Intermediate Business Statistics

SERIES IN QUANTITATIVE METHODS FOR
DECISION MAKING

Robert L. Winkler, Counsulting Editor

Intermediate Business Statistics

Analysis of Variance, Regression, and Time Series

Robert B. Miller
Dean W. Wichern
University of Wisconsin

Holt, Rinehart and Winston

New York   Chicago   San Francisco   Atlanta   Dallas
Montreal   Toronto   London   Sydney

Library of Congress Cataloging in Publication Data

Miller, Robert Burnham, 1942–
Intermediate business statistics.

Bibliography: p. 513
Includes index.
1. Statistics. 2. Business—Mathematical models.
I. Wichern, Dean W., joint author.   II. Title.
HA29.M567   519.5   76-51330

ISBN: 0-03-089101-9

Printed in the United States of America
4567890 038 0987

To Our Parents

PREFACE

In this book we emphasize statistical model building. Although for the most part, the methodology presented is well established and available from many sources, it is not always in a form that can be absorbed by students with rather limited backgrounds in statistics. We have attempted to bring together in a pedagogically sound manner what we regard as the most useful model-building techniques for students (and researchers) in business. Specifically we concentrate on fixed effects analysis of variance (ANOVA) models, multiple regression models, and the time series models developed and popularized by G. E. P. Box and G. M. Jenkins.

We stress the interative nature of model building and the need to investigate the reasonableness of the assumptions made in adopting a particular model. We illustrate how models enable the data analyst to "interrogate" the data so that, where the data are properly formulated, valid and valuable inferences can be made from them.

We assume that people using this book have had a first course in statistics and some exposure to calculus and matrix algebra. For those who are rusty, Appendices 2 and 3 can be used to review some of the fundamentals in these areas. Chapters 2 and 3 of the book are designed to provide a fairly comprehensive review of probability and statistical inference. We discuss statistical inference entirely in terms of normal probability distributions (normal populations). We have not attempted to cover other data-generating mechanisms because, at this point, we are concerned merely with reacquainting the reader with the general concepts of point estimators, confidence intervals, significance tests, and so forth, and this is most easily done in the normal distribution context. Moreover, normality is often taken as a working assumption in later chapters, and some of the results given in these chapters can be directly related to the results in Chapter 3.

Chapters 4–11 represent the main thrust of the text. The material in these chapters, along with Chapters 2 and 3 if necessary, can be organized to satisfy the requirements of several types of courses covering about 45 one-hour lecture periods or fewer. Some possibilities are outlined below.

Course	*Appropriate Chapters*
1. Introduction to statistical model building (ANOVA, regression, and time series fundamentals)	1, 2 and 3 (if necessary), 4, 5, 6, 9; Appendix 3
2. Analysis of variance, regression, and correlation	1, 2 and 3 (if necessary), 4, 5, 6, 7, 8; Appendix 3
3. Forecasting (multiple regression and time series analysis)	1, 2 and 3 (if necessary), 5, 6, 7, 8, 9, 10, 11; Appendix 3 (if necessary)
4. Special topics in regression and time series analysis (Assumes an exposure to multiple regression using matrix algebra; typically requires two courses in statistics)	1, 5, and 6 and Appendix 3 (if necessary), 7, 8, 9, 10, 11
5. Statistical model building	1, 2 and 3 (if necessary), 4–11; Appendices 1, 2, and 3 (if necessary)

Other chapter combinations are possible, and it may be necessary in specific cases only to cover certain sections of various chapters.

We are indebted to our students and colleagues who offered us valuable suggestions for improving the book. We are grateful to typists Jeanne Brooks, Ann Anderson, and Janice Zawacki for the herculean feats they performed in converting handwritten manuscript to typewritten copy. Jim Hickman, Robert Winkler, and other reviewers have contributed greatly to any success this book may enjoy. Finally we must thank the staff at Holt, Rinehart and Winston for their assistance in producing the finished product.

We are also grateful to the Literary Executor of the late Sir Ronald A. Fisher, F.R.S., to Dr. Frank Yates, F.R.S., and to Longman Group Ltd., London, for permission to reprint Tables II, III, IV and VII from their book Statistical *Tables for Biological, Agricultural, and Medical Research* (6th edition, 1974).

We are responsible for the errors that remain and would appreciate having them brought to our attention.

Madison, Wisconsin   R. B. M.
January 1977   D. W. W.

CONTENTS

Chapter Three   ELEMENTS OF STATISTICAL INFERENCE   73

Chapter Four   ANALYSIS OF VARIANCE   124

Chapter Five   SIMPLE LINEAR REGRESSION
                 AND CORRELATION 176

Chapter Nine   TIME SERIES ANALYSIS: AN INTRODUCTION   330

Chapter Ten   TIME SERIES ANALYSIS: MODEL BUILDING AND FORECASTING   353

Intermediate Business Statistics

INTRODUCTION

1.1 A PERSPECTIVE ON STATISTICS

Modern statistical methods are concerned with the generation and analysis of numerical information or data. These methods are perhaps best appreciated in the context of scientific research.

The scope of scientific research is vast, encompassing all realms of human experience. Regardless of the nature of the field of study, a goal of any research effort is to gain understanding of observable phenomena. Given this understanding, a further goal may be to predict or control the end products of these phenomena. The goals above are ordinarily achieved through the iterative process of reconciling theory with certain observable outcomes.

In fact, the scientific method requires that all propositions be "tested against the observable facts." Thus a major element in scientific activity is the conducting of experiments designed to put propositions to the test. However, drawing general conclusions from experimental data (induction) is inherently an uncertain activity, and scientists rarely conclude that a proposition is "true." Rather, they speak in terms of the degree to which propositions agree with the observed facts. A set of propositions about the behavior of some phenomenon is called a "theory" or a "model." A model may range from a very simple to a very complex set of propositions, and the testing of a model or theory may range from a straightforward task to a mammoth undertaking.

Figure 1.1 is a paradigm of the evolution of scientific knowledge. Referring to the figure, we make the following comments:

(1) Data may consist of casual observations, measurements collected by other investigators, or the results of carefully planned experiments or surveys.

1

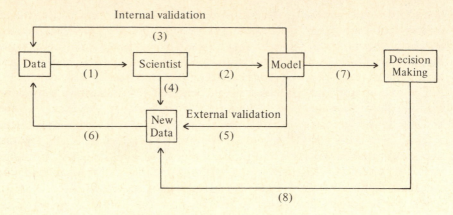

Figure 1.1 A paradigm for scientific activity

(2) After observing the data, the "scientist" (or group of scientists) formulate(s) propositions or hypotheses purporting to explain the particular outcomes observed. This set of propositions constitutes the model of the phenomenon of interest and is the creation of the scientist's imagination. From the model certain consequences may be logically deduced. These consequences are also part of the model.

(3) At the very least, the model must yield consequences that are consistent with the already observed facts. The determination that these consequences are indeed consistent with the data in hand may require a rather sophisticated analysis. We call this step "internal" model validation.

(4) When the scientist is satisfied with the internal model validation, the next question to ask is whether the predictions made using the model are consistent with "new" data. At this point the scientist must plan and carry out an experiment that will yield new data. These data are then used to evaluate model performance.

(5) and (6) The comparison of model predictions with *new* data is called "external" model validation. If the model appears to be inadequate (does not fit the facts), modifications of the model will be made. The revised model is validated using *all* data currently available (branch 6). If the model seems to be making reasonable predictions, it is accepted tentatively as a working hypothesis. The cycle repeats if at any point the hypothesis is inconsistent with new data.

(7) and (8) When a model is accepted as a working hypothesis, it may be shared through publication or private communication. The results produced by an acceptable model are ordinarily used by planners, forecasters, policy makers, and so forth, to make decisions. It must be emphasized that although a model may constitute the "best information currently available," its predictions may prove inconsistent with future observations, so that

decisions based on a particular model may be of little (or negative) value. In fact, a decision-making application may produce precisely the data needed to put a theory to a rigorous test. Social experiments provide a good illustration of this remark. Widespread adoption of a particular medical treatment, for example, will provide data on the treatment's effectiveness.

The role of statistical procedures in a scientific investigation can now be made explicit. As noted above, a model or theory can be empirically based and is empirically tested. Whereas the imagination of the scientist produces a theory, the theory is subject to an empirical validation. It is this feature that makes statistical activity so fundamental to scientific inquiry. Statistics provides guidelines for efficiently generating informative data and for constructing tests of these data against certain propositions.

We might classify the study of statistics as the study of variation or differences in data. We can divide the sources of variation into three main categories: (a) inherent variation, (b) experimentally induced variation, and (c) variation due to errors or mistakes. It is important to distinguish among these categories.

Inherent Variation

Suppose a survey is conducted. We would expect to find that individuals in the survey differ from one another with respect to any number of characteristics, for example, age, height, size of family, and preferences for automobiles, political candidates, and soda pop. This variation is inherent in the collection of individuals.

Experimentally Induced Variation

Suppose subjects are divided into "control" and "experimental" groups, and each group is given a "treatment." If the "treatments" have substantially different effects, we expect the groups to respond differently to them in a detectable way. There will be inherent variation among members of a group, and there will be variation among members of different groups due to the different "treatments." Can we measure the amount of variation coming from the two sources, and can we state whether the variation due to the difference in "treatments" is substantial or inconsequential? The answer to these questions, in certain experimental situations, is yes, as we shall see in Chapter 4. The variation due to treatments mentioned above is an example of experimentally induced variation.

Variation Due to Errors or Mistakes

Variation can be due to mistakes in handling the data. For example, we might transpose digits, or mispunch computer cards, or misplace observations. This kind of variation can be minimized by careful data processing. On a more subtle level, errors of measurement can be due to the imprecision of measuring techniques. For example, in sample surveys in which questionnaires and personal interviews are used to obtain information, a common problem is that of nonresponse or biased response. The effects of nonresponse can be considerable if failure to respond to a questionnaire, for example, is due to the nature of the information sought. Biased response can result from failure of a subject to understand a question or a negative reaction to an interviewer. Furthermore, if an interviewer accepts responses from a secondary respondent (such as a spouse or child) rather than from the actual subject of study, he may receive biased responses either due to the ignorance of the secondary respondent or due to the sensitivity of the information sought.

A statistical analysis often attempts to produce measures of the amount of variation due to various sources. This information is valuable in determining the confidence that can be placed in inferences drawn from the data. In some areas of research, particularly in the physical sciences, rather precise conclusions and predictions can be made. In business and other social sciences, however, we are often forced to be content with rather crude theories. Any ability to make valid assessments of variability is a direct consequence of a careful statistical analysis.

1.2 AN OVERVIEW OF THE BOOK

The overwhelming majority of statistical methods deals with the testing of *quantitative* or *mathematical* propositions. This stems from the fact that scientists have found that propositions usually need to be put in quantitative form before they can be tested empirically. In keeping with this observation, we shall deal with mathematical models in this book. In particular, we shall consider (a) fixed effect analysis of variance models, (b) regression models, and (c) time series models. All of these models are used extensively in business research and decision making. Illustrations of a few of these applications are contained in Chapters 4 through 11. However, our primary purpose is the explication of the statistical methods, rather than the detailed description of the research studies that have been based on the methods.

We primarily consider problems in which the data are numerical *measurements*. We do not consider the analysis of ordinal data in this book. Furthermore we take normality as a working distributional assumption. Not all data generating distributions are normal; however, it is frequently possible to transform the data so that they may be regarded as realizations of a normally distributed random variable. Moreover, inferences based on normal theory methods are often not appreciably affected by mild departures from normality. In summary, nonnormality is often less bothersome than the violation of other assumptions we will be making, specifically, the assumptions of variance homogeneity and independence.

Chapters 2 and 3 constitute a rather extensive review of elementary statistical concepts. We feel such a review is justified because many students taking a second course in statistics are "rusty" and need an extensive review. The review in Chapters 2 and 3 makes the text relatively self-contained. Also, a review section has the additional advantage of presenting the elementary concepts in a format and notation that blends with the application chapters that follow.

Chapter 4 presents some basic analysis of variance (ANOVA) models. These models are useful when the means of several populations are to be compared and functions of the means estimated.

Chapters 5 through 8 are devoted to regression models. Chapter 5 is a detailed exposition of the simple linear regression model. Also included is a discussion of correlation analysis since regression and correlation are closely related. In Chapter 6 is a discussion of multiple regression analysis or regression models with *several* independent variables. In Chapter 7 are discussions of some of the finer points of regression modeling, including such topics as dummy variables, transformation of variables, multicollinearity, variable selection, and nonlinear estimation. Chapter 8 is a case study illustrating a number of regression modeling techniques.

Chapters 9, 10, and 11 are devoted to an exposition of the time series analysis procedures popularized by Box and Jenkins in reference [1]. Chapter 9 is essentially a case study in which the techniques of parametric time series modeling are introduced in the context of an illustrative example. Chapter 10 contains a more general and formal presentation of time series modeling. Finally, in Chapter 11, we consider the application of time series models to seasonal data.

1.3 THE COMPUTER AND STATISTICS

Many large-scale research projects would not be possible without computers to do the necessary data manipulation. Computers are also useful in theoretical studies in which convenient mathematical solutions are unobtainable. All methods in this book can be implemented on computers, and except for the solution of small-scale problems, such implementation would be practically essential. We shall present many examples of calculations done on a computer, and many of the problems will be best solved using a computer.

2

PROBABILITY THEORY

2.1 INTRODUCTION

We all are familiar with odds giving on sporting events, playing games of chance, insuring against risk of loss, and many other activities that evidence the existence of uncertainty in our everyday experience. Although uncertain events cannot be predicted with unerring accuracy, they are often observed to follow certain regular patterns or "laws." (For example, if a fair coin is flipped many times, we expect about half of the flips to result in "heads.") Probability theory is the result of man's attempt to discover the "laws of chance."

Statistical inferences share with the activities mentioned above the element of uncertainty. Statistical inferences are generalizations on the basis of limited information and hence subject to unpredictable amounts of error. It seems natural therefore to study statistical procedures in the context of probability theory. In fact, a fundamental process in statistics is the attempt to evaluate, using probability theory, the precision (or limits on the error) with which an inference can be made in a given set of circumstances. This process naturally leads to the comparison of competing statistical techniques and the development of recommendations for or against using a given technique in a given situation.

The goal of this chapter is to review enough probability theory to provide an adequate context for the development and evaluation of statistical procedures. The basic element in the theory is the probability *distribution*. Thus after briefly introducing some basic probability concepts in Section 2.2, we review the concept of probability distribution in some detail in Section 2.3. In Section 2.4 many results specifically needed for the discussion of statistical inference are presented. These results will be referred to numerous times throughout the book.

6

2.2 BASIC CONCEPTS

Definitions of Probability

Probability is a numerical measure of uncertainty. The need for such a measure was first recognized by gamblers since the ability to calculate odds or probabilities allowed them to develop strategies for minimizing losses in games of chance. The recognition that probability can be used to measure uncertainty in practically every area of human endeavor has led to its widespread use as a tool by decision makers in, for example, government, industry, and medicine.

Various techniques are used to assign probabilities to events. Three of the most popular methods are the symmetry principle, the relative frequency principle, and the method of standard lotteries.

The *symmetry principle* seems to be most useful in simple games of chance. It could be used to assign probability 1/2 to the event "heads up after the flip of a fair coin," or probability 1/6 to the event "a one spot up after the roll of a fair die." In general, if there are n possible outcomes of an experiment, all regarded as "equally likely," then the symmetry principle assigns probability $1/n$ to each outcome.

The *relative frequency principle* is based on a phenomenon that we shall call *statistical regularity*. This phenomenon (and the corresponding probability assessment) is illustrated in the following examples.

EXAMPLE 2.1

We consider a perfectly balanced wheel that is divided into eight equal sectors as shown in Figure 2.1. If the wheel is given a push and allowed to spin freely, it will eventually stop so that the pointer points to one of the eight sectors. (If the pointer falls on a dividing line, we assume that it points to the left-hand sector.)

A wheel like the one in Figure 2.1 is spun 1300 times. After each spin the frequency and relative frequency of "sector 1" is recorded. The frequency of sector 1 is the number of times this sector is obtained, and consequently its *relative* frequency after n spins is

$$\text{relative frequency} = (\text{frequency})/n$$

The relative frequencies are plotted in Figure 2.2. The graph in Figure 2.2 is highly variable when n is small, but this variability decreases as n gets larger. In fact the relative frequencies seem to stabilize around $1/8 = 0.125$. This stabilizing behavior is an example of statistical regularity. Moreover, the limiting value of the relative frequency is taken to be the probability of obtaining sector 1 on one spin of the wheel.

It is interesting to note that an application of the symmetry principle would lead us to assign probability 1/8 to sector 1. In this case the relative frequency argument and the symmetry principle lead to the same probability assignment.

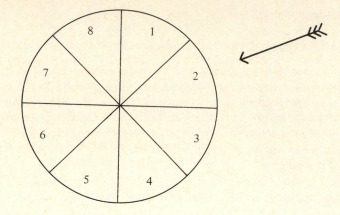

Figure 2.1

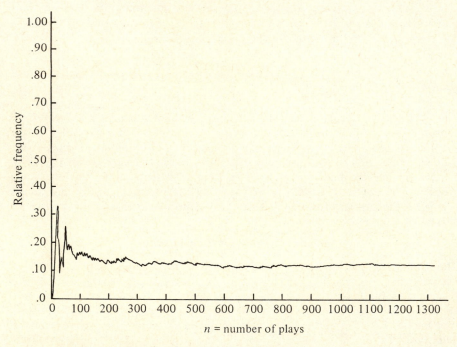

Figure 2.2   Relative frequencies of sector 1

EXAMPLE 2.2

Statistical regularity is observed in innumerable everyday situations. One such example is shown in Figures 2.3 and 2.4. Figure 2.3 shows the total numbers of births and the numbers of males born in the United States during the years 1940 to 1967.[a] Both series of numbers show a steady increase in births through the year 1961 and a steady decrease in the number of births after 1961. If we look at the ratios of male births to total births in Figure 2.4, however, we see an almost constant series of numbers. Since the ratios are based on records of millions of births each year, it is safe to assume that they are indicative of the probability of giving birth to a male baby, at least in the United States.

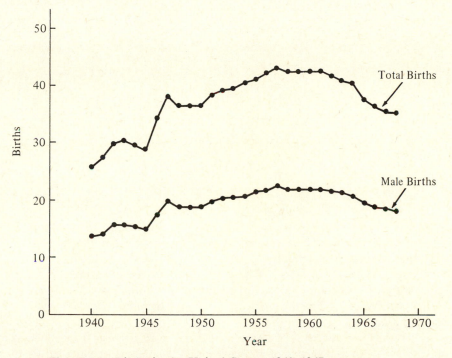

Figure 2.3 Births in the United States, 1940–1967

Both preceding examples of statistical regularity are illustrations of a theorem in probability theory known as the "law of large numbers." The law of large numbers says that if A is an event occurring with probability p and an experiment in which A can occur is repeated n independent times, then the relative frequency of the occurrence of A converges to p with certainty

[a]Source: U.S. Department of Health, Education, and Welfare, Public Health Service, *Vital Statistics of the United States*, Vol. L. *Live Births by Sex and Sex Ratio, United States, 1940–1967*.

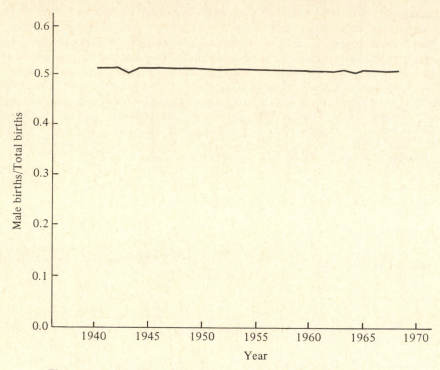

Figure 2.4   Relative frequency of male births, 1940–1967

as $n \to \infty$. Thus the law guarantees that by repeating the appropriate experiments an ever larger number of times, we may approximate probabilities by relative frequencies to any degree of accuracy. Unfortunately it does not answer the question of *how many* times the experiment must be repeated in order to achieve a desired degree of accuracy of approximation. This question can often be answered for a particular problem by more detailed analysis.

In practice the question of "how many" is often irrelevant in the sense that only a limited amount of information is available. In such cases the relative frequency principle states that we should use the relative frequencies that are available *as if* they were probabilities.

The *method of standard lotteries* is used to elicit probabilities from decision makers for events whose probabilities are not obvious by symmetry and which are not capable of extensive repeated observation. The method (or sophisticated variations of it) has been used in cases in which important policy or managerial decisions must be made on the basis of scanty (sometimes nonexistent) empirical evidence. As a simple illustration we consider the decision to undertake a risky financial venture. It is uncertain whether the venture will succeed or fail. In order to assess the odds in favor of success we consider a lottery that pays $1 if the venture succeeds and pays nothing if the venture fails. We are not saying that a successful venture will actually return a dollar. One dollar simply *stands for* the payoff of a successful venture; that is, the actual payoff is *standardized* to allow us to assess the probability of success without allowing the actual amount of payoff to interfere with the assessment.

We now ask the question of the decision maker: "What is the greatest amount of money you would pay to enter the lottery?" His answer to this question is interpreted as his best guess at the probability of the success of the venture. Suppose he says $0.25. We interpret this to mean that he believes the venture has a 25 percent chance of success (that is, probability of success is 1/4). The validity of his statement can be checked by giving him a choice between the following two lotteries:

Lottery 1
Events: Venture succeeds   Venture fails
Payoffs: $1 $0

Lottery 2
A bowl contains 100 chips, 25 black and 75 red. A chip is drawn at random.
Events: Black chip drawn   Red chip drawn
Payoffs: $1 $0

If the decision maker really believes the probability of success is 1/4, he should be indifferent between the two lotteries. If he is not, he should reassess the probability of success.

Mathematical Probability

Probability is ordinarily formally defined in the context of performing an "experiment." The elements of the mathematical probability model are given in the following sections.

Elementary Outcomes   The elementary outcomes of an experiment (of a game, if you wish) are the simplest units in which the results of the experiment are expressed. Elementary outcomes are *mutually exclusive* in the sense that if one of them occurs, each of the others cannot occur, and they are *exhaustive* in the sense that one of them must occur when the experiment is performed.

Sample Space   The elementary outcomes considered as a collection or a set are known as the sample space of an experiment. The sample space will be denoted by the Greek letter Ω.

Events   Events are simply subsets of the sample space Ω and will be denoted by capital letters such as $A, B, C, \ldots$. We note that any set containing a single elementary outcome is an event, since it is a subset of Ω. Likewise the entire sample space Ω and the empty set $\emptyset$ are events.

Since events are simply subsets of Ω, they may be manipulated by the rules of set algebra (which are reviewed in Appendix 1). In addition, events are assumed to satisfy the following properties:

1. If A and B are events such that $B \subset A$, then the difference $A - B$ is an event.
2. If $A_1, A_2, \ldots$ is a sequence of events, then the union and the intersection of these events are events.

Probability   Intuitively speaking, the probability of an event A is a number between 0 and 1 (inclusive) that reflects the (relative) likelihood of the event's occurring when the experiment is performed. (We shall denote the probability of an event A by the symbol $P(A)$, read "probability of A" or "P of A.") Mathematically, probability is a function, P, from the collection of events to the interval $[0, 1]$ that satisfies the following consistency conditions:

1. $P(\emptyset) = 0$ and $P(\Omega) = 1$.
2. $P(A) \geq 0$ for any event A.
3. If A_1 and A_2 are disjoint events (that is, $A_1 \cap A_2 = \emptyset$), then

$$P(A_1 \cup A_2) = P(A_1) + P(A_2)$$

In words, the probability of a union of disjoint events is the sum of the probabilities of the events.

Certain basic rules for calculating probabilities can be derived from the consistency conditions, for example,

1. If $\bar{A}$ denotes the complement of A, then $P(\bar{A}) = 1 - P(A)$.
2. If $A \subset B$, then $P(A) \leq P(B)$.
3. The addition rule: For any events A and B,

$$P(A \cup B) = P(A) + P(B) - P(A \cap B)$$

EXAMPLE 2.3

In a certain region the 1000 employees of a photographic equipment and supplies manufacturing industry are classified by type of work (production or nonproduction) and by size of establishment in which they are employed. The number of employees in each work-size category is given in Table 2.1. The symbols $A_1, A_2, \ldots, A_5$ are used to denote the size categories, and B_1 and B_2 are used to denote "production" and "nonproduction," respectively.

Table 2.1   Number of Employees by Type of Work and Size of Establishment within a Photographic Equipment and Supplies Industry

Number of Employees per Establishment		TYPE OF WORK		Total
		Production (B_1)	Nonproduction (B_2)	
1–4	(A_1)	7	1	8
5–19	(A_2)	14	2	16
20–99	(A_3)	40	39	79
100–499	(A_4)	107	36	143
500 and over	(A_5)	452	302	754
Total		620	380	1000

If an employee is to be selected at random for an interview, what is the probability that he will be a production worker in an establishment of between 20 and 99 workers? A much shorter way to write this question is to ask for the value of $P(A_3 \cap B_1)$. Random selection means that each of the 40 employees in $A_3 \cap B_1$ has the same chance of being selected as any other employee, so the numerical value of $P(A_3 \cap B_1)$ is $40/1000 = 0.04$.

Similar arguments lead to the conclusions that $P(A_3) = 0.079$ and $P(B_1) = 0.620$. Using the addition rule, we can write

$$P(A_3 \cup B_1) = P(A_3) + P(B_1) - P(A_3 \cap B_1) = 0.079 + 0.620 - 0.04 = 0.659$$

In words this says that there is a 65.9 percent chance of randomly selecting an employee who is either a production worker or working in an establishment with 20 to 99 workers, or both.

The event $\bar{A}_5$ is the event "employee works in an establishment with not more than 499 workers." The probability of this event is

$$P(\bar{A}_5) = 1 - P(A_5) = 1 - 0.754 = 0.246$$

The event "employee works in an establishment with not more than 99 workers" is $\overline{A_4 \cup A_5}$, and the probability of this event is

$$P(\overline{A_4 \cup A_5}) = 1 - P(A_4 \cup A_5) = 1 - [P(A_4) + P(A_5)] = 1 - 0.143 - 0.754 = 0.103$$

Conditional Probability

If A and B are any two events with $P(B) > 0$, then the *conditional probability* of A, given B, is denoted by $P(A|B)$ and defined by the equation

$$P(A|B) = \frac{P(A \cap B)}{P(B)} \tag{2.1}$$

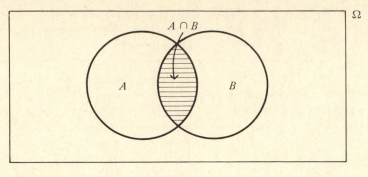

Figure 2.5

The nature of conditional probability may be pictured in a Venn diagram (see Figure 2.5). We suppose the event B has occurred. Then all the elementary outcomes in A but outside of $A \cap B$ could not have occurred. Hence, the conditional probability of A, given B, must be proportional to $P(A \cap B)$. The probability $P(A \cap B)$ is divided by $P(B)$ to form a probability function on the reduced sample space defined by B. In other words, the function defined by equation (2.1) satisfies the following properties:

1. $P(\emptyset|B) = 0$ and $P(B|B) = 1$.
2. $P(A|B) \geq 0$ for any event A.
3.   If A_1 and A_2 are disjoint events, then,

$$P(A_1 \cup A_2|B) = P(A_1|B) + P(A_2|B)$$

All basic probability theorems of this section hold for the conditional probability function defined on subsets of B. Division of $P(A \cap B)$ by $P(B)$ is necessary to insure that all usual consistency conditions are satisfied.

EXAMPLE 2.4

Let us return to the population of employees in Table 2.1. The conditional probability of A_1, given B_1, is

$$P(A_1|B_1) = \frac{P(A_1 \cap B_1)}{P(B_1)} = \frac{0.007}{0.620} = 0.011$$

In words, we are calculating the probability that a production worker works in an establishment with one to four employees. The probability of any A_i, given B_1 or given B_2, may be calculated in a similar way. Table 2.2 displays all such conditional probabilities.

A result that follows immediately from the definition of conditional probability is the *multiplication rule* which says that if $P(B) > 0$, then

$$P(A \cap B) = P(B) \cdot P(A|B) \tag{2.2}$$

Table 2.2 Conditional Probabilities Based on
Table 2.1

i	Event	$P(A_i|B_1)$	$P(A_i|B_2)$
1	A_1	0.011	0.003
2	A_2	0.023	0.005
3	A_3	0.064	0.103
4	A_4	0.173	0.095
5	A_5	0.729	0.794
		1.000	1.000

The events A and B are said to be *independent* if $P(A|B) = P(A)$. This statement implies that knowledge of the occurrence of B has no effect on our assessment of the probability that A occurs. When A and B are independent, equation (2.2) becomes

$$P(A \cap B) = P(B) \cdot P(A) \qquad (2.3)$$

which is the multiplication rule for independent events. Dividing both sides of equation (2.3) by $P(A)$ yields the independence criterion expressed in terms of the conditional and marginal probabilities of B. In this case, $P(B|A) = P(B)$.

EXAMPLE 2.5

Suppose a card is to be drawn randomly from an ordinary 52-card bridge deck and we define the following events:

A = "The card is a spade."
B = "The card is a club."
C = "The card is not a face card and has an even number of spots."
D = "The card is a King or a Jack."

It is easy to see that $P(A) = 13/52 = 1/4$, $P(B) = 1/4$, $P(C) = 20/52 = 5/13$, and $P(D) = 8/52 = 2/13$.
Since A and B are mutually exclusive,

$$P(A \cap B) = P(\emptyset) = 0$$

which immediately implies that A and B are *dependent*. (This is intuitively obvious, since if either of the events occurs, the other cannot occur.) Similarly, C and D are dependent.

Now

$$P(B \cap C) = (5/52) = (1/4) \cdot (5/13) = P(B) \cdot P(C)$$

which implies that B and C are independent. Similar arguments show that the events A and C, A and D, and B and D are independent, as you can easily verify.

We conclude this section on conditional probability with a discussion of Bayes' theorem, a result that has many uses in statistical inference and decision making.

Suppose a sample space Ω is partitioned into k mutually exclusive events, $H_1, H_2, \ldots, H_k$, as shown in Figure 2.6. We assume that the probabilities $P(H_1), P(H_2), \ldots, P(H_k)$ are given. Now suppose the experiment is performed and the event D (also shown in Figure 2.6) occurs. What are the

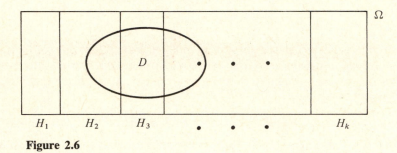

Figure 2.6

conditional probabilities of $H_1, H_2, \ldots, H_k$, given D? Bayes' theorem is a rule for calculating conditional probabilities given the probabilities $P(H_1), P(H_2), \ldots, P(H_k)$ and the conditional probabilities $P(D|H_1)$, $P(D|H_2), \ldots, P(D|H_k)$.

The theorem states that

$$P(H_i|D) = \frac{P(H_i)\, P(D|H_i)}{\sum_{j=1}^{k} P(H_j)\, P(D|H_j)}, \qquad i = 1, 2, \ldots, k \qquad (2.4)$$

(Problem 2.21 develops a proof of the theorem.) The following example illustrates the use of Bayes' theorem.

EXAMPLE 2.6

A manufacturer obtains a certain part from three different suppliers. Fifty percent of the manufacturer's stock of parts comes from supplier 1, 15 percent from supplier 2, and 35 percent from supplier 3. It is known from past experience that 10 percent of supplier 1's parts are defective, 6 percent of supplier 2's parts are defective, and 15 percent of supplier 3's parts are defective. If a part is taken at random from the manufacturer's stock and found to be defective, what is the probability that the part came from supplier 1? Supplier 2? Supplier 3?

Here H_i is "parts come from supplier i," $i = 1, 2, 3$, and $P(H_1) = 0.50$, $P(H_2) = 0.15$, and $P(H_3) = 0.35$. The event D is "part is defective," and $P(D|H_1) = 0.10$, $P(D|H_2) = 0.06$, and $P(D|H_3) = 0.15$. Consequently,

$$\sum_{j=1}^{3} P(H_j) P(D|H_j) = (0.50)(0.10) + (0.15)(0.06) + (0.35)(0.15) = 0.1115$$

and Bayes' theorem gives

$$P(H_1|D) = \frac{(0.50)(0.10)}{0.1115} = \frac{0.050}{0.1115} = 0.45$$

$$P(H_2|D) = \frac{(0.15)(0.06)}{0.1115} = \frac{0.009}{0.1115} = 0.08$$

$$P(H_3|D) = \frac{(0.35)(0.15)}{0.1115} = \frac{0.0525}{0.1115} = 0.47$$

After a defective part has been observed, the chances are slightly greater that the part came from supplier 3 than from supplier 1, and there is less than a 10-percent chance that the part came from supplier 2. Before the experiment yielded a defective part, the chances of a part coming from suppliers 1, 2, and 3 were 0.5, 0.15, and 0.35. Thus a single piece of information has had quite an impact on the probabilities of H_1, H_2, and H_3.

The probabilities $P(H_1), P(H_2), \ldots, P(H_k)$ are called *prior* probabilities because they are assigned before the experiment is run. The probabilities $P(H_1|D), P(H_2|D), \ldots, P(H_k|D)$ are called *posterior* probabilities because they are computed *after* the experiment is run and the datum D observed. Note that the computation of $P(H_i|D)$ depends not only on $P(H_1)$, $P(H_2), \ldots, P(H_k)$, but also on $P(D|H_1)$, $P(D|H_2), \ldots, P(D|H_k)$.

2.3 PROBABILITY DISTRIBUTIONS AND RANDOM VARIABLES

Discrete Distributions

Experiments frequently yield numerical outcomes. A "variable" that assumes the values of the outcomes of an experiment is called a *random variable*. When probabilities are associated with the numerical outcomes, the *probability distribution* of the random variable is formed.

If probability is assigned only to discrete (isolated) values of the random variable, then the probability distribution is said to be *discrete*. If probability is assigned only to intervals of values of the random variable, the probability distribution is said to be *continuous*. We shall give a series of definitions and results for discrete distributions in this section. The next section contains similar definitions and results for continuous distributions.

Let X denote a random variable that assumes values (in ascending numerical order) $x_1, x_2, x_3, \ldots$ with associated probabilities $p(x_1), p(x_2), p(x_3), \ldots$. The function $p(x)$ is called the *probability function* of X and clearly describes its probability distribution. We will sometimes write $p(x) = \Pr[X = x]$ to stand for "the probability that X assumes the value x." The notation $\Pr[X \le x]$ stands for "the probability that X assumes a value no greater than x," and the notation $\Pr[x_i < X \le x_j]$ stands for "the probability that X assumes a value greater than x_i, but no greater than x_j. Clearly

$$\Pr[x_i < X \le x_j] = p(x_{i+1}) + p(x_{i+2}) + \cdots + p(x_j)$$

so this latter probability is a *sum* of values of the probability function of X.

Let X and Y be two discrete random variables, and let x and y be any values of X and Y, respectively. Then the probability that X has value x and Y simultaneously has value y is called the *joint probability function* of X and Y at the point (x, y). This value is denoted by

$$p(x, y) = \Pr[(X = x) \cap (Y = y)] \tag{2.5}$$

We point out that symbols like $[X = x]$, $[Y = y]$, and $[(X = x) \cap (Y = y)]$ represent events and $p(x)$, $p(y)$, and $p(x, y)$ are probabilities that satisfy the axioms of probability given earlier. The same laws of probability hold whether the outcomes are numerical or nonnumerical.

Our notation makes certain facts easy to see. For example, let $y_1, y_2, \ldots, y_n$ denote the values of Y. Since a particular value of X, say x_i, must occur simultaneously with some value of Y, we have

$$p(x_i) = \Pr[X = x_i] = \Pr[X = x_i \cap Y = y_1] + \Pr[X = x_i \cap Y = y_2]$$
$$+ \cdots + \Pr[X = x_i \cap Y = y_n]$$
$$= p(x_i, y_1) + p(x_i, y_2) + \cdots + p(x_i, y_n)$$
$$= \sum_{j=1}^{n} p(x_i, y_j)$$

Similarly, it can be shown that

$$p(y_j) = \sum_{i=1}^{m} p(x_i, y_j)$$

where we assume X can assume the values $x_1, x_2, \ldots, x_m$. The functions $p(x)$ and $p(y)$ are called the *marginal* probability functions of X and Y. They are obtained by forming certain sums of values of the joint probability function.

Continuing with the notation developed above, note that the conditional probability of the event $[Y = y]$, given the event $[X = x]$, is

$$\Pr[Y = y | X = x] = \frac{\Pr[X = x \cap Y = y]}{\Pr[X = x]} \tag{2.6}$$

and therefore letting

$$p(y|x) = \Pr[Y = y | X = x]$$

we have

$$p(y|x) = \frac{p(x, y)}{p(x)} \qquad (2.7)$$

This is called the *conditional probability* function of Y, given $X = x$. Similarly, we can define

$$p(x|y) = \frac{p(x, y)}{p(y)} \qquad (2.8)$$

EXAMPLE 2.7

In attempting to come up with a hiring policy for clerical workers in a department store, the personnel manager referred to past records to obtain data on educational level and length of service of clerical employees. Defining the random variables

$$X = \begin{cases} 0 & \text{if employee has a high school education} \\ 1 & \text{if employee has post-high school educational experience} \end{cases}$$

and

$$Y = \begin{cases} -1 & \text{if employee stayed less than 6 months} \\ 0 & \text{if employee stayed between 6 and 12 months} \\ 1 & \text{if employee stayed over 12 months} \end{cases}$$

he obtained the following joint distribution

$p(x)$	0.80	0.20	
1	0.24	0.05	0.29
0	0.16	0.14	0.30
-1	0.40	0.01	0.41
$y \diagdown x$	0	1	$p(y)$

$$p(x, y) = \Pr[X = x \cap Y = y]$$

The marginal distributions of X and Y are also shown. We are treating relative frequencies as probabilities here. Notice that 41 percent of the employees stayed less than 6 months, and that 97.5 percent ($= 0.40/0.41$) of these employees had only high school education. In terms of random variables we are saying that, given a person stays less than 6 months, the probability that he has only a high school education is 0.975. In symbols we have

$$p(x = 0|y = -1) = \frac{p(x = 0, y = -1)}{p(y = -1)}$$

$$= \frac{0.40}{0.41} = 0.975$$

We are using the conditional distribution of X, given $Y = -1$.

The conditional distributions of Y, given $X = 0$ and $X = 1$ are shown in the following table. These distributions indicate that of the employees who have only a high school education, 50 percent stay less than 6 months, 20 percent stay 6 to 12 months, and 30 percent stay more than 12 months. Of the employees who have post-high school educational experience, only 5 percent stay less than 6 months, 70 percent stay 6 to 12 months, and 25 percent stay over 12 months.

Totals	1	1
1	0.3	0.25
0	0.2	0.70
-1	0.5	0.05
y	$p(y|0)$	$p(y|1)$

The concept of *independent random variables* is an extension of the concept of independent events presented earlier. Let X and Y be two discrete random variables, and let x and y be any values of X and Y, respectively. The events $[X = x]$ and $[Y = y]$ are *independent* if

$$\Pr[X = x \cap Y = y] = \Pr[X = x]\Pr[Y = y] \qquad (2.9)$$

using equation (2.3). Equation (2.9) may be written in terms of probability functions as

$$p(x, y) = p(x)p(y) \qquad (2.10)$$

In defining X and Y to be independent random variables, it seems reasonable to require that equation (2.10) hold for all possible combinations of values x and y, for this would say that no matter what experimental outcomes occurred, the events $[X = x]$ and $[Y = y]$ would be independent.

Our discussion involving two random variables may be extended to any finite number of variables as follows. Let $X_1, X_2, \ldots, X_n$ be a collection of discrete random variables defined on a sample space Ω. Let $x_1, x_2, \ldots, x_n$ be any possible combination of values of the random variables. The association of probabilities with all such combinations is a joint probability function provided only that the total amount of probability associated is equal to one. The joint probability function is denoted by $p(x_1, x_2, \ldots, x_n)$. Let $p(x_1), p(x_2), \ldots, p(x_n)$ denote the marginal probability functions of $X_1, X_2, \ldots, X_n$, respectively. The random variables are said to be mutually independent if

$$p(x_1, x_2, \ldots, x_n) = p(x_1)p(x_2) \ldots p(x_n) \qquad (2.11)$$

for *all* possible combinations of values of $x_1, x_2, \ldots, x_n$.

Independent random variables play an important role in statistics because they provide a model for the collection of data from a population by means of simple random sampling with replacement. Such a model is fundamental to the validity of many statistical inference procedures presented in Chapters 3 through 8 of this book. On the other hand, a significant proportion of practical problems involves dependencies. Notable examples are those in which observations are collected over time, and past observations influence current and future observations. Chapters 9, 10, and 11 are exclusively devoted to methods for handling such observations. Interestingly enough, the models used to describe dependencies also require assumptions about the independence of certain random variables in their formulation. Hence, the concept of independence is seen to play a pervasive role in statistics.

We conclude this section by noting that Bayes' theorem may be translated into the language of discrete random variables. Assuming the random variable X can take one of m values and, making the appropriate identification of symbols, equation (2.4) becomes

$$p(x_i|y) = \frac{p(x_i)p(y|x_i)}{\sum_{j=1}^{m} p(x_j)p(y|x_j)}, \qquad i = 1, 2, \ldots, m \qquad (2.12)$$

The result in equation (2.12) is Bayes' theorem for discrete random variables.

Continuous Distributions

Although discrete distributions assign probabilities to points, continuous distributions assign probabilities to intervals. As an illustration, we consider a very large population of heights. A histogram of such a population might contain many narrow classes because of the great size of the population. Figure 2.7 illustrates the sort of situation we are describing. The histogram is constructed so that the area of any bar is equal to the relative frequency of the

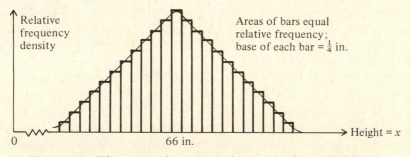

Figure 2.7   Histogram of a hypothetical population of heights

interval that makes up the base of the bar. Such a relative frequency is also the probability that a randomly drawn height will fall in the interval at the base of the bar.

A smooth curve has been drawn through the histogram so that it may be used to approximate probabilities of intervals. Figure 2.8 shows the details of

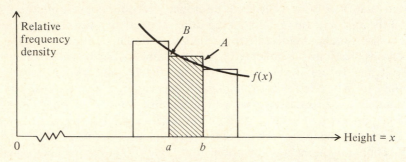

Figure 2.8   Detail of histogram in Figure 2.7

the approximation. We focus on the bar whose base is the interval (a, b). Let $f(x)$ denote the function whose graph is the smooth curve. If we find the area under $f(x)$ over the interval (a, b), we will obtain an approximation to the area of the bar. Although the area under the curve leaves out the area A, it includes the area B, thus making two compensating errors, so to speak. The required area is calculated by evaluating the integral $\int_a^b f(x)\, dx$.

Let X denote a randomly drawn height from the population, and define the symbol $\Pr[a < X < b]$ to be the probability that X falls in the interval (a, b). Then our discussion above allows us to write the approximation

$$\Pr[a < X < b] \doteq \int_a^b f(x)\, dx \tag{2.13}$$

In defining continuous random variables, we will ignore the approximation contained in equation (2.13) and assume that it is a strict equality. The function $f(x)$ is called the probability density function (p.d.f.) of the distribution of X. Since $f(x)$ describes a distribution of probability, it must be a nonnegative function ($f(x) \geq 0$) and its area (integral) over all values of x must be 1, that is, $\int_{-\infty}^{\infty} f(x)\, dx = 1$.

The simplest example of a p.d.f. is the uniform p.d.f.

$$f(x) = 1, \qquad 0 < x < 1 \tag{2.14}$$

This probability function assigns probability density 1 to each point in the interval $(0, 1)$ and probability density 0 to every point outside this interval, as shown in Figure 2.9. Using equation (2.13), we see that

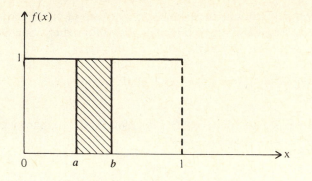

Figure 2.9   Graph of $f(x) = 1$, $0 < x < 1$

$$\Pr[a < X < b] = \int_a^b 1 \, dx = b - a$$

which, as expected, is simply the length of the interval.

The continuous distribution most frequently referred to in statistics is the standard normal distribution. The p.d.f. of this distribution is given by

$$f(z) = \frac{1}{\sqrt{2\pi}} e^{-(z^2/2)}, \qquad -\infty < z < \infty \tag{2.15}$$

We are using Z as a symbol for the standard normal random variable, since this is generally accepted notation. A graph of the p.d.f. in equation (2.15) is shown in Figure 2.10. The p.d.f. is symmetric about $z = 0$. The area under the curve between $z = 0$ and $z = z' > 0$ is given by the integral

$$\int_0^{z'} \frac{1}{\sqrt{2\pi}} e^{-(z^2/2)} \, dz = 0.5 - \int_{z'}^{\infty} \frac{1}{\sqrt{2\pi}} e^{-(z^2/2)} \, dz \tag{2.16}$$

The integral cannot be expressed as an elementary function of z'. Instead, numerical techniques are used to evaluate the integral to any desired

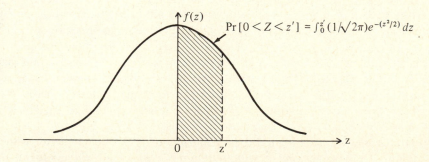

Figure 2.10   Graph of the standardized normal p.d.f.

accuracy. Appendix Table D presents the results of these calculations for selected values of z'. Using this table we may obtain the probability assigned to any interval as the following example illustrates.

EXAMPLE 2.8

Let Z have the standard normal p.d.f. in equation (2.15). The following probabilities may be obtained using Appendix Table D:

1. $\Pr[0 < Z < \infty] = 0.5$ (by symmetry)
2. $\Pr[0 < Z < 1.5] = 0.5 - 0.0668 = 0.4332$
3. $\Pr[Z > 1.5] = 0.0668$ (directly)
4. $\Pr[-2 < Z < -1.5] = \Pr[1.5 < Z < 2]$ (by symmetry)
    $= \Pr[1.5 < Z] - \Pr[2 < Z]$
    $= 0.0668 - 0.0228 = 0.0440$
5. $\Pr[-1.75 < Z < 1.45] = 1 - \Pr[Z > 1.45] - \Pr[Z > 1.75]$
    $= 1 - 0.0735 - 0.0401 = 0.8864$
6. $\Pr[|Z| > 1.96] = \Pr[Z > 1.96] + \Pr[Z < -1.96]$
    $= 2\Pr[Z > 1.96]$ (by symmetry)
    $= 2(0.025)$
    $= 0.05$

Probabilities for other intervals could be obtained by similar calculations.

Continuous approximations of joint distributions involve the same principles outlined for the one-variable case. The idea behind such approximations is pictured in Figure 2.11. The values of two numerical characteristics, X and Y, will fall in the rectangle defined by the inequalities $a < x < b$ and $c < y < d$ with some probability determined by the population under consideration. This probability may be pictured as the volume of a rectangular solid figure of appropriate height. Now, if the base rectangle is sufficiently small, this volume could be approximated by finding the volume of the shaded figure in Figure 2.11. This figure is formed by fitting an appropriate smooth surface $f(x, y)$ over the values of x and y. The volume of the shaded figure is obtained by calculating the double integral of $f(x, y)$ over the rectangular region $a < x < b, c < y < d$. Since a knowledge of double integrals is not assumed in this book, we shall not develop the mathematics any further. We simply state that functions $f(x, y)$ can be found that assign probabilities to regions in the xy plane in such a way that these probabilities satisfy the consistency conditions. Such functions are called joint p.d.f.'s of the continuous random variables X and Y.

Once the notion of joint p.d.f. is understood, it is easy to write down definitions of marginal and conditional p.d.f.'s and the independence of continuous random variables.

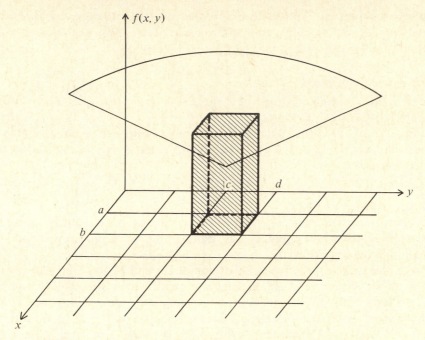

Figure 2.11   Continuous approximation to a joint distribution

The marginal p.d.f.'s of X and Y are given by

$$f(x) = \int_{-\infty}^{\infty} f(x, y)\, dy \tag{2.17}$$

and

$$f(y) = \int_{-\infty}^{\infty} f(x, y)\, dx \tag{2.18}$$

respectively. The conditional p.d.f. of X, given $Y = y$, is

$$f(x|y) = \frac{f(x, y)}{f(y)} \tag{2.19}$$

and the conditional p.d.f. of Y, given $X = x$, is

$$f(y|x) = \frac{f(x, y)}{f(x)} \tag{2.20}$$

From equation (2.20) we have

$$f(x, y) = f(x)f(y|x)$$

and substituting this into equation (2.18) we get

$$f(y) = \int_{-\infty}^{\infty} f(x, y)\, dx = \int_{-\infty}^{\infty} f(x)f(y|x)\, dx$$

Finally, inserting the expression on the right for $f(y)$ into equation (2.19) we can write

$$f(x|y) = \frac{f(x)f(y|x)}{\displaystyle\int_{-\infty}^{\infty} f(x')f(y|x')\,dx'} \qquad (2.21)$$

which is the continuous form of Bayes' theorem. Note the similarity of equation (2.21) to equation (2.12).

Let X and Y be continuous random variables with joint p.d.f. $f(x, y)$ and marginal p.d.f.'s $f(x)$ and $f(y)$, respectively. Then X and Y are said to be independent if

$$f(x, y) = f(x)f(y) \qquad (2.22)$$

for all values of x and y.

A more complete definition of independence would have required also that $f(x, y)$ be positive over a *rectangular* region in the xy plane, but we shall not pursue this point here.

Note the similarity between equations (2.22) and (2.10). Generally such similarities exist between the discrete and continuous cases. Thus it should not be hard for you to conceive of defining joint p.d.f.'s for any finite collection of continuous random variables and the independence of such random variables by extending equation (2.11) to the continuous case. The condition for the independence of a collection of continuous random variables $X_1, X_2, \ldots, X_n$ is

$$f(x_1, x_2, \ldots, x_n) = f(x_1)f(x_2) \ldots f(x_n) \qquad (2.23)$$

where $f(x_1, x_2, \ldots, x_n)$ is the joint p.d.f. and $f(x_1), f(x_2), \ldots, f(x_n)$ are the marginal p.d.f.'s of the random variables.

Functions of Random Variables and Mathematical Expectations

A function of a random variable is simply a coding of the values of the random variable. For example, if X is the (uncertain) number of sales of a product, and each unit of the product sells for $1.50, then the dollar volume of sales is $g(X) = (\$1.50)X$. Other examples of simple functions of the random variable X are $g(X) = X$, $g(X) = X^2$, and $g(X) = 1/X$.

A function of a collection of random variables, $X_1, X_2, \ldots, X_n$, is a set of numbers $g(X_1, X_2, \ldots, X_n)$ assigned to the possible combinations of values of the random variables. For example, $g(X_1, X_2, \ldots, X_n)$ might be the sum, average, or some other linear combination of the X_i's, or $g(X_1, X_2, \ldots, X_n)$ might be the product of the X_i's.

If X is a discrete random variable with probability function $p(x)$ and $g(X)$ is a function of X, then the mathematical expectation of $g(X)$ is denoted by $E[g(X)]$ and is given by

$$E[g(X)] = \sum_x g(x)p(x) \qquad (2.24)$$

If X is a continuous random variable with p.d.f. $f(x)$, and $g(X)$ is a function of X, then the mathematical expectation of $g(X)$ is denoted by $E[g(X)]$ and is given by

$$E[g(X)] = \int_{-\infty}^{\infty} g(x)f(x)\,dx \qquad (2.25)$$

Mathematical expectations of functions of collections of variables are defined by formulas analogous to equations (2.24) and (2.25), except that they involve multiple sums and integrals.

Table 2.3 provides a list of some of the more commonly used mathematical expectations of functions of a single random variable.

Table 2.3   Some Common Mathematical Expectations

$g(X)$	$E[g(X)]$	*Common symbols*
X	$E(X)$: mean	μ_X
$(X - \mu_X)^2$	$E(X - \mu_X)^2$: variance	σ_X^2, or $\text{Var}(X)$
$[(X - \mu_X)/\sigma_X]^3$	$E[(X - \mu_X)/\sigma_X]^3$: skewness	γ_1
$[(X - \mu_X)/\sigma_X]^4$	$E[(X - \mu_X)/\sigma_X]^4$: kurtosis	γ_2

Referring to the entries in Table 2.3, we note that the positive square root of the variance $\sigma_X = +\sqrt{\sigma_X^2}$ is called the *standard deviation* of X. Moreover, the function $h(X) = (X - \mu_X)/\sigma_X$ appearing in the skewness and kurtosis is called the *standardized version* of X. The mean and variance of a standardized random variable are always 0 and 1, respectively.

Let $g(X) = a + bX$ be any linear function of the random variable X; then it can be shown that

$$E(a + bX) = a + bE(X) \qquad (2.26)$$

and

$$\text{Var}(a + bX) = b^2\,\text{Var}(X) \qquad (2.27)$$

If X and Y are random variables, the mathematical expectation of the function $g(X, Y) = (X - \mu_X)(Y - \mu_Y)$ is called the *covariance* of X and Y and is denoted by the symbol σ_{XY} or the symbol $\text{Cov}(X, Y)$; that is,

$$\sigma_{XY} = \text{Cov}(X, Y) = E[(X - \mu_X)(Y - \mu_Y)] = E(XY) - \mu_X\mu_Y \qquad (2.28)$$

The *correlation coefficient* of X and Y, denoted by ρ_{XY}, is simply the covariance of the standardized versions of X and Y, or the ratio of the covariance of X and Y to the product of their standard deviations.

Specifically,

$$\rho_{XY} = \text{Cov}\left[\left(\frac{X - \mu_X}{\sigma_X}\right), \left(\frac{Y - \mu_Y}{\sigma_Y}\right)\right] \qquad (2.29)$$

$$= E\left[\left(\frac{X - \mu_X}{\sigma_X}\right)\left(\frac{Y - \mu_Y}{\sigma_Y}\right)\right]$$

$$= \frac{E[(X - \mu_X)(Y - \mu_Y)]}{\sigma_X \sigma_Y}$$

$$= \frac{\sigma_{XY}}{\sigma_X \sigma_Y} \qquad (2.30)$$

It can be shown that $-1 \leq \rho_{XY} \leq 1$, and that X and Y are linear functions of one another if and only if ρ_{XY} is either $+1$ or -1. If $\rho_{XY} = 0$, X and Y are said to be uncorrelated. Moreover, $\rho_{XY} = 0$ for independent random variables.[b]

EXAMPLE 2.9

We shall calculate the correlation coefficient of the variables X and Y given in Example 2.7. To do this we must first compute the means and variances of X and Y. The calculation of the variances is simplified by noting that

$$\text{Var}(X) = \text{Cov}(X, X) = E(X^2) - \mu_X^2 \qquad (2.31)$$

using equation (2.28). The complete set of calculations follows:

x	$p(x)$	$xp(x)$	$x^2p(x)$
0	0.8	0	0
1	0.2	0.2	0.2
	1.0	$\overline{0.2} = \mu_X$	$\overline{0.2} = E(X^2)$

$\sigma_X^2 = E(X^2) - \mu_X^2 = 0.2 - (0.2)^2 = 0.16$
$\sigma_X = \sqrt{0.16} = 0.4$

y	$p(y)$	$yp(y)$	$y^2p(y)$
-1	0.41	-0.41	0.41
0	0.30	0	0
1	0.29	0.29	0.29
	1.00	$\overline{-0.12} = \mu_Y$	$\overline{0.70} = E(Y^2)$

$\sigma_Y^2 = 0.7 - (-0.12)^2 = 0.6856$
$\sigma_Y = 0.828$

$$E(XY) = (0)(-1)(0.4) + (0)(0)(0.16) + (0)(1)(0.24)$$
$$+(1)(-1)(0.01) + (1)(0)(0.14) + (1)(1)(0.05) = 0.04$$

$$\sigma_{XY} = E(XY) - \mu_X\mu_Y = (0.04) - (0.2)(-0.12) = 0.04 + 0.024 = 0.064$$

$$\rho_{XY} = \frac{\sigma_{XY}}{\sigma_X \sigma_Y} = \frac{0.064}{(0.4)(0.828)} = 0.19$$

[b]The converse of this statement is not true in general; that is, $\rho_{XY} = 0$ does not necessarily imply that X and Y are independent.

Cumulative Distribution Functions and Percentage Points

The cumulative distribution function (c.d.f.) of a random variable X is denoted by $F(x)$ and is defined by

$$F(x) = \Pr[X \le x] = \begin{cases} \sum_{t \le x} p(t) & \text{(discrete case)} \\ \int_{-\infty}^{x} f(t)\, dt & \text{(continuous case)} \end{cases} \tag{2.32}$$

A c.d.f. is a never-decreasing function whose values are bounded by 0 and 1.

As an example, consider the uniform distribution, $f(x) = 1$, $0 \le x \le 1$. Thus when x is less than 0, $F(x) = 0$. For x such that $0 \le x \le 1$ we have

$$F(x) = \int_{0}^{x} 1\, dt = x \tag{2.33}$$

and, finally, $F(x) = 1$ for x greater than or equal to 1. The c.d.f. of the uniform distribution is graphed in Figure 2.12.

If Z has the standard normal distribution with p.d.f. given in equation (2.15), then the c.d.f. of Z is

$$\Pr[Z \le z'] = \int_{-\infty}^{z'} \frac{1}{\sqrt{2\pi}} e^{-(z^2/2)}\, dz \tag{2.34}$$

which is the shaded area shown in Figure 2.13. Thus we see that Appendix

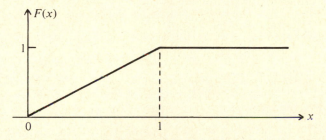

Figure 2.12 The c.d.f. of the p.d.f. $f(x) = 1$, $0 < x < 1$

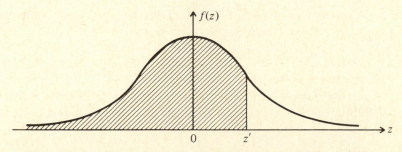

Figure 2.13 The p.d.f. of the standard normal random variable

Table D can be used to provide values of the c.d.f. of the standard normal random variable.

The c.d.f. can be used to define other characteristics of probability distributions. For example, if X is a random variable and p is a number between 0 and 1, the 100 pth *percentile* of the distribution of X is a number η_p such that $\Pr[X < \eta_p] \leq p$ and $\Pr[X \leq \eta_p] \geq p$. Twenty-fifth, 50th, and 75th percentiles of a distribution are called *quartiles* of the distribution, and a 50th percentile is also called a *median*.

EXAMPLE 2.10

Let Z be the standard normal variable. The 60th percentile of Z is the number $\eta_{0.6}$ which has 60 percent of the area under the p.d.f. to its left and 40 percent to its right, as shown in Figure 2.14. From Appendix Table D, $\Pr[Z \leq \eta_{0.6}] = 0.6$, provided

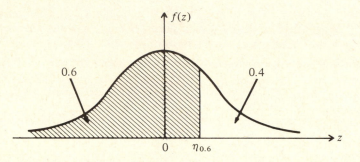

Figure 2.14 The 60th percentile of the standard normal distribution

$\eta_{0.6} = 0.2534$. Thus 0.2534 is the 60th percentile of the standard normal distribution. Similar arguments show that the 25th, 50th, and 75th percentiles of the standard normal distribution are -0.6723, 0, and 0.6723, respectively.

Normal Distributions

We have already suggested that the normal (or Gaussian) distribution is the most important and widely used distribution in statistical analysis. One reason for this is that many everyday phenomena appear to be approximately normally distributed. Another is that many commonly used statistics have approximate normal distributions. The normal p.d.f. curve is symmetric about its mean μ and extends from $-\infty$ to $+\infty$ as indicated in Figure 2.15.

Formally the random variable X has a normal distribution if its p.d.f. is

$$f(x) = \frac{1}{\sqrt{2\pi\sigma^2}} e^{-(x-\mu)^2/2\sigma^2}, \qquad -\infty < x < \infty \qquad (2.35)$$

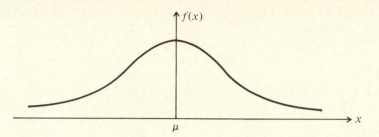

Figure 2.15 The normal curve

The parameters μ and σ^2 are the mean and variance of the distribution, respectively, and although μ may have any numerical value whatsoever, σ^2 must be positive. In our usual notation,

$$\mu_X = \mu \tag{2.36}$$

and

$$\sigma_X^2 = \sigma^2 \tag{2.37}$$

A common practice among statisticians is to refer to a normal distribution with mean μ and variance σ^2 by the symbol $N(\mu, \sigma^2)$. This is a convenient convention, for it eliminates the need to write the lengthy phrase, "normal distribution with mean μ and variance σ^2." We will adopt this notation in this book.

The standardized version of a $N(\mu, \sigma^2)$ random variable is $Z = (X - \mu)/\sigma$, and Z has the standard normal distribution whose p.d.f. is displayed in equation (2.15). In fact, the p.d.f. in equation (2.15) can be obtained from the p.d.f. in equation (2.35) by setting $\mu = 0$, $\sigma^2 = 1$, and $x = z$. This relationship between X and Z allows us to compute probabilities for X using Appendix Table D. The technique for doing this is illustrated in the following example.

EXAMPLE 2.11

Suppose X has an $N(68, (1.5)^2)$ distribution. The p.d.f. of X is plotted in Figure 2.16. The shaded area in this figure is $\Pr[68.5 < X < 71.5]$. Now, the event $(68.5 < X < 71.5)$ is also the event $(68.5 - 68 < X - 68 < 71.5 - 68)$, and this is also the event

$$\frac{68.5 - 68}{1.5} < \frac{X - 68}{1.5} < \frac{71.5 - 68}{1.5}$$

or $(\tfrac{1}{3} < Z < 2\tfrac{1}{3})$, where Z is the standardized version of X. Thus we have

$$\Pr[68.5 < X < 71.5] = \Pr[\tfrac{1}{3} < Z < 2\tfrac{1}{3}] \tag{2.38}$$

But the right-hand side of equation (2.38) can be evaluated using Appendix Table D.

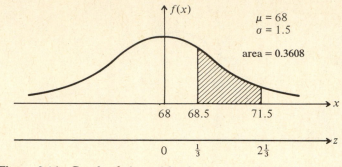

$$\mu = 68$$
$$\sigma = 1.5$$

area = 0.3608

Figure 2.16   Graph of the p.d.f. of a normal distribution on original and standardized scales

This gives $\Pr[\frac{1}{3} < Z < 2\frac{1}{3}] = 0.3608$. Figure 2.16 shows the change of scale effected by standardizing X.

Table 2.4 displays some commonly used probabilities for the normal distribution.

Table 2.4   Probabilities for an $N(\mu, \sigma^2)$ distribution

k	$Pr[|X - \mu| \leq k\sigma]$
0.5	0.3830
1	0.6826
1.645	0.9000
1.96	0.9500
2	0.9545
2.576	0.9900
3	0.9974

The *bivariate* normal distribution is an interesting example of a joint distribution of two continuous random variables. It is the model for several important statistical techniques. Among these are estimation of a correlation coefficient and certain regression problems.

Let the five parameters μ_1, μ_2, σ_1^2, σ_2^2, and ρ be such that $-\infty < \mu_i < \infty$ and $\sigma_i^2 > 0$, $i = 1, 2$, and $-1 \leq \rho \leq 1$. The continuous random variables X and Y have a bivariate normal distribution with the above parameters if the joint p.d.f. of X and Y is

$$f(x, y) = \frac{1}{2\pi\sigma_1\sigma_2\sqrt{1-\rho^2}} \exp\left\{-\frac{1}{2(1-\rho^2)}\left[\left(\frac{x-\mu_1}{\sigma_1}\right)^2\right.\right.$$

$$\left.\left. - 2\rho\left(\frac{x-\mu_1}{\sigma_1}\right)\left(\frac{y-\mu_2}{\sigma_2}\right) + \left(\frac{y-\mu_2}{\sigma_2}\right)^2\right]\right\} \quad (2.39)$$

for $-\infty < x < \infty$ and $-\infty < y < \infty$.

It can be shown that (a) the marginal distributions of X and Y are $N(\mu_1, \sigma_1^2)$ and $N(\mu_2, \sigma_2^2)$, respectively; (b) the correlation coefficient of X and Y is $\rho_{X,Y} = \rho$, and X and Y are independent if and only if $\rho = 0$; (c) the conditional distribution of Y, given $X = x$, is $N(\mu_2 + \rho(\sigma_2/\sigma_1)(x - \mu_1), (1 - \rho^2)\sigma_2^2)$; and (d) the conditional distribution of X, given $Y = y$, is $N(\mu_1 + \rho(\sigma_1/\sigma_2)(y - \mu_2), (1 - \rho^2)\sigma_1^2)$.

Note that (c) states that the conditional mean of Y, given $X = x$, is a linear function of x; that is,

$$E(Y|X = x) = \alpha + \beta x$$

where $\alpha = \mu_2 - \rho(\sigma_2/\sigma_1)\mu_1$ and $\beta = \rho(\sigma_2/\sigma_1)$. [The symbol $E(Y|X = x)$ stands for the mean of the conditional distribution of Y given $X = x$.] Furthermore, the conditional variance of Y, given $X = x$, is $(1 - \rho^2)\sigma_2^2$, the same value for every x. It should be pointed out that the conditional variance of Y, given $X = x$, is *never larger* than the marginal variance of Y, σ_2^2, because the factor $(1 - \rho^2)$ is a fraction between 0 and 1, inclusive.

EXAMPLE 2.12

The preceding remarks may be illustrated numerically by consideration of a bivariate normal distribution with parameters $\mu_1 = \mu_2 = 0$, $\sigma_1^2 = 4$, $\sigma_2^2 = 9$, and $\rho = 0.6$. The conditional distributions of Y, given $X = 1$ and $X = 2$, are $N[(0.6)(3/2)(1), (1 - 0.36)(9)] = N(0.9, 5.76)$ and $N[(0.6)(3/2)(2), (1 - 0.36)(9)] = N(1.8, 5.76)$, respectively. The conditional means $E(Y|X = 1) = 0.9$ and $E(Y|X = 2) = 1.8$ fall on the line

$$E(Y|X = x) = \rho\frac{\sigma_2}{\sigma_1}x = (0.6)\frac{3}{2}(x)$$

whereas the conditional variances for any given set of values of X have the same value $(1 - 0.36)(9) = 5.76$.

EXAMPLE 2.13

Suppose that the joint distribution of a population of fathers' heights (X) and heights of oldest adult daughters (Y) is bivariate normal with parameters $\mu_1 = 68$ in., $\mu_2 = 64$ in., $\sigma_1 = 2.5$ in., $\sigma_2 = 2.7$ in., and $\rho = 0.5$. The conditional mean of daughters' heights given fathers' heights is

$$E(Y|X = x) = \mu_2 + \rho \frac{\sigma_2}{\sigma_1}(x - \mu_1)$$

$$= 64 + (0.5)\left(\frac{2.7}{2.5}\right)(x - 68)$$

$$= 64 + (0.54)(x - 68)$$

$$= 27.28 + (0.54)x \qquad (2.40)$$

Equation (2.40) is interpreted as follows. Consider two populations of fathers, those who are 68 in. tall and those who are 69 in. tall. According to equation (2.40) the *average* height of the oldest daughters of the taller fathers is 0.54 of an inch greater than the *average* height of the oldest daughters of the shorter fathers. Furthermore, because of the assumption of bivariate normality, the *distribution* of heights of daughters whose fathers are 69 in. tall is $N[27.28 + (0.54)(69), (1-0.25)(2.7)^2] = N(64.54, 5.4675)$, whereas the *distribution* of heights of daughters whose fathers are 68 in. tall is $N[27.28 + (0.54)(68), (1-0.25)(2.7)^2] = N(64, 5.4675)$. Similar statements may be made about daughters whose fathers are any specified height.

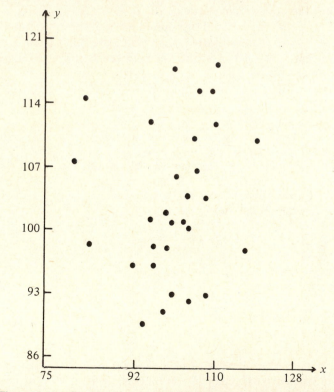

Figure 2.17 Scatter diagram of 32 observations from a bivariate normal distribution $\mu_1 = 100$, $\mu_2 = 105$, $\sigma_1^2 = 100$, $\sigma_2^2 = 100$, $\rho = 0.5$

EXAMPLE 2.14

The generation of observations from a bivariate normal distribution is easily accomplished using a computer. A technique for generating bivariate normal samples is presented in Section 2.4, under the discussion of Simulating Random Samples. In this example we will interpret the results of three sampling experiments that make use of this technique.

Figure 2.17 shows a scatter diagram of 32 observations generated from a bivariate normal distribution with parameters $\mu_1 = 100$, $\mu_2 = 105$, $\sigma_1^2 = 100$, $\sigma_2^2 = 100$, and $\rho = 0.5$, and Figure 2.18 shows 35 obsevations from a bivariate normal distribution with parameters $\mu_1 = 100$, $\mu_2 = 105$, $\sigma_1^2 = 100$, $\sigma_2^2 = 100$, and $\rho = 0.9$. The appearance of linearity is much stronger in the second figure than in the first. Figure 2.19 shows a scatter diagram of 35 observations from a bivariate normal distribution with parameters $\mu_1 = 100$, $\mu_2 = 105$, $\sigma_1^2 = 100$, $\sigma_2^2 = 100$, and $\rho = 0$. In this case the random variables X and Y are independent (discussed previously) and the scatter diagram shows no linear tendencies.

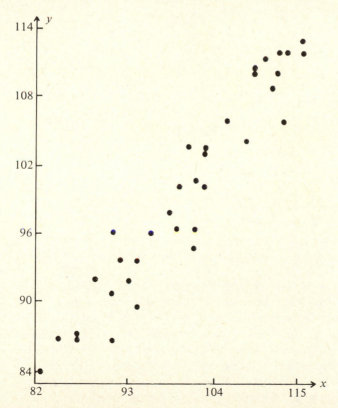

Figure 2.18 Scatter diagram of 35 observations from a bivariate normal distribution $\mu_1 = 100$, $\mu_2 = 105$, $\sigma_1^2 = 100$, $\sigma_2^2 = 100$, $\rho = 0.9$

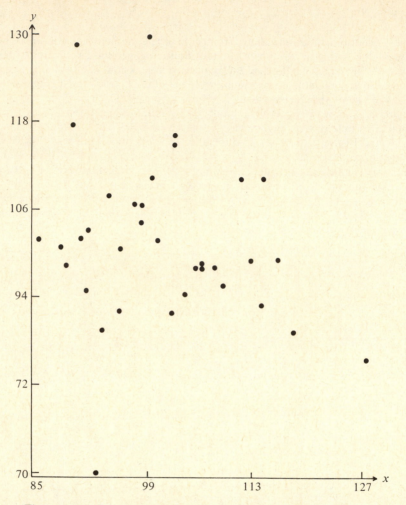

Figure 2.19   Scatter diagram of 35 observations from a bivariate normal distribution $\mu_1 = 100$, $\mu_2 = 105$, $\sigma_1^2 = 100$, $\sigma_2^2 = 100$, $\rho = 0$

Another way to study the bivariate normal distribution is by means of contours of constant density. A contour of constant density is the set of all (x, y) values for which the p.d.f. $f(x, y)$ in equation (2.39) assumes some fixed value c. Contours of the bivariate normal distribution are always ellipses with center at (μ_1, μ_2). This is also the point at which the p.d.f. in equation (2.39) is greatest in value. As c is made smaller, the contour ellipses become wider. These remarks are illustrated in Figure 2.20 in which contour plots of a number of bivariate normal distributions are shown. These plots also show the effects of varying the parameters μ_1, μ_2, σ_1^2, σ_2^2, and ρ. Notice in particular the different orientations as ρ is positive or negative.

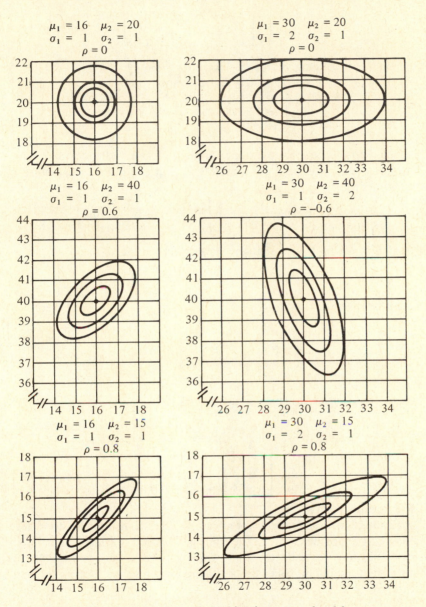

Figure 2.20 Contour plots of the bivariate normal p.d.f.

The bivariate normal p.d.f. in equation (2.39) may be written very simply in terms of matrices.[c] Let the 2×1 column vectors $\mathbf{x}$ and $\mathbf{\mu}$ be defined by

$$\mathbf{x} = \begin{bmatrix} x \\ y \end{bmatrix} \quad \text{and} \quad \mathbf{\mu} = \begin{bmatrix} \mu_1 \\ \mu_2 \end{bmatrix}$$

Then the 2×1 column vector $(\mathbf{x} - \mathbf{\mu})$ is $(x - \mu_1, y - \mu_2)'$.

Let the 2×2 symmetric matrix Σ be defined by

$$\Sigma = \begin{bmatrix} \sigma_1^2 & \rho\sigma_1\sigma_2 \\ \rho\sigma_1\sigma_2 & \sigma_2^2 \end{bmatrix} = \begin{bmatrix} \sigma_X^2 & \sigma_{X,Y} \\ \sigma_{X,Y} & \sigma_Y^2 \end{bmatrix}$$

Σ is called the variance–covariance matrix of the distribution.

The result that we shall demonstrate is that the p.d.f. in equation (2.39) may be written

$$f(x, y) = \frac{1}{2\pi \sqrt{|\Sigma|}} \exp\left[-\frac{1}{2}(\mathbf{x} - \mathbf{\mu})'\Sigma^{-1}(\mathbf{x} - \mathbf{\mu})\right]$$

To see this, we first note that the determinant of Σ is

$$|\Sigma| = \sigma_1^2\sigma_2^2 - \rho^2\sigma_1^2\sigma_2^2 = \sigma_1^2\sigma_2^2(1 - \rho^2)$$

and the inverse of Σ is

$$\Sigma^{-1} = \frac{1}{\sigma_1^2\sigma_2^2(1 - \rho^2)} \begin{bmatrix} \sigma_2^2 & -\rho\sigma_1\sigma_2 \\ -\rho\sigma_1\sigma_2 & \sigma_1^2 \end{bmatrix} = \frac{1}{1 - \rho^2} \begin{bmatrix} \dfrac{1}{\sigma_1^2} & -\dfrac{\rho}{\sigma_1\sigma_2} \\ -\dfrac{\rho}{\sigma_1\sigma_2} & \dfrac{1}{\sigma_2^2} \end{bmatrix}$$

Thus the quadratic form $(\mathbf{x} - \mathbf{\mu})'\Sigma^{-1}(\mathbf{x} - \mathbf{\mu})$ is

$$\frac{1}{1 - \rho^2}\left[\frac{1}{\sigma_1^2}(x - \mu_1)^2 - 2\frac{\rho}{\sigma_1\sigma_2}(x - \mu_1)(y - \mu_2) + \frac{1}{\sigma_2^2}(y - \mu_2)\right]$$

$$= \frac{1}{1 - \rho^2}\left[\left(\frac{x - \mu_1}{\sigma_1}\right)^2 - 2\rho\left(\frac{x - \mu_1}{\sigma_1}\right)\left(\frac{y - \mu_2}{\sigma_2}\right) + \left(\frac{y - \mu_2}{\sigma_2}\right)^2\right]$$

and the result follows.

In closing, we remark that joint normal distributions may be defined in any number of dimensions, say k. The joint p.d.f.'s of k-variate normal distributions would have exactly the same form as above with the understanding that the vectors and matrices are k dimensional and the factor (2π) is raised to the power $k/2$.

[c] You may wish to review Appendix 3 at this point.

2.4 SAMPLING DISTRIBUTIONS

General Discussion

Statistical *inference* is concerned with using a sample to draw conclusions about a population. To achieve this purpose, functions of the sample items (such as the mean and variance) are often formed. If the sample is random, then functions of the sample items are random variables. Such random variables are called *statistics*, and their probability distributions are called the *sampling distributions* of the statistics. Using a relative frequency interpretation of probability, we say that sampling distributions describe the behavior of the statistics when random samples are repeatedly drawn from the population of interest.

Suppose we wish to draw a random sample of size n from a population of N numerical elements that will be denoted by $\omega_1, \omega_2, \ldots, \omega_N$. We note, for later consideration, that the mean and variance of this population will be denoted by μ and σ^2 and are given by the equations

$$\mu = \frac{1}{N} \sum_{i=1}^{N} \omega_i \tag{2.41}$$

and

$$\sigma^2 = \frac{1}{N} \sum_{i=1}^{N} (\omega_i - \mu)^2 \tag{2.42}$$

A random drawing from the population is a drawing made in such a way that each item in the population has the same chance of being selected. Let us denote the result of a random drawing by the random variable X. Then, of course, the probability distribution of X assigns probability $1/N$ to each item ω_i, $i = 1, 2, \ldots, N$. [This is why the mean and variance of X are given by equations (2.41) and (2.42), respectively.]

Now suppose that n random drawings are to be taken from the population *with replacement*; that is, after each drawing, the drawn item is replaced before the next drawing occurs. This means that the drawings are independent, and all are governed by the *same* distribution, the distribution of X defined in the preceding paragraph. Symbolically we could denote the results of the n drawings by the n random variables, $X_1, X_2, \ldots, X_n$ where X_i is the outcome of the ith draw, for $i = 1, 2, \ldots, n$. The assumption of sampling with replacement implies that the X_i's are mutually independent, and that each X_i has the same distribution as X. In other words, in each draw each item in the population has the same chance of being selected.

Since the X_i's have the same distribution, they have equal means and variances—those given by equations (2.41) and (2.42). We may write this as

$$\mu_{X_i} = E(X_i) = \mu \quad \text{and} \quad \sigma^2_{X_i} = \sigma^2, \quad i = 1, 2, \ldots, n \tag{2.43}$$

If the sampling from the population is done without replacement, the sample items, $X_1, X_2, \ldots, X_n$, cannot be independent, and detailed analysis of the sampling scheme is more complicated than in the "with replacement" case. In both cases the most commonly used estimator of the population mean is the *sample* mean

$$\bar{X} = \frac{1}{n}(X_1 + X_2 + \cdots + X_n)$$

$$= \left(\frac{1}{n}\right) X_1 + \left(\frac{1}{n}\right) X_2 + \cdots + \left(\frac{1}{n}\right) X_n \tag{2.44}$$

The sample mean, $\bar{X}$, is a *linear* function of the sample items, and this fact allows us to compute the mean and variance of its sampling distribution rather easily. We now state some important general results about linear functions of random variables and then apply them to the sample mean.

If $X_1, X_2, \ldots, X_n$ are random variables with means and variances $\mu_{X_1}, \mu_{X_2}, \ldots, \mu_{X_n}$, $\sigma^2_{X_1}, \sigma^2_{X_2}, \ldots, \sigma^2_{X_n}$, and covariances $\sigma_{X_1,X_2}, \ldots, \sigma_{X_1,X_n}$, $\sigma_{X_2,X_3}, \ldots, \sigma_{X_2,X_n}, \ldots, \sigma_{X_{n-1},X_n}$, and $L = b_0 + b_1 X_1 + b_2 X_2 + \cdots + b_n X_n$ is any linear combination of the X_i's, then the mean of L is

$$E(L) = b_0 + b_1 E(X_1) + b_2 E(X_2) + \cdots + b_n E(X_n) \tag{2.45}$$

and the variance of L is given by

$$\sigma_L^2 = \sum_{i=1}^{n} b_i^2 \sigma^2_{X_i} + 2 \sum \sum_{i<j} b_i b_j \sigma_{X_i,X_j} \tag{2.46}$$

Although equation (2.46) is written in compact form using summation notation, it actually involves a great many terms. To illustrate, let us take $n = 3$. Then

$$\sigma_L^2 = b_1^2 \sigma^2_{X_1} + b_2^2 \sigma^2_{X_2} + b_3^2 \sigma^2_{X_3} + 2b_1 b_2 \sigma_{X_1,X_2} + 2b_1 b_3 \sigma_{X_1,X_3} + 2b_2 b_3 \sigma_{X_2,X_3}$$

If we take $n = 4$, the number of terms in equation (2.46) grows to

$$4 + \binom{4}{2} = 4 + 6 = 10$$

and so forth. Thus the variance of a linear combination of dependent random variables is quite a complicated expression. If the variables are independent, however, then all the covariance terms are zero, and equation (2.46) takes on the much simplified form

$$\sigma_L^2 = b_1^2 \sigma^2_{X_1} + b_2^2 \sigma^2_{X_2} + \cdots + b_n^2 \sigma^2_{X_n} \tag{2.47}$$

The sample mean is a linear function of $X_1, X_2, \ldots, X_n$ with $b_0 = 0$, and $b_1 = b_2 = \cdots = b_n = 1/n$. Thus equations (2.45) and (2.46) may be used to derive the mean and variance of $\bar{X}$. In the sampling with replacement case, statement (2.43) holds, and $\sigma_{X_i,X_j} = 0$ because the X's are independent. Thus we get

$$\mu_{\bar{X}} = E(\bar{X}) = \mu \quad \text{and} \quad \sigma_{\bar{X}}^2 = \text{Var}(\bar{X}) = \frac{\sigma^2}{n} \tag{2.48}$$

Interestingly enough, statement (2.43) also holds in the sampling without replacement case, but

$$\sigma_{X_i,X_j} = \frac{1}{N^2(N-1)} \left(\sum_{i=1}^{N} \sum_{j=1}^{N} \omega_i \omega_j - N \sum_{i=1}^{N} \omega_i^2 \right) = -\frac{\sigma^2}{N-1} \qquad (2.49)$$

Substituting these relationships into equations (2.45) and (2.46) yields

$$\mu_{\bar{x}} = E(\bar{X}) = \mu \qquad \text{and} \qquad \sigma_{\bar{x}}^2 = \text{Var}(\bar{X}) = \frac{\sigma^2}{n} \left(\frac{N-n}{N-1} \right) \qquad (2.50)$$

Thus the variance of $\bar{X}$ when sampling without replacement is smaller by the multiplicative factor $[(N-n)/(N-1)]$ than the variance of $\bar{X}$ when sampling with replacement.

If N is very large compared to n, the factor $[(N-n)/(N-1)]$ is very close to 1, so that the without replacement modification is negligible. Furthermore, if N is very large compared to σ^2, then the covariance between any pair of observations X_i and X_j, which is $-[\sigma^2/(N-1)]$, will be small. In practical terms, this means that if we are sampling without replacement from a very large population, the sample mean behaves essentially as if the sampling were being done *with* replacement. Thus for relatively small samples from large populations the mathematical model that assumes sampling with replacement and hence independent draws from the population may be used for either of these sampling methods. This is convenient because the model assuming independence is easier to work with.

The foregoing discussion is not intended to be a justification for assuming independent draws when such an assumption is unwarranted (that is, when N is relatively small). It is intended to show why the independent sampling model is so widely used in practice and extensively discussed in textbooks.

In this book we shall assume that the items of a random sample are independent random variables unless the contrary is explicitly stated.

Sampling from Normal Populations

If the number of distinct items in a population is very large, it is often convenient to *assume* the population is infinite and to describe the population by a continuous p.d.f. Suppose the population can be described by a normal distribution. When this assumption is made, certain mathematical results can be derived which strengthen the statements made in the previous section.

Let $X_1, X_2, \ldots, X_n$ denote a random sample from an $N(\mu, \sigma^2)$ population, and let $L = b_0 + b_1 X_1 + \cdots + b_n X_n$ denote a linear combination of the X_i's. Then it can be shown that the sampling distribution of L is *normal* with mean $b_0 + (b_1 + b_2 + \cdots + b_n)\mu$ and variance $(b_1^2 + b_2^2 + \cdots + b_n^2)\sigma^2$. In particular, if L is the sample mean, then $\bar{X}$ has an $N(\mu, \sigma^2/n)$ distribution.

EXAMPLE 2.15

Suppose that a population of accounts receivable has an $N[\mu = \$68, \sigma^2 = (\$2.50)^2]$ distribution, and a simple random sample of size $n = 9$ is to be drawn from the population. The sampling distribution of the sample mean $\bar{X}$ will be normal with mean $\mu = \$68$ and variance $\sigma_{\bar{X}}^2 = \sigma^2/n = (\$2.50)^2/9 = (\$0.833)^2$. Table 2.5 displays 10 independent samples of size 9 from the population along with the sample means. (The samples were generated using the techniques described in Section 2.4 under the discussion of Simulating Random Samples.)

Table 2.5 Ten Random Samples and Sample Means
from an $N[68, (2.5)^2]$ Population

Sample number	x_1	x_2	x_3	x_4	x_5	x_6	x_7	x_8	x_9	MEAN $\bar{x}$
1	68.3	66.9	64.8	64.9	69.0	67.8	67.2	66.7	67.6	67.1
2	67.2	66.1	68.5	63.6	70.3	72.4	66.1	68.4	67.7	67.8
3	65.7	69.5	73.4	70.2	68.6	68.2	70.7	64.7	65.1	68.5
4	64.7	70.4	68.3	64.4	70.5	71.3	68.2	68.9	69.7	68.5
5	68.1	68.1	71.0	65.6	65.9	64.7	65.0	71.0	66.0	67.3
6	66.1	62.5	68.8	62.6	66.4	65.1	68.8	69.5	71.3	66.8
7	73.0	65.9	63.8	64.8	63.1	63.9	70.9	61.5	65.3	65.8
8	71.0	65.9	69.0	70.4	69.9	67.1	69.7	68.5	63.8	68.3
9	65.2	71.0	69.0	70.4	69.9	67.1	69.7	68.5	63.8	68.3
10	63.4	65.6	71.4	63.8	68.0	66.7	69.1	68.8	69.8	67.4

Figure 2.21 is a plot of the p.d.f. of the population, the p.d.f. of the sampling distribution of $\bar{X}$, and a dot diagram of the 10 means in Table 2.5. Notice how the sample means tend to scatter about $\mu = \$68$ in a fashion consistent with their theoretical sampling distribution. A dot diagram of the 90 sample items in Table 2.5 would scatter about $\mu = \$68$ in a fashion consistent with the population distribution. This illustrates the fact that although both the population and the sampling distribution of $\bar{X}$ are centered on the same mean, the variability in the former is greater than the variability in the latter.

The facts describing the behavior of the sample average are fundamental to the theory of statistical inference and will prove useful when we consider the estimation of population means.

The Central Limit Theorem

In the previous two sections we have placed great emphasis on properties of the sample mean $\bar{X}$. Our justification for this is that many statistical problems involve the estimation of population means, and $\bar{X}$ is a reasonable estimator of the population mean.

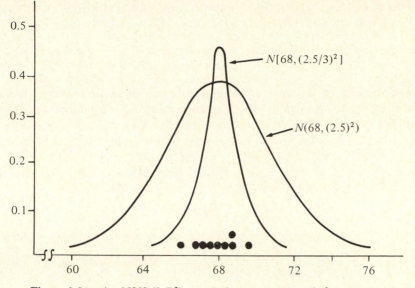

Figure 2.21   An $N[68,(2.5)^2]$ population, an $N[68(2.5/3)^2]$ sampling distribution, and a dot diagram of 10 sample means

Another interesting property of sample averages is that their sampling distributions can often be approximated by a normal distribution regardless of the form of the parent population. This result is known as the central limit theorem (CLT).

The central limit theorem is actually a mathematical statement about the standardized version of $\bar{X}$, namely,

$$Z_n = \sqrt{n}\,\frac{(\bar{X} - \mu)}{\sigma} \tag{2.51}$$

(Here μ and σ are the mean and standard deviation of the population, and n is the size of the random sample taken from the population.) The central limit theorem says that as n increases, the sampling distribution of Z_n may be more and more closely approximated by a standard normal distribution. In other words, when the sample size n is large, then probabilities like $\Pr[Z_n \le z']$ are very close to the probability $\Pr[Z \le z']$ where Z is a standard normal random variable.

Consider the events,

$$Z_n \le z' \tag{2.52}$$

$$\sqrt{n}\,\frac{(\bar{X} - \mu)}{\sigma} \le z' \tag{2.53}$$

$$\bar{X} \le \mu + z'\,\frac{\sigma}{\sqrt{n}} \tag{2.54}$$

As the reader can verify, these events are equivalent. Thus the events in expressions (2.52), (2.53), and (2.54) all have the same probability; that is,

$$\Pr[Z_n \le z'] = \Pr\left[\frac{\sqrt{n}\,(\bar{X} - \mu)}{\sigma} \le z'\right] = \Pr\left[\bar{X} \le \mu + z'\frac{\sigma}{\sqrt{n}}\right] \quad (2.55)$$

When n is large, however, the central limit theorem tells us that this probability is approximately the probability $\Pr[Z \le z']$ obtainable from the standard normal table. We can use the standard normal distribution to approximate probabilities whose exact values are obtainable only from the actual sampling distribution of $\bar{X}$ *provided the sample size is large*. The practical implication of this result is that we can often calculate probabilities for $\bar{X}$ without having to know its exact sampling distribution.

EXAMPLE 2.16

Suppose we anticipate sampling from a population of accounts with standard deviation $\sigma = \$100$ whose mean μ is unknown. For a given sample size n, what is the probability that the sample mean $\bar{X}$ will differ from the population mean by no more than $10? Strictly speaking, we cannot answer this question without knowing the sampling distribution of $\bar{X}$, but if n is sufficiently large, we may obtain an approximate answer by appealing to the central limit theorem.

We are seeking the probability of the event,

$$-\$10 \le \bar{X} - \mu \le \$10 \quad (2.56)$$

that is, the difference between $\bar{X}$ and μ is no more than $10. According to the central limit theorem, however, we may use the standard normal distribution to approximate the probability of events of the form,

$$-z' \le \frac{\sqrt{n}(\bar{X} - \mu)}{\sigma} \le z'$$

which can be written in the equivalent form

$$-z'\frac{\sigma}{\sqrt{n}} \le \bar{X} - \mu \le z'\frac{\sigma}{\sqrt{n}} \quad (2.57)$$

Now, expression (2.57) would be the same as expression (2.56) if we set $z'(\sigma/\sqrt{n}) = \$10$, or

$$z' = \frac{\$10\sqrt{n}}{\sigma} = \frac{\$10\sqrt{n}}{\$100} = \frac{\sqrt{n}}{10}$$

Thus for given values of n, we can find the value of z' and evaluate $\Pr[-z' \le Z \le z']$ using Appendix Table D. For example, take $n = 25$. Then

$$z' = \frac{\sqrt{25}}{10} = \frac{5}{10} = \frac{1}{2} \quad \text{and} \quad \Pr\left[-\frac{1}{2} \le Z \le \frac{1}{2}\right] = 0.3830$$

(using Appendix Table D). Table 2.6 shows the results of a number of such

Table 2.6 Probabilities That $-\$10 \leq \bar{X} - \mu \leq \$10$ Using the Normal Approximation for Various Values of n, $\sigma = \$100$

n	$z' = \sqrt{n}/10$	$\Pr[-z' \leq Z \leq z] \doteq \Pr[-\$10 \leq \bar{X} - \mu \leq \$10]$
25	$\frac{1}{2}$	0.3830
64	$\frac{4}{5}$	0.5762
225	$1\frac{1}{2}$	0.8664
400	2	0.9545

calculations. For the larger sample sizes especially, $\Pr[-z' \leq Z \leq z']$ may be taken as an accurate approximation of $\Pr[-\$10 \leq \bar{X} - \mu \leq \$10]$.

Table 2.6 indicates that if we want the probability of the event in expression (2.56) to be at least 0.8664, we must take a sample of at least $n = 225$ items from the population. To obtain a probability of at least 0.9545, a sample of size $n \geq 400$ is required.

The argument in Example 2.16 may be modified to provide a solution to the following problem. Suppose we wish to take a random sample from a population with known standard deviation σ and unknown mean μ. Suppose further that we want the probability of the event

$$-L \leq \bar{X} - \mu \leq L$$

to be no less than some specified value, say p. How large must the sample size n be? Using the normal approximation and expression (2.57), we set $L = z'(\sigma/\sqrt{n})$. Here z' will be determined by the preassigned value of p, and σ and L are given, so we may solve for n as follows:

$$\sqrt{n} = \frac{z'\sigma}{L}$$

or

$$n = \frac{(z')^2\sigma^2}{L^2} \tag{2.58}$$

As an illustration, take $\sigma = 10$, $L = \frac{1}{2}$, and $p = 0.9545$. Then we must have $z' = 2$ because for a standard normal random variable $\Pr[-2 \leq Z \leq 2] = 0.9545$. This leads to

$$n = \frac{(2)^2(10)^2}{(\frac{1}{2})^2} = (16)(100) = 1600$$

as the required minimum sample size.

In practice, the value of σ is not known, but equation (2.58) might still be applied using a reasonable guess at σ or by obtaining an estimate of σ from, perhaps, a pilot study.

The Chi-Square Distribution

Although the estimation of means is an important statistical problem, variances are also important parameters, and their estimation can lead to the consideration of a sampling distribution known as the "χ-square distribution."

Assume that $X_1, X_2, \ldots, X_n$ is a random sample from an $N(\mu, \sigma^2)$ population. We know that the sample mean

$$\bar{X} = \frac{1}{n} \sum_{i=1}^{n} X_i$$

has an $N(\mu, \sigma^2/n)$ distribution. Now consider the distribution of the *sample variance*

$$S^2 = \frac{1}{n} \sum_{i=1}^{n} (X_i - \bar{X})^2$$

Rather than study the distribution of S^2 directly, it is customary to study the distribution of

$$\frac{nS^2}{\sigma^2} = \sum_{i=1}^{n} \frac{(X_i - \bar{X})^2}{\sigma^2}$$

The distribution of this random variable depends on a single parameter, called "degrees of freedom,"[d] and in this case the degrees of freedom parameter has value $n - 1$. Thus the *distribution* of nS^2/σ^2 is completely known once the value of n is known regardless of whether or not the value of σ^2 is known. It is this property that makes the *distribution* of nS^2/σ^2 useful for statistical inference. The distribution of nS^2/σ^2 is called χ-square with $n - 1$ degrees of freedom and is abbreviated $\chi^2(n - 1)$.

The discussion in the previous paragraph outlines one of the more important ways the χ-square distribution can arise. There are other applications of the χ-square distribution, however, where the degrees of freedom are determined differently. Thus our definition of the χ-square distribution will incorporate a general notation for degrees of freedom that can be specialized to any particular application. We shall denote degrees of freedom by the symbol ν and we always assume $\nu > 0$.

The p.d.f. of the χ-square distribution with ν degrees of freedom,

---

[d] The terminology "degrees of freedom" comes from the following considerations. Suppose we have $n = 3$ sample items, $x_1 = 5$, $x_2 = 3$, $x_3 = 7$. Their mean is $\bar{x} = (5 + 3 + 7)/3 = 5$. Now, the standard deviation is computed from the deviations $x_i - \bar{x}$, that is, $5 - 5 = 0$, $3 - 5 = -2$, and $7 - 5 = 2$. We see that

$$\sum_{i=1}^{3} (x_i - \bar{x}) = 0 + (-2) + (2) = 0$$

and hence knowing two of the deviations we can always obtain the third deviation by subtraction. In this sense, only two deviations are "free" to vary; the third will always be determined by the other two. Hence, there are two "degrees of freedom." In general, if there are n deviations from the sample mean, only $n - 1$ of them are "free" to vary, and hence only $n - 1$ degrees of freedom.

abbreviated $\chi^2(\nu)$, is

$$f(y) = k_1(\nu)y^{(\nu/2)-1}\,e^{-(y/2)}, \qquad y > 0 \tag{2.59}$$

where $k_1(\nu)$ is a constant that insures that the integral of $f(y)$ over the interval $(0, \infty)$ is one.

Remarks   1. The χ^2 random variable Y takes on only positive values with positive probability. (This would necessarily be so for the application in the first paragraph of this section since nS^2/σ^2 could never be a negative quantity.)

2. The constant $k_1(\nu)$ is a rather complicated function of ν which will not be displayed here.

3. Graphs of four χ^2 distributions, for $\nu = 1, 2, 8,$ and 11 are shown in Figure 2.22. Appendix Table F displays values of y for which $\Pr[Y \geq y] = p$ for various choices of the probability p and degrees of freedom ν.

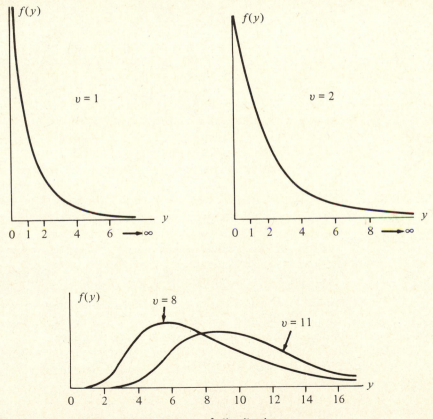

Figure 2.22   Graphs of four χ^2 distributions

EXAMPLE 2.17

Let $X_1, X_2, \ldots, X_{10}$ denote a random sample of size $n = 10$ from an $N(\mu, \sigma^2)$ distribution, and let

$$Y = \frac{(10)S^2}{\sigma^2} = \sum_{i=1}^{10} \frac{(X_i - \bar{X})^2}{\sigma^2}$$

Then according to the discussion in the first paragraph of this section, the random variable Y has a $\chi^2(9)$ distribution (here $\nu = n - 1 = 10 - 1 = 9$). From Appendix Table F, $\Pr[Y > 8.34] = 0.50$, $\Pr[Y > 14.7] = 0.10$, and $\Pr[Y > 21.7] = 0.01$.

The mean and variance of the χ^2 distribution are

$$\mu_Y = \nu \tag{2.60}$$

and

$$\sigma_Y^2 = 2\nu \tag{2.61}$$

Moreover, when ν is large, the approximate distribution of the random variable

$$Z_\nu = \sqrt{2Y} - \sqrt{2\nu - 1} \tag{2.62}$$

is $N(0, 1)$.

There is an important relationship between the normal and χ^2 distributions. Specifically, if X has an $N(\mu, \sigma^2)$ distribution, so that $Z = (X - \mu)/\sigma$ has an $N(0, 1)$ distribution, then the distribution of $Z^2 = (X - \mu)^2/\sigma^2$ is $\chi^2(1)$. Furthermore, if $X_1, X_2, \ldots, X_n$ is a random sample from an $N(\mu, \sigma^2)$ population, so that $Z_1 = (X_1 - \mu)/\sigma$, $Z_2 = (X_2 - \mu)/\sigma, \ldots, Z_n = (X_n - \mu)/\sigma$ are independent $N(0, 1)$ random variables, then

$$\sum_{i=1}^{n} Z_i^2 = \sum_{i=1}^{n} \frac{(X_i - \mu)^2}{\sigma^2}$$

has a $\chi^2(n)$ distribution.

The t Distribution

Another pervasive sampling distribution is the t distribution (often called "Student's t"). This distribution arises in many statistical applications, particularly those involving the estimation of means. In this section we shall formally define the t distribution, discuss some of its properties, and then consider a situation in which the distribution arises in practice.

Like the χ-square distribution, the t distribution depends on a single parameter called "degrees of freedom" which we shall denote by ν. The p.d.f. of the t distribution with ν degrees of freedom is

$$f(t) = k_2(\nu)\left(1 + \frac{t^2}{\nu}\right)^{-(\nu+1)/2}, \qquad -\infty < t < \infty \tag{2.63}$$

where $k_2(\nu)$ is a constant that insures that the integral of $f(t)$ over the interval $(-\infty, \infty)$ will be 1. Like the normalizing constant for the χ^2 distribution, the form of $k_2(\nu)$ is unimportant for our purposes and will not be displayed. The t distribution with ν degrees of freedom will often be denoted by the symbol $t(\nu)$.

The p.d.f. for $\nu = 16$ is shown in Figure 2.23 along with the p.d.f. of the $N(0, 1)$ distribution. Note that the t distribution is symmetric about the

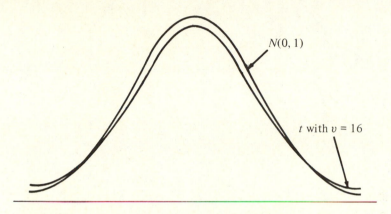

Figure 2.23   The $t(16)$ and $N(0, 1)$ p.d.f.'s

origin, but that it has fatter tails than the normal distribution. Figure 2.24 illustrates these facts in a different way. The central intervals which contain 95 percent of the values in a t distribution with 16 degrees of freedom and an $N(0, 1)$ distribution are shown in the figure. It is clear that one must go farther into the tails of the t distribution to capture 95 percent of the values than in the tails of the normal distribution. Appendix Table E gives intervals of selected probability content for t distributions with degrees of freedom ranging from $\nu = 1$ through $\nu = 30$ and for selected values of ν larger than 30. The $\nu = \infty$ row at the bottom of the table is particularly interesting because it displays the intervals appropriate for the $N(0, 1)$ distribution [for example, the 95-percent interval for $\nu = \infty$ is $(-1.96, 1.96)$]. This is the consequence of the fact that the limit of the p.d.f. in equation (2.63) as $\nu \to \infty$ is the p.d.f. of the $N(0, 1)$ distribution. Informally we say that the t distribution approaches the normal distribution as the degrees of freedom increase.

For $\nu > 2$ the mean and variance of the $t(\nu)$ distribution are

$$\mu_{t(\nu)} = 0 \qquad (2.64)$$

and

$$\sigma^2_{t(\nu)} = \frac{\nu}{\nu - 2} \qquad (2.65)$$

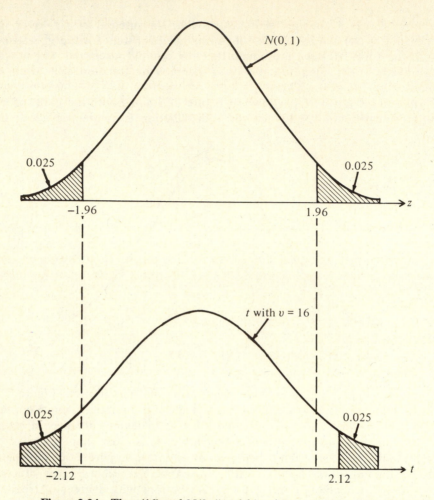

Figure 2.24   The $t(16)$ and $N(0, 1)$ p.d.f.'s with central 95-percent intervals

A random variable that has the p.d.f. in equation (2.63) may be defined in the following way. Let Z be an $N(0, 1)$ random variable, and let Y be a $\chi^2(\nu)$ random variable that is *independent* of Z. Then the random variable defined by

$$t = \frac{Z}{\sqrt{Y/\nu}}$$
(2.66)

has a $t(\nu)$ distribution. Equation (2.66) says we can form a $t(\nu)$ random variable by dividing an $N(0, 1)$ random variable by the square root of an independent $\chi^2(\nu)$ random variable divided by its degrees of freedom.

The definition in equation (2.66) stems from the following considerations. Let $X_1, X_2, \ldots, X_n$ denote a random sample from an $N(\mu, \sigma^2)$ population, and let $\bar{X}$ and S^2 be the sample mean and variance. We have already seen that $Z = \sqrt{n}(\bar{X} - \mu)/\sigma$ has an $N(0, 1)$ distribution, and $Y = nS^2/\sigma^2$ has a $\chi^2(n-1)$ distribution. In addition, it is possible to show that the random variables Z and Y defined here are *independent*. This is a fact that is not easy to prove, and it depends on our assumption of a normal population. However, given these assumptions, it is clear that we can use equation (2.66) to form a $t(n-1)$ random variable, namely,

$$t = \frac{Z}{\sqrt{Y/\nu}} = \left[\frac{\sqrt{n}(\bar{X} - \mu)}{\sigma}\right] \bigg/ \sqrt{\left(\frac{nS^2}{\sigma^2}\right)\bigg/(n-1)}$$

$$= \sqrt{n}(\bar{X} - \mu) \bigg/ \sqrt{\frac{1}{n-1}\sum_{i=1}^{n}(X_i - \bar{X})^2} \qquad (2.67)$$

Neither the random variable in equation (2.67) nor its distribution depends on σ^2. This means that the sample mean and population mean may be related in a distribution that does not depend on σ^2. This fact allows us to make inferences about the population mean without knowing the population variance.

The F Distribution

Suppose U and V are *independent* random variables such that U has a $\chi^2(\nu_1)$ distribution and V has a $\chi^2(\nu_2)$ distribution. The random variable defined by the ratio

$$F = \frac{U/\nu_1}{V/\nu_2} \qquad (2.68)$$

is said to have an F distribution with ν_1 and ν_2 degrees of freedom. This F distribution will be denoted by the symbol $F(\nu_1, \nu_2)$ and its p.d.f. is

$$f(x) = k_3(\nu_1, \nu_2)x^{(\nu_1 - 2)/2}(\nu_2 + \nu_1 x)^{-(\nu_1 + \nu_2)/2} \qquad 0 \le x < \infty \qquad (2.69)$$

where the normalizing constant, $k_3(\nu_1, \nu_2)$, is a complicated expression involving ν_1 and ν_2.

The shape of the F distribution changes as either ν_1 or ν_2 or both change. A typical F distribution is pictured at the top of Appendix Table G. We note that ν_1 is the number of degrees of freedom associated with the numerator of the F ratio, whereas ν_2 is the number of degrees of freedom associated with the denominator. The $100(1 - \alpha)$ percentile of the $F(\nu_1, \nu_2)$ distribution is denoted by $F_\alpha(\nu_1, \nu_2)$. Appendix Table G gives 95th and 99th percentiles of the F distribution for a number of values of ν_1 and ν_2.

We will have numerous occasions to refer to the F distribution in later chapters. Here we simply introduce a situation in which the F distribution arises in statistical inference.

Let $X_{11}, X_{12}, \ldots, X_{1m}$ and $X_{21}, X_{22}, \ldots, X_{2n}$ denote *independent* random samples from $N(\mu_1, \sigma_1^2)$ and $N(\mu_2, \sigma_2^2)$ populations, respectively. We know from the discussion of the χ-square distribution that the random variables

$$\sum_{i=1}^{m} \frac{(X_{1i} - \bar{X}_1)^2}{\sigma_1^2} \quad \text{and} \quad \sum_{j=1}^{n} \frac{(X_{2j} - \bar{X}_2)^2}{\sigma_2^2}$$

have $\chi^2(m-1)$ and $\chi^2(n-1)$ distributions, respectively. Furthermore, these two χ-square random variables are independent because they are functions of independent random samples. Thus the ratio

$$F = \frac{\displaystyle\sum_{i=1}^{m} \frac{(X_{1i} - \bar{X}_1)^2}{\sigma_1^2(m-1)}}{\displaystyle\sum_{j=1}^{n} \frac{(X_{2j} - \bar{X}_2)^2}{\sigma_2^2(n-1)}} \tag{2.70}$$

has an $F(m-1, n-1)$ distribution. This fact can be used to make inferences about the variance ratio σ_2^2/σ_1^2. Further discussion of this situation is presented in Chapter 3.

Simulating Random Samples

Simulation occurs in many areas of applied mathematics. In statistics the need for simulation usually arises from the need for studying the sampling distribution of a statistic. Often purely analytical tools are not adequate for the study of a very complicated statistic. On the other hand, the observed behavior of the statistic computed from simulated samples can yield important insights relatively quickly. Sometimes such insights even lead to more analytical work, as they suggest "results" that were not evident before.

In this section we shall give a fairly detailed introduction to the simulation of random samples from *continuous* probability distributions. A basic result needed in this discussion follows.

Theorem   Let X be a continuous random variable with c.d.f. $F(x)$. Define the random variable $U = F(X)$. Then U has the uniform distribution on $(0, 1)$.

Proof   The c.d.f. of U is

$$F_U(u) = \Pr[U \le u] = \Pr[F(X) \le u] \tag{2.71}$$

where u is a number between 0 and 1. Now let x_u denote the unique value of X such that $F(x_u) = u$. Figure 2.25 shows that such an x must exist because $F(x)$ is an increasing function of x. But $F(x) \le u$ whenever $x \le x_u$ (see Figure 2.25), so

$$\Pr[F(X) \le u] = \Pr[X \le x_u] = F(x_u) = u$$

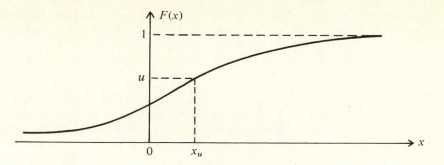

Figure 2.25   The c.d.f. of the continuous random variable X showing the solution of the equation $F(x_u) = u$

Combining this fact with equation (2.71), we have

$$F_U(u) = F(x_u) = u$$

for $0 < u < 1$. This is the c.d.f. of the uniform distribution on $(0, 1)$, however, so U must have this distribution.

The theorem suggests a very useful technique for generating random samples from continuous populations. This technique will be illustrated in the following example.

EXAMPLE 2.18

Let T be a random variable with the c.d.f.

$$F(t) = 1 - e^{-t}, \qquad t > 0 \tag{2.72}$$

This c.d.f. is graphed in Figure 2.26. In order to draw an item at random from this distribution, we first draw an item u at random from the interval $(0, 1)$. This item corresponds to a value of T, t_u, using the equation $u = 1 - e^{-t_u}$. Solving this equation for t_u, we get $t_u = -\ln(1 - u)$. From the theorem we have $\Pr[T \le t_u] = \Pr[U \le u] = u$, so t_u is the $100u$th percentile of this distribution.

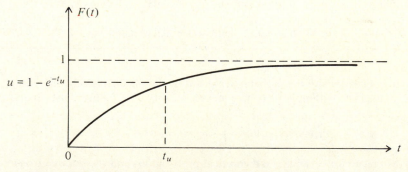

Figure 2.26   The c.d.f. $F(t) = 1 - e^t, t > 0$

The suggested procedure may be implemented only if we can draw items at random from the continuous uniform distribution. In practice, we can accurately measure to only a finite number of decimal places, and hence we can only hope to approximate sampling from a continuous uniform distribution. By using a table of random digits, like Appendix Table A, we may "measure" a uniform variable to any desired number of decimal places.

Suppose we want five-place accuracy. We can arbitrarily pick a starting point in Appendix Table A, read off five consecutive digits, and form a five-place decimal. If we choose line 27, column 21 of the table, we obtain the number $u = 0.38670$. Now, corresponding to $u = 0.38670$ there is a value $t_{0.38670} = -\ln(1 - 0.38670) = 0.48890$, and this is the sampled value of T. Continuing in the table of random digits we obtain another random fraction $u' = 0.88243$, so $t_{u'} = -\ln(1 - 0.88243) = 2.14072$. In this way, a random sample of any size may be generated.

The procedure outlined above was used to generate a sample of 100 items from the distribution whose c.d.f. is given in Figure 2.26. A histogram of the sample is displayed in Figure 2.27, and Table 2.7 compares the relative frequencies of the

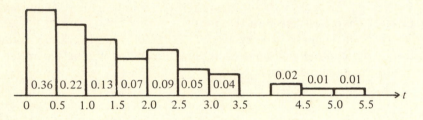

Figure 2.27   Histogram of 100 observations generated from the distribution with c.d.f. $F(t) = 1 - e^{-t}$, $t > 0$

Table 2.7   Comparison of Probabilities and Relative Frequencies from a Sampling Experiment

Class	0–0.5	0.5–1.0	1.0–1.5	1.5–2.0	2.0–2.5	2.5–3.0	3.0–3.5	3.5–4.0	4.0–4.5	4.5–5.0	5.0–5.5
Probability	0.39	0.24	0.14	0.09	0.05	0.03	0.02	0.01	0.01	0	0
Relative frequency	0.36	0.22	0.13	0.07	0.09	0.05	0.04	0	0.02	0.01	0.01

histogram classes with the probabilities of those classes obtained by integrating the p.d.f. of the distribution. For example, the probability of T falling in the second class is

$$\Pr[0.5 < T < 1.0] = \int_{0.5}^{1} e^{-t}\, dt = e^{-0.5} - e^{-1} = 0.24$$

The discrepancies between the theoretical probabilities and the relative frequencies are due to sampling variability.

The procedure outlined in Example 2.18 could be used to generate samples from any continuous distribution, but it is not always the most efficient method. The next example presents a very efficient method for generating random samples from an $N(0, 1)$ population. The method is due to Box and Muller, reference [19].

EXAMPLE 2.19

Let u_1 and u_2 be two items independently generated from a uniform distribution on $(0, 1)$. Then

$$z_1 = \sqrt{-2 \ln u_1} \cos (2\pi u_2)$$

$$z_2 = \sqrt{-2 \ln u_2} \sin (2\pi u_1)$$

(2.73)

are two independent items from an $N(0, 1)$ distribution.

The equations in (2.73) can be solved for z_1, z_2 given u_1 and u_2 and this procedure is easier than solving the equation $F(z) = u$. Table 2.8 shows how 20 pairs of independent

Table 2.8 Generation of 40 Independent $N(0, 1)$ Items[a]

Pair No.	u_1	u_2	z_1	z_2
1	0.644642	0.701380	−0.28183	−0.66436
2	0.335852	0.080943	1.29026	1.92393
3	0.078680	0.771056	0.29745	0.34214
4	0.467546	0.293248	−0.33097	0.31719
5	0.254436	0.070989	1.49264	2.29920
6	0.285633	0.246192	0.03787	1.63251
7	0.775011	0.403185	−0.58590	−1.33125
8	0.526138	0.640749	−0.71828	−0.15426
9	0.272030	0.184547	0.64505	1.82081
10	0.797893	0.101837	0.53905	−2.04142
11	0.033454	0.306802	−0.91072	0.32075
12	0.784079	0.583375	−0.60395	−1.01449
13	0.093892	0.728301	−0.29564	0.44299
14	0.367762	0.828140	0.66688	0.45354
15	0.653190	0.770659	0.11946	−0.59235
16	0.235196	0.485799	−1.69461	1.19644
17	0.657691	0.130158	0.62600	−1.68917
18	0.806037	0.926460	0.58783	−0.36688
19	0.857939	0.645951	−0.87179	0.72802
20	0.899488	0.282561	−0.09351	−0.93864

[a] For each pair (u_1, u_2) of uniform variables,

$$z_1 = \sqrt{-2 \ln u_1} \cos (2\pi u_2)$$

$$z_2 = \sqrt{-2 \ln u_2} \sin (2\pi u_1)$$

uniform numbers can be transformed into 20 pairs of independent $N(0, 1)$ numbers, and Table 2.9 presents a comparison of the relative frequency distribution of the 40 numbers with the theoretical $N(0, 1)$ probability distribution. The agreement between the two is reasonably good.

Table 2.9 Comparison of Probabilities and Relative Frequencies for a Sample of Size 40 from an $N(0, 1)$ Population

Class	Below −1.5	−1.5 to −0.5	−0.5 to 0.5	0.5 to 1.5	Above 1.5
Probability	0.067	0.242	0.383	0.242	0.067
Relative frequency	0.075	0.250	0.350	0.225	0.100

Example 2.19 illustrates the generation of independent items from a single $N(0, 1)$ population. Once a sample of $N(0, 1)$ items has been generated, it may be converted to a sample of $N(\mu, \sigma^2)$ items by the transformation

$$x = \sigma z + \mu \qquad (2.74)$$

This technique was used to generate the numbers in Table 2.5.

We now turn to the generation of samples from a bivariate normal distribution. Suppose the values of the parameters μ_1, μ_2, σ_1^2, σ_1^2, and ρ are fixed, and we wish to generate a sample of size n from the bivariate normal distribution with these parameter values. Each (x, y) pair may be generated sequentially by first generating x according to the marginal distribution $N(\mu_1, \sigma_1^2)$ and then generating y from the conditional distribution, given the value of x, $N(\mu_2 + \rho(\sigma_2/\sigma_1)(x - \mu_1), (1 - \rho^2)\sigma_2^2)$.

To illustrate the procedure, suppose we wish to generate an observation from the bivariate normal distribution with parameters $\mu_1 = 100$, $\mu_2 = 105$, $\sigma_1^2 = 100$, $\sigma_2^2 = 100$, and $\rho = 0.5$. Let (z_1, z_2) be a pair of independent $N(0, 1)$ items generated by the procedure outlined in Example 2.19 (or some other suitable method). For illustration, we take $z_1 = -0.586$ and $z_2 = 0.359$. The value of x is calculated by equation (2.74) as

$$x = \sigma_1 z_1 + \mu_1 = (10)(-0.586) + 100 = 94.14$$

The value of y corresponding to x is generated from the distribution

$$N\left(\mu_2 + \rho\frac{\sigma_2}{\sigma_1}(x - \mu_1), (1 - \rho^2)\sigma_2^2\right) = N\left(\mu_2 + \rho\sigma_2 z_1, (1 - \rho^2)\sigma_2^2\right)$$
$$= N(105 + (0.5)(10)(-0.586), (1 - 0.25)(100))$$
$$= N(102.07, 75)$$

Thus to obtain the value of y, we calculate

$$y = \sqrt{75}\, z_2 + 102.07$$

$$= (8.66025)(0.359) + 102.07$$

$$= 105.18$$

This procedure may be repeated until a sample of n values is drawn. The items displayed in Figures 2.17, 2.18, and 2.19 were obtained in this way.

In closing, we remark that simulations are usually done with computers, as they involve quite a bit of tedious arithmetic. Most computers have routines to generate uniform random numbers, and from these, random samples from other continuous populations can usually be obtained.

2.5   FINAL COMMENTS

Our goal in this chapter has been to develop a probability model for experiments whose outcomes are governed by the laws of chance. The basic element in this model is the probability *distribution*. A distribution is characterized by its probability function if it is discrete, by its probability density function if it is continuous, or by its cumulative distribution function in either case. Using the probability function or the probability density function, useful mathematical expectations (weighted averages) may be defined. Examples are the mean and the variance. Using the cumulative distribution function, concepts such as medians and percentiles may be defined.

In practice, it is common to use characteristics of samples (statistics) to estimate analogous characteristics of populations (parameters). Such practice naturally leads to the study of the behavior of statistics. In random samples the behavior of statistics will be governed by the laws of chance, and the probability distributions of statistics are called *sampling distributions*.

The sampling distribution of the sample mean has been presented in detail because the sample mean is such a commonly used statistic. Furthermore, the central limit theorem illustrates the large sample approximation of a sampling distribution by a normal distribution, an often-encountered technique in statistics.

Other sampling distributions that will be referred to often in this book (the χ^2, t, and F) were defined and discussed briefly in Section 2.4. These distributions play a central role in the theory of statistical inference.

Finally, we have discussed in some detail the generation of random samples from theoretical probability distributions. This technique is a useful device for both the theoretician and the practitioner.

PROBLEMS

2.1 For each of the experiments listed below write down the sample space Ω and state how many events (subsets) there are.
   (a) Tossing a coin twice (denote "heads" by H and "tails" by T).
   (b) Tossing a coin three times.
   (c) Rolling a die.
   (d) Rolling a die twice.
   (e) Drawing a card from a bridge deck until you get a spade.

2.2 What probabilities would you assign to the elementary outcomes in each part of Problem 2.1?

2.3 Mr. M and Mr. W are playing a game that involves flipping a fair coin. If the coin shows "heads" Mr. W wins a point, if "tails," Mr. M wins a point. The first player to obtain three points is the winner. Write down the sample space, and beside each elementary outcome record the winner.

2.4 Which of the following cannot be probability functions regardless of the probabilities assigned to other possible events? Explain your answers.
   (a) $P(\{a, b\}) = 1/2$, $P(\{a\}) = 2/3$
   (b) $P(\{a, b\}) = 1/4$, $P(\{c, d\}) = 1/2$
   (c) $P(\{a, b, c\}) = 0$
   (d) $P(\{b\}) = P(\{c\}) = P(\{d\}) = 1/3$
   (e) $P(\{a, b, c\}) = 2/3$, $P(\{a, b\}) = P(\{b, c\}) = 1/4$

2.5 Repeat experiments in Problem 2.1, parts (a) and (c), 200 times and record the frequencies and relative frequencies of the elementary outcomes. What do you conclude?

2.6 Find a sports story in a daily newspaper, and measure the word lengths in letters. Record the numbers of words of lengths 1, 2, 3, etc., and calculate the relative frequencies. Repeat your experiment for several other sports stories. Do your data suggest any conclusions? How might you compare word lengths in sports writing with word lengths in, say, the stories in the *Wall Street Journal*?

2.7 Choose a page in a residential telephone book and record the frequencies of the last digits in the telephone numbers. Calculate the relative frequencies. Do these data suggest any conclusions?

2.8 Cite three examples of statistical regularity that you have observed.

2.9 Given $P(A) = 0.4$, $P(B) = 0.5$, and $P(A \cap B) = 0.3$, find (a) $P(A \cup B)$; (b) $P(A \cap \bar{B})$ and $P(\bar{A} \cap B)$; and (c) $P(\bar{A} \cap \bar{B})$.

2.10 If A and B represent mutually exclusive events, $P(A) = 0.25$, and $P(B) = 0.40$, find each of the following probabilities: (a) $P(\bar{A})$; (b) $P(\bar{B})$; (c) $P(A \cup B)$; (d) $P(A \cap B)$; and (e) $P(\bar{A} \cap \bar{B})$.

2.11   Prove that

$$P(A \cup B \cup C) = P(A) + P(B) + P(C) - P(A \cap B) - P(A \cap C)$$
$$- P(B \cap C) + P(A \cup B \cup C)$$

2.12   Consider the game in Problem 2.3. Suppose one player is ahead two points to one point and the players must stop playing without finishing the game. What is the probability that the person with two points would have eventually won the game? How should the stakes be divided?

2.13   If the eight possible outcomes for three flips of a coin have equal probabilities of 1/8, what is the probability of getting one head and two tails (not necessarily in that order)?

2.14   If the numbers from 1 through 100 are written on slips of paper and mixed in a goldfish bowl, and each slip of paper has an equal chance of being drawn, what is the probability of drawing a number that is divisible by 13?

2.15   A man owns a house in town and a cabin in the mountains. In any one year the probability of the house being burglarized is 0.01, and the probability of the cabin being burglarized is 0.05. For any one year what is the probability that
(a) Both will be burglarized?
(b) One or the other (but not both) will be burglarized?
(c) Neither will be burglarized?
(d) Did you assume the two events were independent? Do you think this a reasonable assumption?

2.16   The interval $(0, 1)$ is to be divided into two parts at random. What is the probability that one of the parts will be at least twice as long as the other?

2.17   A box contains six white and three green chips. A second box contains four white and five green chips. A chip is selected at random from the first box and placed in the second box. Then one is drawn from the second. What is the probability that the second chip is green?

2.18   A box contains two red and three blue chips. What is the probability that two chips drawn from the box simultaneously will both be red?

2.19   The probability that a certain door is locked is 1/2. The key to the door is one of 12 keys in a cabinet. If a person selects two keys at random from the cabinet and takes them to the door with him, what is the probability that he can open the door without returning for another key?

2.20   Playing five-card draw poker, Mr. Jones is dealt a pair and decides to flip a coin: Heads he keeps the pair and draws three cards, tails he keeps the pair and a "kicker" to bluff his opponents and draws only two cards. If he draws three cards, the probability of Mr. Jones' improving his hand is 0.287; if he draws only two cards the probability is 0.260. What is the probability that he will improve his hand?

2.21   (Bayes' Theorem): Suppose the sample space Ω is divided into three mutually exclusive subsets, H_1, H_2, H_3, as pictured in the accompanying figure, and suppose $P(H_1)$, $P(H_2)$, and $P(H_3)$ are given. Now suppose the experiment is performed, and you are told (1) that the event D occurred (Figure P2.21) and (2) the values of $P(D|H_i)$, $i = 1, 2, 3$.

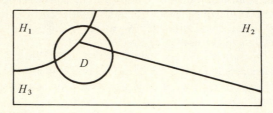

Figure P2.21

(a)   Show that

$$P(H_i|D) = \frac{P(H_i)P(D|H_i)}{\sum_{j=1}^{3} P(H_j)P(D|H_j)}, \qquad i = 1, 2, 3$$

(b)   Generalize the result above to the case in which Ω is divided into k mutually exclusive events (called a "partition of Ω"), where k is an arbitrary positive integer; that is, show that

$$P(H_i|D) = \frac{P(H_i)P(D|H_i)}{\sum_{j=1}^{k} P(H_j)P(D|H_j)}, \qquad i = 1, 2, \ldots, k$$

This result is known as Bayes' theorem. The probabilities $P(H_i)$, $i = 1, 2, \ldots, k$ are called *a priori*, or prior, probabilities, that is, probabilities assigned *before* the experiment is performed. The probabilities $P(H_i|D)$, $i = 1, 2, \ldots, k$ are called *a posteriori*, or posterior, probabilities, that is, those probabilities assigned *after* the experiment is performed and D is observed.

2.22   (Bayes' Theorem): Suppose the experiment in Problem 2.21(a) is performed a *second* time, and you are told (1) that the event E occurred and (2) the values of $P(E|H_i)$, $i = 1, 2, 3$.
(a)   Show that

$$P(H_i|D \cap E) = \frac{P(H_i)P(D \cap E|H_i)}{\sum_{j=1}^{3} P(H_j)P(D \cap E|H_j)}$$

$$= \frac{P(H_i|D)P(E|H_i \cap D)}{\sum_{j=1}^{3} P(H_j|D)P(E|H_j \cap D)}, \qquad i = 1, 2, 3$$

(b)   Generalize this result to the case in which Ω is partitioned into k mutually exclusive events and the experiment is repeated n times.

2.23   (Bayes' Theorem): Two bowls, numbered I and II, are hidden behind a screen. Bowl I contains seven red and three blue chips, and bowl II contains five red and five blue chips. A person behind the screen flips a fair coin. If he obtains a "head" he draws a chip at random from bowl I, and if he obtains a "tail" he draws a chip at random from bowl II. In either case he shows you the chip he has drawn. If he shows you a red chip, what is the probability that he drew the chip from bowl I? From bowl II?

2.24   Consider the same setup as in Problem 2.23, but suppose the person draws two chips without replacement from the appropriate bowl. Assuming the person shows you two red chips, calculate the *a posteriori* probabilities of bowl I and bowl II using the two methods in Problem 2.22.

2.25   (Bayes' Theorem): A manufacturer obtains a certain part from three different suppliers. Fifty percent of the manufacturer's stock of parts comes from supplier I, 15 percent from supplier II, and 35 percent from supplier III. It is known from past experience that 10 percent of supplier I's parts are defective, 6 percent of supplier II's parts are defective, and 15 percent of supplier III's parts are defective. If a part is taken at random from the manufacturer's stock and found to be defective, what is the probability that the part came from supplier I? Supplier II? Supplier III?

2.26   If the probability that a married man will vote in a given election is 0.50, the probability that a married woman will vote in the election is 0.60, and the probability that a woman will vote in the election given that her husband votes is 0.90, find
   (a)   The probability that a husband and wife both vote in the election.
   (b)   The probability that a man will vote in the election given that his wife will vote.
   (c)   The probability that at least one of a married couple will vote in the election.

2.27   Suppose that among three indistinguishable boxes one contains two pennies, one contains a penny and a dime, and one contains two dimes. Selecting one of the boxes at random, a coin is drawn from it at random. If it is a penny, what is the probability that the other coin in this box is also a penny?

2.28   A certain gentleman has two possible routes driving home from the office: a more scenic route through the countryside which he takes 60 percent of the time, and a faster route on the expressway which he takes 40 percent of the time. If he takes the scenic route, the probability he will be late for dinner is 0.8; if he takes the expressway, the probability he will be late for dinner is 0.1. If he is late for dinner, what is the probability that he took the scenic route?

2.29   Half of all the articles bought at an expensive women's accessory shop are later returned. The manager has noticed that if an item is bought by a woman who is accompanied by at least one other woman, it is invariably returned. Items bought under other circumstances are returned 37.5 percent of the time. If an item is returned, what is the probability it was bought by a woman accompanied by at least one other woman?

2.30   A student has studied two sets of questions in preparation for an examination. One set contains six questions of which he can answer five; the other set contains four questions of which he can answer two. The examination consists of three questions selected at random from one of the two sets, the set having been selected at random, and two questions selected at random from the other set. What is the probability that the student can answer all five questions on the examination?

2.31   If A and B are independent, $P(A) = 0.20$ and $P(B) = 0.45$, find (a) $P(A|B)$; (b) $P(A \cap B)$; (c) $P(A \cup B)$; (d) $P(\bar{A} \cap \bar{B})$.

2.32   Two dice are thrown successively; what is the probability that the first will show an odd number and the sum of the two throws will be even? Are these events independent?

2.33   In a hypothetical business school, the 1000 students were classified by sex and by whether or not they smoked more or less than one pack of cigarettes a day. The results are given in the following table:

SMOKE

Sex	≤ 1 *pack/day*	>1 *pack/day*	*Total*
Male	400	300	700
Female	150	150	300
Total	550	450	

If a student is drawn at random, what is the probability that
(a)   The student is a male?
(b)   The student smokes more than one pack of cigarettes per day?
(c)   The student is a female *and* smokes less than one pack of cigarettes per day?
(d)   The student smokes less than one pack of cigarettes per day, given that he is a male?
Are the events "male" and "smokes more than one pack of cigarettes per day" independent? Explain.

2.34   The students in Problem 2.33 were classified according to sex and declared major. The results were as follows:

MAJOR

Sex	Accounting	Finance	Quantitative Methods	Other
Male	360	200	70	70
Female	140	100	30	30

(a) Find two pairs of independent events.
(b) Find two pairs of dependent events.

2.35 Three events, A_1, A_2, A_3, are said to be *mutually independent* if

$$P(A_i \cap A_j) = P(A_i)P(A_j), \qquad i, j = 1, 2, 3, i \neq j$$

and

$$P(A_1 \cap A_2 \cap A_3) = P(A_1)P(A_2)P(A_3)$$

(a) Generalize this definition to n events.
(b) A fair die is to be thrown two independent times. Define the events,
$A_1 = $ "odd number of spots on first toss"
$A_2 = $ "odd number of spots on second toss"
$A_3 = $ "sum of the spots on the two tosses is odd"
   (1) Show that A_1 and A_2, A_1 and A_3, and A_2 and A_3 are pairs of independent events.
   (2) Show that

$$P(A_1 A_2 A_3) \neq P(A_1)P(A_2)P(A_3)$$

and hence that A_1, A_2, and A_3 are not mutually independent.

2.36 (a) Prove by induction that

$$P(A_1 \cap A_2 \ldots \cap A_n) = P(A_1)P(A_2|A_1)P(A_3|A_1 \cap A_2)$$
$$\cdots \cdot \ P(A_n|A_1 \cap \ldots \cap A_{n-1})$$

(b) Specialize part (a) to the case in which $A_1, A_2, \ldots, A_n$ are mutually independent.

2.37 Suppose you toss a fair coin three independent times. Show, using Problem 2.36, that you must assign probability 1/8 to each of the elementary outcomes. Suppose the coin is biased, so that the probability of "heads" on each toss is 3/4. What probabilities must be assigned to the elementary outcomes?

2.38 If possible, give examples of two events that are mutually exclusive and independent, mutually exclusive but not independent, not mutually exclusive but independent, not mutually exclusive and not independent.

2.39 (a) Suppose a two-card hand is drawn at random from an ordinary bridge deck. Find (1) P(two Aces), (2) P(Ace of hearts and Jack of clubs), (3) P(at least one Ace), (4) P(two Aces|at least one Ace), (5) P(Ace of hearts and Jack of clubs|at least one Ace).
(b) Suppose the experiment consisting of drawing a two-card hand is performed five independent times. What is the probability that all five hands have two Aces? What is the probability that at least one of the five hands does not have two Aces?

2.40 In a triangle taste test the taster is presented with three samples, two of which are alike, and is asked to pick the odd one by tasting. If a taster has no well-developed sense of taste and can only pick the odd one by chance, what is

the probability that in five trials he will make
(a)   Five correct decisions?
(b)   No correct decisions?
(c)   At least one correct decision?

2.41   Independent tosses of a coin are repeated until a head appears. What is the probability that six tosses will be required?

2.42   A perfectly balanced wheel is divided into three unequal parts corresponding to the intervals [0, 1/3), [1/3, 1/2), and [1/2, 1). The wheel is to be spun until the pointer points to the interval [0, 1/3). What is the probability that the number of spins is less than or equal to 5? Less than or equal to 10? Less than or equal to n, for any positive integer n? Answer similar questions if the wheel is to spin until the interval [0, 1/3) is pointed out twice.

2.43   (a)   Write down all the samples of size 2 without replacement from the numbers 1, 2, 3, 4, 5. For each sample calculate the arithmetic average of the sample items. Construct a frequency distribution of these averages (that is, for each possible value of the average count the number of times this value occurs in the course of sampling). Now calculate the arithmetic average of all the arithmetic averages you obtained. Compare this average with the arithmetic average of the numbers 1, 2, 3, 4, 5.
       (b)   Repeat part (a) with samples of size 3.

2.44   We can simulate sampling chips from a bowl with replacement using the random digits in Appendix Table A. For example, suppose we wanted to draw four chips with replacement from a bowl containing 200 red and 800 blue chips. This can be done by going through the list of random digits and recording a red chip for a 0 or 1 and a blue chip for a 2, 3, 4, 5, 6, 7, 8, or 9. Thus the first four digits of the last line of Appendix Table A would yield the sample BRBR, that is, a sample with two red and two blue chips.
       (a)   Repeat the sampling experiment described above 100 times and record the number of samples containing 0, 1, 2, 3, or 4 red chips. Compute the relative frequencies of the samples containing 0, 1, 2, 3, or 4 red chips.
       (b)   Draw 100 samples of size 4 with replacement from a bowl containing 400 red and 600 blue chips, a bowl containing 600 red and 400 blue chips, and a bowl containing 800 red and 200 blue chips, calculating in each case the relative frequencies of samples containing 0, 1, 2, 3, or 4 red chips. (Note that the composition of the bowl influences the behavior of the relative frequencies.)

2.45   Sampling without replacement may also be simulated using random numbers. Take, for example, the drawing of a sample of size 4 from the bowl with 200 red and 800 blue chips. Go through the list of random digits, grouping the digits three at a time. Let the numbers from 000 to 199 represent red chips. Let the numbers from 200 to 999 represent blue chips. If the same number is obtained twice, it is disregarded the second time, and the next three-digit number is used instead. For example, line 32 of Appendix Table A yields BBRB, but note that the number 6 was disregarded once in obtaining this sample.

(a) Repeat parts (a) and (b) of Problem 2.44 sampling without replacement.

(b) Compare sampling with and without replacement for the cases mentioned in this problem and in Problem 2.44. How would the two methods have compared if there had been only 10 chips in the bowl, but the ratios of red to blue chips had been the same in each case?

2.46 The number of oil tankers, say X, arriving at a certain refinery each day has the probability function,

x	0	1	2	3	4	5	6
$p(x)$	0.14	0.27	0.27	0.18	0.09	0.04	0.01

The existing service facility of the refinery can handle three tankers a day. If more than three tankers arrive in a day the excess must be sent to another port.

(a) On a given day what is the probability of having to send a tanker to another port?

(b) To what extent must the present facilities be increased so that at least 90 percent of the time *all* the tankers arriving on a given day could be served?

(c) Compute the mean and most probable numbers of tankers arriving per day.

(d) Let Y denote the number of tankers serviced daily. Find the probability function and mean of Y.

2.47 You are given the following joint probability function of the random variables X_1 and X_2.

X_2			
3	0.15	0.20	0.05
2	0.10	0.05	0.20
1	0.05	0.15	0.05
X_1	1	2	3

$p(x_1, x_2)$

(a) Place the marginal probability functions in the table.

(b) Find the probability functions of the random variables $Y_1 = X_1 + X_2$ and $Y_2 = X_1 X_2$.

(c) Find the means and variances of X_1, X_2, Y_1, and Y_2.

(d) Are X_1 and X_2 independent? Explain.

2.48 Repeat Problem 2.47 with the joint probability function of X_1 and X_2 given below.

X_2			
3	0.120	0.160	0.120
2	0.105	0.140	0.105
1	0.075	0.100	0.075
X_1	1	2	3

Compare the answers in the two problems.

2.49   The joint probability function of X_1, X_2, and X_3 is given below.

(x_1, x_2, x_3)	$p(x_1, x_2, x_3)$
0 0 0	0.125
0 0 1	0.125
0 1 0	0.100
1 0 0	0.080
0 1 1	0.150
1 0 1	0.120
1 1 0	0.090
1 1 1	0.210

(a)   Find the joint probability functions of X_1 and X_2. Of X_1 and X_3. Of X_2 and X_3.

(b)   Find the marginal probability functions of X_1, X_2, and X_3.

(c)   Find the conditional probability function (c.p.f.) of X_3, given $X_1 = 0$, and compute the mean and variance of this probability function.

(d)   Define $T = X_1 + X_2 + X_3$. Find the probability function of T. Find the c.p.f. of T, given $X_1 = 0$. Compute the means and variances of these two probability functions.

2.50   Calculate the covariance and correlation coefficient for the pair of random variables X_1 and X_2 in Problems 2.47, 2.48, and 2.49.

2.51   Suppose that the random variable X has the folllowing probability distribution:

x	-2	-4	6	4	2	1
$p(x)$	0.1	0.2	0.3	0.2	0.1	0.1

(a)   Verify that the above is a probability distribution.

(b)   Find $E(X)$, σ_X^2, and σ_X.

(c)   Find $E(2X)$, σ_{2X}^2, $E(X/2 - 3)$, and $\sigma_{(x/2-3)}^2$.

2.52   A finance company receives 10 applications for loans. Of these 10, 4 are good risks and 6 are bad risks. But the company does not know this. Also the company does not have enough resources to extend loans to all 10 applicants. Thus the company decided to extend loans to 3 applicants chosen at random from the 10 who applied. If X denotes the number of bad risks selected, find $E(X)$ and σ_X^2. If on an average loan the company figures to gain \$50 on a good risk and lose \$70 on a bad risk, what is the expected return from these 3 loans?

2.53   (Simple inventory problem): A store owner with an amount I of a certain product on hand at the end of a day must decide how much of the product to order for delivery the following morning to meet an uncertain demand. Let

I = inventory on hand at close of business day

D = random variable representing demand for the product the next business day

A = amount of the product to be ordered for delivery before the store opens the next day

L = loss due to excess stock or excess demand

α = constant of proportionality

Thus $I + A$ is the quantity of product on hand at the beginning of the new day. In addition suppose the loss incurred by the store keeper is proportional to the square of the difference between the quantity on hand and the demand; that is

$$L = \alpha [(I + A) - D]^2$$

(a)   Show that the expected loss is given by

$$\mu_L = E(L) = \alpha \{ [(I + A) - \mu_D]^2 + \sigma_D^2 \}$$

where $\mu_D = E(D)$ and $\sigma_D^2 = \mathrm{Var}(D)$.

(b)   Suppose the probability function for demand is

d	0	1	2	3	4	5
$p(d)$	0.1	0.1	0.2	0.3	0.2	0.1

(1)   Calculate the mean, median, mode (most probable value), and variance associated with this probability distribution.

(2)   Let $I = 2$, $\alpha = 2$, and suppose that A is chosen to be the "most probable" demand less the inventory on hand. Evaluate the expected loss. Evaluate the expected loss for

$$A = \text{median} - I = \eta_5 - I \qquad \text{and} \qquad A = \text{mean} - I = \mu_D - I$$

Observe that $A = \mu_D - I$ minimizes the expected loss.

2.54   Suppose demand for a certain product is independent from week to week. Let X_1 and X_2 be demands for two consecutive weeks, and assume $\mu_{X_i} = 400$ and $\sigma_{X_i}^2 = 625$, $i = 1, 2$.

(a)   Find the mean and variance of total demand over the two-week period.

(b)   The profit on each item sold is $1.50. Find the mean and variance of total profit over the two-week period.

(c)   Answer parts (a) and (b) assuming that $\sigma_{X_1, X_2} = -125$ instead of $\sigma_{X_1, X_2} = 0$. What effect does this change in the covariance have on expected profit? What effect would the assumption $\sigma_{X_1, X_2} = 125$ have on expected profit?

2.55   A shipment of 1000 packages of light bulbs contains 100 packages with broken bulbs. Suppose an inspector plans to choose three packages at random and *without replacement* from the shipment. Let X_i denote the number of defective packages obtained on the ith draw, $i = 1, 2, 3$.

(a)   Develop a probability model for the joint distribution of X_1, X_2, X_3.

(b)   Compute σ_{X_1, X_2}, σ_{X_1, X_3}, and σ_{X_2, X_3}.

(c)   Find the probability distribution of $T = X_1 + X_2 + X_3$, and compute the mean and variance of T.

2.56   An auditor wishes to sample a set of 500 accounts in order to estimate the number of accounts past due. He plans to select four accounts at random without replacement. Let X_i denote the number of past due accounts obtained on the ith draw for $i = 1, 2, 3, 4$.

(a)   Assuming there are 50 past due accounts in the set, what is the joint probability function of X_1, X_2, X_3, and X_4? What is the distribution of $T = X_1 + X_2 + X_3 + X_4$? What is the probability that T will be at least 1, that is, $\Pr[T \geq 1]$?

(b)   Answer part (a) assuming the number of past due accounts in the set is 100. Compare the value of $\Pr[T \geq 1]$ with that obtained in part (a).

(c)   The auditor plans to use the value of $125T$ as an estimate of the number of past due accounts in the set. Find the probability function of $125T$, and calculate its mean and variance assuming 50 past due accounts in the set.

(d)   Repeat part (c) assuming 100 past due accounts in the set.

2.57   Let X and Y be random variables with means μ_X, μ_Y, variances σ_X^2, σ_Y^2, and covariance of $\sigma_{X,Y}$. Find expressions for the mean and variance of each of the following random variables.

(a) $X + Y$ (b) $X - Y$

(c) $3X - 2Y$ (d) $\frac{1}{3}X + \frac{2}{3}Y$

(e) $\frac{3}{4}X + \frac{1}{4}Y$ (f) $\frac{1}{3}X - \frac{2}{3}Y$

2.58   Suppose that the annual returns on a certain investment are independent from year to year. Let R_i, $i = 1, 2, 3, \ldots, h$, denote the returns in the next h years, and let μ_i and σ_i^2 denote the mean and variance of R_i, $i = 1, 2, \ldots, h$. The present value of the investment, assuming a constant discount factor of $(1 + j)^{-1}$, is

$$R = (1+j)^{-1}R_1 + (1+j)^{-2}R_2 + \cdots + (1+j)^{-h}R_h$$

Find the mean and variance of R in terms of the μ_i and σ_i^2, $i = 1, 2, \ldots, h$.

2.59   Let $X_1, X_2, \ldots, X_{10}$ denote the outcomes of 10 random drawings from a population with mean μ and variance σ^2. Let $\bar{X}$ denote the mean of the X_i's. Find the mean and variance of $\bar{X}$ if

(a) $\mu = 100, \sigma^2 = 100$ (b) $\mu = 100, \sigma^2 = 49$

(c) $\mu = 0, \sigma^2 = 10$ (d) $\mu = -50, \sigma^2 = 100$

(e) $\mu = -1, \sigma^2 = 0.1$ (f) $\mu = -50, \sigma^2 = 0.1$.

2.60   Let Z be the standardized normal random variable. Find the following probabilities using Appendix Table D.

(a) $\Pr[Z \leq 1.75]$ (b) $\Pr[Z \leq -0.68]$

(c) $\Pr[-1.96 \leq Z \leq 1.96]$ (d) $\Pr[-2 \leq Z \leq 2]$

(e) $\Pr[Z > 1.645]$ (f) $\Pr[Z > -1.7]$

(g) $\Pr[|Z| > 2]$ (h) $\Pr[|Z| < 1]$

2.61   Let $\bar{X}$ denote the mean of a random sample of size 16 from a population of heights with mean $\mu = 68$ in. and standard deviation $\sigma = 2$ in. Use the central limit theorem to approximate the following probabilities:

(a) $\Pr[\bar{X} \leq 69 \text{ in.}]$
(b) $\Pr[\bar{X} \leq 67 \text{ in.}]$
(c) $\Pr[67 \text{ in.} \leq \bar{X} \leq 69 \text{ in.}]$
(d) $\Pr[\bar{X} > 68.5 \text{ in.}]$
(e) $\Pr[\bar{X} \leq 68.75 \text{ in.}]$

2.62 If T is the sum of a random sample of size 100 from a population of accounts with mean \$400 and variance $(\$25)^2$, use the central limit theorem to approximate the following probabilities: (a) $\Pr[T \leq \$400,500]$; (b) $\Pr[T > \$400,250]$; and (c) $\Pr[\$399,510 \leq T \leq \$400,490]$.

2.63 Let $\bar{X}$ be the mean of a random sample from a population with mean μ and variance $\sigma^2 = 100$. Of what size should the sample be if we want the probability of the event

$$-1 \leq \bar{X} - \mu \leq 1$$

to be 0.90?

2.64 Rework Problem 2.63 assuming $\sigma^2 = 25$. Compare the answers to the two problems.

2.65 Let X have a normal distribution with mean 2 and variance 11. State the distribution of $Y = a + bX$ for each of the following combinations of a and b:
(a) $a = 2, b = 1$
(b) $a = 3, b = 1$
(c) $a = 2, b = 2$
(d) $a = 2, b = -1$
(e) $a = 2, b = -2$
(f) $a = -2, b = 1$
(g) $a = -2, b = 2$
(h) $a = -2, b = -2$.

2.66 Suppose X has a normal distribution with mean 4 and variance 11. Find
(a) $\Pr[X < 9]$
(b) $\Pr[X < -1]$
(c) $\Pr[X > 8]$
(d) $\Pr[X > -5]$
(e) $\Pr[7 < X < 9]$
(f) $\Pr[1 < X < 10]$
(g) $\Pr[-2 < X < 10]$
(h) $\Pr[1 < X < 7]$.

2.67 Suppose X has a normal distribution with mean -5 and variance 25. Find
(a) $\Pr[-10 < X < 0]$
(b) $\Pr[-4 < X < 1]$
(c) $\Pr[-14.80 < X < 4.80]$
(d) $\Pr[X > -3]$

2.68 Suppose that the heights (in feet) of a certain population of people are normally distributed with mean $5\frac{5}{6}$ ft and standard deviation $\frac{1}{3}$ ft. Let X denote the height (in feet) of a person drawn at random from the population. If Y denotes the height of the person in inches, what is the functional relationship between X and Y? What are the mean and standard deviations of Y? What distribution does Y have? Calculate the probability that the person's height is greater than 72 in.

2.69 Suppose that the diameters of cardboard disks produced for frozen pizza packages are normally distributed with mean 12 in. and standard deviation 0.05 in. To be used in a pizza package, a disk must have a diameter no less than 11.8 in. and no greater than 12.1 in.

(a) What is the probability that a randomly drawn disk will meet the criteria stated above?

(b) If 10 disks are selected independently, what is the probability that at least 1 disk will not meet the criteria?

2.70 A company has found that the length of time a newly hired agent stays with the company is approximately normally distributed with mean 18 months and standard deviation 3 months.

(a) Find the probability that a newly hired agent will stay with the company at least two years.

(b) What is the probability that two new agents hired independently will both stay with the company at least two years?

(c) What is the probability that the *average* length of stay of the two agents is greater than two years?

2.71 A manufacturing process is known to produce widgets whose lengths (X) are normally distributed with parameters $\mu = 0.5$ in. and $\sigma = 0.01$ in. A widget is selected at random.

(a) Evaluate the probabilities $\Pr[0.48 \leq X \leq 0.53]$ and $\Pr[X \geq 0.53]$.

(b) If 10 widgets are drawn independently and measured, what is the probability that their *average* length will be no more than 0.505 in.?

2.72 A company has found that its profits Y are related to its sales X by the formula $Y = 0.16X + 500$. If next year's sales have an $N(7{,}500; 800^2)$ distribution, what is the distribution of next year's profit? What is the probability that next year's profit will be between 1500 and 1800?

2.73 The lifetimes of two brands of tires were tested with the following results:

Class Number	Lifetimes in Thousands of Miles	Frequency for Brand 1	Frequency for Brand 2
1	23–25	1	0
2	25–27	7	4
3	27–29	15	20
4	29–31	10	32
5	31–33	15	30
6	33–35	17	12
7	35–37	13	2
8	37–39	9	0
9	39–41	8	0
10	41–43	5	0
		100	100

(a) Draw histograms of tire life for both brands, and depict on each histogram the interval $\bar{x} \pm s$, where $\bar{x}$ and s are the sample mean and standard deviation.

(b) The event "lifetime less than 31,000 miles" corresponds to the classes numbered 1, 2, 3, and 4. Calculate the relative frequencies of this event for the two brands.

(c) Suppose you approximated a given tire lifetime distribution with a normal distribution with $\mu = \bar{x}$ and $\sigma = s$. Using such a normal approximation, calculate the probability of the event "lifetime less than 31,000 miles" for each of the brands.

(d) If you owned a fleet of trucks, which brand of tire would you prefer to use? If the law forbade truck tires to be used for more than 31,000 miles, would you change your answer?

2.74 Let $X_1, X_2, \ldots, X_n$ denote a random sample of size n from an $N(100, 25)$ population. Calculate $\Pr[|\bar{X} - 100| < 1.5]$ assuming $n = 10$, 25, 64, and 100. Explain in words what is happening here.

2.75 Find the quartiles and the 95th percentile of the χ-square distributions with $\nu = 3$, 9, and 20. Compare the three distributions on the basis of these characteristics. Find the means and standard deviations of each of these distributions.

2.76 If $X_1, X_2, \ldots, X_{12}$ is a random sample from an $N(\mu, \sigma^2)$ population, and $Y = \Sigma_{i=1}^{12} (X_i - \bar{X})^2/\sigma^2$, find (a) $\Pr[Y \le 19.7]$; (b) $\Pr[3.82 \le Y \le 21.9]$; and (c) $\Pr[Y > 24.7]$.

2.77 Suppose Y has a $\chi^2(100)$ distribution. Use the normal approximation in equation (2.62) to find the value y such that $\Pr[Y \le y] = 0.95$. Compare this value to that given in Appendix Table F.

2.78 Find the 95-percent percentiles of the t distributions with $\nu = 3, 9$, and 20. Find the means and variances of each of these distributions.

2.79 Let $X_1, X_2, \ldots, X_n$ denote a random sample of size n from an $N(\mu, \sigma^2)$ population. Find $t'(n)$ such that

$$\Pr\left[\left|\frac{\sqrt{n}(\bar{X} - \mu)}{\sqrt{(1/(n - 1))\Sigma_{i=1}^{n} (X_i - \bar{X})^2}}\right| < t'(n)\right] = 0.95$$

for $n = 4, 10, 25$, and 31. To what number will the sequence $[t'(n)]$ converge as $n \to \infty$?

2.80 Let X and Y have a bivariate normal distribution with parameters μ_1, μ_2, σ_1^2, σ_2^2, and ρ. For each of the cases below, determine the marginal distributions of X and Y, and the conditional distributions of X, given $Y = y$, and of Y, given $X = x$.

(a) $\mu_1 = 10$, $\mu_2 = -1$, $\sigma_1^2 = 25$, $\sigma_2^2 = 16$, $\rho = 0.4$
(b) $\mu_1 = 10$, $\mu_2 = -1$, $\sigma_1^2 = 25$, $\sigma_2^2 = 10$, $\rho = -0.7$
(c) $\mu_1 = -10$, $\mu_2 = 10$, $\sigma_1^2 = 16$, $\sigma_2^2 = 25$, $\rho = 0.7$
(d) $\mu_1 = -10$, $\mu_2 = 10$, $\sigma_1^2 = 16$, $\sigma_2^2 = 25$, $\rho = -0.4$

2.81   Generate random samples of size 3 from the distributions with the following c.d.f.'s:

(a) $F(x) = \begin{cases} 0, & x < -1 \\ \dfrac{x+1}{2}, & -1 \le x < 1 \\ 1, & x \ge 1 \end{cases}$

(b) $F(x) = \begin{cases} 0, & x < -1 \\ \dfrac{(1+x)^2}{2}, & -1 \le x < 0 \\ 1 - \dfrac{(1-x)^2}{2}, & 0 \le x < 1 \\ 1, & x \ge 1 \end{cases}$

(c) $F(x) = \begin{cases} 0, & x < -1 \\ \dfrac{x^3+1}{2}, & -1 \le x < 1 \\ 1, & x \ge 1 \end{cases}$

2.82   Select a continuous probability distribution and generate a random sample of size 100 from it. Compare the sample with the population.

3

ELEMENTS OF
STATISTICAL INFERENCE

3.1 INTRODUCTION

In Chapter 2 we stated that "statistical inferences are generalizations on the basis of limited information." Before we can apply probability theory to a study of methods of statistical inference, we must be precise about the type of information that is available and the sort of generalization that is desired.

In elementary statistical theory, we assume that an experimenter wishes to know the numerical value of a characteristic (parameter) of a population, but he cannot observe all the elements of the population. We also assume that the experimenter can draw a random sample from the population and make measurements on the items of the sample that yield information about the parameter of interest. The assumption of random sampling is often hard to defend because the experimenter lacks control over the population. Techniques for dealing with nonrandom samples will be presented in later chapters of this book. In this chapter we retain the classical assumption of random sampling for purposes of reviewing elementary inference concepts.

Once the assumption of random sampling is accepted, a host of statistical methods is available to solve a wide variety of problems. The methods that follow will focus on the problems of estimating means and variances of normal populations and means of arbitrary populations when sample sizes are large. The inference tools we shall review extensively are the confidence interval and the hypothesis test.

Suppose that an experimenter wishes to know the value of a parameter, θ, of a well-defined population, and that a random sample of some predetermined size, n, may be drawn from the population. The items of the random sample are denoted by $X_1, X_2, \ldots, X_n$, and each item is assumed to be generated by the population p.d.f.[a] denoted by $f(x|\theta)$. The parameter θ is important in this discussion and its presence in the p.d.f. is noted explicitly. Since the X_i's are independent, their joint p.d.f. is given by

$$f(x_1, x_2, \ldots, x_n|\theta) = f(x_1|\theta) \ f(x_2|\theta) \ \cdots \ f(x_n|\theta) \tag{3.1}$$

[see equation (2.23)].

A *statistic* is a function of the sample items. An important statistical problem is the choice of a statistic for making inferences about θ. This problem is discussed extensively in Section 3.8. For the purposes of the present discussion we assume that such a statistic has been chosen and denote it by $\hat{\theta}$. Note that $\hat{\theta}$ is a random variable because it is a function of the items of the random sample, and hence $\hat{\theta}$ has a sampling distribution. (For example, $\hat{\theta}$ could be the sample mean and have an approximate normal distribution if n is large.) Thus it makes sense to make probability statements about $\hat{\theta}$. Furthermore, if $\hat{\theta}$ is to be useful in making inferences about θ, its sampling distribution must link $\hat{\theta}$ and θ in some way. (For example, if θ is the mean of the population, $\hat{\theta}$ is the sample mean, and n is large, then $\hat{\theta}$ has an approximate $N(\theta, \sigma^2/n)$ distribution, so that the sampling distribution of $\hat{\theta}$ depends on θ.)

3.2 CONFIDENCE INTERVALS

It is often possible to find two functions of $\hat{\theta}$, call them $\phi_1(\hat{\theta})$ and $\phi_2(\hat{\theta})$, such that

$$\Pr[\phi_1(\hat{\theta}) < \theta < \phi_2(\hat{\theta})] = 1 - \alpha \tag{3.2}$$

where α is some preassigned number (which is frequently close to zero). It is important to note that $\phi_1(\hat{\theta})$ and $\phi_2(\hat{\theta})$ do *not* depend on the value of θ. If such functions are found and numerical values of $X_1, X_2, \ldots, X_n$ are obtained by sampling the population, then the numerical values of $\hat{\theta}, \phi_1(\hat{\theta})$, and $\phi_2(\hat{\theta})$ can be calculated. The interval $[\phi_1(\hat{\theta}), \phi_2(\hat{\theta})]$, whose endpoints are the numerical values of $\phi_1(\hat{\theta})$ and $\phi_2(\hat{\theta})$, is called a $100(1 - \alpha)$—percent confidence interval for θ. In view of equation (3.2), we should be "$100(1 - \alpha)$ percent confident" that the value of θ, although unknown, lies somewhere in the numerical interval. We admit the possibility that the numerical interval does not contain θ, and we measure the risk of this being the case by α. The larger the value of $(1 - \alpha)$, the larger the chance that a given interval produced by evaluating $\phi_1(\hat{\theta})$ and $\phi_2(\hat{\theta})$ has of containing the numerical

---

[a] If the X_i's are discrete, then they are governed by a probability function. Although we treat the continuous case here, the ideas and methods are analogous for the discrete case.

value of the population parameter θ. We now present a number of applications of this theory.

Estimation of the Mean of an $N(\mu, \sigma^2)$ Population with σ^2 Known

Here we let $\theta = \mu$, and we let $\hat{\theta} = \bar{X}$, the sample mean. From the discussion of normal populations in Section 2.4 we know that $\bar{X}$ has an $N(\mu, \sigma^2/n)$ sampling distribution, and hence that $Z = (\bar{X} - \mu)/(\sigma/\sqrt{n})$ has a standard normal distribution. Letting $z_{\alpha/2}$ denote the $100(1 - \alpha/2)$th percentile of the standard normal distribution, we can write

$$\Pr\left[-z_{\alpha/2} < \frac{\bar{X} - \mu}{\sigma/\sqrt{n}} < z_{\alpha/2}\right] = 1 - \alpha \tag{3.3}$$

or

$$\Pr\left[-z_{\alpha/2}\left(\frac{\sigma}{\sqrt{n}}\right) < \bar{X} - \mu < z_{\alpha/2}\left(\frac{\sigma}{\sqrt{n}}\right)\right] = 1 - \alpha \tag{3.4}$$

This latter expression is symmetric in $\bar{X}$ and μ; consequently,

$$\Pr\left[-z_{\alpha/2}\left(\frac{\sigma}{\sqrt{n}}\right) < \mu - \bar{X} < z_{\alpha/2}\left(\frac{\sigma}{\sqrt{n}}\right)\right] = 1 - \alpha \tag{3.5}$$

and from this it follows that

$$\Pr\left[\bar{X} - z_{\alpha/2}\left(\frac{\sigma}{\sqrt{n}}\right) < \mu < \bar{X} + z_{\alpha/2}\left(\frac{\sigma}{\sqrt{n}}\right)\right] = 1 - \alpha \tag{3.6}$$

We see from equation (3.6) that

$$\phi_1(\bar{X}) = \bar{X} - z_{\alpha/2}\left(\frac{\sigma}{\sqrt{n}}\right) \quad \text{and} \quad \phi_2(\bar{X}) = \bar{X} + z_{\alpha/2}\left(\frac{\sigma}{\sqrt{n}}\right)$$

Table 3.1 displays selected values of $1 - \alpha$ and $z_{\alpha/2}$ for this case.

Table 3.1   Values of $z_{\alpha/2}$ such that $\Pr[\bar{X} - z_{\alpha/2}(\sigma/\sqrt{n}) < \mu < \bar{X} + z_{\alpha/2}(\sigma/\sqrt{n})] = 1 - \alpha$

$1 - \alpha$	0.85	0.90	0.95	0.9545	0.98	0.99	0.995
$z_{\alpha/2}$	1.440	1.645	1.960	2.000	2.326	2.576	2.807

EXAMPLE 3.1

To illustrate the preceding remarks, let us look at the results of a sampling experiment. Each repetition of the experiment consists of the drawing of $n = 7$ items at random from an $N(100, 100)$ population. The random interval in equation (3.6) will be in the form

$$\left[\bar{X} - 1.96\left(\frac{10}{\sqrt{7}}\right), \bar{X} + 1.96\left(\frac{10}{\sqrt{7}}\right)\right] = [\bar{X} - 7.41, \bar{X} + 7.41]$$

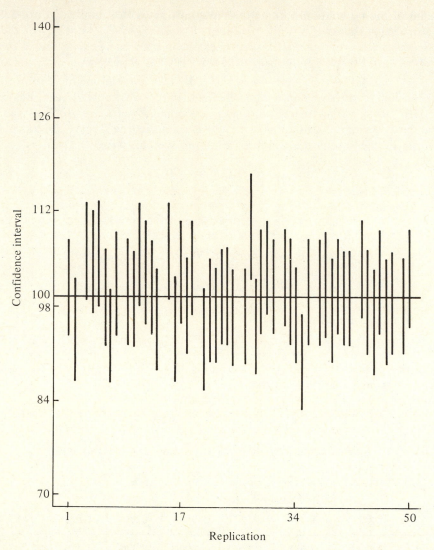

Figure 3.1   Fifty 95-percent confidence intervals for the mean of an $N(100, 100)$ population based on $n = 7$ observations using the normal distribution

Figure 3.1 shows the intervals resulting from 50 repetitions of the experiment. Although the number of repetitions is not large, we would expect roughly 95 percent of the intervals to contain $\mu = 100$, the mean of the sampled population. We see that 48 out of 50 (96 percent) of them do contain $\mu = 100$, a result that is certainly consistent with the theoretical prediction.

In practice we will not know the value of μ, and we can rarely afford the luxury of repetitions of an experiment. But Example 3.1 suggests that the computation of an interval like that in equation (3.6) may be a rational way to express an inference about μ on the basis of a single sample. Given the demonstrated behavior of the interval in repeated samples, it seems reasonable to believe that the interval computed from a *single* sample is likely to contain μ.

EXAMPLE 3.2

A management instructor routinely administers a standardized management aptitude test to his students. Scores from the test are approximately normally distributed with standard deviation $\sigma = 10$. Suppose that the instructor plans to administer the test to a group of $n = 49$ students just entering management training with the idea that their scores represent a random sample from the population of scores made by people with little or no previous management training. His purpose is to estimate the average score for this population which we shall denote by μ.

Let us suppose the instructor administers the test and finds the average of the 49 scores to be $\bar{X} = 102.9$. A 95-percent confidence interval for μ is (using $z_{0.025} = 1.96$)

$$\left[102.9 - 1.96 \left(\frac{10}{7} \right), 102.9 + 1.96 \left(\frac{10}{7} \right) \right] = [100.1, \ 105.7]$$

A 99-percent confidence interval for μ is (using $z_{0.005} = 2.576$)

$$\left[102.9 - 2.576 \left(\frac{10}{7} \right), 102.9 + 2.576 \left(\frac{10}{7} \right) \right] = [99.2, \ 106.6]$$

The 99-percent interval is wider than the 95-percent interval because

$$z_{0.005} = 2.576 > 1.96 = z_{0.025}$$

This makes intuitive sense since it is reasonable to expect that the larger interval is more likely to contain μ.

Estimation of the Mean of an $N(\mu, \sigma^2)$ Population with σ^2 Unknown

Let $X_1, X_2, \ldots, X_n$ denote a random sample from an $N(\mu, \sigma^2)$ population whose parameters are unknown. Our primary interest is in the estimation of μ, and, in particular, we would like to be able to construct a confidence interval for μ despite the fact that σ^2 is unknown. Interestingly enough, this can be done by using the t distribution defined in Section 2.4. In that section, we saw that the random variable

$$\frac{\sqrt{n}(\bar{X} - \mu)}{\sqrt{\sum_{i=1}^{n} (X_i - \bar{X})^2 / (n-1)}} \tag{3.7}$$

has a t distribution with $\nu = n - 1$ degrees of freedom. Referring to

Appendix Table E we find a number $t_{\alpha/2}(\nu)$ such that

$$\Pr\left[-t_{\alpha/2}(\nu) \le \frac{\sqrt{n}\,(\bar{X}-\mu)}{\sqrt{\sum_{i=1}^{n}(X_i-\bar{X})^2/(n-1)}} \le t_{\alpha/2}(\nu)\right] = 1-\alpha \qquad (3.8)$$

This is a rather cumbersome statement to write, but it can be usefully simplified by defining the symbol

$$\bar{s} = \sqrt{\sum_{i=1}^{n}\frac{(X_i-\bar{X})^2}{(n-1)}} \qquad (3.9)$$

Then equation (3.8) becomes

$$\Pr\left[-t_{\alpha/2}(\nu) \le \frac{\sqrt{n}\,(X-\mu)}{\bar{s}} \le t_{\alpha/2}(\nu)\right] = 1-\alpha \qquad (3.10)$$

which can be written

$$\Pr\left[-t_{\alpha/2}(\nu)\left(\frac{\bar{s}}{\sqrt{n}}\right) \le \bar{X}-\mu \le t_{\alpha/2}(\nu)\left(\frac{\bar{s}}{\sqrt{n}}\right)\right] = 1-\alpha \qquad (3.11)$$

or as

$$\Pr\left[\bar{X}-t_{\alpha/2}(\nu)\left(\frac{\bar{s}}{\sqrt{n}}\right) \le \mu \le \bar{X}+t_{\alpha/2}(\nu)\left(\frac{\bar{s}}{\sqrt{n}}\right)\right] = 1-\alpha \qquad (3.12)$$

Comparison of equations (3.12) and (3.6) is instructive. In the latter equation σ is known, and the factor $z_{\alpha/2}$ enters into the endpoints of the interval. In equation (3.12) σ is replaced by the estimator $\bar{s}$, and $z_{\alpha/2}$ is replaced by $t_{\alpha/2}(\nu)$. Since $t_{\alpha/2}(\nu)$ is larger than $z_{\alpha/2}$, it may be thought of as the adjustment that must be made because σ is estimated from the data.

EXAMPLE 3.3

Figure 3.2 shows the intervals resulting from 50 repetitions of an experiment that consists of drawing $n = 7$ items at random from an $N(100, 100)$ population. Since there are six degrees of freedom, we have $t_{0.025}(6) = 2.447$; so each interval is a realization of the random interval

$$\left[\bar{X}-2.447\frac{\bar{s}}{\sqrt{7}}, \bar{X}+2.447\frac{\bar{s}}{\sqrt{7}}\right]$$

The widths of the intervals vary from repetition to repetition because they depend on $\bar{s}$, which is a random variable. The *average* of the widths of the 50 intervals in Figure 3.2 is 18.58. This is greater than the constant width, 14.82, of the intervals in Figure 3.1.

To summarize our discussion, if $X_1, X_2, \ldots, X_n$ is a random sample from an $N(\mu, \sigma^2)$ population, then equation (3.12) is true, and for a given sample the

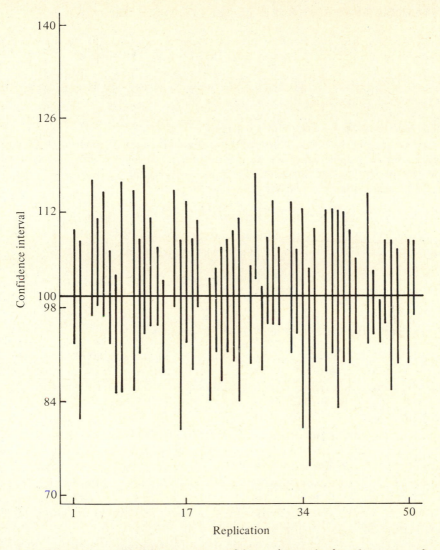

Figure 3.2 Fifty 95-percent confidence intervals for the mean of an $N(100, 100)$ population based on $n = 7$ observations using the t distribution

value of the interval

$$\left[\bar{X} - t_{\alpha/2}(\nu) \frac{\bar{s}}{\sqrt{n}}, \bar{X} + t_{\alpha/2}(\nu) \frac{\bar{s}}{\sqrt{n}} \right] \tag{3.13}$$

with $\nu = n - 1$ is a $100(1 - \alpha)$-percent confidence interval for μ.

EXAMPLE 3.4

Eighteen consecutive monthly rates of return[b] were computed for the common stock of an anonymous American company. They are listed below.

−0.097,	−0.019,	−0.272,	−0.068,	0.041,	0.188,
−0.029,	−0.219,	−0.050,	−0.077,	−0.127,	−0.125,
0.395,	0.335,	−0.066,	−0.088,	−0.288,	0.256

For these data, $\bar{X} = -0.017$ and $\bar{s} = 0.193$.

Let us accept for a moment the proposition that these monthly rates of return come from a normal population. Then we may use the t distribution with $\nu = 18 - 1 = 17$ degrees of freedom to construct confidence intervals for μ. With $\nu = 17$, $t_{0.025}(17) = 2.11$; so a 95-percent confidence interval for μ is

$$\left[-0.017 - 2.11 \frac{0.193}{\sqrt{18}}, -0.017 + 2.11 \frac{0.193}{\sqrt{18}}\right] = [-0.113, 0.079]$$

Similarly, $t_{0.005}(17) = 2.898$, and a 99-percent confidence interval for μ is

$$\left[-0.017 - 2.898 \frac{0.193}{\sqrt{18}}, -0.017 + 2.898 \frac{0.193}{\sqrt{18}}\right] = [-0.149, 0.115]$$

Estimation of the Variance of an $N(\mu, \sigma^2)$ Population

Although the mean of a population is often of great interest, the variance may be equally interesting. For example, an instructor who observes a class's scores on a series of tests usually keeps track of the variances as well as the means of the scores on successive tests. Assuming the relative difficulty of the tests remains constant, an increase in the variances may indicate a widening gap between the better and average students. Decreasing variances might be caused by the dropping of the course by the poorer students and a resulting homogenization of the class. As another example, a quality control engineer knows that a disturbance in the production process can frequently show up as a change in variability rather than a simple change in level or mean. These remarks are intended to motivate the study of the estimation of a population variance.

Let $X_1, X_2, \ldots, X_n$ denote a random sample from an $N(\mu, \sigma^2)$ population whose mean and variance are unknown. Confidence intervals for σ^2 may be obtained as follows. In the discussion of the χ-square distribution in Section 2.4 we saw that the random variable

---

[b] "Monthly rate of return" is defined to be the monthly rate of change of the price of the company's common stock on the open market. Thus, for example, the rate of return for the month of March would be

$$\frac{\text{(stock price at end of March)} - \text{(stock price at end of February)}}{\text{(stock price at end of February)}}$$

This calculation assumes no drastic modifications in the company's operation such as a stock split or a merger.

$$\frac{nS^2}{\sigma^2} = \sum_{i=1}^{n} \frac{(X_i - \bar{X})^2}{\sigma^2} = \frac{(n-1)\bar{s}^2}{\sigma^2}$$

has a $\chi^2(n-1)$ distribution. Using Appendix Table F we can find numbers $\chi^2_{1-\alpha/2}(n-1)$ and $\chi^2_{\alpha/2}(n-1)$ such that

$$\Pr\left[\frac{(n-1)\bar{s}^2}{\sigma^2} < \chi^2_{1-(\alpha/2)}(n-1)\right] = \Pr\left[\frac{(n-1)\bar{s}^2}{\sigma^2} > \chi^2_{\alpha/2}(n-1)\right] = \frac{\alpha}{2} \quad (3.14)$$

(see Figure 3.3). This means that

$$\Pr\left[\chi^2_{1-(\alpha/2)}(n-1) \le \frac{(n-1)\bar{s}^2}{\sigma^2} \le \chi^2_{\alpha/2}(n-1)\right] = 1 - \alpha \quad (3.15)$$

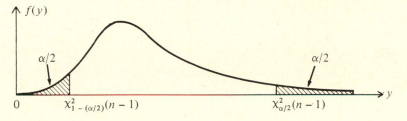

Figure 3.3 The $\chi^2(n-1)$ p.d.f., $n-1 \ge 3$

Now the event

$$\chi^2_{1-(\alpha/2)}(n-1) \le \frac{(n-1)\bar{s}^2}{\sigma^2} \le \chi^2_{\alpha/2}(n-1) \quad (3.16)$$

may be written

$$\frac{1}{\chi^2_{1-(\alpha/2)}(n-1)} \ge \frac{\sigma^2}{(n-1)\bar{s}^2} \ge \frac{1}{\chi^2_{\alpha/2}(n-1)} \quad (3.17)$$

or

$$\frac{(n-1)\bar{s}^2}{\chi^2_{\alpha/2}(n-1)} \le \sigma^2 \le \frac{(n-1)\bar{s}^2}{\chi^2_{1-(\alpha/2)}(n-1)} \quad (3.18)$$

Thus

$$\Pr\left[\frac{(n-1)\bar{s}^2}{\chi^2_{\alpha/2}(n-1)} \le \sigma^2 \le \frac{(n-1)\bar{s}^2}{\chi^2_{1-(\alpha/2)}(n-1)}\right] = 1 - \alpha \quad (3.19)$$

and this expression can be used as a basis for a $100(1-\alpha)$-percent confidence interval for σ^2.

EXAMPLE 3.5

Let us estimate the variance of the population of monthly rates of return discussed in Example 3.4. The sample of $n = 18$ rates yielded $\bar{s} = 0.193$. Thus

$$(n-1)\bar{s}^2 = (17)(0.193)^2 = 0.633$$

To construct a 95-percent confidence interval for σ^2, we must determine $\chi^2_{0.975}(17)$ and $\chi^2_{0.025}(17)$ from Appendix Table F. If Y is a $\chi^2(17)$ variable, we see from Appendix Table F that

$$\Pr[Y < 7.56] = 0.025 \qquad \text{and} \qquad \Pr[Y > 30.19] = 0.025$$

so that

$$\chi^2_{0.975}(17) = 7.56 \qquad \text{and} \qquad \chi^2_{0.025}(17) = 30.19$$

Thus a 95-percent confidence interval for σ^2 is

$$\left[\frac{(0.633)}{(30.19)}, \frac{(0.633)}{(7.56)} \right] = [0.02, 0.08]$$

It is interesting to note that the confidence interval suggested by equation (3.19) is *not* formed by adding and subtracting a factor from the estimate $\bar{s}^2$. This can be contrasted with the intervals for μ that were formed by adding and subtracting factors from $\bar{X}$.

3.3 HYPOTHESIS TESTS

Simple Hypotheses

In this section we assume the parameter of interest, θ, may be equal to one of only two values, say θ_0 or θ_1. We shall denote the hypothesis that θ has the value θ_0 by the symbol H_0: $\theta = \theta_0$, and the hypothesis the θ has the value θ_1 by the symbol H_a: $\theta = \theta_1$. It is customary to call H_0 the *null hypothesis* and H_a the *alternative hypothesis*. This terminology is rather arbitrary, and it should be clear to the reader that an experimenter is free to choose any values he wishes to be null and alternative hypotheses.

The two hypotheses can be "tested" for plausibility using sample evidence. The test is based on the value of a statistic $\hat{\theta}$, which is a function of the sample items and whose sampling distribution depends on θ. (If this were not the case, the statistic could not be informative about θ.) The possible values of the statistic are divided into two regions, one of which is favorable to H_0 and the other one favorable to H_a. The consequences of using such a test are shown in Table 3.2. Concluding H_a: $\theta = \theta_1$ when H_0: $\theta = \theta_0$ is correct is called a *type* I *error*, and concluding H_0: $\theta = \theta_0$ when H_a: $\theta = \theta_1$ is correct is called a *type* II *error*.

The precision of a test is judged by an examination of the error probabilities. The probability of a type I error is called the *significance level* of the test and is traditionally denoted by α, although this notation proves to be inconvenient in some contexts. The probability of a type II error is denoted by β. This probability is not usually given a name, but $1 - \beta$, the probability of concluding H_a when H_a is correct, is called the *power* of the

Table 3.2   Consequences of a Test of Hypotheses

| | | CORRECT HYPOTHESIS | |
		$H_0: \theta = \theta_0$	$H_a: \theta = \theta_1$
CONCLUSION	$H_0: \theta = \theta_0$	correct conclusion	type II error
	$H_a: \theta = \theta_1$	type I error	correct conclusion

test when $\theta = \theta_1$. The term "power" refers to the probability of accepting the alternative hypothesis, and this probability can be calculated when H_0 is correct as well as when H_a is correct. The power of the test when $\theta = \theta_0$ is the probability of accepting H_a when H_0 is true and is thus the significance level of the test.

The *power function* of a test is the function that associates with each value of θ the probability of accepting H_a, assuming that value of θ is correct. The power function will be denoted by $1 - \beta(\theta)$. In this notation the significance level of the test is $\alpha = 1 - \beta(\theta_0)$, and the power of the test at $\theta = \theta_1$ is $1 - \beta(\theta_1)$. Table 3.3 shows the probabilities of the consequences in Table 3.2.

The consequence probabilities in Table 3.2 may be used to develop criteria for choosing tests. Three possibilities are listed in the following paragraphs.

Table 3.3   Probabilities of Consequences in Table 3.2

| | | CORRECT HYPOTHESIS | |
		$H_0: \theta = \theta_0$	$H_a: \theta = \theta_1$
CONCLUSION	$H_0: \theta = \theta_0$	$\beta(\theta_0)$	$\beta(\theta_1)$
	$H_a: \theta = \theta_1$	$\alpha = 1 - \beta(\theta_0)$	$1 - \beta(\theta_1)$

Relativity Small Error Probabilities   A relatively small error probability is the requirement that for any fixed hypothesis the probability of committing an error is less than the probability of drawing a correct conclusion. In symbols we require $\beta(\theta_0) > 1 - \beta(\theta_0)$ and $\beta(\theta_1) < 1 - \beta(\theta_1)$. This requirement cannot always be met.

Unbiasedness   A test is unbiased if the significance level is less than the power at $\theta = \theta_1$, that is, if $\alpha = 1 - \beta(\theta_0) < 1 - \beta(\theta_1)$, or $\beta(\theta_0) > \beta(\theta_1)$.

Most Powerful for Fixed α   Let the significance level α be fixed. The best test of significance level α is the test whose significance level is α and whose power at $\theta = \theta_1$ is greater than the power at $\theta = \theta_1$ of any other test. Although this criterion may seem like a very strong one, it does lead to practical results. In fact, it is possible to develop the form that a most powerful test must take. Thus it is always possible, at least in theory, to derive the best test for any fixed significance level α.

Some of the theory for developing most powerful tests is presented in Section 3.8. Generally we shall not derive tests in this book, but we shall always use most powerful tests where this is possible.

We shall now confine the theory just presented to some important cases and illustrate each case with an example. The data will always be regarded as a *random sample* of size n from some population.

The first case we consider is that of testing the mean of a $N(\mu, \sigma^2)$ population, where σ^2 is assumed to be known. The most powerful test of H_0: $\mu = \mu_0$ against H_a: $\mu = \mu_1$ is based on the sample mean $\bar{X}$.

If $\mu_0 < \mu_1$, the test accepts H_0 if $\bar{X}$ is *less* than some critical value c and accepts H_a if $\bar{X}$ is *greater* than c. The significance level of the test is

$$\alpha = 1 - \beta(\mu_0) = \Pr[\bar{X} > c\,;\, \mu = \mu_0] = \Pr\left[Z > \frac{c - \mu_0}{\sigma/\sqrt{n}}\right] \qquad (3.20)$$

where Z is the standard normal random variable. The power of the test at $\mu = \mu_1$ is

$$1 - \beta(\mu_1) = \Pr[\bar{X} > c\,;\, \mu = \mu_1] = \Pr\left[Z > \frac{c - \mu_1}{\sigma/\sqrt{n}}\right] \qquad (3.21)$$

If $\mu_0 > \mu_1$, the test accepts H_0 if $\bar{X}$ is *greater* than some critical value c and accepts H_a if $\bar{X}$ is *less* than c. The significance level of the test is

$$\alpha = 1 - \beta(\mu_0) = \Pr[\bar{X} < c\,;\, \mu = \mu_0] = \Pr\left[Z < \frac{c - \mu_0}{\sigma/\sqrt{n}}\right] \qquad (3.22)$$

and the power of the test at $\mu = \mu_1$ is

$$1 - \beta(\mu_1) = \Pr[\bar{X} < c\,;\, \mu = \mu_1] = \Pr\left[Z < \frac{c - \mu_1}{\sigma/\sqrt{n}}\right] \qquad (3.23)$$

EXAMPLE 3.6

Suppose a random sample can be selected from a population of heights whose mean is unknown but whose variance is known to be $\sigma^2 = (2\text{ in.})^2$. Let us further assume that the population is normally distributed. There is some uncertainty about the value of μ. If the population is "typical," it should have a mean around, say, 68 in. There is a feeling, however, that the mean height might be 69 in. We thus set forth two hypotheses about μ, namely, H_0: $\mu = 68$ in. and H_a: $\mu = 69$ in.

Now suppose the sample size is $n = 9$, and we decide to set the significance level at $\alpha = 0.05$. The sample mean $\bar{X}$ will have an $N(\mu, 4/9)$ distribution, so using equation (3.20) we have

$$0.05 = \Pr\left[\bar{X} > c \; ; \mu = 68\right] = \Pr\left[Z > \frac{c - 68}{2/3}\right]$$

where Z has the standard normal distribution. Appendix Table D gives $\Pr[Z > 1.645] = 0.05$, so we must have $(c - 68)/\frac{2}{3} = 1.645$, or

$$c = 68 + (2/3)(1.645) = 69.097$$

Once c has been determined, we can calculate the probability of concluding H_a when H_a is correct. It is, using equation (3.21),

$$1 - \beta(69) = \Pr[\bar{X} > 69.097; \mu = 69] = \Pr\left[Z > \frac{69.097 - 69}{2/3}\right] = 0.4424$$

The arrangements of various consequences and probabilities are displayed in Table 3.4.

Table 3.4   Consequences and Probabilities in Testing $\mu = 68$ and $\mu = 69$ (when $n = 9$ and one error probability is set at 0.05)

CONSEQUENCES

| | | Correct hypothesis | |
		$\mu = 68$	$\mu = 69$
Conclusion	$\mu = 68$	correct conclusion	error
	$\mu = 69$	error	correct conclusion

PROBABILITIES

| | | Correct hypothesis | |
		$\mu = 68$	$\mu = 69$
Conclusion	$\mu = 68$	0.95	0.5576
	$\mu = 69$	0.05	0.4424

The reader may think it a bit odd to construct a test in which some values of $\bar{X}$ greater than 69 are considered more favorable to the hypothesis $\mu = 68$ than to the hypothesis $\mu = 69$. Yet this can be a consequence of arbitrarily fixing an error probability, as we have seen. The anomaly may be remedied by taking a larger sample size. To illustrate this, consider fixing the same error probability as before at 0.05 and using a sample of size $n = 16$. Then the critical value of the test is 68.823, and the consequence probabilities are shown in Table 3.5. Note that when $\mu = 69$ is the correct hypothesis, the probability of a correct conclusion is now greater than the probability of error.

Table 3.5 Consequences and Probabilities in Testing $\mu = 68$ and $\mu = 69$ (when $n = 16$ and one error probability is set at 0.05)

CONSEQUENCES

| | | Correct hypothesis | |
		$\mu = 68$	$\mu = 69$
Conclusion	$\mu = 68$	correct conclusion	error
	$\mu = 69$	error	correct conclusion

PROBABILITIES

| | | Correct hypothesis | |
		$\mu = 68$	$\mu = 69$
Conclusion	$\mu = 68$	0.95	0.3613
	$\mu = 69$	0.05	0.6387

We might feel intuitively that the most natural test would have equal error probabilities, and yet we have considered tests with unequal probabilities. Two lines of argument have been advanced to justify the use of such tests, and these will be outlined in the next two paragraphs.

The Weight of Evidence Argument   Suppose we believe that the hypothesis $\mu = 68$ is *not* correct, and we want to demonstrate this in the strongest possible way. If we really believe the evidence from the sample will be strongly against $\mu = 68$, then we believe this hypothesis will be rejected even if a test is used that strongly favors the acceptance of $\mu = 68$ a priori. Just such a test is shown in Table 3.4. If the sample mean turns out to be greater than 69.097, then we reject the hypothesis $\mu = 68$ under circumstances very favorable to it, and consequently we may argue that the hypothesis has been rejected decisively. In other words the weight of the experimental evidence is overwhelmingly against $\mu = 68$.

The More Serious Error Argument   Suppose some action is to be based on the conclusion drawn from a test. If we conclude $\mu = 68$, action A will be taken. If we conclude $\mu = 69$, action B will be taken. If it is more serious to take action B when $\mu = 68$ is correct than it is to take action A when $\mu = 69$, then there would be justification for making one error probability smaller than the other.

To be more specific, let us consider a scenario from production. Periodically samples of a product are taken from an assembly line and checked for quality. Assume that one of the product characteristics checked is supposed to have a mean of 68 in., and deviations from this specification of 1 in. or more are unacceptable. Then it would be of interest to know if the mean is 69 in. The actions to be taken are clear. If the conclusion is $\mu = 68$ in., the assembly line continues. If the conclusion is $\mu = 69$ in., the assembly line is stopped and remedial action is taken. Clearly the cost of stoppage could be very great, so that this action would only be taken if the sample evidence strongly favored $\mu = 69$ in. Again we are led to a test like that pictured in Table 3.4.

The second case we consider is that of testing the mean of an $N(\mu, \sigma^2)$ population whose variance is unknown. The most powerful test of H_0: $\mu = \mu_0$ against H_a: $\mu = \mu_1$ is based on the t statistic

$$t = \frac{\bar{X} - \mu_0}{s/\sqrt{n}} \tag{3.24}$$

When the hypothesis H_0: $\mu = \mu_0$ is correct, then the statistic in equation (3.24) has the t distribution with $n - 1$ degrees of freedom (see discussion of the t distribution in Section 2.4).

If $\mu_0 < \mu_1$, the test accepts H_a if the value of t is greater than some critical value c and accepts H_0 if the value of t is less than c.

The significance level of the test is

$$\alpha = \Pr\left[\frac{\bar{X} - \mu_0}{s/\sqrt{n}} > c; \mu_0\right] \tag{3.25}$$

which can be evaluated using tabled values of the $t(n - 1)$ distribution. The calculation of the power of the test requires the ability to find

$$\Pr\left[\frac{\bar{X} - \mu_0}{s/\sqrt{n}} > c; \mu_1\right]$$

and this is not an elementary problem.

If $\mu_0 > \mu_1$, the test accepts H_a if t is less than the critical value c and accepts H_0 if t is greater than c. The significance level is

$$\alpha = \Pr\left[\frac{\bar{X} - \mu_0}{s/\sqrt{n}} < c; \mu_0\right] \tag{3.26}$$

Again discussion of power is omitted.

EXAMPLE 3.7

An assembly line produces rods whose average length is supposed to be 36 in., and all rods more than 36.5 in. in length must be reworked. Periodically random samples of $n = 20$ rods are taken from the line and measured in order to check the average length of rods being produced. The null hypothesis is H_0: $\mu = 36$ in., and the alternative hypothesis is H_a: $\mu = 36.5$ in. If H_a is accepted, the assembly line is stopped and checked, and this is costly; so a significance level of $\alpha = 0.01$ is used. This means that occasionally the line will be allowed to continue when it is producing unacceptable rods, as this is considered less serious than stopping the line when it is producing acceptable rods.

If we assume the population of rod lengths is normal, we may use the t test as a guide in deciding whether the line should be stopped or allowed to continue. Since the significance level $\alpha = 0.01$ is given we may use equation (3.25) to solve for c. We have $n - 1 = 19$ degrees of freedom and $\Pr[t(19) > 2.539] = 0.01$, so the critical value is $c = 2.539$.

To illustrate, suppose a sample of 20 items has $\bar{X} = 36.1$ in. and $\bar{s} = 0.2$ in. Then the value of t is

$$\frac{36.1 \text{ in.} - 36 \text{ in.}}{(0.2 \text{ in.})/\sqrt{20}} = 2.24 < 2.539$$

so the line would not be stopped. On the other hand a sample with $\bar{X} = 36.2$ in. and $\bar{s} = 0.25$ in. yields a t of

$$\frac{36.2 \text{ in.} - 36 \text{ in.}}{(0.25 \text{ in.})/\sqrt{20}} = 3.58$$

and the line would be stopped.

The third case we consider is the testing of the variance of an $N(\mu, \sigma^2)$ population. The most powerful test of H_0: $\sigma^2 = \sigma_0^2$ against H_a: $\sigma^2 = \sigma_1^2$ is based on the statistic

$$\sum_{i=1}^{n} (X_i - \bar{X})^2 = nS^2$$

When H_0 is correct the random variable

$$\frac{nS^2}{\sigma_0^2} = \sum_{i=1}^{n} \frac{(X_i - \bar{X})^2}{\sigma_0^2}$$

has a $\chi^2(n - 1)$ distribution, as discussed in Section 2.4.

If $\sigma_0^2 < \sigma_1^2$, the test accepts H_0 if nS^2/σ_0^2 is less than a critical value c and accepts H_a if nS^2/σ_0^2 is greater than c. The significance level of the test is

$$\alpha = \Pr\left[\frac{nS^2}{\sigma_0^2} > c; \sigma^2 = \sigma_0^2\right] \tag{3.27}$$

If $\sigma_0^2 > \sigma_1^2$, the test accepts H_0 if nS^2/σ_0^2 is greater than a critical value c and accepts H_a if nS^2/σ_0^2 is less than c. The significance level of the test is

$$\alpha = \Pr\left[\frac{nS^2}{\sigma_0^2} < c; \sigma^2 = \sigma_0^2\right] \tag{3.28}$$

Discussions of the power of the tests above are omitted.

EXAMPLE 3.8

Let us use the data presented in Example 3.5 to test the hypotheses H_0: $\sigma^2 = 0.04$ and H_a: $\sigma^2 = 0.03$. We set the significance level at $\alpha = 0.10$. Since $n = 18$, we have 17 degrees of freedom and find in Appendix Table F that

$$\Pr\left[\frac{nS^2}{\sigma_0^2} < 10.1; \sigma^2 = \sigma_0^2 = 0.04\right] = 0.1$$

so the critical value of the test is 10.1. From Example 3.5, $\bar{s} = 0.193$, so

$$\sum_{i=1}^{18} (X_i - \bar{X})^2 = nS^2 = (n - 1)\bar{s}^2 = (17)(0.193)^2 = 0.633$$

and

$$\frac{nS^2}{0.04} = \frac{(0.633)}{(0.04)} = 15.83 > 10.1$$

Thus we accept H_0: $\sigma^2 = 0.04$.

The Concept of P Value

Different experimenters may wish to use different significance levels and hence apply different tests to the same data. This leaves open the possibility that experimenters *could* draw different conclusions from the same facts. A technique for reporting experimental results that takes this possibility into account is the *P* value. The *P* value is the *smallest* significance level at which the observed data would have caused rejection of the null hypothesis.

For example, suppose we wish to test the hypotheses H_0: $\mu = 68$ and H_a: $\mu = 69$, where μ is the mean of a normal population with variance $\sigma^2 = 4$. Suppose further that a sample of size $n = 9$ yields a mean $\bar{X} = 68.8$. If a critical value of $c = 68.8$ had been used, the significance level of the test would have been

$$\Pr[\bar{X} > 68.8; \mu = 68] = \Pr\left[Z > \frac{68.8 - 68}{2/3}\right] = \Pr[Z > 1.2] = 0.1151$$

and this is the smallest significance level at which the null hypothesis would be rejected by a most powerful test. Thus if a significance level of 0.05 had been chosen a priori, the observation of $\bar{X} = 68.8$ would not have led to a rejection of H_0. On the other hand, if an a priori choice of $\alpha = 0.15$ had been made, $\bar{X} = 68.8$ would have led to the rejection of H_0. It is clear from this discussion that if the *P* value is reported, then experimenters with different a priori significance levels may decide immediately whether they accept or reject H_0. Thus we see the *P* value as a useful technique for reporting the results of a test.

As a final example let us reconsider the problem in the previous paragraph without assuming that σ^2 is known. Then the test will be based on the value of the statistic $t = (\bar{X} - 68)/(\bar{s}/\sqrt{9})$. Now if $\bar{X} = 69$ and $\bar{s} = 1.5$, the t value is

$(69 - 68)/(1.5/\sqrt{9}) = 2.00$. Appendix Table E does not allow us to calculate the probability $\Pr[t(8) > 2.00]$, which would be the P value in this case. The table does, however, show that

$$\Pr[t(8) > 2.306] = 0.025 \qquad \text{and} \qquad \Pr[t(8) > 1.86] = 0.05$$

Thus the P value of the test is between 0.025 and 0.05.

Composite Hypotheses and Significance Testing

In practice experimenters often find it difficult to specify hypotheses as precisely as is required by the theory presented in the discussion of simple hypotheses of this section. In other words a hypothesis of the form H_a: $\theta = \theta_1$ is too specific. On the other hand it may be possible to set up a hypothesis specifying a *range* of values of θ. For example, in an acceptance sampling problem we may be able to say that shipments with no more than 2 percent defectives are acceptable. A null hypothesis describing an acceptable lot would be H_0: $p \leq 0.02$, where p is the fraction of defectives in a lot. The alternative hypothesis is H_a: $p > 0.02$, and this alternative describes the main feature of an unacceptable shipment, namely, its fraction of defectives is greater than 0.02. A shipment with 5 percent defectives is unacceptable, as is a shipment with 10 percent defectives. Both of these cases are covered by the alternative H_a: $p > 0.02$. We thus see the possibility of setting up hypotheses containing *more than one value of the parameter*. Such hypotheses are called *composite*. Our purpose in this section is to show how the theory discussed previously may be extended to cover the testing of composite hypotheses. We begin with an example.

EXAMPLE 3.9

A typewriter manufacturer claims that unique features of its machines guarantee increased typing speed. On current equipment typists in the typing pool at the Travis Manufacturing Company type an average of 40 words per minute (wpm) with a standard deviation of $\sigma = 5$ wpm.

Travis management may be skeptical of the typewriter manufacturer's claim, believing that the new machines do not really increase speed. Hence management would wish to test the null hypothesis

$$H_0: \mu = 40$$

where μ is the average words per minute of the typing pool. Management might consider buying new machines if they indeed proved superior to the present equipment. Thus the alternative hypothesis of interest is

$$H_a: \mu > 40$$

If H_0 cannot be abandoned in favor of H_a, then the new equipment will not be considered.

Suppose $n = 10$ typists are to be randomly selected from the typing pool and asked to type a typical passage. The words per minute of each typist are to be recorded and averaged to obtain $\bar{X}$. If we assume that the population of words per minute for all typists in the pool is $N(\mu, 25)$, then $\bar{X}$ has an $N(\mu, 25/10)$ sampling distribution. The question we pose is this: What values of $\bar{X}$ will be large enough to cast doubt on the null hypothesis?

Despite the fact that we are not testing H_0 against any particular value of μ in H_a, we might still apply the theory of the simple hypothesis discussed earlier by setting a significance level and finding a critical value. Note that this process does not depend on any particular value of μ in H_a. To illustrate, let us take $\alpha = 0.05$. Since we are interested in alternatives greater than 40, the calculation of the significance level is

$$0.05 = \Pr[\bar{X} > c\,;\, \mu = 40] = \Pr\left[Z > \frac{c - 40}{(5/\sqrt{10})}\right] \qquad (3.29)$$

and this equation yields a critical value c. In fact, since $\Pr[Z > 1.645] = 0.05$, we have

$$\frac{c - 40}{(5/\sqrt{10})} = 1.645$$

or

$$c = 40 + \left(\frac{5}{\sqrt{10}}\right)(1.645) = 42.6$$

Now what are the implications of using this test? The power of the test for any specific alternative in H_a can be calculated. For example, the power when $\mu = 41$ is

$$1 - \beta(41) = \Pr[\bar{X} > 42.6\,;\, \mu = 41] = \Pr\left[Z > \frac{42.6 - 41}{(5/\sqrt{10})}\right] = \Pr[Z > 1.01] = 0.1562$$

whereas the power when $\mu = 42$ is

$$1 - \beta(42) = \Pr[\bar{X} > 42.6\,;\, \mu = 42] = \Pr\left[Z > \frac{42.6 - 42}{(5/\sqrt{10})}\right] = \Pr[Z > 0.38] = 0.3520$$

Table 3.6 Power for Selected Values of μ of the Test Proposed in Example 3.9

μ	40.5	41	41.5	42	42.5	43	44	45
$1 - \beta(\mu)$	0.0918	0.1562	0.2420	0.3520	0.4761	0.5987	0.8133	0.9357

Table 3.6 shows powers for a number of alternatives. The greater the value of the alternative μ, the greater the power of the test to accept that alternative over the null hypothesis. Values of μ close to 40 are not very likely to be chosen over the null hypothesis. For example, if in fact $\mu = 41$, the probability of rejecting H_0 is only 0.1562. We conclude that if management uses this test as a basis for decision making, management is not anxious to reject H_0 unless alternatives like $\mu = 43$ or $\mu = 44$ are correct. This may be quite sensible because only a substantial increase in words per minute could justify the expense of purchasing a number of new typewriters.

To illustrate the application of the test, suppose the experiment is run and the value of $\bar{X}$ is 43. This value calls for the rejection of H_0 at the 5-percent significance level since $43 > 42.6$. The P value is

$$\Pr[\bar{X} > 43;\ \mu = 40] = \Pr\left[Z > \frac{43-40}{(5/\sqrt{10})}\right] = \Pr[Z > 1.9] = 0.0287$$

Thus H_0 would have been rejected by any test whose significance level was bigger than 0.0287.

Example 3.9 brings out some of the important issues in testing composite hypotheses. It is worth repeating that (a) the critical value of the test is found by referring solely to the null hypothesis and a preassigned significance level; (b) no particular alternative in H_a is accepted in favor of H_0, but certain alternatives are much more likely to lead to rejection of H_0 than others.

These properties are characteristic of tests called by statisticians *significance tests*. The term "significance" comes from the practice of calling data that lead to the rejection of H_0 significant. This is a technical use of the term and should be interpreted carefully. A statistically significant result need not be of great "practical significance." Significance tests are popular with practitioners because they do not require the delineation of specific alternative hypotheses. There are some criticisms of significance tests, however, that should be made explicit.

The first criticism is that because the test can be formed by specifying only the null hypothesis and the significance level, there is a strong temptation to forget about power considerations. In fact, many practitioners have fallen into the habit of using standard significance levels like 0.1, 0.05, and 0.01 regardless of the specific properties of their experiments, such as the sample size and the variance of the population being sampled. Such practice could lead to setting up a test of very limited power for an alternative that the experimenter considers rather important to detect. The practice of choosing standard significance levels is often rationalized by the argument that the type I error is much more important than type II errors, but this claim is rarely supported with concrete evidence.

The second criticism is that if the experimenter is so uncertain about the possible alternatives to H_0, he should take an estimation approach to the problem from the start. To illustrate this, consider again the situation in Example 3.9. Management could have constructed confidence intervals for μ on the basis of the data and found a set of plausible values of μ. This information would have been more explicit than the information that the hypothesis H_0: $\mu = 40$ was rejected. The 95-percent confidence interval for μ is

$$\left[\bar{X} - 1.96\left(\frac{5}{\sqrt{10}}\right),\ \bar{X} + 1.9\left(\frac{5}{\sqrt{10}}\right)\right] = [43 - 3.1,\ 43 + 3.1] = [39.9,\ 46.1]$$

and the 90-percent confidence interval for μ is

$$\left[\bar{X} - 1.645\left(\frac{5}{\sqrt{10}}\right), \bar{X} + 1.645\left(\frac{5}{\sqrt{10}}\right)\right] = [43 - 2.6, 43 + 2.6] = [40.4, 45.6]$$

Since plausible values of μ could range anywhere from about 40 to about 46, it seems likely that more precise information is needed before a decision can be made. Further experimentation appears warranted.

In Example 3.9 the hypotheses were H_0: $\mu = 40$ and H_a: $\mu > 40$. This alternative hypothesis is called a *one-sided* alternative because it contains only values greater than 40. In general if a null hypothesis is of the form H_0: $\theta = \theta_0$, then it is possible to test one of two one-sided alternative hypotheses, namely, H_a: $\theta > \theta_0$ or H_a: $\theta < \theta_0$. Another type of composite hypothesis is the two-sided alternative H_a: $\theta \neq \theta_0$. This alternative hypothesis contains values both smaller and greater than θ_0. The following example shows how to conduct a test for a two-sided alternative hypothesis.

EXAMPLE 3.10

A management instructor routinely gives his beginning management classes an aptitude test. The scores on the test are normally distributed and have variance $\sigma^2 = 100$. "Typical" students should obtain an average score of about 100 on the test. In order to see if his students are "typical" the instructor proposes to test H_0: $\mu = 100$ against the alternative H_a: $\mu \neq 100$. The data are to be the scores obtained by the $n = 49$ students in a beginning class. The assumption is made that these students may be treated as a random sample from the population of all students who attend the instructor's classes. This assumption could be faulty if the quality of students is changing over time, but given the assumption, inferences about μ may be drawn from the current class.

An intuitively reasonable test of H_0 against H_a can be based on the sample mean $\bar{X}$. The test says to accept H_0 if the difference between $\bar{X}$ and 100 does not fall too far below or too far above 0; that is, accept H_0 if

$$-c \leq \bar{X} - 100 \leq c \tag{3.30}$$

To see why this is a reasonable test look at Figure 3.4. The figure shows the sampling

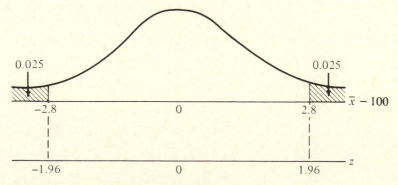

Figure 3.4   The $N(0, 100/49)$ distribution

distribution of $\bar{X} - 100$ when H_0: $\mu = 100$ is the correct hypothesis. When H_0 is correct, $\bar{X} - 100$ has an $N(0, 100/49)$ distribution. Thus a value of $\bar{X}$ too far from 100, or equivalently a value of $\bar{X} - 100$ too far from 0, will constitute evidence against H_0. Since the alternative hypothesis contains all values of μ on either side of 100, it is reasonable to reject H_0 when the value of $\bar{X} - 100$ falls outside of an interval whose endpoints are symmetric about 0, namely, $(-c, c)$. Using this convention the value of c can be determined by fixing the significance level. Figure 3.4 shows that $c = 2.8$ when the significance level is fixed at $\alpha = 0.05$. In general, H_0 is rejected when $\bar{X} - 100 < -c$ or $\bar{X} - 100 > c$ and the significance level is

$$\alpha = \Pr[(\bar{X} - 100 < -c) \cup (\bar{X} - 100 > c); \mu = 100]$$
$$= 1 - \Pr[-c \leq \bar{X} - 100 \leq c; \mu = 100]$$
$$= 1 - \Pr\left[\frac{-c}{(10/7)} \leq \frac{\bar{X} - 100}{(10/7)} \leq \frac{c}{(10/7)}; \mu = 100\right]$$
$$= 1 - \Pr\left[-\frac{c}{(10/7)} \leq Z \leq \frac{c}{(10/7)}\right]$$

where Z has the standard normal distribution. This follows from the fact that when H_0: $\mu = 100$ is correct, $Z = (\bar{X} - 100)/(10/7)$ has the standard normal distribution. With $\alpha = 0.05$ we have

$$0.05 = 1 - \Pr\left[-\frac{c}{(10/7)} \leq Z \leq \frac{c}{(10/7)}\right]$$

or

$$\Pr\left[-\frac{c}{(10/7)} \leq Z \leq \frac{c}{(10/7)}\right] = 0.95$$

Thus we must have

$$\frac{c}{(10/7)} = 1.96$$

or

$$c = \left(\frac{10}{7}\right)(1.96) = 2.8$$

as in Figure 3.4.

If we had chosen $\alpha = 0.1$, the equation for c would have been $c/(10/7) = 1.645$, or $c = (10/7)(1.645) = 2.35$.

Table 3.7 shows the power function of the test with $\alpha = 0.05$ for a number of alternative values of μ. As is intuitively reasonable, the power function is symmetric about $\mu = 100$. To illustrate the necessary calculations, we compute $1 - \beta(102)$ as follows:

$$1 - \beta(102) = 1 - \Pr[-2.8 \leq \bar{X} - 100 \leq 2.8; \mu = 102]$$
$$= 1 - \Pr[-4.8 \leq \bar{X} - 102 \leq 0.8; \mu = 102]$$
$$= 1 - \Pr\left[\frac{-4.8}{(10/7)} \leq \frac{\bar{X} - 102}{(10/7)} \leq \frac{0.8}{(10/7)}; \mu = 102\right]$$
$$= 1 - \Pr[-3.36 \leq Z \leq 0.56]$$
$$= 1 - (0.7123 - 0.0004) = 1 - 0.7119 = 0.2881$$

Notice that when $\mu = 102$ is assumed to be correct, the standardized version of $\bar{X}$ is $(\bar{X} - 102)/(10/7)$.

Table 3.7 Power Function of the Test with $\alpha = 0.05$

μ	95	96	97	98	99	100	101	102	103	104	105
$1 - \beta(\mu)$	0.9382	0.7995	0.5557	0.2881	0.1077	0.05	0.1077	0.2881	0.5557	0.7995	0.9382

Table 3.7 shows that the power of the test does not rise above 0.5 until alternatives such as 103 and 97 are considered. A population of students whose average score is 102, for example, is not likely to be detected by the test. If it were deemed important to detect such a population and to maintain a significance level of 0.05, then more students would have to be tested.

The concept of P value may be applied to tests of two-sided alternatives. To illustrate this, suppose the average of $n = 49$ test scores is $\bar{X} = 103$. We might think that the P value is

$$\Pr[\bar{X} - 100 > 3; \mu = 100] = \Pr[Z > 2.1] = 0.0179 \qquad (3.31)$$

It must be remembered, however, that the P value is the *smallest significance level* at which H_0 could have been rejected by the given data. Thus the P value when $\bar{X} = 103$ is observed must be calculated by the formula

$$1 - \Pr[-3 \le \bar{X} - 100 \le 3; \mu = 100] = 1 - \Pr[-2.1 \le Z \le 2.1] = 1 - 0.9642 = 0.0358$$

Thus the P value is *twice* the value given in equation (3.31).

Various significance tests for the mean of an $N(\mu, \sigma^2)$ population with σ^2 known are displayed in Table 3.8.

Significance tests for the mean of an $N(\mu, \sigma^2)$ population whose variance is unknown may be based on the t statistic. Table 3.9 displays the possible tests. We assume that $X_1, X_2, \ldots, X_n$ is a random sample from the population,

Table 3.8 Tests for the Mean of an $N(\mu, \sigma^2)$ Population Whose Variance Is Known

Null Hypothesis	Alternative Hypothesis	Reject H_0 if:	Power $1 - \beta(\mu_1)$
$H_0: \mu = \mu_0$	$H_a: \mu \ne \mu_0$	$\lvert \bar{X} - \mu_0 \rvert > c$	$1 - \Pr\left[\dfrac{\mu_0 - \mu_1 - c}{(\sigma/\sqrt{n})} \le Z \le \dfrac{\mu_0 - \mu_1 + c}{(\sigma/\sqrt{n})}\right]$
$H_0: \mu = \mu_0$	$H_a: \mu > \mu_0$	$\bar{X} - \mu_0 > c$	$\Pr\left[Z > \dfrac{\mu_0 - \mu_1 + c}{(\sigma/\sqrt{n})}\right]$
$H_0: \mu = \mu_0$	$H_a: \mu < \mu_0$	$\bar{X} - \mu_0 < -c$	$\Pr\left[Z < \dfrac{\mu_0 - \mu_1 - c}{(\sigma/\sqrt{n})}\right]$

Table 3.9   Tests for the Mean of an $N(\mu, \sigma^2)$ Population When the Variance Is Unknown

Null Hypothesis	Alternative Hypothesis	Reject H_0 if:
$H_0: \mu = \mu_0$	$H_a: \mu < \mu_0$	$t < -c$
$H_0: \mu = \mu_0$	$H_a: \mu > \mu_0$	$t > c$
$H_0: \mu = \mu_0$	$H_a: \mu \neq \mu_0$	$|t| > c$

that $\bar{X} = (1/n) \sum_{i=1}^{n} X_i$, that $\bar{s} = \sqrt{\sum_{i=1}^{n} (X_i - \bar{X})^2 / (n-1)}$, and that

$$t = \frac{\bar{X} - \mu_0}{(\bar{s}/\sqrt{n})} \qquad (3.32)$$

When $H_0: \mu = \mu_0$ is correct, t has a $t(n-1)$ distribution. Using this fact critical values of the tests can be determined once a significance level is fixed. The power calculations for this test are beyond the scope of this book.

EXAMPLE 3.11

Let us use the data in Example 3.4 to test the hypotheses $H_0: \mu = 0$ and $H_a: \mu \neq 0$. In words, we are testing whether or not the average monthly rate of return on the company's stock is different from zero. We shall use a 5-percent significance level. This means that

$$0.05 = 1 - \Pr\left[-c < \frac{\bar{X} - 0}{\bar{s}/\sqrt{18}} < c\right]$$

$$= 1 - \Pr[-c < t(17) < c]$$

or

$$\Pr[-c < t(17) < c] = 0.95$$

and this means that $c = 2.11$. The computed value of t is

$$t = \frac{-0.017 - 0}{(0.193)/\sqrt{18}} = -0.374 \qquad \text{and} \qquad |t| = 0.374 < 2.11$$

Thus we do not reject $H_0: \mu = 0$ at the 5-percent significance level.

We note that $\mu = 0$ was contained in the 95-percent confidence interval for μ (see Example 3.4) and consequently must be interpreted as a plausible value for μ from an estimation point of view. The implied correspondence between values in the confidence interval and acceptable hypotheses is not accidental. A more thorough explanation of the relationship between confidence intervals and hypothesis tests is given in Section 3.4.

In closing this section we mention one final warning about significance tests. This has to do with "acceptance" of the null hypothesis. The power calculations in this section have demonstrated that even when the null hypothesis is not correct, the chances of rejecting it may be very small. This will happen if the correct hypothesis is close to the null hypothesis. For this reason it is common to replace the statement "accept the null hypothesis" with the statement "do not reject the null hypothesis." This language recognizes the fact that the null hypothesis may be, and most likely is, false, but the test may not be powerful enough to reject the false null hypothesis. There are even situations in which this would be considered a desirable feature of the test, situations in which alternatives close to the null hypothesis are to be deliberately ignored. (See Example 3.9 for an illustration of this remark.)

3.4 THE RELATIONSHIP BETWEEN CONFIDENCE INTERVALS AND SIGNIFICANCE TESTS

It is intuitively plausible that a confidence interval for a parameter contains the values of the parameter that cannot be rejected by certain significance tests. This is, in fact, the case. We will illustrate this basic principle with an example.

Let $X_1, X_2, \ldots, X_n$ denote a random sample from an $N(\mu, \sigma^2)$ population with σ^2 known. Then

$$\Pr\left[\bar{X} - 1.96\left(\frac{\sigma}{\sqrt{n}}\right) \leq \mu \leq \bar{X} + 1.96\left(\frac{\sigma}{\sqrt{n}}\right)\right] = 0.95$$

as shown in the first part of Section 3.2. Thus when $\mu = \mu_0$,

$$\Pr\left[\bar{X} - 1.96\left(\frac{\sigma}{\sqrt{n}}\right) \leq \mu_0 \leq \bar{X} + 1.96\left(\frac{\sigma}{\sqrt{n}}\right)\right] = 0.95$$

and this is equivalent to the statement

$$\Pr\left[\mu_0 - 1.96\left(\frac{\sigma}{\sqrt{n}}\right) \leq \bar{X} \leq \mu_0 + 1.96\left(\frac{\sigma}{\sqrt{n}}\right)\right] = 0.95$$

The equivalence of the events in the above statements shows that μ_0 will lie in the 95-percent confidence interval for μ if and only if $\bar{X}$ falls in the interval $[\mu_0 - 1.96(\sigma/\sqrt{n}), \mu_0 + 1.96(\sigma/\sqrt{n})]$, and this latter condition means that H_0: $\mu = \mu_0$ will not be rejected at the 5-percent significance level by the first test described in Table 3.8. We see that another way to perform the significance test is to construct the confidence interval and to reject H_0 only if μ_0 does *not* lie in the interval.

As might be expected, tests of one-sided alternatives are related to one-sided confidence intervals. Again, assuming an $N(\mu, \sigma^2)$ population, a one-sided 95-percent confidence interval for μ can be based on the statement

$$\Pr\left[\bar{X} - 1.645\frac{\sigma}{\sqrt{n}} \le \mu\right] = 0.95$$

Now μ_0 falls in the interval $(\bar{X} - 1.645(\sigma/\sqrt{n}), \infty)$ if and only if $\bar{X} - \mu_0 < 1.645(\sigma/\sqrt{n})$. This, however, is precisely the condition under which H_0: $\mu = \mu_0$ is accepted and H_a: $\mu > \mu_0$ is rejected at the 5-percent significance level by the second test listed in Table 3.8.

Similar demonstrations can be made to show the equivalence of the other confidence interval and testing procedures developed in this chapter.

3.5 DETERMINING SAMPLE SIZE IN EXPERIMENTAL DESIGN

Sometimes we are in the position of having to design an experiment with a specified level of precision. We may take either an estimation or testing approach to this problem. Both these approaches will be illustrated on the assumption that we wish to make inferences about the mean of a normal population whose variance is known.

We consider the estimation approach first. A confidence interval for μ has the form

$$\left[\bar{X} - z_{\alpha/2}\frac{\sigma}{\sqrt{n}}, \bar{X} + z_{\alpha/2}\frac{\sigma}{\sqrt{n}}\right]$$

and it is natural to measure the *precision* of the interval by its *half-width* $z_{\alpha/2}(\sigma/\sqrt{n})$. This is because the probability is $1 - \alpha$ that $\bar{X}$ and μ differ by no more than $z_{\alpha/2}(\sigma/\sqrt{n})$. In designing the experiment we fix $1 - \alpha$ and the half-width $z_{\alpha/2}(\sigma/\sqrt{n})$. Then we can solve for the sample size n needed to guarantee the specified level of precision. Let h denote the specified half-width. Then

$$h = z_{\alpha/2}\frac{\sigma}{\sqrt{n}}$$

and

$$n = \left(\frac{z_{\alpha/2}}{h}\right)^2 \sigma^2 \qquad (3.33)$$

Thus the required sample size is proportional to the population variance σ^2. Table 3.10 displays values of the proportionality factor $(z_{\alpha/2}/h)^2$ corresponding to selected values of $1 - \alpha$ and h. Equation (3.33) can be used only if a value for σ^2 is specified. If an acceptable value for σ^2 is not known a priori, it may be possible to get a rough estimate by conducting a small pilot study.

In some significance testing problems it may be possible to specify a deviation from the null hypothesis that is important to detect. If a value of

Table 3.10 Values[a] of $(z_{\alpha/2}/h)^2$ for Selected Values of $1-\alpha$ and h

$1-\alpha$ \ h	0.01	0.10	0.50	1.0
0.85	20,736	208	9	2
0.90	27,061	271	11	3
0.95	38,416	385	16	4
0.99	66,358	664	27	7

[a] Values are rounded to the nearest integer so as to show the magnitude of n when $\sigma^2 = 1$.

the power function is specified at that deviation, along with a significance level, values of n and c can be determined. Let us illustrate this with an example.

EXAMPLE 3.12

We return to the situation in Example 3.10. There the population distribution was $N(\mu, 100)$, and the hypotheses to be tested were H_0: $\mu = 100$ and H_a: $\mu \neq 100$. The power function of the test using $n = 49$ and $\alpha = 0.05$ is given in Table 3.7. Suppose it is deemed important to detect a deviation from $\mu = 100$ of two points or more, and the probability of detecting a two-point deviation is to be 0.75. The significance level is to remain at 0.05. These two requirements are

$$1 - \beta(100) = 0.05 \tag{3.34}$$

$$1 - \beta(102) = 0.75 \tag{3.35}$$

$$1 - \beta(98) = 0.75 \tag{3.36}$$

Only one of the conditions of equations (3.35) and (3.36) needs to be imposed due to the symmetry of the power function about 100. Now,

$$1 - \beta(100) = 1 - \Pr[-c \leq \bar{X} - 100 \leq c; \mu = 100] = 1 - \Pr\left[-\frac{c}{(10/\sqrt{n})} \leq Z \leq \frac{c}{(10/\sqrt{n})}\right] \tag{3.37}$$

and

$$1 - \beta(102) = 1 - \Pr[-c < \bar{X} - 100 < c; \mu = 102] = 1 - \Pr\left[\frac{-c-2}{(10/\sqrt{n})} \leq Z \leq \frac{c-2}{(10/\sqrt{n})}\right] \tag{3.38}$$

Setting equation (3.37) equal to 0.05 and equation (3.38) equal to 0.75, we get

$$\Pr\left[-\frac{c}{(10/\sqrt{n})} < Z < \frac{c}{(10/\sqrt{n})}\right] = 0.95 \tag{3.39}$$

and

$$\Pr\left[\frac{-c-2}{(10/\sqrt{n})} < Z < \frac{c-2}{(10/\sqrt{n})}\right] = 0.25 \tag{3.40}$$

Figure 3.5 illustrates these two equations. The illustration of equation (3.40) contains

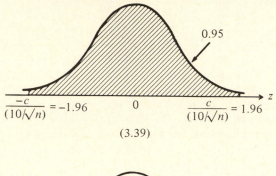

(3.39)

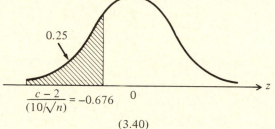

(3.40)

Figure 3.5 Illustrations of equations (3.39) and (3.40)

an approximation because we are replacing equation (3.40) by equation (3.41),

$$\Pr\left[Z \leq \frac{c-2}{(10/\sqrt{n})}\right] = 0.25 \tag{3.41}$$

This approximation is generally used in practice and yields acceptable results in most cases. We shall check its accuracy for this example in a moment.

Figure 3.5 shows that we must have

$$\frac{c}{(10/\sqrt{n})} = 1.96 \tag{3.42}$$

and

$$\frac{c-2}{(10/\sqrt{n})} = -0.676 \tag{3.43}$$

These equations lead to the relationship

$$-0.676 = \frac{c}{(10/\sqrt{n})} - \frac{2}{(10/\sqrt{n})} = 1.96 - \frac{\sqrt{n}}{5} \qquad (3.44)$$

which gives

$$\sqrt{n} = 5(1.96 + 0.676) = 13.18 \qquad \text{and} \qquad n = 174 \qquad (3.45)$$

Substituting equation (3.45) into equation (3.42) yields

$$c = 1.96 \frac{10}{\sqrt{n}} = \frac{19.6}{13.18} = 1.49 \qquad (3.46)$$

Our calculations have led to the following test: Take a random sample of size $n = 174$ from the population and calculate $\bar{X}$. If

$$-1.49 \le \bar{X} - 100 < 1.49$$

accept H_0; otherwise reject H_0. This test is such that the significance level $\alpha = 0.05$ and the power function at 102 and 98 is

$$1 - \beta(98) = 1 - \beta(102) = 0.75$$

Note that this test can be applied only if a sample of size 174 is available.

In concluding this example, let us check the approximation in equation (3.41) by substituting $n = 174$ and $c = 1.49$ into the probability statement in equation (3.40). The resulting probability should be close to 0.25. Now

$$\Pr\left[\frac{-1.49-2}{(10/\sqrt{174})} \le Z \le \frac{1.49-2}{(10/\sqrt{174})}\right] = \Pr[-4.60 \le Z \le -0.673]$$

$$= 0.2505 - 0.00002 = 0.25048$$

and this is sufficiently close to 0.25 for practical purposes.

3.6 LARGE SAMPLE AND ROBUST PROCEDURES BASED ON THE t STATISTIC

In the previous sections the assumption of a normal population allowed us to present a number of elegant examples of statistical inference. Quite often in practice, however, the normality assumption cannot be justified. Much effort has gone into the development of techniques that are not sensitive to assumptions about the population, and such techniques are called by different writers "nonparametric," "distribution free," or "robust." The property of being distribution free or nonparametric is characteristic of many statistical methods *when the sample size n is large*. Our purpose in this section is to present important large sample robust techniques for making statistical inferences about a population mean. These techniques are based on the t statistic

$$t = \frac{\bar{X} - \mu}{\bar{s}/\sqrt{n}} \qquad (3.47)$$

where $\bar{s} = \sqrt{\sum_{i=1}^{n}(X_i - \bar{X})^2/(n-1)}$. Here we assume $X_1, X_2, \ldots, X_n$ is a random sample from an arbitrary population with mean μ and finite variance σ^2. The value of σ^2 is not assumed to be known. We now quote a fundamental theorem from mathematical statistics.

Theorem 3.1   Given the assumptions made above, the random variable in equation (3.47) has an approximate $N(0, 1)$ distribution when n is large. Consequently, when n is large, $\bar{X}$ has an approximate $N(\mu, \bar{s}^2/n)$ distribution.

Theorem 3.1 is remarkable, for it allows the population to be completely arbitrary except for the assumption of a finite variance. Of course, the question of when n is "large enough" naturally arises. A popular rule of thumb is that for practical purposes an n larger than 30 is "large enough." This rule may be acceptable as long as the population is roughly symmetric, but many applications would seem to require larger n. A specific example will be noted below. Any rule of thumb should be thought of as only a benchmark with which to compare one's own requirements, as what is "large enough" can and will vary with circumstances.

The t statistic may be used to develop large sample tests of the hypothesis H_0: $\mu = \mu_0$ against various alternatives. Table 3.11 summarizes the possibilities. Since we are assuming that n is large, we use Theorem 3.1 to assert that

$$\Pr\left[\frac{\bar{X}-\mu_0}{\bar{s}/\sqrt{n}} < c \, ; \mu_1\right] = \Pr\left[\bar{X} < \mu_0 + c\left(\frac{\bar{s}}{\sqrt{n}}\right); \mu_1\right]$$

$$= \Pr\left[\frac{\bar{X}-\mu_1}{\bar{s}/\sqrt{n}} < \frac{\mu_0-\mu_1+c(\bar{s}/\sqrt{n})}{\bar{s}/\sqrt{n}}; \mu_1\right]$$

$$\doteq \Pr\left[Z < \frac{\mu_0-\mu_1+c(\bar{s}/\sqrt{n})}{\bar{s}/\sqrt{n}}\right] \tag{3.48}$$

Thus "power" calculations for the test can be made, but only *after* the data, and in particular $\bar{s}$, have been observed. This is the reason we have put "power" in quotation marks. Equations such as those in the display of (3.48) are useful for computing significance levels before and P values after the experiment has been performed.

EXAMPLE 3.13

Let us return to the situation in Example 3.10. There a management instructor contemplated drawing a random sample of $n = 49$ scores from a population of management aptitude test scores. In that example we assumed that the population of scores was $N(\mu, 100)$; but because $n = 49$, it is possible for us to relax both the assumption of normality and the assumption that σ^2 is known to be 100.

If the instructor wished to test the hypothesis H_0: $\mu = 100$ against the alternative

Table 3.11 Large Sample Tests for a Population Mean

Null Hypothesis	Alternative Hypothesis	Reject H_0 if	Power $1 - \beta(\mu_1)$
$H_0: \mu = \mu_0$	$H_a: \mu \neq \mu_0$	$\left\| \dfrac{\bar{X} - \mu_0}{\bar{s}/\sqrt{n}} \right\| > c$	$1 - \Pr\left[\dfrac{\mu_0 - \mu_1 - c\bar{s}/\sqrt{n}}{\bar{s}/\sqrt{n}} \leq Z \leq \dfrac{\mu_0 - \mu_1 + c\bar{s}/\sqrt{n}}{\bar{s}/\sqrt{n}} \right]$
$H_0: \mu = \mu_0$	$H_a: \mu > \mu_0$	$\dfrac{\bar{X} - \mu_0}{\bar{s}/\sqrt{n}} > c$	$\Pr\left[Z > \dfrac{\mu_0 - \mu_1 + c\bar{s}/\sqrt{n}}{\bar{s}/\sqrt{n}} \right]$
$H_0: \mu = \mu_0$	$H_a: \mu < \mu_0$	$\dfrac{\bar{X} - \mu_0}{\bar{s}/\sqrt{n}} < -c$	$\Pr\left[Z < \dfrac{\mu_0 - \mu_1 - c\bar{s}/\sqrt{n}}{\bar{s}/\sqrt{n}} \right]$

H_a: $\mu \neq 100$ at the 5-percent significance level, he could calculate the value of $t = (\bar{X} - 100)/(\bar{s}/\sqrt{49})$ and see if it falls in the interval $(-1.96, 1.96)$. In this example, suppose $\bar{X} = 102.9$ and $\bar{s} = 11.2$; consequently,

$$t = \frac{102.9 - 100}{(11.2/7)} = 1.81$$

and H_0 is not rejected at the 5-percent significance level. In Example 3.10 the value $\bar{X} = 102.9$ would lead to the rejection of H_0 at the 5-percent significance level, but there we assumed the instructor knew $\sigma^2 = 100$. On average the test based on σ^2 unknown should be less sensitive to departures from H_0 than the test based on σ^2 known. Thus the result obtained in this example is not surprising.

To compute the P value associated with $\bar{X} = 102.9$, we calculate

$$2\Pr[\bar{X} > 102.9; 100] \doteq 2\Pr\left[Z > \frac{102.9 - 100}{(11.2/7)}\right] = 2\Pr[Z > 1.81] = 0.0704$$

Thus only a test with a significance level larger than about 0.07 would have rejected H_0: $\mu = 100$.

Theorem 3.1 yields a statement that can be used as a basis of a confidence interval for μ. Since for large n, $\bar{X}$ is approximately a $N(\mu, \bar{s}^2/n)$ random variable, we have

$$\Pr\left[\mu - z_{\alpha/2}\frac{\bar{s}}{\sqrt{n}} \leq \bar{X} \leq \mu + z_{\alpha/2}\frac{\bar{s}}{\sqrt{n}}\right] = 1 - \alpha \tag{3.49}$$

where $z_{\alpha/2}$ is the $100(1 - \alpha/2)$th percentile of the standard normal distribution. But equation (3.49) may be written

$$\Pr\left[\bar{X} - z_{\alpha/2}\frac{\bar{s}}{\sqrt{n}} \leq \mu \leq \bar{X} + z_{\alpha/2}\frac{\bar{s}}{\sqrt{n}}\right] = 1 - \alpha \tag{3.50}$$

and this statement will yield an approximate $100(1 - \alpha)$-percent confidence interval for μ if n is sufficiently large.

EXAMPLE 3.14

Using the data $n = 49$, $\bar{X} = 102.9$, and $\bar{s} = 11.2$ from Example 3.13, we see that an approximate 95-percent confidence interval for μ is

$$\left[\bar{X} - 1.96\frac{\bar{s}}{\sqrt{n}}, \bar{X} + 1.96\frac{\bar{s}}{\sqrt{n}}\right] = \left[102.9 - 1.96\left(\frac{11.2}{\sqrt{49}}\right), 102.9 + 1.96\left(\frac{11.2}{\sqrt{49}}\right)\right]$$
$$= (99.76, 106.04)$$

The t statistic often yields quite robust methods of inference even when the sample sizes are small. We may loosely summarize the results as follows: As long as the population does not stray too far from symmetry, tests and confidence intervals based on the t statistic will have significance levels and confidence coefficients very close to those obtained when the populations are normal. That is, the best statistical methods for making inferences about the mean of a *normal* population are valid (but not necessarily most precise), in many cases, for making inferences about the mean of a nonnormal population. Thus many practitioners routinely use the t statistic with the confidence that they are not erring greatly in doing so. Of course, the routine application of such methods to samples from highly skewed populations could produce misleading results, so some care must be taken to check the assumption of symmetry.

3.7 COMPARING MEANS AND VARIANCES OF TWO NORMAL POPULATIONS

Comparison of Means Using Independent Samples

A common statistical problem is the comparison of the means of two populations. In this section we assume the two populations are $N(\mu_1, \sigma_1^2)$ and $N(\mu_2, \sigma_2^2)$, respectively, and we assume the data are *independent* random samples from the populations. It is convenient to make inferences about the parameter $\delta = \mu_2 - \mu_1$. Let $\bar{X}_1$ and $\bar{X}_2$ denote the means of independent random samples of size m and n from the $N(\mu_1, \sigma_1^2)$ and $N(\mu_2, \sigma_2^2)$ populations, respectively. Then we have

$$E(\bar{X}_1) = \mu_1 \qquad\qquad \text{Var}(\bar{X}_1) = \frac{\sigma_1^2}{m}$$

$$E(\bar{X}_2) = \mu_2 \qquad\qquad \text{Var}(\bar{X}_2) = \frac{\sigma_2^2}{n}$$

$$E(\bar{X}_2 - \bar{X}_1) = \mu_2 - \mu_1 = \delta \qquad \text{Var}(\bar{X}_2 - \bar{X}_1) = \frac{\sigma_1^2}{m} + \frac{\sigma_2^2}{n}$$

The last result follows from the assumption that the samples are independent and hence that the sample means are independent. Furthermore, since $\bar{X}_1$ and $\bar{X}_2$ are normal random variables, so is $\bar{X}_2 - \bar{X}_1$.

We now present inference procedures under the assumption that σ_1^2 and σ_2^2 are equal; that is, $\sigma_1^2 = \sigma_2^2 = \sigma^2$. Let $\bar{s}_1^2$ and $\bar{s}_2^2$ be the two sample variances, and let

$$\bar{s}_p^2 = \frac{(m-1)\bar{s}_1^2 + (n-1)\bar{s}_2^2}{m+n-2}$$

be an estimator of the common population variance σ^2 constructed as a weighted average of $\bar{s}_1^2$ and $\bar{s}_2^2$. Then the random variable

$$t = \frac{(\bar{X}_2 - \bar{X}_1) - (\mu_2 - \mu_1)}{\bar{s}_p \sqrt{(1/m) + (1/n)}} \tag{3.51}$$

has a $t(m + n - 2)$ distribution. This fact can be used to make inferences about $\delta = \mu_2 - \mu_1$.

EXAMPLE 3.15

Let us suppose we want to evaluate a new method of teaching calculus to business students, and assume we have a standardized exam that adequately measures proficiency in calculus. The new teaching method is to be compared to a traditional one. Two groups of students are selected *at random* from prebusiness majors who are to take calculus. One group (called the "control" group) receives traditional instruction, whereas the other group (called the "experimental" group) receives instruction under the new method. At the end of the course both groups take the proficiency test. The test scores are to be used to measure the difference in average

Table 3.12   Test Scores for Two Methods
of Instruction

i	X_{1i}	i	X_{2i}
1	59.16	1	95.56
2	73.18	2	86.47
3	73.67	3	83.29
4	60.08	4	68.12
5	75.29	5	70.83
6	72.78	6	84.14
7	83.92	7	81.07
8	74.09	8	75.03
9	70.61	9	85.01
10	60.36	10	66.18
11	75.07		
12	55.86		

$m = 12$ $\qquad\qquad\qquad$ $n = 10$

$\bar{X}_1 = 69.51$ $\qquad\qquad\qquad$ $\bar{X}_2 = 79.57$

$\Sigma_{i=1}^{12}(X_{1i} - \bar{X}_1)^2 = 801.71$ $\qquad$ $\Sigma_{i=1}^{10}(X_{2i} - \bar{X}_2)^2 = 772.21$

$\bar{s}_1 = \sqrt{\frac{1}{11}\Sigma_{i=1}^{12}(X_{1i} - \bar{X}_1)^2} = 8.54$ $\qquad$ $\bar{s}_2 = \sqrt{\frac{1}{9}\Sigma_{i=1}^{10}(X_{2i} - \bar{X}_2)^2} = 9.29$

$\bar{X}_2 - \bar{X}_1 = 10.06$ $\qquad\qquad$ $\bar{s}_p = \sqrt{\frac{801.71 + 772.21}{20}} = 8.87$

$t = \dfrac{(\bar{X}_2 - \bar{X}_1) - 0}{\bar{s}_p \sqrt{(1/12) + (1/10)}} = \dfrac{10.06}{(8.87)\sqrt{11/60}} = 2.65$

proficiency gained under the two methods of instruction. We let μ_2 denote average proficiency using the new method and μ_1 denote the average proficiency using the traditional method. One approach to this problem would be to test H_0: $\mu_2 - \mu_1 = 0$ (no difference in the methods) against H_a: $\mu_2 - \mu_1 > 0$ (new method superior to the traditional method).

If $m = 12$, $n = 10$, and the null hypothesis is assumed to be true, then the statistic

$$t = \frac{\bar{X}_2 - \bar{X}_1}{\bar{s}_p \sqrt{(1/12) + (1/10)}}$$

has a $t(20)$ distribution. Since $\Pr[t(20) > 1.725] = 0.05$, we would reject H_0 at the 5-percent significance level if the value of t exceeds 1.725. Table 3.12 displays some hypothetical data and the associated statistics. Since the computed value of t is $2.65 > 1.75$, we reject at the 5-percent significance level the hypothesis that the two methods yield the same average performance.

A 90-percent confidence interval for $\mu_2 - \mu_1$ is based on the fact that

$$\Pr\left[-1.725 < t = \frac{(\bar{X}_2 - \bar{X}_1) - (\mu_2 - \mu_1)}{\bar{s}_p \sqrt{(1/12) + (1/10)}} \le 1.725\right] = 0.90$$

or

$$\Pr[(\bar{X}_2 - \bar{X}_1) - 1.725\bar{s}_p \sqrt{(1/12) + (1/10)} \le \mu_2 - \mu_1 < (\bar{X}_2 - \bar{X}_1)$$
$$+ 1.725\bar{s}_p \sqrt{(1/12) + (1/10)}] = 0.90$$

Thus the 90-percent confidence interval for $\mu_2 - \mu_1$ is

$$[(\bar{X}_2 - \bar{X}_1) - 1.725\bar{s}_p \sqrt{(1/12) + (1/10)}, (\bar{X}_2 - \bar{X}_1) + 1.725\bar{s}_p \sqrt{(1/12) + (1/10)}]$$

or, for our example,

$$[10.06 - (1.725)(8.87)(0.4281), 10.06 + (1.725)(8.87)(0.4281)] = (3.51, 16.61)$$

The half-length of the interval is 6.55.

Although the distribution of the t random variable in equation (3.51) is derived from the assumption of normal populations, two-sample t statistic methods are quite robust. For this reason they are widely used in practice even when populations are nonnormal.

In presenting the two-sample t statistic we assumed that the two populations had equal variances. When this assumption is not correct, the t statistic can lead to incorrect inferences. We now present a hypothesis test that is appropriate in such cases. The test statistic is[c]

$$\tau = \frac{(\bar{X}_2 - \bar{X}_1)}{\sqrt{\dfrac{\bar{s}_1^2}{m} + \dfrac{\bar{s}_2^2}{n}}} \tag{3.52}$$

[c] The test statistic τ is due to Welch (see reference [45]).

and the null hypothesis H_0: $\mu_1 = \mu_2$ is rejected at the 100α-percent signifi-
cance level if $|\tau|$ is greater than $t_{\alpha/2}(\nu')$, where ν' is computed in the
following steps:

1. Evaluate: $a = (\bar{s}_1^2/m)[(\bar{s}_1^2/m) + (\bar{s}_2^2/n)]^{-1}$ (3.53)

2. Evaluate: $1/\nu' = [a^2/(m-1)] + (1-a)^2/(n-1)$ (3.54)

3. Evaluate: $t_{\alpha/2}(\nu')$ using the linear interpolation:

$$t_{\alpha/2}(\nu') = t_{\alpha/2}(\nu_U) - \left[\frac{(1/\nu_U) - (1/\nu')}{(1/\nu_U) - (1/\nu_L)}\right][t_{\alpha/2}(\nu_U) - t_{\alpha/2}(\nu_L)]$$

where $\nu_L = ([\nu'] + 1)$, $\nu_U = [\nu']$, and $[\nu']$ is the greatest integer in ν'.

EXAMPLE 3.16

Suppose $m = 8$, $\bar{X}_1 = 100$, $\bar{s}_1^2 = 400$, $n = 10$, $\bar{X}_2 = 115.9$, and $\bar{s}_2^2 = 100$. The value of
τ is

$$\tau = \frac{(115.9 - 100)}{\sqrt{(400/8) + (100/10)}} = 2.05$$

To see if this value is significant at the 5-percent level, we must calculate $t_{0.025}(\nu')$.
Now

$$a = \frac{400}{8}\left(\frac{400}{8} + \frac{100}{10}\right)^{-1} = \frac{50}{60} = 0.8333$$

and

$$\frac{1}{\nu'} = \frac{(0.8333)^2}{7} + \frac{(0.1667)^2}{9} = 0.1023$$

Thus $\nu' = 9.78$, $\nu_U = [\nu'] = 9$, and $\nu_L = [\nu'] + 1 = 10$. The interpolation scheme yields

$$t_{0.025}(\nu') = 2.262 - \left(\frac{0.1111 - 0.1023}{0.1111 - 0.1000}\right)(2.262 - 2.228) = 2.235$$

since

$$t_{0.025}(\nu_U) = t_{0.025}(9) = 2.262 \quad \text{and} \quad t_{0.025}(\nu_L) = t_{0.025}(10) = 2.228$$

Hence H_0: $\mu_1 = \mu_2$ is not rejected at the 5-percent significance level because
$2.05 < 2.235$.

Now suppose we had used the two-sample t statistic. The value of $\bar{s}_p^2$ is

$$\bar{s}_p^2 = (\tfrac{7}{16})\bar{s}_1^2 + (\tfrac{9}{16})\bar{s}_2^2 = (\tfrac{7}{16})(400) + (\tfrac{9}{16})(100) = 231.25$$

and $s_p = 15.21$. The value of the two-sample t statistic is

$$t = \frac{115.9 - 100}{(15.21)\sqrt{(1/8) + (1/10)}} = 2.20$$

But $t_{0.025}(16) = 2.12$, so the value of t would lead to the rejection of H_0 at the 5-percent
significance level *if we assumed the two population variances equal.*

Since the two methods of analysis disagree, and since it is reasonable to believe that the population variances are not equal here, we must conclude that the two-sample t statistic is not an appropriate test statistic in this case.

Comparison of Means Using Paired Samples

Let (X, Y) denote a randomly selected pair of observations which, because they are paired, come from some joint distribution with joint p.d.f. $f(x, y)$ or joint probability function $p(x, y)$. Let μ_X, μ_Y, σ_X^2, σ_Y^2, and σ_{XY} denote the means, variances, and covariance of X and Y. We are interested in the distribution of the difference $D = Y - X$. This distribution may be interpreted as the population of differences between all paired observations X and Y. From equations (2.45) and (2.46) we see that

$$\mu_D = E(D) = E(Y - X) = E(Y) - E(X) = \mu_Y - \mu_X \tag{3.55}$$

and

$$\sigma_D^2 = \sigma_{Y-X}^2 = \sigma_Y^2 + \sigma_X^2 - 2\sigma_{XY} \tag{3.56}$$

Thus the mean difference between Y and X is the difference of their means, but the variance of $Y - X$ is a rather complicated function. In order to make inferences about μ_D we must have a random sample from the population of differences. Such a sample is obtained by selecting n independent random pairs (X_1, Y_1), $(X_2, Y_2), \ldots, (X_n, Y_n)$ and forming $D_1 = Y_1 - X_1$, $D_2 = Y_2 - X_2, \ldots, D_n = Y_n - X_n$. Then

$$\bar{D} = \frac{1}{n} \sum_{i=1}^{n} D_i$$

is the mean of the sample of differences, and from equation (2.48) we know that

$$\mu_{\bar{D}} = \mu_D \quad \text{and} \quad \sigma_{\bar{D}}^2 = \frac{\sigma_D^2}{n} \tag{3.57}$$

If we further assume that X and Y have the bivariate normal distribution defined in equation (2.39), it can be shown that $\bar{D}$ has an $N(\mu_D, \sigma_D^2/n)$ distribution; that is, $\bar{D}$ behaves like the mean of a random sample from an $N(\mu_D, \sigma_D^2)$ population (the population of differences), and hence inferences about μ_D may be handled according to the methods set out in Sections 3.2 and 3.3. To test $H_0: \mu_D = \mu_D^0$, say, calculate $\bar{D}$ and

$$\bar{s}_D = \sqrt{\frac{1}{n-1} \sum_{i=1}^{n} (D_i - \bar{D})^2}$$

and form the t statistic,

$$t = \frac{\bar{D} - \mu_D^0}{\bar{s}_D/\sqrt{n}}$$

This statistic can then be compared in the usual way with critical values of a

t distribution with $(n-1)$ degrees of freedom to determine whether H_0 should be rejected. A $100(1-\alpha)$-percent confidence interval for μ_D is

$$\left[\bar{D} - t_{\alpha/2}(n-1) \frac{\bar{s}_D}{\sqrt{n}}, \ \bar{D} + t_{\alpha/2}(n-1) \frac{\bar{s}_D}{\sqrt{n}} \right] \qquad (3.58)$$

where $t_{\alpha/2}(n-1)$ is the $100(1-\alpha/2)$th percentile of the t distribution with $n-1$ degrees of freedom.

EXAMPLE 3.17

We wish to test the durability of two brands of tires, brand A and brand B. An experiment that yields paired data may be performed as follows. Assume we have 20 tires of each brand. We select 20 drivers at random and install a pair of tires (one brand A, one brand B) on the rear axle of each car. Ten of the 20 drivers are selected at random, and the brand A tire is installed on the left side of their cars; the remaining 10 drivers have the brand A tire installed on the right side. (This procedure tends to eliminate the effects of different wear characteristics on right and left sides of cars.) Each driver is to drive 10,000 miles under normal circumstances and then bring the tires in for inspection. One possible measure of durability is tread depth. Table 3.13 shows hypothetical tread depths for the tires in our experiment. The data can be used to test the hypothesis that the two brands experienced the same tread wear over 10,000 miles against the alternative hypothesis that the two brands differed in tread wear. Since $t_{0.025}(19) = 2.093$, we reject $H_0: \mu_D = 0$ at the 5-percent level given these data. A 95-percent confidence interval for the average difference in tread depth is

$$[-0.26 - (2.093)(0.10), -0.26 + (2.093)(0.10)] = [-0.47, -0.05]$$

It appears that the average depth of brand A is somewhere between 0.47 and 0.05 32nd of an inch greater than the average depth of brand B.

Comparison of Variances of Two Normal Populations

We noted in the first part of this section (Comparison of Means Using Independent Samples) that an assumption implicit in the routine use of the two-sample t statistic was the equality of the two population variances. Thus it would be helpful to have a technique for checking this assumption. *If the populations are normal,* an "F test" provides such a technique. Let $X_{11}, X_{12}, \ldots, X_{1m}$ and $X_{21}, X_{22}, \ldots, X_{2n}$ denote independent random samples from $N(\mu_1, \sigma_1^2)$ and $N(\mu_2, \sigma_2^2)$ populations, respectively. We wish to test the null hypothesis $H_0: \sigma_1^2 = \sigma_2^2$ against $H_a: \sigma_1^2 \neq \sigma_2^2$. It is traditional to estimate the ratio σ_1^2/σ_2^2 and reject H_0 if this estimate strays "too far" from 1. To illustrate, define the statistics,

$$\bar{X}_1 = \frac{1}{m} \sum_{i=1}^{m} X_{1i}$$

Table 3.13 Analysis of Paired Data

Subject	TREAD DEPTH (1/32 IN.) Brand A	Brand B	$D = Brand B - Brand A$
1	6.73	6.32	−0.41
2	6.47	5.95	−0.52
3	6.63	6.18	−0.45
4	6.60	6.14	−0.46
5	6.61	5.21	−0.60
6	6.06	5.40	−0.66
7	5.77	6.33	+0.56
8	6.62	6.17	−0.45
9	6.23	6.66	+0.43
10	6.31	5.73	−0.58
11	6.51	6.02	−0.49
12	5.84	6.38	+0.54
13	6.40	5.86	−0.54
14	6.10	6.57	+0.47
15	6.68	6.25	−0.43
16	6.52	6.03	−0.49
17	6.32	6.73	+0.41
18	6.48	5.98	−0.50
19	6.26	5.67	−0.59
20	6.50	6.00	−0.50

$\bar{D} = -0.26$, $\bar{s}_D = 0.45$, $\bar{s}_D/\sqrt{20} = 0.10$

$t = \bar{D}/(\bar{s}_D/\sqrt{20}) = -0.26/(0.10) = -2.6$

$$\bar{X}_2 = \frac{1}{n}\sum_{i=1}^{n} X_{2i}$$

and

$$F = \frac{\sum_{i=1}^{m}(X_{1i} - \bar{X}_1)^2/(m-1)}{\sum_{i=1}^{n}(X_{2i} - \bar{X}_2)^2/(n-1)}$$

Note that F is the ratio of an estimator of σ_1^2 to an estimator of σ_2^2. Now when H_0 is true, we see from equation (2.70) that the distribution of this ratio is the F distribution with $m-1$ and $n-1$ degrees of freedom. Upper 5 and 1 percentage points of the F distribution are shown in Appendix Table G. These tables allow us to determine whether a computed F value is "too large." Of course it is conceivable that F might be "too small," but tables of lower percentage points are not provided. To get around this difficulty in practice we always associate σ_1^2 with the population that produces the *larger sample variance*. There is no loss of generality in this convention.

EXAMPLE 3.18

Let us use the data in Table 3.12 to test the equality of the two variances. The computed value of F is

$$F = \frac{(9.29)^2}{(8.54)^2} = \frac{86.3041}{72.9316} = 1.18$$

Now according to Appendix Table G an F variable with 9 and 11 degrees of freedom exceeds 2.90 with probability 0.05. Hence the value of F is not "too large," and we cannot conclude from the data in Table 3.12 that a difference between σ_1^2 and σ_2^2 exists.

Unfortunately, the F test is *not* robust against departures from normality in the populations. Hence it is not always a "safe" method to use in practice. A robust test of the equality of variances is presented in Chapter 4.

3.8   METHODS FOR DERIVING ESTIMATORS AND TESTS

Statistical inference is the act of drawing conclusions from numerical data about the parameters of a probability model. Typically the model is assumed to be a description of the mechanism that generates the data, but certain parameters of the model are in doubt. The data are observed in hopes that they will provide some information about the unknown values of these parameters. Thus the estimation of parameters is a fundamental goal of statistical inference.

The extraction of information from numerical data may be done efficiently or inefficiently. Statistical estimation theory attempts to ascertain the efficiency of estimation methods and thus to establish guidelines for the proper choice of statistical methods in practice. A discussion of the many techniques for determining the efficiency of statistical procedures lies outside the scope of this book. Instead we shall briefly describe some of the widely used methods of generating estimators and tests. A thorough critique of these methods involves concepts introduced in courses on mathematical statistics.

In this section we assume the notation presented in Section 3.1, but sometimes θ will stand for a collection of parameters.

The Method of Least Squares

The *method of least squares* may be explained in the following way. Let X be a random item drawn from a population with unknown parameter θ, and suppose that the mean of X is a function of θ. We write $E(X) = h(\theta)$. Let $X_1, X_2, \ldots, X_n$ be a random sample of items from the population. Then each X_i has mean $h(\theta)$, and the squared deviation of X_i from $h(\theta)$ is

$$[X_i - h(\theta)]^2, \qquad i = 1, 2, \ldots, n \tag{3.59}$$

A measure of the total variation of the X_i's about $h(\theta)$ is the sum of the squared deviations in equation (3.59), namely,

$$S(\theta) = \sum_{i=1}^{n} [X_i - h(\theta)]^2 \qquad (3.60)$$

The method of least squares says to choose the estimator $\hat{\theta}$ to minimize the function in equation (3.60). This is a plausible suggestion because it estimates θ with that value $\hat{\theta}$ which makes $h(\hat{\theta})$ behave most like the center of gravity of the X_i's. This in turn makes sense because $h(\theta)$ itself is the center of gravity (mean) of X.

EXAMPLE 3.19

Let $X_1, X_2, \ldots, X_n$ be a random sample from a population with mean μ and variance σ^2, and assume we want to estimate μ. Since $E(X_i) = \mu$, $i = 1, 2, \ldots, n$, the method of least squares would have us minimize

$$S(\mu) = \sum_{i=1}^{n} (X_i - \mu)^2 = \sum_{i=1}^{n} X_i^2 - 2\mu \sum_{i=1}^{n} X_i + n\mu^2$$

Since the function $S(\mu)$ is a quadratic function of μ with positive coefficient on μ^2, it can be minimized by setting its first derivative equal to zero and solving the resulting linear equation for the minimizing value $\hat{\mu}$. Differentiating $S(\mu)$ with respect to μ, we have

$$S'(\hat{\mu}) = -2 \sum_{i=1}^{n} X_i + 2\hat{\mu} n$$

and setting this equal to zero we get

$$\hat{\mu} = \frac{1}{n} \sum_{i=1}^{n} X_i = \bar{X}$$

the sample mean.

We remark that the method of least squares is applicable in very general situations. For example, suppose $E(X_i) = h_i(\theta)$, a different function of θ for each $i = 1, 2, \ldots, n$. If the variance of each X_i is the same, that is, if

$$\sigma_{X_i}^2 = E[X_i - h_i(\theta)]^2 = \sigma^2 \qquad \text{for every } i = 1, 2, \ldots, n$$

then the method of least squares is still sensible. This is because the function

$$S(\theta) = \sum_{i=1}^{n} [X_i - h_i(\theta)]^2$$

will still be a measure of the total variation in the X_i's. If, on the other hand, the variances of the X_i's are different, then the criterion function in $S(\theta)$ is no longer appropriate because it does not account for the changing variability of the X_i's. An example of the application of this general form is regression analysis, which will be discussed in Chapters 5 and 6.

The Method of Maximum Likelihood

The *method of maximum likelihood* assumes that an expression for the joint distribution of the sample items, such as equation (3.1), is given. Once the numerical values of $X_1, X_2, \ldots, X_n$ are known as the result of sampling the population, equation (3.1) is treated as a function of the parameter(s) θ. This function is called the *likelihood function* of the parameter(s). The term "likelihood" comes from the fact that for each value of θ the likelihood function is (roughly) proportional to the probability of observing the data actually obtained by sampling. Thus finding the parameter value(s) that maximize the likelihood function may be interpreted as finding those value(s) "most likely" to have generated the observations on hand. Such maximizing value(s) are called "maximum" likelihood estimates.

EXAMPLE 3.20

Let $X_1, X_2, \ldots, X_n$ be a random sample from an $N(\mu, \sigma^2)$ population with σ^2 known and μ unknown. The joint p.d.f. of $X_1, X_2, \ldots, X_n$ is

$$f(x_1, \ldots, x_n) = (2\pi\sigma^2)^{-n/2} \exp\left[-(1/2\sigma^2) \sum_{i=1}^{n} (x_i - \mu)^2\right] \tag{3.61}$$

For fixed $x_1, x_2, \ldots, x_n$ the expression in equation (3.61), considered as a function of μ, is the likelihood function of μ. This function will be maximized if $\sum_{i=1}^{n} (x_i - \mu)^2$ is made as small as possible. However from Example 3.19 we know that the maximum occurs at $\hat{\mu} = (1/n) \sum_{i=1}^{n} x_i$. Thus the maximum likelihood estimator of μ is the sample mean $\bar{X}$, which was also the least squares estimator.

The method of maximum likelihood can be extended to populations with more than one unknown parameter. A particular example follows.

EXAMPLE 3.21

Let $X_1, X_2, \ldots, X_n$ be a random sample from an $N(\mu, \sigma^2)$ population with both μ and σ^2 unknown. The joint p.d.f. of $X_1, X_2, \ldots, X_n$ is the joint p.d.f. in equation (3.61), but the likelihood function is now a function of the two parameters μ and σ^2. It can be shown that this likelihood function is maximized when μ is taken to be $\bar{x} = (1/n) \sum_{i=1}^{n} x_i$ and σ^2 is taken to be

$$s^2 = \frac{1}{n} \sum_{i=1}^{n} (x_i - \bar{x})^2$$

Thus the maximum likelihood estimators of μ and σ^2 are the mean $\bar{X}$ and variance S^2 of the sample, respectively.

The preceding examples illustrate some of the unique features of the method of maximum likelihood.

1.   In order to use the method we must make an assumption about the

form of the population distribution. This can be a rather severe restriction if little is known a priori about the population.

2. The method often yields intuitively reasonable estimators.

One problem frequently encountered in trying to apply the method of maximum likelihood is the difficulty in solving for the values that maximize the likelihood function. This can require a great deal of effort, and frequently we must resort to computer algorithms designed to find the maximum of a function of several variables. In addition, a plot of the likelihood function may reveal that *several* values of the population parameter yield plausible explanations of the sample results, and yet the mathematical procedure will select only one of them as the *maximum* likelihood estimate; that is, the likelihood function may have several "almost" maximums, and these should be investigated and reported.

Most Powerful Tests

The theory of most powerful tests is based on the concept of likelihood function which was defined in the previous section. We shall use the notation $L(\theta; x_1, x_2, \ldots, x_n)$ for this function. Values of θ that make $L(\theta; x_1, x_2, \ldots, x_n)$ *large* are said to be *supported by the data*. The best supported value of θ, the one that maximizes the likelihood function, is the maximum likelihood estimate of θ, as we noted in the previous section.

Now, if two hypotheses about θ, H_0: $\theta = \theta_0$ and H_a: $\theta = \theta_1$ are put forward, we may see which of them is better supported by the data by comparing the values of the likelihood function evaluated at θ_0 and θ_1. The most obvious thing to do is to compute the so-called *likelihood ratio*

$$\lambda(x_1, x_2, \ldots, x_n) = \frac{L(\theta_1; x_1, \ldots, x_n)}{L(\theta_0; x_1, \ldots, x_n)} \tag{3.62}$$

If this ratio is less than 1, then θ_0 is better supported than θ_1, and if the ratio is greater than 1, then θ_1 is better supported than θ_0.

As an example, let $f(x_i; \theta)$ be $N(\theta, \sigma^2)$, where σ^2 is known. Then the likelihood function is [see equation (3.61) with $\mu = \theta$]

$$L(\theta; x_1, \ldots, x_n) = \left(\frac{1}{2\pi\sigma^2}\right)^{n/2} \exp\left[\frac{-1}{2\sigma^2} \sum_{i=1}^{n} (x_i - \bar{x})^2\right] \exp\left[\frac{-n}{2\sigma^2} (\bar{x} - \theta)^2\right] \tag{3.63}$$

and the likelihood ratio is

$$\lambda(x_1, \ldots, x_n) = \exp\left\{-\frac{n}{2\sigma^2} [(\bar{x} - \theta_1)^2 - (\bar{x} - \theta_0)^2]\right\}$$

$$= \exp\left\{-\frac{n}{2\sigma^2} [-2\bar{x}(\theta_1 - \theta_0) + \theta_1^2 - \theta_0^2]\right\}$$

$$= \exp\left[-\frac{n(\theta_1^2 - \theta_0^2)}{2\sigma^2}\right] \exp\left[\frac{n\bar{x}}{\sigma^2}(\theta_1 - \theta_0)\right] \tag{3.64}$$

Now this ratio is greater than 1 if and only if

$$\exp\left[\frac{n\bar{x}}{\sigma^2}(\theta_1 - \theta_0)\right] > \exp\left[\frac{n(\theta_1{}^2 - \theta_0{}^2)}{2\sigma^2}\right] \tag{3.65}$$

and this is true if and only if

$$\frac{n\bar{x}}{\sigma^2}(\theta_1 - \theta_0) > \left[\frac{n(\theta_1{}^2 - \theta_0{}^2)}{2\sigma^2}\right] = \frac{n(\theta_1 - \theta_0)(\theta_0 + \theta_1)}{2\sigma^2} \tag{3.66}$$

because the exponential function is monotone.

We must consider two cases: $\theta_0 < \theta_1$ and $\theta_0 > \theta_1$. If $\theta_0 < \theta_1$, then $\theta_1 - \theta_0 > 0$ and equation (3.66) holds if and only if

$$\bar{x} > \frac{\theta_0 + \theta_1}{2} \tag{3.67}$$

If $\theta_0 > \theta_1$, then $\theta_1 - \theta_0 < 0$ and equation (3.66) holds if and only if

$$\bar{x} < \frac{\theta_0 + \theta_1}{2} \tag{3.68}$$

In either case, we see that the test is based on the value of $\bar{X}$, the sample mean, and the critical value of the test is $c = (\theta_0 + \theta_1)/2$, the point halfway between θ_0 and θ_1.

It is intuitively satisfying to compare the likelihood ratio with 1 in order to derive a test, but in order to achieve a preassigned significance level, we must consider comparing the ratio to some other value which we shall denote by K. Using this notation, we consider the values $x_1, x_2, \ldots, x_n$ to be evidence against the null hypothesis when

$$\lambda(x_1, x_2, \ldots, x_n) > K \tag{3.69}$$

An argument similar to that given above shows that this condition is equivalent to the condition

$$\bar{x} > c \tag{3.70}$$

if $\theta_0 < \theta_1$, and to the condition

$$\bar{x} < c \tag{3.71}$$

if $\theta_0 > \theta_1$, where c is a function of θ_0, θ_1, n, σ, and K.

In general it can be shown that if α is a preassigned significance level, then the test whose significance level is α and whose critical value is found from the condition given in equation (3.69) has the greatest possible power at $\theta = \theta_1$. This is known as the "Neyman-Pearson lemma."

3.9 FINAL COMMENTS

In this chapter we have reviewed some elementary statistical inference procedures by concentrating on inferences about normal population means and variances. These inferences were based on random samples from the populations. Specifically, we considered the construction of confidence intervals and hypothesis tests and their relationship to one another. In Section 3.8, we discussed briefly some of the more commonly employed justifications for our procedures.

PROBLEMS

3.1   Let $X_1, X_2, \ldots, X_n$ be independent random variables such that $E[X_i] = \mu$ and $\sigma_{X_i}^2 = \sigma_i^2$, $i = 1, 2, \ldots, n$. Define

$$S(\mu) = \sum_{i=1}^{n} \left(\frac{X_i - \mu}{\sigma_i} \right)^2$$

and show that $S(\mu)$ is minimized by

$$\hat{\mu} = \sum_{i=1}^{n} w_i X_i$$

where

$$w_i = \left[\sum_{j=1}^{n} \frac{1}{\sigma_j^2} \right]^{-1} \left(\frac{1}{\sigma_i^2} \right)$$

$\hat{\mu}$ is known as the *weighted least squares estimator* of μ. The weights are needed because the X_i's do not have constant variance.

3.2   Let $X_1, X_2, \ldots, X_n$ be a random sample from a population with p.d.f.

$$f(x) = \mu e^{-\mu x}, \qquad x > 0$$

where μ is a positive number.
(a)   Show that the joint p.d.f. of $X_1, X_2, \ldots, X_n$ is

$$f(x_1, x_2, \ldots, x_n) = \mu^n \exp \left[-\mu \sum_{i=1}^{n} x_i \right]$$

(b)   Show that the maximum likelihood estimator of μ is $\hat{\mu} = n / \sum_{i=1}^{n} X_i = 1/\bar{X}$.

3.3   Suppose the weights of newborn infants are normally distributed with standard deviation $\sigma = 15$ oz. The birth weights of the last 100 infants born in a certain hospital were recorded. The average of these weights is $\bar{X} = 114$ oz. Construct a 97-percent confidence interval for μ, the mean of the population of weights.

3.4   The lifetimes of a certain brand of movie floodlights are normally distributed with standard deviation $\sigma = 56$ hours. Ten floodlights are purchased and left on until they all burn out. The lengths of lifetimes are 276, 119, 269, 167, 177, 299, 244, 246, 230, 178 hours. Use these data to construct a 95-percent confidence interval for μ, the average length of life for the lights. The manufacturer guarantees the lights will burn 210 hours, on the average. Is this a plausible guarantee?

3.5   The salary of junior executives in a large retailing firm is normally distributed with standard deviation $\sigma = \$1500$. If a random sample of 25 junior executives yields an average salary of $16,400, what is the 95-percent confidence interval for μ, the average salary of all junior executives?

3.6   Let $X_1, X_2, \ldots, X_n$ denote a random sample from an $N(\mu, \sigma^2)$ population where σ^2 is known. Show that

$$\Pr\left[\frac{\sqrt{n}(\bar{X} - \mu)}{\sigma} < 1.645\right] = 0.95$$

and hence that in a large number of repeated samples the interval $[\bar{X} - 1.645(\sigma/\sqrt{n}), \infty)$ will contain μ approximately 95 percent of the time. This interval can be used as the basis for a so-called *semi-infinite* 95-percent confidence interval for μ.

3.7   (Continuation) Given the situation in Problem 3.6 and a sample of size $n = 16$ with $\bar{X} = 68$ and $\sigma = 2$, what is the semi-infinite 95-percent confidence interval for μ?

3.8   (Continuation) How would you construct a 99-percent semi-infinite confidence interval for μ of the form $[\bar{X} - z(\sigma/\sqrt{n}), \infty)$?

3.9   (Continuation) Show how a confidence interval for μ of the form $(-\infty, \bar{X} + z(\sigma/\sqrt{n})]$ could be formed.

3.10   Redo Problem 3.3 without assuming that the population standard deviation is known. You are given $\bar{s} = 14.8$ oz.

3.11   Redo Problem 3.4 *without* assuming that the population standard deviation is known.

3.12   Redo Problem 3.5 *without* assuming that the population standard deviation is known. You are given that $\bar{s} = \$1575$.

3.13   Use the data in Problem 3.10 to estimate the variance of the population of weights.

3.14   Use the data in Problem 3.4 to estimate the variance of the population of lifetimes.

3.15   Use the data in Problem 3.12 to estimate the variance of the population of salaries.

3.16   If $X_1, X_2, \ldots, X_n$ is a random sample of size n from an $N(\mu, \sigma^2)$ population, and $nS^2/\sigma^2 = \sum_{i=1}^{n} (X_i - X)^2/\sigma^2$, show that there is a number χ_u^2 such that $\Pr[nS^2/\sigma^2 < \chi_u^2] = 0.95$ and hence that in a large number of repeated samples the interval $[nS^2/\chi_u^2, \infty)$ contains σ^2 about 95 percent of the time. This interval may be used to construct semi-infinite 95-percent confidence intervals for σ^2.

3.17   (Continuation) Construct 95-percent semi-infinite confidence intervals for σ^2 for the populations described in Problems 3.10, 3.4, and 3.12.

3.18   A cereal packaging machine is supposed to turn out boxes that contain 20 oz of cereal on the average. Past experience indicates that the standard deviation of the content weights of the packages turned out by the machine is $\sigma = 0.25$ oz. Boxes containing less than 19.8 oz of cereal, on the average, are considered unacceptable. Thus the manufacturer wishes to test H_0: $\mu = 20$ and H_a: $\mu = 19.8$.

   (a)   If samples of size $n = 20$ are drawn from current production to test the hypotheses, find the power function of the test assuming a significance level of $\alpha = 0.025$.

   (b)   If a sample of $n = 20$ items has mean weight 19.9, what conclusion is drawn about μ at the 2.5-percent significance level? Compute the P value for this case, and interpret it.

   (c)   Find the power function of a test based on $n = 36$ items with $\alpha = 0.025$.

   (d)   If you were not given the value of σ, what test would you use? Could you compute the power function of the test?

3.19   Suppose the variance of the package weights in Problem 3.18 is not known, and it is desired to test H_0: $\sigma^2 = 0.0625$ against H_a: $\sigma^2 = 0.09$. Construct the appropriate test using a 5-percent significance level. If a sample size of $n = 20$ yields $\sum_{i=1}^{n}(X_i - \bar{X})^2 = 1.52$, what hypothesis do you accept at the 5-percent significance level? What can you say about the P value here?

3.20   An automobile manufacturer claims that a certain make of car averages 16 miles to a gallon of gas. A consumer group disputes the claim and says that the true figure is about 14 miles to the gallon. The hypotheses to be tested are H_0: $\mu = 16$ and H_a: $\mu = 14$. Assume the gas mileage population is $N(\mu, 9)$. A sample of $n = 18$ cars, all with exactly 2000 miles and properly tuned, is selected to test the hypotheses. Each car is to be driven until three tanks of gas are used up. Then the miles per gallon for each car will be recorded.

   (a)   Using the critical value $c = 15$, compute the power function of the test.

   (b)   If a sample of $n = 25$ had been chosen, what would be the power function of the test in part (a)?

3.21   Consider the situation in Problem 3.20, but suppose interest centers on testing H_0: $\sigma^2 = 9$ against H_a: $\sigma^2 = 12$. Construct the appropriate test using a 10-percent significance level.

3.22   Collect 51 consecutive weekly closing prices for a stock of your choice, and calculate the 50 weekly *changes* in closing price. (If p_{t-1} is the price at the end of week $t - 1$, and p_t is the price at the end of week t, then $p_t - p_{t-1}$ is the change in closing price during week t.) Let μ denote the average weekly change in closing price.

   (a)   Test H_0: $\mu = 0$ against H_a: $\mu \neq 0$ using your data. Take $\alpha = 0.05$.

   (b)   Compute the P value using your data.

(c) Assume that the $\bar{s}$ from your data is equal to the population variance σ^2. What test would you have used if you had been given a priori that σ^2 had this value? Compute the power function of this test.

3.23 Let $X_1, X_2, \ldots, X_{25}$ denote a random sample of size 25 from an $N(\mu, 100)$ population, and suppose the null hypothesis H_0: $\mu = 100$ is to be tested.
(a) If the alternative hypothesis is H_a: $\mu \neq 100$, compute the power function of the appropriate test. Use $\alpha = 0.05$.
(b) If the alternative hypothesis is H_a: $\mu > 100$, compute the power function of the appropriate test. Use $\alpha = 0.05$.
(c) For the test in (a) compute the P value associated with $\bar{X} = 103.5$.
(d) For the test in (b) compute the P value associated with $\bar{X} = 103.5$.

3.24 Suppose you plan to test the mean of an $N(\mu, \sigma^2)$ distribution where the null and alternative hypotheses are H_0: $\mu = \mu_0$ and H_a: $\mu > \mu_0$. You want the significance level of the test to be $\alpha = 0.05$, and you want the power of the test at $\mu_1 = \mu_0 + \frac{1}{2}\sigma$ to be 0.85. Show that these requirements lead to the equations

$$c - \mu_0 = 1.645 \frac{\sigma}{\sqrt{n}}$$

and

$$c - \mu_0 = -1.03 \frac{\sigma}{\sqrt{n}} + 0.5\sigma$$

and hence to the result that $n = 29$. Thus n may be determined without knowledge of σ.

3.25 Two appraisers are to be compared to see if they assign the same average assessed value to houses in a certain city. Sixty-four houses are to be selected at random from the city, and each appraiser is to value each of the houses. Let the random variable X_i denote the *difference* in the two appraisers' assessments of the ith house, and suppose the population of assessment differences is $N(\mu, \sigma^2)$. The hypotheses to be tested are H_0: $\mu = 0$ and H_a: $\mu \neq 0$.
(a) Develop a test of the hypotheses given above.
(b) If a sample of size $n = 64$ yields $\bar{X} = \$200$ and $\bar{s} = \$500$, what conclusion do you draw? What is the P value here?

3.26 (a) Describe at least two "before and after" experiments that would lead to paired data.
(b) Describe at least two experiments that would lead to comparison of means using independent samples.

3.27 Suppose that a paired data experiment yields $n = 15$, $\bar{D} = -9.8$, and $\bar{s}_D = 19.4$.
(a) Test H_0: $\mu_D = 0$ against H_a: $\mu_D \neq 0$ at the 5-percent significance level.
(b) Test H_0: $\mu_D = 0$ against H_a: $\mu_D < 0$ at the 5-percent significance level.
(c) Construct 90- and 95-percent confidence intervals for μ_D. Do these intervals contain 0? Comment.

3.28   The data below are annual rates of return based on Standard and Poor's index of 425 stocks for the years 1951 through 1972. One column contains rates of return ignoring dividends paid on the stocks, and the other column shows rates based on the market values of the stocks *and* dividends paid.

(a)   Construct a scatter diagram of these data and note how the two columns of figures tend to "move together."

(b)   Form the differences D_i, $i = 1, 2, \ldots, 22$, and calculate $\bar{D}$ and $\bar{s}_D$.

(c)   Test H_0: $\mu_D = 0$ against H_a: $\mu_D \neq 0$.

(d)   Calculate 90- and 95-percent confidence intervals for μ_D.

(e)   Interpret the results in parts (c) and (d) verbally.

Annual Rates of Return ($\times 100$) Based on Standard and Poor's Index

Year	Ignoring Dividends	Including Dividends
1951	17.7	24.8
1952	11.0	16.9
1953	−7.5	−2.1
1954	49.7	56.0
1955	30.1	34.6
1956	3.4	7.1
1957	−14.4	−10.7
1958	37.6	41.8
1959	9.4	12.6
1960	−4.7	−1.6
1961	23.1	26.5
1962	−12.8	−9.9
1963	20.1	23.7
1964	13.1	16.4
1965	9.9	13.1
1966	−13.4	−10.4
1967	23.3	26.8
1968	7.5	10.5
1969	−10.2	−7.3
1970	−0.6	2.6
1971	11.7	14.9
1972	17.0	19.8

3.29   The data in this problem are similar to the data in Problem 3.28, except that the rates of return are based on the Dow Jones Industrial Average. Redo Problem 3.28 using these data.

Annual Rates of Return ($\times 100$) Based on Dow Jones
Industrial Average

Year	Ignoring Dividends	Including Dividends
1951	14.4	21.3
1952	8.4	14.1
1953	-3.8	1.7
1954	44.0	50.2
1955	20.8	26.1
1956	2.3	7.0
1957	-12.8	-8.4
1958	34.0	38.5
1959	16.4	19.9
1960	-9.3	-6.2
1961	18.7	22.4
1962	-10.8	-7.6
1963	17.0	20.6
1964	14.6	18.7
1965	10.9	14.2
1966	-18.9	-15.7
1967	15.2	19.0
1968	4.3	7.7
1969	-15.2	-11.6
1970	4.8	8.8
1971	6.1	9.8
1972	14.6	18.2

3.30 Suppose an experiment yields $m = 9$, $n = 12$, $\bar{X}_1 = 68$, $\bar{X}_2 = 60$, $\bar{s}_1^2 = 25$, and $\bar{s}_2^2 = 30$.

   (a)   Use the two-sample t statistic to test H_0: $\mu_2 - \mu_1 = 0$ against H_a: $\mu_2 - \mu_1 \neq 0$ at a significance level of your choice.

   (b)   Construct 90- and 95-percent confidence intervals for $\mu_2 - \mu_1$.

   (c)   Assuming the populations are normal, test the hypothesis H_0: $\sigma_1^2 = \sigma_2^2$ against H_a: $\sigma_1^2 \neq \sigma_2^2$.

3.31 Suppose an experiment yields $m = 20$, $n = 20$, $\bar{X}_1 = 72$, $\bar{X}_2 = 71$, $\bar{s}_1^2 = 4$, and $\bar{s}_2^2 = 4$.

   (a)   Use the two-sample t statistic to test H_0: $\mu_2 - \mu_1 = 0$ against H_a: $\mu_2 - \mu_1 \neq 0$.

   (b)   Construct a 95-percent confidence interval for $\mu_2 - \mu_1$.

3.32 At the beginning of the school year 67 sixth graders were randomly divided into two groups of size $m = 32$ and $n = 35$. Then a coin was flipped to assign a male teacher to one class and a female teacher to the other class. The teachers were instructed to follow a prescribed syllabus in order to hold the content of the two classes as constant as possible. At the end of the year the students were given an achievement test covering reading and mathematical skills. The composite scores for the students are shown below.

Class with Male Teacher

Student No.	1	2	3	4	5	6	7	8	9	10	11
Score	103	114	86	106	109	108	94	98	103	113	100
Student No.	12	13	14	15	16	17	18	19	20	21	22
Score	109	102	95	103	118	100	123	120	100	106	104
Student No.	23	24	25	26	27	28	29	30	31	32	
Score	121	114	98	93	91	100	116	95	118	113	

Class with Female Teacher

Student No.	1	2	3	4	5	6	7	8	9	10	11	12
Score	111	103	112	106	97	112	110	109	110	104	93	90
Student No.	13	14	15	16	17	18	19	20	21	22	23	24
Score	106	87	112	98	92	104	87	87	92	100	100	113
Student No.	25	26	27	28	29	30	31	32	33	34	35	
Score	95	94	87	94	96	96	101	84	91	97	110	

Use these data to test the hypothesis that students respond equally well to instruction by the male and the female teachers. What generalizations, if any, would you be willing to make on the basis of the test?

4

ANALYSIS OF VARIANCE

4.1 INTRODUCTION

In previous chapters we have been concerned with making inferences about populations from simple random samples selected from these populations. We have discussed inferences about population means. In this chapter we shall consider the problem of comparing two or more population means and, in the process, introduce a useful analytical technique known as the "analysis of variance."

We have tacitly assumed in our earlier discussions of statistical inference that it is actually possible to select a simple random sample or, at the very least, that routinely recorded observations may be legitimately regarded as a simple random sample.[a] We shall refer to the act of selecting a random sample from an existing, well-defined population as the "sampling experiment" and to the plan for generating the random sample as the "sampling, or experimental, design." The problem of selecting an appropriate experimental design is an important one, and a brief discussion of this topic is essential for understanding much of the material included in later parts of this chapter. We will discuss experimental design in the context of particular examples. The reader should consult references [7], [8], and [9] for more complete discussions.

Measurements are often recorded under different sets of conditions to see if the general level of the observations differs significantly over these sets. For example, in order to study the effects of age and sex on the time required

---

[a] Here we are referring to the observations themselves as the random sample rather than a set of independent and identically distributed random variables. However, the reader should have no trouble with these subtleties at this point.

There are, of course, situations in which it is not sensible to regard the observations as a random sample, and, in fact, it is precisely the nature of the dependence in the observations that is of interest. This is particularly true for observations recorded over time.

to learn a new task, the time may be recorded for individuals of different ages and sex. Here a set of conditions corresponds to a sex-age combination, and differences in the times associated with the different combinations can be used to measure the "effects" of age and sex on the variable of interest (learning time). It is common practice to refer to the quantities specifying the conditions under which the measurements are taken (in this example, age and sex) as "factors," and to the measured variable itself (in this case, learning time) as the "response."

As an additional example, an accountant may be interested in examining the effects of interim reporting methods (produced by different accounting theories) and years of experience on the predictions of common stock price made by a group of security analysts. Here the factors under investigation are "interim reporting method" and "experience," and the response is "predicted stock price." The objective in this study would, presumably, be to see if different accounting methods and differing years of experience, and perhaps their "interaction," lead to substantially different stock price predictions.

These examples will allow us to discuss various aspects of experimental design. We will concentrate in particular on several important contributions to the theory of experimental design which lead to an increase in the "power" of the subsequent analysis.

Any good plan for generating observations starts with a clearly specified objective and a list of constraints within which the investigator must operate. The constraints may include economic constraints (a budget), time constraints (have the results next week!), and equipment constraints (we cannot take measurements out there!). Questions such as "What is it I'm interested in?" "What physical resources do I have available?" and "How much money and time do I have?" must be considered initially. The experimental design ordinarily specifies:

1. The set of factors included in the study, and their settings.
2. The set of experimental units (the *size* of the experiment).
3. The rules and procedures by which the experimental units are assigned to combinations of factor settings (or vice versa).
4. The nature of the measurements on the experimental units.

In the learning time example, the factors chosen for study were age and sex. There are obviously two "settings" of sex (male and female), and there are (let us suppose) three "settings" of age: 10–20 years, 21–35 years, and over 35 years. The experimental units are individuals, and the measurements (observations) are times in minutes needed to learn the task. Suppose it is decided to have two experimental units for every combination of factor settings. Then there will be a total of 12 observations. The only element of the experimental design left unspecified at this point is the method that will be used to assign experimental units to the combinations of factor settings.

Assume that a large pool of individuals is willing to participate in the learning time experiment. One way to carry out the assignment of experimental units is to divide (stratify) the pool into smaller groups corresponding to the sex and age classifications and then select two individuals at random (using a table of random digits) from each group for the actual experiment. Alternatively, the investigator may simply pick two females, for example, between the ages of 10 and 20 years who he feels are "representative" of this age group, two whose ages fall in the 21- to 35-year category, and two who are over 35 years of age—repeating the procedure for the males.

The method used to assign experimental units to combinations of factor settings is a very important aspect of experimental design. Of the two methods just discussed, the random sampling procedure is better than the judgmental selection process because with the former method the individuals selected may be regarded as a random sample from the parent population of individuals (of a particular set and age classification) that might have been chosen. Statistical techniques based on the assumption of random samples can then be employed to make inferences about the populations of learning times. This is not true for the judgmental method. Another disadvantage of the judgmental method is that the investigator unconsciously or otherwise may select individuals with characteristics that bias the experimental results. For example, some agility may be required to learn the task and the investigator may select "agile looking" females, thereby masking any real difference in learning time due to sex.

Consider the assignment of experimental units in the stock price example. Suppose that there are six interim reporting methods and three classifications for years of experience. The security analysts are to be presented with a sequence of historical stock prices and quarterly and annual reports (of a hypothetical company) varying with the interim accounting method for a particular set of economic conditions. With this information, they must predict the common stock price for the next period.

In this example the experimental units are the analysts and the observations are predicted stock prices. A collection of security analysts can be stratified according to years of experience and then assigned at random to the different accounting methods. Equivalently, each analyst could receive one of the six different collections of historical reports at random. This use of randomization, called *restricted randomization* because the randomizing is done only over the accounting method factor, allows the investigator to regard each group of predicted stock prices corresponding to a given experience-accounting method combination as a random sample from a (hypothetical) parent population of values that might have been observed. Again, statistical techniques based on random samples can be employed to make inferences about the "populations." Here, inferences about the population means, for example, would represent inferences about the effects of the experience and accounting method factors since, presumably, these are the only factors that distinguish the populations from one another.

To summarize, randomization allows us to make valid inferences about populations from the subsequent observations, and since it tends to produce homogeneous groups (that is, groups whose characteristics are roughly the same other than those due to the factors of interest), systematic effects due to factors not under the direct control of the experimenter (biases) are eliminated. In our later examples we shall assume that randomization is employed over the settings of *all* the factors considered in the investigation. This kind of randomization is called *complete randomization*.

We will conclude our abbreviated discussion of experimental design with a comment on factor selection. Suppose in the stock price example the analysts were not classified according to years of experience; that is, the only factor considered was an interim accounting method. If, in fact, experience is an important variable in explaining differences in stock price predictions, its effect will be *confounded* with the effect produced by the accounting method. Any observed significant differences among the populations of stock price predictions may be due to differences in accounting methods, or differences in years of experience, or to an interaction between these two factors, and there is no way to distinguish among the three possibilities. Thus the experimenter must give the factor selection problem careful consideration. Subject to any external constraints, the factors under investigation should represent those variables likely to have an important effect on the response.

It is rather interesting that under suitable conditions the effects of factors on a response can be determined by analyzing the *variability* of the observations; that is, significant differences in response level over the sets of conditions peculiar to the investigation can be uncovered by an *analysis of variance*. To do this it is necessary to postulate a model that links the observations with the combinations of factor settings, called "factor levels." Some of these models are discussed in forthcoming sections of this chapter. The reader should keep in mind that in many cases the validity of these models depends in part on the experimental design considerations just discussed.

4.2 ONE-FACTOR ANALYSIS OF VARIANCE— AN INTRODUCTION

The ideas involved in one-factor analysis of variance are best illustrated in the context of an example.

EXAMPLE 4.1

Suppose that typists in a typing pool are selected to evaluate four comparable typewriters produced by different manufacturers. Five typists are randomly assigned to each machine and asked to type the same material for 10 minutes. At the end of the time period, the words per minute (wpm) are recorded. The data are presented in Table 4.1.

Table 4.1   Samples of the Output from Four Different Typewriters

Sample (Typewriter) Number	Output from Typewriter j (wpm)					$\bar{Y}_j$
$j = 1$	69	62	70	67	62	66
2	67	72	76	69	71	71
3	76	70	71	66	77	72
4	60	64	67	58	66	63
						$\bar{Y} = 68$

Assuming that the typists are of roughly equal ability, the only apparent factor that influences output is the "brand of typewriter." However, it is clear from inspection of the data and some careful thought that the outputs from a given typewriter will not be exactly the same due to the effect (hopefully small) of extraneous factors that have not been specifically accounted for. (We know, for example, that the typists are not *precisely* of equal ability.) Therefore we regard the five observations for each typewriter in Table 4.1 as a random sample from the hypothetical population of possible outputs that might have been observed. The population mean represents the long-run performance level for a particular typewriter at a certain level of competence, and any deviation from this mean is due to the effects of the extraneous factors. A comparison of typewriters (with respect to output) is then provided by a comparison of population means.

The analysis of variance argument proceeds in the following manner. We examine the variability *between* samples relative to the variability *within* samples. If the samples (when represented by their sample averages) are "spread out" relative to the spread of the observations within each sample, the hypothesis of equal population means is rejected since such an event is unlikely if the population means are identical. On the other hand, if the spread between sample averages is roughly the same as the spread of the observations within samples, then the hypothesis of equal population means is not rejected since this event is not unlikely if we are sampling from populations with the same mean.[b] Our comments are illustrated in Figure 4.1

[b] This statement and the previous one are not quite right, since clearly the population *variance* will influence the variability within and between samples. However, later we will assume that the populations are *exactly* the same except for a possible shift in location. In this case what we have said is correct.

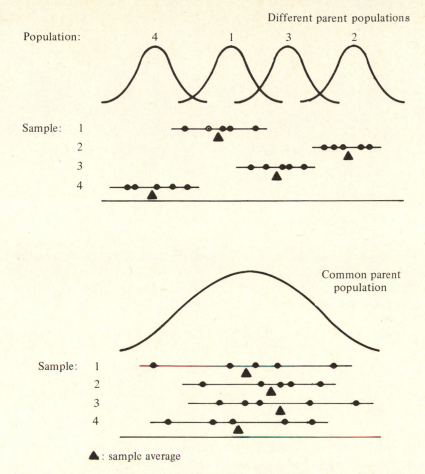

Figure 4.1 Variation within samples in relation to variation between samples for two alternatives

where four samples of size 5 are plotted under each alternative. It is almost inconceivable that the four samples pictured in the top half of the figure come from the same parent population. Note that the variation within each sample is small relative to the dispersion between samples (that is, the variability in the sample averages). Yet it is quite conceivable that the four samples pictured in the bottom half of the figure come from a common parent population. Note in this case that the variation within samples is large compared with the spread of the sample averages.

Suppose that we have J samples of the same size, n, from J populations with the same variance but possibly different means. Let Y_{ij} denote the ith observation of the j sample, $i = 1, 2, \ldots, n$ and $j = 1, 2, \ldots, J$. Let $\bar{Y}_j$ denote the jth sample average, $j = 1, 2, \ldots, J$, and $\bar{\bar{Y}}$ denote the overall average of

Table 4.2 Sample Number

1	2	3	$\cdots$	J
Y_{11}	Y_{12}	Y_{13}		Y_{1J}
Y_{21}	Y_{22}	Y_{23}		Y_{2J}
Y_{31}	Y_{32}	Y_{33}		Y_{3J}
$\vdots$	$\vdots$	$\vdots$		$\vdots$
Y_{n1}	Y_{n2}	Y_{n3}		Y_{nJ}

$$\bar{Y}_1 = \frac{1}{n}\sum_{i=1}^{n} Y_{i1} \qquad \bar{Y}_2 = \frac{1}{n}\sum_{i=1}^{n} Y_{i2} \qquad \bar{Y}_3 = \frac{1}{n}\sum_{i=1}^{n} Y_{i3} \qquad \bar{Y}_J = \frac{1}{n}\sum_{i=1}^{n} Y_{iJ}$$

$$\bar{Y} = \frac{1}{Jn}\sum_{j=1}^{J}\sum_{i=1}^{n} Y_{ij} = \frac{1}{J}\sum_{j=1}^{J} \bar{Y}_j$$

(4.1)

the *pooled* samples. The observations and the sample averages can then be set out as in Table 4.2.

A measure of the variability *between* samples is provided by the sum of weighted squared deviations

$$SS_b = \sum_{j=1}^{J} n(\bar{Y}_j - \bar{Y})^2 \tag{4.2}$$

However, the quantity actually employed in the test for the equality of population means (assuming a common parent population) is

$$\frac{SS_b}{J-1} = \frac{1}{J-1}\sum_{j=1}^{J} n(\bar{Y}_j - \bar{Y})^2 = n\frac{1}{J-1}\sum_{j=1}^{J} (\bar{Y}_j - \bar{Y})^2 = ns_{\bar{Y}}^2 \tag{4.3}$$

where $s_{\bar{Y}}^2$ is recognized as the sample variance of the *sample averages*. Since the theoretical variance of sample averages is σ^2/n (the population variance divided by the sample size), $s_{\bar{Y}}^2$ may be interpreted as an estimator of σ^2/n, and hence $ns_{\bar{Y}}^2$ is an estimator[c] of σ^2.

We take as a measure of the variation *within* samples the average "within sample variance." Thus let

$$s_j^2 = \frac{1}{(n-1)}\sum_{i=1}^{n} (Y_{ij} - \bar{Y}_j)^2, \qquad j = 1, 2, \ldots, J \tag{4.4}$$

be the sample variance for the jth sample. Then the average within sample variance (the "pooled" sample variance) is given by

$$s_p^2 = \frac{1}{J}\sum_{j=1}^{J} s_j^2 \tag{4.5}$$

The pooled variance, s_p^2, measures the within sample variability and is an

---

[c] In fact, assuming a *common* parent population, $ns_{\bar{Y}}^2$ is an *unbiased* estimator of σ^2; that is, $E(ns_{\bar{Y}}^2) = \sigma^2$ (see Section 4.4).

unbiased estimator of the assumed common population variance σ^2 *regard-less of whether the population means are identical.* Since it is the relative variability that is important (we must *compare* the variability between samples to the variability within samples), the ratio

$$F = \frac{ns_{\bar{Y}}^2}{s_p^2} \qquad (4.6)$$

is examined. We would expect the value of this ratio to be in the neighborhood of 1 if the samples all come from the same parent population because, under these circumstances, the numerator and denominator of the F ratio are estimators of the same quantity, σ^2. Alternatively, if the samples come from populations with the same variance but different means, the numerator will tend to be larger than the denominator, and hence the F ratio will tend to be much greater than 1. Thus when the value of the F ratio is "large," the notion of a common parent population is rejected. The theoretical justification for this procedure is presented in Section 4.4. At this point we are simply arguing that it is a reasonable way to proceed.

EXAMPLE 4.1 (*continued*)

Using the typewriter data presented in Table 4.1, the sample variances are

$$s_1^2 = \frac{1}{5-1}[(69-66)^2 + (62-66)^2 + \cdots + (62-66)^2] = 14.5$$

$$s_2^2 = \frac{1}{5-1}[(67-71)^2 + (72-71)^2 + \cdots + (71-71)^2] = 11.5$$

and similarly,

$$s_3^2 = 20.5$$

$$s_4^2 = 15.0$$

Their average value is

$$s_p^2 = \frac{1}{4}\sum_{j=1}^{4} s_j^2 = \frac{1}{4}(14.5 + 11.5 + 20.5 + 15.0) = 15.375$$

Using the entries in the last column of Table 4.1,

$$s_{\bar{Y}}^2 = \frac{1}{4-1}[(66-68)^2 + (71-68)^2 + \cdots + (63-68)^2] = 18$$

and

$$ns_{\bar{Y}}^2 = 5(18) = 90$$

Therefore the F ratio becomes

$$F = \frac{90}{15.375} = 5.854$$

Interpretation of this value will be considered in Section 4.4. We simply note at this point that given the assumptions of the next section, the probability of observing an F ratio at least as large as the one we observed if the samples come from identical normal populations is less than 0.01.

4.3 THE ONE-FACTOR MODEL AND ASSUMPTIONS

Suppose we have J samples from J populations (either actual populations or hypothetical populations), and we are interested in comparing population means. Then each observation—regarded initially as the random variable Y_{ij}—can be represented by the linear model

$$Y_{ij} = \mu_j + e_{ij}, \qquad i = 1, 2, \ldots, n$$
$$j = 1, 2, \ldots, J \qquad (4.7)$$

where μ_j is the mean of the jth population, e_{ij} is a random error term representing the deviation of Y_{ij} from the population mean, and n is the size of the jth sample. (We point out that, in general, the sample sizes need not be the same. The case of unequal sample sizes is treated in Section 4.5.) We now make the following assumptions: For all i and j

1. $E(e_{ij}) = 0$ and $\mathrm{Var}(e_{ij}) = \sigma^2$.
2. e_{ij} are normally distributed.   (4.8)
3. e_{ij} are independent random variables.

The assumptions set out in equation (4.8) imply that the Y_{ij}'s are independent normal variables with a common variance σ^2 but possibly different means. The hypothesis of interest is then H_0: $\mu_1 = \mu_2 = \cdots = \mu_J$. The assumptions are summarized in Table 4.3. Thus statistically speaking the problem is one of comparing the means of several normal populations with the same variance.

Table 4.3 Assumptions Underlying the One-Factor Analysis of Variance

Population Number	Assumed Distribution	Sample
1	$N(\mu_1, \sigma^2)$	$Y_{11}, Y_{21}, \ldots, Y_{n1}$
2	$N(\mu_2, \sigma^2)$	$Y_{12}, Y_{22}, \ldots, Y_{n2}$
3	$N(\mu_3, \sigma^2)$	$Y_{13}, Y_{23}, \ldots, Y_{n3}$
$\vdots$	$\vdots$	$\vdots$
J	$N(\mu_J, \sigma^2)$	$Y_{1J}, Y_{2J}, \ldots, Y_{nJ}$

Reparameterization

In many instances it is convenient to write the model in equation (4.7) in a somewhat different form. This is particularly true if the differences between population means are due to the "effects" of an experimental variable (controllable factor) like the accounting method variable introduced in Section 4.1. Therefore we write the jth mean μ_j as the sum of an overall mean or level μ and a differential effect τ_j. An alternative form of the one-factor linear model is then

$$Y_{ij} = \mu + \tau_j + e_{ij}, \qquad i = 1, 2, \ldots, n$$
$$j = 1, 2, \ldots, J \qquad (4.9)$$

where μ is defined such that $\sum_{j=1}^{J} \tau_j = 0$, and again, e_{ij} are independent $N(0, \sigma^2)$ variables. In this form a test of the equality of population means is exactly a test of the hypothesis H_0: $\tau_1 = \tau_2 = \cdots = \tau_J = 0$. The model in equation (4.9) is appealing because the factor effect (τ_j) is explicitly identified. Moreover, it can be easily and naturally extended to accommodate the effects of additional factors (see Section 4.9).

Comparing equations (4.7) and (4.9) with μ defined as

$$\mu = \frac{1}{J} \sum_{j=1}^{J} \mu_j \qquad (4.10)$$

we see that

$$\tau_j = \mu_j - \mu, \qquad j = 1, 2, \ldots, J \qquad (4.11)$$

Thus the assumptions underlying equation (4.9) are exactly those displayed in Table 4.3 with $\mu_j = \mu + \tau_j$, $j = 1, 2, \ldots, J$.

EXAMPLE 4.2

In the typewriter experiment introduced in Example 4.1 the factor is typewriter brand. The "levels" of the factor correspond to a particular brand of typewriter. The typewriters are arbitrarily indexed from 1 to 4. If Y_{ij} is the output in words per minute for the jth typewriter brand, then μ_j is the mean output for the jth brand, and τ_j is the differential effect of the jth brand on output.

EXAMPLE 4.3

In the spinning of synthetic fibers chemicals are extruded through rectangular orifices (of the same size) in an annular spinnerette. The width of a rectangular orifice is well determined, but the length can be varied. Suppose five spinnerettes are manufactured with orifices of the same width but different lengths. The lengths of the orifices for the five spinnerettes are 360, 370, 380, 390, and 400 microns, respectively.

In this case the factor is orifice length. The five levels of the factor correspond to the five lengths with 360 microns the first level, and 400 microns the fifth level. If Y_{ij} is a measure of the "shape" of the fibers produced at the jth length (level), $j = 1, 2, \ldots, 5$, then μ_j is the mean shape of the population of fibers extruded through orifices of length j, and τ_j is the differential effect of the jth length on fiber shape.

Notice that in both examples one factor is being considered. In the first example the factor levels correspond to categories, whereas in the second example the factor levels have a quantitative significance.

The linear models in equations (4.7) and (4.9), given the underlying assumptions we have imposed, are referred to as *fixed effects* models[d] since the conclusions reached pertain *only* to those factor levels included in the study. Since the models in equations (4.7) and (4.9) can be derived from one another, inferences about the parameters in one model are easily converted into inferences about the parameters in the other model.

Least Squares Estimators of the Model Parameters

We can estimate the parameters in either analysis of variance model using the principle of least squares.[e] Specifically, we choose as estimators of the parameters those functions of the observations that minimize the error sum of squares $\sum_{j=1}^{J} \sum_{i=1}^{n} e_{ij}^2$. The subsequent estimators are called the "least squares" estimators. If we denote the least squares estimator of μ_j by $\hat{\mu}_j$, the least squares estimators of μ and τ_j are $\hat{\mu} = (1/J)\sum_{j=1}^{J}\hat{\mu}_j$ [recall equation (4.10)]; and $\hat{\tau}_j = \hat{\mu}_j - \hat{\mu}$, $j = 1, 2, \ldots, J$ [see equation (4.11)]. Conversely, we can obtain the least squares estimators $\hat{\mu}$ and $\hat{\tau}_j$ and then get $\hat{\mu}_j$ from the relationship

$$\hat{\mu}_j = \hat{\mu} + \hat{\tau}_j, \qquad j = 1, 2, \ldots, J$$

It can be shown that subject to the condition $\sum \tau_j = 0$, the unique least squares estimators are

$$\hat{\mu} = \bar{Y} = \frac{1}{Jn} \sum_{j=1}^{J} \sum_{i=1}^{n} Y_{ij} \tag{4.12}$$

and

$$\hat{\tau}_j = \bar{Y}_j - \bar{Y} = \frac{1}{n} \sum_{i=1}^{n} Y_{ij} - \bar{Y}, \qquad j = 1, 2, \ldots, J \tag{4.13}$$

---

[d] Other analysis of variance models exist. Models in which the factor levels are regarded as a sample of factor levels that might have been chosen for study are called *random effects* models. For these models, conclusions extend to the population of factor levels. *Mixed* models are those in which the levels of some of the factors are regarded as fixed and the levels of other factors are regarded as a sample from a population of possible levels. In this book we shall only be concerned with fixed effects models. References [8] and [9] contain discussions of the random and mixed effects models.

[e] Given the assumptions underlying the analysis of variance model, the least squares estimators are also unbiased estimators of the model parameters and the minimum mean square error estimators.

Consequently

$$\hat{\mu}_j = \hat{\mu} + \hat{\tau}_j = \bar{Y}_j, \qquad j = 1, 2, \dots, J \tag{4.14}$$

The least squares estimates are then given by substituting the observations for the corresponding random variables Y_{ij} in equations (4.12), (4.13), and (4.14). The reader will notice that the estimator of the population mean μ_j, $j = 1, 2, \dots, J$ is the sample average $\bar{Y}_j$ and the estimator of the differential effect τ_j, $j = 1, 2, \dots, J$ is the difference between the jth sample average $\bar{Y}_j$ and the overall sample average $\bar{Y}$.

4.4 THE *F* RATIO AND THE ANALYSIS OF VARIANCE TABLE

If the assumptions set out in Table 4.3 hold and the null hypothesis H_0: $\mu_1 = \mu_2 = \cdots = \mu_J$ (or equivalently H_0: $\tau_1 = \tau_2 = \cdots = \tau_J = 0$) is true, then the ratio in equation (4.6) has the F distribution with $\nu_1 = J - 1$ and $\nu_2 = J(n - 1)$ degrees of freedom (see the discussion of the F distribution in Section 2.4). As we shall point out shortly, $J - 1$ is the number of degrees of freedom associated with the numerator of the F ratio and $J(n - 1)$ is the number of degrees of freedom associated with the denominator.

Percentage points of the F distribution for various numbers of degrees of freedom are given in Appendix Table G. These points will allow us to test H_0 at specified significance levels and, as we shall demonstrate in Section 4.6, allow us to construct simultaneous interval estimates of linear functions of the population means.

The calculations leading to the F ratio can be conveniently set out in the form of a table known as the *analysis of variance* (ANOVA) table. This table is based on the following algebraic identity (see Problem 4.5):

$$\sum_{j=1}^{J} \sum_{i=1}^{n} (Y_{ij} - \bar{Y})^2 = \sum_{j=1}^{J} n(\bar{Y}_j - \bar{Y})^2 + \sum_{j=1}^{J} \sum_{i=1}^{n} (Y_{ij} - \bar{Y}_j)^2 \tag{4.15}$$

that is,

$$\begin{pmatrix} \text{total variation in} \\ \text{the observations} \\ \text{about the overall} \\ \text{average, denoted} \\ \text{by } SS_t \end{pmatrix} = \begin{pmatrix} \text{variation between} \\ \text{samples, denoted} \\ \text{by } SS_b \end{pmatrix} + \begin{pmatrix} \text{variation within} \\ \text{samples, denoted} \\ \text{by } SS_w \end{pmatrix}$$

We note that SS_b has already been introduced in equation (4.2), and using equations (4.4) and (4.5),

$$SS_w = J(n - 1)s_p^2$$

Each of the terms in equation (4.15) has a number associated with it that represents the number of independent pieces of information needed to calculate it. This number is known as the degrees of freedom (d.f.) and the values corresponding to the variation breakup in equation (4.15) are presented below.

$$\text{variation:} \qquad SS_t = SS_b + SS_w$$

$$\text{d.f.:} \qquad (Jn - 1) = (J - 1) + J(n - 1) \qquad (4.16)$$

In particular, the sum of the numbers of degrees of freedom associated with the variation between samples and the variation within samples is the number of degrees of freedom associated with the total variation.

The ANOVA table for the one-factor model is given in Table 4.4(a). The "sum of squares" and "degrees of freedom" columns contain the identities in equations (4.15) and (4.16). The "mean squares" column contains the between samples and within samples variations divided by their respective degrees of freedom. The F ratio is simply the ratio of the mean squares and is precisely the ratio displayed in equation (4.6). It is this quantity that is referred to an appropriate percentage point of an F distribution with $\nu_1 = J - 1$ and $\nu_2 = J(n - 1)$ degrees of freedom to test the null hypothesis

$$H_0: \mu_1 = \mu_2 = \cdots = \mu_J \qquad (\text{or } H_0: \tau_1 = \tau_2 = \cdots = \tau_J = 0)$$

Let $F_\gamma(\nu_1, \nu_2)$ denote the upper γ percentage point of an F distribution with ν_1 and ν_2 degrees of freedom; then at the γ level of significance[f] a test of H_0 is provided by the rule[g]

$$\text{Reject } H_0: \mu_1 = \mu_2 = \cdots = \mu_J \qquad \text{if } F > F_\gamma(J - 1, J(n - 1))$$

$$\text{Do not reject } H_0: \qquad\qquad\qquad \text{if } F \le F_\gamma(J - 1, J(n - 1))$$

The alternative hypothesis is $H_a: \mu_i \ne \mu_j$ for some $i \ne j$; that is, at least one of the population means is different from the others. For the model in equation (4.9) the hypothesis being tested is

$$H_0: \tau_1 = \tau_2 = \cdots = \tau_J = 0$$

versus

$$H_a: \tau_j \ne 0 \qquad \text{for some } j$$

---

[f] From this point on we will use γ to denote the significance level. This is done to avoid notational difficulties in Chapters 5 and 6.

[g] Given the assumptions underlying the one-factor analysis of variance model,

$$E(MS_b) = E(ns_{\bar{y}}^2) = \sigma^2 + \frac{n}{(J-1)} \sum_{j=1}^{J} \tau_j^2$$

and

$$E(MS_w) = E(s_p^2) = \sigma^2 \qquad (\text{always})$$

Thus if $H_0: \tau_1 = \tau_2 = \cdots = \tau_J = 0$ is true, $E(MS_b) = \sigma^2$ and we would expect the F ratio, $F = MS_b/MS_w$, to be near 1. On the other hand, if H_0 is false, the numerator in the F ratio will tend to be larger than the denominator. (Its expected value is σ^2 plus a positive constant.) Thus *large* values of the F ratio tend to reinforce the alternative hypothesis, $H_a: \tau_j \ne 0$ for some j. This is the reason the F test is one sided, and in particular we reject H_0 for large F values.

Table 4.4 The One-Factor Analysis of Variance Table (equal sample sizes)

(a) GENERAL ANOVA TABLE

Source of Variation	Sum of Squares (SS)	Degrees of Freedom (d.f.)	Mean Squares (MS)	F Ratio
Between samples	$SS_b = \sum_{j=1}^{J} n(\bar{Y}_j - \bar{Y})^2$ $= n \sum_{j=1}^{J} (\bar{Y}_j - \bar{Y})^2$	$J - 1$	$MS_b = \dfrac{SS_b}{(J-1)} = ns_{\bar{Y}}^2$	$F = \dfrac{MS_b}{MS_w} = \dfrac{ns_{\bar{Y}}^2}{s_p^2}$
Within samples	$SS_w = \sum_{j=1}^{J} \sum_{i=1}^{n} (Y_{ij} - \bar{Y}_j)^2$	$J(n-1)$	$MS_w = \dfrac{SS_w}{J(n-1)} = s_p^2$	
Total (about mean)	$SS_t = \sum_{j=1}^{J} \sum_{i=1}^{n} (Y_{ij} - \bar{Y})^2$	$Jn - 1$		

(b) ANOVA TABLE FOR THE OBSERVATIONS IN TABLE 4.1

Source of Variation	Sum of Squares (SS)	Degrees of Freedom (d.f.)	Mean Squares (MS)	F Ratio
Between samples	$SS_b = 270$	$J - 1 = 3$	$MS_b = ns_{\bar{Y}}^2 = 90$	$F = \dfrac{MS_b}{MS_w} = \dfrac{90}{15.375}$
Within samples	$SS_w = 246$	$J(n-1) = 16$	$MS_w = s_p^2 = 15.375$	$= 5.854$
Total (about mean)	$SS_t = 516$	$Jn - 1 = 19$		

EXAMPLE 4.4

The ANOVA table for the typewriter data given in Example 4.1 is set out in Table 4.4(b). The F ratio $F = 5.854$ exceeds $F_{0.01}(3, 16) = 5.29$, the upper 1-percent point of an F distribution with 3 and 16 degrees of freedom. Therefore we would reject the null hypothesis H_0: $\mu_1 = \mu_2 = \mu_3 = \mu_4$ in favor of H_a: $\mu_i \neq \mu_j$ for some $i \neq j$ at the 1-percent level of significance; that is, it appears as if there are real differences between the brands of typewriters.

Moreover, using equation (4.14) we have the point estimates,

$$\hat{\mu}_1 = \bar{Y}_1 = 66 \qquad \hat{\mu}_3 = \bar{Y}_3 = 72$$

$$\hat{\mu}_2 = \bar{Y}_2 = 71 \qquad \hat{\mu}_4 = \bar{Y}_4 = 63$$

Since, from equation (4.12), $\hat{\mu} = \bar{Y} = 68$, the brand (differential) effects are estimated to be [see equation (4.13)],

$$\hat{\tau}_1 = \bar{Y}_1 - \bar{Y} = 66 - 68 = -2 \qquad \hat{\tau}_3 = \bar{Y}_3 - \bar{Y} = 72 - 68 = 4$$

$$\hat{\tau}_2 = \bar{Y}_2 - \bar{Y} = 71 - 68 = 3 \qquad \hat{\tau}_4 = \bar{Y}_4 - \bar{Y} = 63 - 68 = -5$$

4.5   UNEQUAL SAMPLE SIZES

The argument leading to the analysis of the variance table and the F test set out in previous sections needs little modification for unequal sample sizes. If we let n_j, $j = 1, 2, \ldots, J$, denote the size of the jth sample, then the calculations leading to the F ratio are set out in the ANOVA table displayed in Table 4.5. Defining the total number of observations to be $N = \sum_{j=1}^{J} n_j$, a test[h] of the null hypothesis

$$H_0: \mu_1 = \mu_2 = \cdots = \mu_J \qquad (\text{or } H_0: \tau_1 = \tau_2 = \cdots = \tau_J = 0)$$

can be carried out by referring the F ratio to an appropriate percentage point of an F distribution with $\nu_1 = J - 1$ and $\nu_2 = N - J$ degrees of freedom.

Point estimators of the model parameters are given by equations (4.12), (4.13), and (4.14) with n_j replacing n, $j = 1, 2, \ldots, J$, except in the divisor of equation (4.12) were n is replaced by N. The τ_j's are defined such that $\sum_{j=1}^{J} n_j \tau_j = 0$.

The reader should note that the entries in Table 4.5 reduce to those in Table 4.4(a) when $n_j = n$ for all j. We have concentrated on the equal sample

[h] For unequal sample sizes,

$$E(MS_b) = \sigma^2 + \frac{1}{(J-1)} \sum_{j=1}^{J} n_j \tau_j^2$$

and

$$E(MS_w) = \sigma^2 \qquad (\text{always})$$

These results can be compared with those in the previous footnote. Again values of the F ratio much greater than 1 discredit the null hypothesis and reinforce the alternative hypothesis.

Table 4.5 The One-Factor Analysis of Variance Table (unequal sample sizes)

Source of Variation	Sum of Squares (SS)	Degrees of Freedom (d.f.)	Mean Squares (MS)	F Ratio[a]
Between samples	$SS_b = \sum_{j=1}^{J} n_j (\bar{Y}_j - \bar{Y})^2$	$J - 1$	$MS_b = \dfrac{SS_b}{(J-1)}$	$F = \dfrac{MS_b}{MS_w}$
Within samples	$SS_w = \sum_{j=1}^{J} \sum_{i=1}^{n_j} (Y_{ij} - \bar{Y}_j)^2$	$\left(\sum_{j=1}^{J} n_j\right) - J$	$MS_w = \dfrac{SS_w}{[(\Sigma n_j) - J]}$	
Total (about mean)	$SS_t = \sum_{j=1}^{J} \sum_{i=1}^{n_j} (Y_{ij} - \bar{Y})^2$	$\left(\sum_{j=1}^{J} n_j\right) - 1$		

[a] The F ratio can be referred to an F distribution with $\nu_1 = J - 1$ and $\nu_2 = (\Sigma_{j=1}^{J} n_j) - J = N - J$ degrees of freedom.

size case since, if at all possible, we should arrange to take samples of the same size. The reason for this will be explored more fully in Section 4.7 when we discuss the effects of departures from the assumptions underlying the one-factor model.

4.6 COMPARISON OF MEANS

The F test in Section 4.4 is only concerned with the question of whether the population means are different. If the answer is yes (the hypothesis of equal means is rejected), then it is natural to ask how much different are the population means. In this section we shall be concerned with the latter question and indicate how means can be compared given the assumptions in Table 4.3.

An immediate possibility is to compare pairs of means. Using arguments like those in Chapter 3, we can show that a $100 \times (1 - \gamma)$-percent confidence interval for the difference $\mu_i - \mu_j$, $i \neq j$, is given by

$$(\bar{Y}_i - \bar{Y}_j) \pm t_{\gamma/2}(N - J)s_p \sqrt{\frac{1}{n_i} + \frac{1}{n_j}} \qquad (4.17)$$

where $t_{\gamma/2}(N - J)$ is the upper $\gamma/2$ percentage point of a t distribution with $N - J = (\sum_{j=1}^{J} n_j) - J$ degrees of freedom, and s_p is the square root of the residual mean square s_p^2 in the ANOVA table.[i] Using this method, pairs of means may be compared, and each statement will carry a confidence coefficient of $(1 - \gamma)$. However, we frequently are interested in relating the results of one statement to the results of another statement; that is, we would like to interpret several statements simultaneously. Unfortunately we can be less certain about simultaneous statements than about individual statements.

For example, suppose we estimate the individual differences given below by the 95-percent confidence intervals shown in parentheses on the right.

$$\mu_1 - \mu_2: \qquad (4.7, 11.9)$$

$$\mu_2 - \mu_3: \qquad (-4.0, 3.2)$$

$$\mu_1 - \mu_3: \qquad (4.3, 11.5)$$

If we now ask how confident are we that all three statements (intervals) hold *simultaneously*, the answer is *less than* 95 percent. The reason is that if any

---

[i]We use an estimator, s_p^2, of the assumed common population variance σ^2 obtained from pooling the information in *all* the samples. Since this quantity has $N - J$ degrees of freedom associated with it, the confidence interval is based on a t random variable with $N - J$ degrees of freedom.

To achieve generality we have presented the unequal sample size result. We will continue to do this throughout the remainder of this section. The reader should have no difficulty obtaining the corresponding result for equal sample sizes by simply replacing n_j with n, $j = 1, 2, \ldots, J$.

one of the individual statements is in error, the collection of statements is in error.

Even if the statements above were *independent* of one another, the simultaneous confidence level would be reduced to $(0.95)^3 = 0.857$. In fact, the statements are dependent since, for example, they all involve the common term s_p. It is possible to exploit this dependence and obtain a set of statements with an associated *simultaneous* confidence coefficient. The method we will introduce is not only applicable to simple comparisons of means, like differences; it can also be applied to more complicated "contrasts" in the population means that may be of interest.

Definition 4.1   A linear function of the form

$$c_1\mu_1 + c_2\mu_2 + \cdots + c_J\mu_J = \sum_{j=1}^{J} c_j\mu_j$$

where $\sum_{j=1}^{J} c_j = 0$ is called a *contrast* in the means, $\mu_1, \mu_2, \ldots, \mu_J$.

Any linear function of the J population means such that the coefficients sum to zero is called a *contrast*. Suppose $J = 3$. Then some examples of contrasts are

$$(0)\mu_1 + (1)\mu_2 + (-1)\mu_3 = \mu_2 - \mu_3$$

$$(1)\mu_1 + (0)\mu_2 + (-1)\mu_3 = \mu_1 - \mu_3$$

$$\left(\frac{1}{2}\right)\mu_1 + \left(\frac{1}{2}\right)\mu_2 + (-1)\mu_3 = \left(\frac{\mu_1 + \mu_2}{2}\right) - \mu_3$$

$$(-2)\mu_1 + (-2)\mu_2 + (4)\mu_3 = 4\mu_3 - 2(\mu_1 + \mu_2)$$

The next theorem, due to Scheffé [8], will allow us to make simultaneous statements with a given confidence coefficient.

Theorem 4.1   With $100 \times (1 - \gamma)$-percent confidence *all* contrasts in the J population means of the form $\sum_{j=1}^{J} c_j\mu_j$ are bracketed by the bounds,

$$\sum_{j=1}^{J} c_j\bar{Y}_j \pm \sqrt{F_\gamma(J-1, N-J)}\, s_p \sqrt{(J-1)\sum_{j=1}^{J}\left(\frac{c_j^2}{n_j}\right)} \qquad (4.18)$$

where $F_\gamma(J - 1, N - J)$ denotes the upper γ percentage point of an F distribution with $\nu_1 = (J - 1)$ and $\nu_2 = (N - J)$ degrees of freedom, $N = \sum_{j=1}^{J} n_j$, and s_p^2 is the within mean square.

Corollary   If $n_j = n$ for all j, then with $100 \times (1 - \gamma)$-percent confidence *all* contrasts of the form $\sum_{j=1}^{J} c_j\mu_j$ are bracketed by the bounds

$$\sum_{j=1}^{J} c_j\bar{Y}_j \pm \sqrt{F_\gamma[J-1, J(n-1)]}\, s_p \sqrt{\frac{(J-1)}{n}\sum_{j=1}^{J} c_j^2} \qquad (4.19)$$

In essence, Scheffé's procedure, set out in equations (4.18) and (4.19), increases the widths of the individual intervals so that all are simultaneously true with a prescribed level of confidence.[j]

EXAMPLE 4.5

Consider the typewriter data of Table 4.1. We have seen from Table 4.4(b) that the underlying population means appear to be different from one another. Suppose we compare the means in pairs. Then the individual 95-percent confidence intervals based on the t distribution [see equation (4.17)] are shown in Table 4.6 in the left-hand column and the 95-percent simultaneous confidence intervals [see equation (4.19)] are shown on the right. The reader will notice that the simultaneous intervals are wider than the individual intervals. On the other hand we can be 95 percent confident that the intervals in the right-hand column (along with any other intervals that might be calculated for contrasts) *simultaneously* contain the indicated differences, whereas we can be much less confident that the intervals in the left-hand column simultaneously contain these differences.

Table 4.6   Comparison of Means—Typewriter Data

INDIVIDUAL INTERVALS	SIMULTANEOUS INTERVALS
$\mu_i - \mu_j$: $(\bar{Y}_i - \bar{Y}_j) \pm t_{0.025}(16) s_p \sqrt{\frac{1}{5} + \frac{1}{5}}$	$\mu_i - \mu_j$: $(\bar{Y}_i - \bar{Y}_j) \pm \sqrt{F_{0.05}(3, 16)}\, s_p\, \sqrt{\frac{2}{5}(2)}$
$\mu_1 - \mu_2$: $-5 \pm (2.12)\sqrt{15.375}\,\sqrt{\frac{2}{5}}$ $= -5 \pm 5.3$	$\mu_1 - \mu_2$: $-5 \pm \sqrt{3.24}\,\sqrt{15.375}\sqrt{\frac{2}{5}}$ $= -5 \pm 7.7$
$\mu_1 - \mu_3$: -6 ± 5.3	$\mu_1 - \mu_3$: -6 ± 7.7
$\mu_1 - \mu_4$: 3 ± 5.3	$\mu_1 - \mu_4$: 3 ± 7.7
$\mu_2 - \mu_3$: -1 ± 5.3	$\mu_2 - \mu_3$: -1 ± 7.7
$\mu_2 - \mu_4$: 8 ± 5.3	$\mu_2 - \mu_4$: 8 ± 7.7
$\mu_3 - \mu_4$: 9 ± 5.3	$\mu_3 - \mu_4$: 9 ± 7.7

The results in Table 4.6 indicate that typewriters 1 and 4 are essentially indistinguishable. Similarly, typewriters 2 and 3 are indistinguishable. However, there appear to be real differences (on average) between typewriters 2 and 4 and 3 and 4. This suggests that typewriters 2 and 3 may be regarded as a group and typewriters 1 and 4 may be regarded as a group. In fact, we can compare the averages of the two groups[k] by estimating the contrast,

$$\frac{1}{2}\mu_2 + \frac{1}{2}\mu_3 - \frac{1}{2}\mu_1 - \frac{1}{2}\mu_4 = \frac{(\mu_2 + \mu_3)}{2} - \frac{(\mu_1 + \mu_4)}{2}$$

[j] If only differences in pairs of means are of interest and the sample sizes are equal, then simultaneous intervals that are shorter than those produced by Scheffé's procedure can be obtained. These intervals are derived from a result due to Tukey (see Scheffé [8], p. 73).

[k] It may be of interest to do this if, for example, typewriters 2 and 3 are manufactured in the United States and typewriters 1 and 4 are produced outside the United States.

using Scheffé's procedure. With 95-percent confidence this contrast is bracketed by

$$\left[\frac{(\bar{Y}_2 + \bar{Y}_3)}{2} - \frac{(\bar{Y}_1 + \bar{Y}_4)}{2}\right] \pm \sqrt{F_{0.05}(3, 16)} \, s_p \sqrt{\frac{3}{5}} \quad (1)$$

that is, by

$$(71.5 - 64.5) \pm \sqrt{3.24} \, \sqrt{15.375} \, \sqrt{\tfrac{2}{5}} = 7 \pm 5.5$$

It appears that the difference between the group means is real and in particular the average output of typewriters 2 and 3 (as a group) is larger than the average output of typewriters 1 and 4 (as a group).

In summary, we point out that Scheffé's procedure is well suited for "data snooping"—that is, making inferences about contrasts that are suggested by the data as well as contrasts that are formulated before the experiment is performed.

4.7 DEPARTURES FROM ASSUMPTIONS AND THE EXAMINATION OF RESIDUALS

The inferential procedures we have developed for the one-factor model depend on a number of underlying assumptions. Specifically, we assume that the observations are normally distributed, independent, and have a common variance. Clearly, there will be cases in which one or more of these assumptions is violated. In fact, in the real world a situation in which *all* of the assumptions are *exactly* true would be nonexistent. Thus it is of interest to examine the effect of departures from the assumptions on inferences about the population means. We shall comment on each of the three assumptions in turn.

Normality

If the *number of observations* in each group is *reasonably large,* inferences made about means assuming normality remain valid for nonnormal populations, all other things being equal. Once again, the effect of nonnormality is slight if the sample sizes are reasonably large and inferences are being made about population means. In particular, the F test is essentially unaffected if the underlying populations are all nonnormal, provided they are of the same form. Clearly, it is always a good idea to check the normality assumption and this can be done using methods presented in Section 5.6.

Equality of Variances

If the *number of observations* in each sample (group) *is the same,* inferences made about means assuming a common variance are not seriously affected by unequal population variances, all other things being equal.

Thus the experimenter should try to arrange for equal sample sizes. (This is why we concentrated on the equal sample size situation in previous sections.) If we couple this requirement with our requirement for minimizing the effect of nonnormality, we should try to arrange for equal *and* reasonably large sample sizes. In general for the case of equal sample sizes, inequality of population variances causes the true significance level associated with the *F* test to *exceed* the nominal value, but the difference is ordinarily slight.

A test for the equality of population variances can and should be performed. Unfortunately some of the popular tests designed for this purpose are severely affected by nonnormality (see Box [13]). A large sample test for homogeneity (equality) of variances that is not appreciably affected by nonnormality is given in Supplement 4A.

If nonhomogeneity of variance is evident and equal sample sizes cannot be arranged, then it is frequently possible to transform the observations to new observations that can be regarded as drawings from populations with (approximately) equal variances. In cases in which the variability appears to increase with the level of the observations, the logarithmic transformation is often appropriate; that is, the logarithms of the observations are represented by the one-factor model, and the usual analysis is carried out with these quantities.

Independence

Of the three assumptions, independence is the most crucial. If this assumption is violated, the effect on inferences about population means can be severe; that is, serious errors in inference can result. Ordinarily, lack of independence will occur *within* samples. In some experimental situations independence can be assured by *randomly* assigning experimental units to the different levels of the factor, as we pointed out in our discussion in Section 4.1. However, it is not always possible to employ randomization. For situations in which randomization is impossible and lack of independence is indicated, the only recourse appears to be to determine (estimate) the nature of the dependence and allow for it in the subsequent analysis.

Independence can be checked within samples, provided they are reasonably large, by computing the sample autocorrelation coefficients [see Section 5.9, equation (5.69)].

Examination of Residuals

Consider the one-factor model

$$Y_{ij} = \mu + \tau_j + e_{ij}, \qquad i = 1, 2, \ldots, n_j; \quad j = 1, 2, \ldots, J$$

subject to the restriction $\Sigma n_j \tau_j = 0$ where e_{ij} are independent $N(0, \sigma^2)$ variables. Then, clearly,

$$e_{ij} = Y_{ij} - \mu - \tau_j, \qquad i = 1, 2, \ldots, n_j; \quad j = 1, 2, \ldots, J \qquad (4.20)$$

And if the model *is correct*, the error components e_{ij} can be *estimated* by the deviations,

$$\hat{e}_{ij} = Y_{ij} - \hat{\mu} - \hat{\tau}_j = Y_{ij} - (\bar{Y}) - (\bar{Y}_j - \bar{Y}) \qquad i = 1, 2, \ldots, n_j; \quad j = 1, 2, \ldots, J$$
$$= Y_{ij} - \bar{Y}_j, \qquad (4.21)$$

obtained by substituting the least squares estimates $\hat{\mu} = \bar{Y}$ and $\hat{\tau}_j = \bar{Y}_j - \bar{Y}$ into equation (4.20) and replacing the random variables Y_{ij} by their observed values. The deviations $\hat{e}_{ij}$ are referred to as the *residuals*. Since the residuals are just the estimates of the errors on the assumption that the model is correct, an analysis of the residuals can reveal potential discrepancies between the data and the model (with accompanying assumptions) used to explain it. In particular, if the assumptions of the model are met, the residuals should vary in a haphazard and unsystematic way and should look like drawings from a normal population with mean zero and variance σ^2.

Therefore to check on the model and the underlying assumption we can

1. Plot all the residuals as a dot diagram. The result should look like a random sample from an $N(0, \sigma^2)$ population as indicated below. (Note that σ^2 can be estimated by the residual mean square s_p^2.) Extreme observations should be investigated.

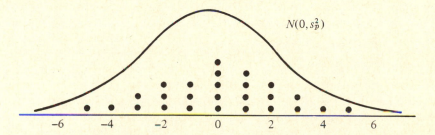

2. Plot the residuals of *each* sample as dot diagrams and compare the spreads. They should be more or less the same (like the example below). If not, the homogeneity of variance assumption may be invalid.

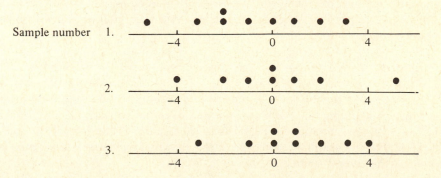

3. Plot the residuals in the order in which the observations occurred. There should be no systematic behavior apparent over time (such as the example below). If a systematic pattern over time exists, the assumption of independence must be questioned.

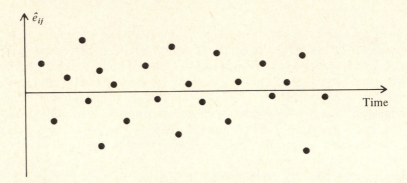

Although any particular phenomenon can be subjected to more sophisticated statistical tests, a series of residual plots like the ones we have presented provide simple and important ways of detecting abnormalities. The reader is referred to Sections 5.6 and 6.6 for more complete discussions of residual analysis.

4.8 TWO-FACTOR ANALYSIS OF VARIANCE— AN INTRODUCTION

In this section we extend the results of the previous sections and consider models for analyzing the effects of two factors, say 1 and 2. We now assume that there are I levels of factor 1 and J levels of factor 2, and that the two factors are "crossed"; that is, all possible combinations of the levels of the factors are present in the experimental design. Moreover, at the ith level of factor 1 and the jth level of factor 2, referred to as the ijth "cell," $i = 1, 2, \ldots, I$, $j = 1, 2, \ldots, J$, there will be *at least* one observation. Thus the observations and their averages, assuming an equal number of observations, say, K per cell, can be set out in a two-way table as shown on the opposite page.

Remarks 1. We denote the kth observation, $k = 1, 2, \ldots, K$ in the ijth cell by Y_{ijk}, $i = 1, 2, \ldots, I$, $j = 1, 2, \ldots, J$. The arithmetic average of the observations at the ith level of factor 1 is denoted by $\bar{Y}_{i..}$, the dots in the subscript indicating that we have summed over the subscripts j and k; that is, to get $\bar{Y}_{i..}$ we sum the observations in each cell over the J levels of factor 2. Similarly, the arithmetic average at the jth level of factor 2 is denoted by $\bar{Y}_{.j.}$ and the overall arithmetic average by $\bar{Y}_{...}$.

FACTOR 2

Level \\ Level	1	2	...	j	...	J	Row average
1	Y_{111} ⋮ Y_{11K}	Y_{121} ⋮ Y_{12K}	...	Y_{1j1} ⋮ Y_{1jK}	...	Y_{1J1} ⋮ Y_{1JK}	$\bar{Y}_{1..}$
2	Y_{211} ⋮ Y_{21K}	Y_{221} ⋮ Y_{22K}	...	Y_{2j1} ⋮ Y_{2jK}	...	Y_{2J1} ⋮ Y_{2JK}	$\bar{Y}_{2..}$
⋮	⋮	⋮		⋮		⋮	⋮
i	Y_{i11} ⋮ Y_{i1K}	Y_{i21} ⋮ Y_{i2K}	...	Y_{ij1} ⋮ Y_{ijK}	...	Y_{iJ1} ⋮ Y_{iJK}	$\bar{Y}_{i..}$
⋮	⋮	⋮		⋮		⋮	
I	Y_{I11} ⋮ Y_{I1K}	Y_{I21} ⋮ Y_{I2K}	...	Y_{Ij1} ⋮ Y_{IjK}	...	Y_{IJ1} ⋮ Y_{IJK}	$\bar{Y}_{I..}$
Column average	$\bar{Y}_{.1.}$	$\bar{Y}_{.2.}$	...	$\bar{Y}_{.j.}$	...	$\bar{Y}_{.J.}$	$\bar{Y}_{...}$

FACTOR 1

overall average

2. Although it is possible to have the case $K = 1$—that is, there is one observation for each of the IJ combinations of levels of the two factors[1]—it is ordinarily desirable to replicate the experiment several times. If the number of replications is K, then we are faced with the more general situation indicated in the two-way table above. The analysis of the replicated two-way layout is considered in Section 4.10.

3. As we shall see shortly, we are still interested in making inferences about population means. For the two-factor situation we will assume that the locations of the underlying populations can deviate from one another because of the separate effects of factors 1 and 2 (or both), or because of the joint effect of factors 1 and 2 (called the "interaction").

4. It frequently happens that only one of the two factors is of immediate

[1] When there is only one observation per cell, we shall drop the third subscript, k. Thus the observation in the ijth cell is Y_{ij}, the arithmetic average of the observations at the ith level of factor 1 is $\bar{Y}_{i.}$, the overall arithmetic average is $\bar{Y}_{..}$, and so forth.

interest. The other factor is included so that its (potential) effect on the response can be allowed for, thereby increasing the precision (sensitivity) of the analysis.

EXAMPLE 4.6

Suppose, in the context of the stock price example introduced in Section 4.1, the security analysts are divided into three experience classes, 0–1 year, 1–5 years, and more than 5 years, and then randomly assigned to interim reporting methods 1, $2, \ldots, 6$, say. Suppose there were 12 analysts in each experience class, two of whom received reporting method 1, two reporting method 2, and so forth. Let factor 1 be "experience" and factor 2 "interim accounting method." Then factor 1 has $I = 3$ levels; $i = 1$ corresponds to 0–1 year, $i = 2$ to 1–5 years, and $i = 3$ to more than 5 years. Factor 2 has $J = 6$ levels; $j = 1$ corresponds to method 1, $j = 2$ to method $2, \ldots, j = 6$ to method 6. In each of the ij cells, $i = 1, 2, 3, j = 1, 2, \ldots, 6$, there are $K = 2$ observations. If the observed response is $Y =$ "predicted stock price," then, for example, Y_{231} is the predicted stock price associated with the first analyst in the second experience group (that is, 1–5 years) receiving accounting method number 3.

Note that in this example the effect of accounting method on predicted stock price may be of primary interest; the effect of experience may be of secondary importance. Experience is included to account for some of the variation in predicted stock price, thus increasing the sensitivity of the analysis.

EXAMPLE 4.7

Consider the typewriter problem given in Example 4.1. Suppose that the output from each typewriter was produced by the same five typists; that is, each typist produced output on all four typewriters. Under these conditions we can allow for an effect due to differences in typists by analyzing the data in Table 4.1 as a two-way layout with "typist" as the first factor and "brand of typewriter" as the second factor. Thus the $I = 5$ levels of factor 1 correspond to the five typists, and the $J = 4$ levels of factor 2 correspond to the four typewriter brands. There is $K = 1$ observation in the ijth cell, $i = 1, 2, \ldots, 5, j = 1, 2, \ldots, 4$. Finally Y_{ij} represents the output (in words per minute) of the ith typist on the jth typewriter. Notice that we have deleted the subscript k here since we do not have to distinguish between observations *within a cell*—there is only one observation per cell.

4.9   THE TWO-FACTOR MODEL AND ASSUMPTIONS— WITH AND WITHOUT REPLICATION

It is most convenient to write the two-factor models as generalizations of the one-factor model given by equation (4.9). Thus if Y_{ijk} is the random variable representing the eventual value of the kth replicate in the ijth cell,

$i = 1, 2, \ldots, I$, $j = 1, 2, \ldots, J$, and $k = 1, 2, \ldots, K$, the two-factor fixed effects model is

$$Y_{ijk} = \mu + \beta_i + \tau_j + (\beta\tau)_{ij} + e_{ijk} \tag{4.22}$$

where μ is the overall mean; β_i is the ith differential, or "main," effect of factor 1; τ_j is the jth differential, or "main," effect of factor 2; $(\beta\tau)_{ij}$ is called the "interaction" effect[m] at the ith level of factor 1 and the jth level of factor 2; and e_{ijk} are independent $N(0, \sigma^2)$ random variables [see equation (4.8)]. If $(\beta\tau)_{ij} = 0$ for all i and j, the model in equation (4.22) is said to be *additive*. (As we shall see in Section 4.10, we may test for additivity.) Thus the two-factor model above says that a given observed value is accounted for by an overall level, an effect due to factor 1, an effect due to factor 2, an effect due to the interaction of factors 1 and 2, and some error representing, say, measurement error and the effects of other factors not explicitly included in the model.

In order to obtain unique (and meaningful) least squares estimators of the model parameters, the following side conditions on the differential effects are ordinarily imposed:

$$\sum_{i=1}^{I} \beta_i = \sum_{j=1}^{J} \tau_j = 0$$

$$\sum_{i=1}^{I} (\beta\tau)_{ij} = 0, \qquad j = 1, 2, \ldots, J \tag{4.23}$$

$$\sum_{j=1}^{J} (\beta\tau)_{ij} = 0, \qquad i = 1, 2, \ldots, I$$

For *unreplicated* experiments the two-factor model is somewhat different. Thus if we have two factors 1 and 2 with I and J levels, respectively, and $K = 1$ observation per cell, the two-factor fixed effects model is

$$Y_{ij} = \mu + \beta_i + \tau_j + e_{ij} \tag{4.24}$$

with μ, β_i, and τ_j, $i = 1, 2, \ldots, I$, $j = 1, 2, \ldots, J$, defined as before and e_{ij} independent $N(0, \sigma^2)$ random variables. To obtain unique least squares estimators of the parameters μ, β_i, and τ_j for all i and j, the side conditions

$$\sum_{i=1}^{I} \beta_i = \sum_{j=1}^{J} \tau_j = 0 \tag{4.25}$$

are imposed.

It is immediately apparent that the model in equation (4.24) does not contain an interaction term. In order to test for factor 1 effects and factor 2 effects in the unreplicated case, it is necessary to assume $(\beta\tau)_{ij} = 0$ for all i

---

[m]As an example, consider the learning time example introduced in Section 4.1. If the task to be learned requires a good deal of digital dexterity, it may be that sex has no effect on learning time for young persons (dexterity is roughly the same for males and females) but has a substantial effect for old persons (males are perhaps less dexterous than females). This differential influence of sex, depending on age, means that the age and sex factors interact.

and j. However, if the experimenter doubts the adequacy of the additive two-factor model, a nonadditive model can be postulated and a test for interaction carried out. The appropriate procedure is due to Tukey [44] and is discussed briefly in Section 4.11.

Least Squares Estimators of the Model Parameters

Given the restrictions set out in equations (4.23) and (4.25) on the differential factor effects, least squares estimators of the model parameters can be derived. These estimators are given below.

Replicated Two-Factor Layout

$$\hat{\mu} = \bar{Y}_{...} = \frac{1}{IJK} \sum_{i=1}^{I} \sum_{j=1}^{J} \sum_{k=1}^{K} Y_{ijk}$$

$$\hat{\beta}_i = \bar{Y}_{i..} - \bar{Y}_{...} \quad \text{where} \quad \bar{Y}_{i..} = \frac{1}{JK} \sum_{j=1}^{J} \sum_{k=1}^{K} Y_{ijk}, \qquad i = 1, 2, \dots, I$$

$$\hat{\tau}_j = \bar{Y}_{.j.} - \bar{Y}_{...} \quad \text{where} \quad \bar{Y}_{.j.} = \frac{1}{IK} \sum_{i=1}^{I} \sum_{k=1}^{K} Y_{ijk}, \qquad j = 1, 2, \dots, J$$

$$(\hat{\beta\tau})_{ij} = \bar{Y}_{ij.} - \bar{Y}_{i..} - \bar{Y}_{.j.} + \bar{Y}_{...} \quad \text{where} \quad \bar{Y}_{ij.} = \frac{1}{K} \sum_{k=1}^{K} Y_{ijk}, \qquad \begin{aligned} i &= 1, 2, \dots, I \\ j &= 1, 2, \dots, J \end{aligned}$$

(4.26)

Unreplicated Two-Factor Layout

$$\hat{\mu} = \bar{Y}_{..} = \frac{1}{IJ} \sum_{i=1}^{I} \sum_{j=1}^{J} Y_{ij}$$

$$\hat{\beta}_i = \bar{Y}_{i.} - \bar{Y}_{..} \quad \text{where} \quad \bar{Y}_{i.} = \frac{1}{J} \sum_{j=1}^{J} Y_{ij}, \qquad i = 1, 2, \dots, I \qquad (4.27)$$

$$\hat{\tau}_j = \bar{Y}_{.j} - \bar{Y}_{..} \quad \text{where} \quad \bar{Y}_{.j} = \frac{1}{I} \sum_{i=1}^{I} Y_{ij}, \qquad j = 1, 2, \dots, J$$

The expected values of several important sample quantities, given that $E(e_{ijk}) = E(e_{ij}) = 0$ for all i, j, k and the restrictions imposed on the differential effects, are presented in Table 4.7. In this table we have set

$$\mu_i = \mu + \beta_i, \qquad i = 1, 2, \dots, I$$
and
$$\mu_j = \mu + \tau_j, \qquad j = 1, 2, \dots, J$$

(4.28)

These quantities will be referred to as the underlying population means associated with the ith level of factor 1 and the jth level of factor 2, respectively.

Using the results in Table 4.7 it is easy to show that the least squares estimators in equations (4.26) and (4.27) are unbiased estimators of the corresponding model parameters.

Table 4.7 Expected Values for the Two-Factor
Fixed Effects Models

$Y_{ijk} = \mu + \beta_i + \tau_j + (\beta\tau)_{ij} + e_{ijk}$	$Y_{ij} = \mu + \beta_i + \tau_j + e_{ij}$

$E(Y_{ijk}) = \mu + \beta_i + \tau_j + (\beta\tau)_{ij}$ $\qquad E(Y_{ij}) = \mu + \beta_i + \tau_j$

$E(\bar{Y}_{i..}) = \mu + \beta_i = \mu_i$ $\qquad\qquad E(\bar{Y}_{i.}) = \mu + \beta_i = \mu_i$

$E(\bar{Y}_{.j.}) = \mu + \tau_j = \mu_j$ $\qquad\qquad E(\bar{Y}_{.j}) = \mu + \tau_j = \mu_j$

$E(\bar{Y}_{ij.}) = \mu + \beta_i + \tau_j + (\beta\tau)_{ij}$ $\qquad E(\bar{Y}_{..}) = \mu$

$E(\bar{Y}_{...}) = \mu$

4.10 ANOVA TABLE AND *F* TESTS FOR THE REPLICATED TWO-WAY LAYOUT

For the replicated two-way layout the hypotheses:

1. H_0: $\beta_i = 0$, for all i (no factor 1 main effects)
2. H_0: $\tau_j = 0$, for all j (no factor 2 main effects)
3. H_0: $(\beta\tau)_{ij} = 0$, for all i and j (no interaction effects)

are of interest and can be tested given the assumptions imposed on the error terms and the restrictions on the differential effects. The tests are based on observed values of random variables having F probability distributions. Like the one-factor case, the calculations leading to the "F tests" are conveniently set out in an analysis of variance table.

The ANOVA table is given in Table 4.8. The entries in the "sum of squares" and "degrees of freedom" columns are derived from the algebraic identities,

$$\sum_{i=1}^{I}\sum_{j=1}^{J}\sum_{k=1}^{K}(Y_{ijk} - \bar{Y}_{...})^2 = JK\sum_{i=1}^{I}(\bar{Y}_{i..} - \bar{Y}_{...})^2 + IK\sum_{j=1}^{J}(\bar{Y}_{.j.} - \bar{Y}_{...})^2 +$$

$$\begin{pmatrix}\text{total variation about}\\ \text{sample average}\end{pmatrix} = \begin{pmatrix}\text{variation due}\\ \text{to factor 1}\end{pmatrix} + \begin{pmatrix}\text{variation due}\\ \text{to factor 2}\end{pmatrix} +$$

$$\hfill (4.29)$$

$$K\sum_{i=1}^{I}\sum_{j=1}^{J}(\bar{Y}_{ij.} - \bar{Y}_{i..} - \bar{Y}_{.j.} + \bar{Y}_{...})^2 + \sum_{i=1}^{I}\sum_{j=1}^{J}\sum_{k=1}^{K}(Y_{ijk} - \bar{Y}_{ij.})^2$$

$$\begin{pmatrix}\text{variation due to interac-}\\ \text{tion of factors 1 and 2}\end{pmatrix} + \begin{pmatrix}\text{residual (error)}\\ \text{variation}\end{pmatrix}$$

and

d.f.: $IJK - 1 = (I - 1) + (J - 1) + (I - 1)(J - 1) + IJ(K - 1)$

The remaining entries are calculated as indicated in the table. The F ratios may be used to test the hypotheses (1), (2), and (3) above, and these tests are summarized in Table 4.9.

Table 4.8   The Two-Factor Analysis of Variance Table (replicated two-way layout)

Source of Variation	Sum of Squares (SS)	Degrees of Freedom (d.f.)	Mean Squares (MS)	F Ratio
Factor 1	$SS_1 = JK \sum_{i=1}^{I} (\bar{Y}_{i..} - \bar{Y}_{...})^2$	$(I-1)$	$MS_1 = \dfrac{SS_1}{(I-1)}$	$F = \dfrac{MS_1}{s^2}$
Factor 2	$SS_2 = IK \sum_{j=1}^{J} (\bar{Y}_{.j.} - \bar{Y}_{...})^2$	$(J-1)$	$MS_2 = \dfrac{SS_2}{(J-1)}$	$F = \dfrac{MS_2}{s^2}$
Interaction	$SS_{12} = K \sum_{i=1}^{I} \sum_{j=1}^{J} (\bar{Y}_{ij.} - \bar{Y}_{i..} - \bar{Y}_{.j.} + \bar{Y}_{...})^2$	$(I-1)(J-1)$	$MS_{12} = \dfrac{SS_{12}}{(I-1)(J-1)}$	$F = \dfrac{MS_{12}}{s^2}$
Residual	$SS_R = \sum_{i=1}^{I} \sum_{j=1}^{J} \sum_{k=1}^{K} (Y_{ijk} - \bar{Y}_{ij.})^2$	$IJ(K-1)$	$MS_R = \dfrac{SS_R}{IJ(K-1)} = s^2$	
Total (about mean)	$\sum_{i=1}^{I} \sum_{j=1}^{J} \sum_{k=1}^{K} (Y_{ijk} - \bar{Y}_{...})^2$	$IJK - 1$		

Table 4.9   Tests of Hypothesis for the Replicated Two-Factor Classification (fixed effects, γ level of significance)

	HYPOTHESIS		
	H_0: all $\beta_i = 0$ (no factor 1 main effects) H_a: not all $\beta_i = 0$	H_0: all $\tau_j = 0$ (no factor 2 main effects) H_a: not all $\tau_j = 0$	H_0: all $(\beta\tau)_{ij} = 0$ (no interaction effects) H_a: not all $(\beta\tau)_{ij} = 0$
Test statistic	$F = \dfrac{MS_1}{s^2}$	$F = \dfrac{MS_2}{s^2}$	$F = \dfrac{MS_{12}}{s^2}$
Degrees of freedom	$\nu_1 = (I-1); \ \nu_2 = IJ(K-1)$	$\nu_1 = (J-1); \ \nu_2 = IJ(K-1)$	$\nu_1 = (I-1)(J-1); \ \nu_2 = IJ(K-1)$
Critical value	$F_\gamma(\nu_1, \nu_2)$	$F_\gamma(\nu_1, \nu_2)$	$F_\gamma(\nu_1, \nu_2)$
Test	Reject H_0 if observed $F > F_\gamma(\nu_1, \nu_2)$; otherwise do not reject H_0		

Large values of the F ratios are regarded as evidence against the appropriate null hypotheses, since if H_0 is not true, the numerators of these ratios will tend to be larger than the denominators. This can be seen by examining the expected mean squares in Table 4.10.

Table 4.10   Expected Mean Squares (EMS) for the Replicated Two-Factor Classification [fixed effects, $Var(e_{ijk}) = \sigma^2$]

Source of Variation	Expected Mean Squares (EMS)
Factor 1	$E(MS_1) = \sigma^2 + \left(JK \sum_{i=1}^{I} \beta_i^2 \right) \Big/ (I-1)$
Factor 2	$E(MS_2) = \sigma^2 + \left(IK \sum_{j=1}^{J} \tau_j^2 \right) \Big/ (J-1)$
Interaction	$E(MS_{12}) = \sigma^2 + \left(K \sum_{i=1}^{I} \sum_{j=1}^{J} (\beta\tau)_{ij}^2 \right) \Big/ (I-1)(J-1)$
Residual	$E(MS_R) = E(s^2) = \sigma^2$

EXAMPLE 4.8

A food-packaging firm is interested in the effects of "training" and "motivation" on employee output. An experiment is designed whereby 24 employees responsible for packaging a certain meat product are divided at random into six groups of four people each.

Each person is subjected to one of two types of training programs designed to increase packaging speed and one of three types of monetary reward systems. Table 4.11 gives the output per two-hour period in number of packages under the various combinations of training and motivation.

Table 4.11   Hypothetical Packaging Data

		MOTIVATION 1	2	3	
TRAINING	1	17 21 49 54 35.25	64 49 34 52.5 63	62 72 61 71.5 91	53.08
	2	33 37 40 16 31.5	41 64 34 64 50.15	56 62 57 72 61.75	48
		33.375	51.625	66.625 50.54	

I = 2
J = 3
K = 4

Let Y_{ijk} be the output in number of packages. We assume the two-factor fixed effects model where β_i, $i = 1, 2$, represents the differential training effect, τ_j, $j = 1, 2, 3$, represents the differential motivation effect, and $(\beta\tau)_{ij}$ represents the training-motivation interaction. The ANOVA table is given in Table 4.12 (note that there are $K = 4$ replications of the experiment).

Now $F_{0.05}(1, 18) = 4.41$, and $F_{0.05}(2, 18) = 3.55$. Therefore we would conclude (at the 5-percent level) that

1. There are no effects due to type of training; that is, H_0: all $\beta_i = 0$ is not rejected ($F = 0.8 < F_{0.05}(1, 18) = 4.41$).
2. There are differential motivation effects; that is, H_0: all $\tau_j = 0$ is rejected ($F = 11.4 > F_{0.05}(2, 18) = 3.55$); the "level" of motivation makes a difference.
3. There are no interaction effects; that is, H_0: all $(\beta\tau)_{ij} = 0$ is not rejected ($F = 0.2 < F_{0.05}(2, 18) = 3.55$).

4.11   ANOVA TABLE AND *F* TESTS FOR THE UNREPLICATED TWO-WAY LAYOUT

For the unreplicated two-way layout the hypotheses:

1. H_0: $\beta_i = 0$, for all i (no factor 1 main effects)
2. H_0: $\tau_j = 0$, for all j (no factor 2 main effects)

can be tested. Moreover, although an additive model (no interaction term) is usually assumed, it is possible to test for the presence of interaction effects using a procedure due to Tukey[44]. Tukey's test is considered at the end of this section.

The ANOVA table for the unreplicated two-way layout is given in Table 4.13. The entries in the table are based on the algebraic identities:

$$\sum_{i=1}^{I}\sum_{j=1}^{J}(Y_{ij} - \bar{Y}_{..})^2 = J\sum_{i=1}^{I}(\bar{Y}_{i.} - \bar{Y}_{..})^2 + I\sum_{j=1}^{J}(\bar{Y}_{.j} - \bar{Y}_{..})^2 + \sum_{i=1}^{I}\sum_{j=1}^{J}(Y_{ij} - \bar{Y}_{i.} - \bar{Y}_{.j} + \bar{Y}_{..})^2$$

$$\begin{pmatrix} \text{total variation} \\ \text{about sample} \\ \text{average} \end{pmatrix} = \begin{pmatrix} \text{variation} \\ \text{due to} \\ \text{factor 1} \end{pmatrix} + \begin{pmatrix} \text{variation} \\ \text{due to} \\ \text{factor 2} \end{pmatrix} + \begin{pmatrix} \text{residual (error)} \\ \text{variation} \end{pmatrix} \quad (4.30)$$

and

$$\text{d.f.:} \quad IJ - 1 = (I - 1) + (J - 1) + (I - 1)(J - 1)$$

The *F* tests are summarized in Table 4.14, and the expected mean squares given the usual assumptions are set out in Table 4.15. Examining the latter table it can be seen that if the null hypotheses H_0: all $\beta_i = 0$ and H_0: all $\tau_j = 0$ are true, MS_1 and MS_2 are unbiased estimators of σ^2, and hence the corresponding *F* ratios will tend to have values near 1. If the null hypotheses are not true, however, the quantities MS_1 and MS_2 tend to be larger than σ^2,

Table 4.12 ANOVA Table for Packaging Example

Source of Variation	Sum of Squares (SS)	Degrees of Freedom (d.f.)	Mean Squares (MS)	F Ratio
Factor 1 (training)	155.0	1	155.0	$F = 0.8$
Factor 2 (motivation)	4436.3	2	2218.2	$F = 11.4$
Interaction (training-motivation)	69.4	2	34.7	$F = 0.2$
Residual	3487.3	18	193.7	
Total (about mean)	8148.0	23		

Table 4.13 The Two-Factor Analysis of Variance Table (unreplicated two-way layout)

Source of Variation	Sum of Squares (SS)	Degrees of Freedom (d.f.)	Mean Squares (MS)	F Ratio
Factor 1	$SS_1 = J\sum_{i=1}^{I}(\bar{Y}_{i.} - \bar{Y}_{..})^2$	$(I-1)$	$MS_1 = \dfrac{SS_1}{(I-1)}$	$F = \dfrac{MS_1}{s^2}$
Factor 2	$SS_2 = I\sum_{j=1}^{J}(\bar{Y}_{.j} - \bar{Y}_{..})^2$	$(J-1)$	$MS_2 = \dfrac{SS_2}{(J-1)}$	$F = \dfrac{MS_2}{s^2}$
Residual	$SS_R = \sum_{i=1}^{I}\sum_{j=1}^{J}(Y_{ij} - \bar{Y}_{i.} - \bar{Y}_{.j} + \bar{Y}_{..})^2$	$(I-1)(J-1)$	$MS_R = \dfrac{SS_R}{(I-1)(J-1)} = s^2$	
Total (about mean)	$\sum_{i=1}^{I}\sum_{j=1}^{J}(Y_{ij} - \bar{Y}_{..})^2$	$IJ - 1$		

Table 4.14   Tests of Hypotheses for the Unreplicated
Two-Factor Classification (fixed effects, γ
level of significance)

HYPOTHESIS

	H_0: all $\beta_i = 0$ (no factor 1 main effects)	H_0: all $\tau_j = 0$ (no factor 2 main effects)
	H_a: not all $\beta_i = 0$	H_a: not all $\tau_j = 0$
Test statistic	$F = \dfrac{MS_1}{s^2}$	$F = \dfrac{MS_2}{s^2}$
Degrees of freedom	$\nu_1 = (I - 1)$; $\nu_2 = (I - 1)(J - 1)$	$\nu_1 = (J - 1)$; $\nu_2 = (I - 1)(J - 1)$
Critical value	$F_\gamma(\nu_1, \nu_2)$	$F_\gamma(\nu_1, \nu_2)$
Test	Reject H_0 if observed $F > F_\gamma(\nu_1, \nu_2)$; otherwise do not reject	

Table 4.15   Expected Mean Squares for the
Unreplicated Two-Factor
Classification [fixed effects,
$\mathrm{Var}(e_{ij}) = \sigma^2$]

Source of Variation	Expected Mean Squares (EMS)
Factor 1	$E(MS_1) = \sigma^2 + \dfrac{J}{(I-1)} \sum\limits_{i=1}^{I} \beta_i^2$
Factor 2	$E(MS_2) = \sigma^2 + \dfrac{I}{(J-1)} \sum\limits_{j=1}^{J} \tau_j^2$
Residual	$E(MS_R) = E(s^2) = \sigma^2$

and hence the F ratios will tend to be much greater than 1. Consequently, large F ratios are regarded as evidence against the null hypotheses.

EXAMPLE 4.9

Recall the typewriter data of Table 4.1. Suppose that the output was produced by five different typists, each typist using all four typewriters. The data along with row and column averages can be displayed as in Table 4.16.

We now postulate the two-factor fixed effects model

$$Y_{ij} = \mu + \beta_i + \tau_j + e_{ij}, \qquad i = 1, 2, \ldots, 4; \quad j = 1, 2, \ldots, 5$$

with e_{ij} independent $N(0, \sigma^2)$ variables and

$$\sum_{i=1}^{I} \beta_i = \sum_{j=1}^{J} \tau_j = 0$$

as a description of the data generating process. Here β_i, $i = 1, 2, \ldots, 4$, represent the incremental mean outputs (main effects) due to differences in typewriters and τ_j, $j = 1, 2, \ldots, 5$, represent the incremental mean outputs (main effects) due to differences in typists.

The ANOVA table is given in Table 4.17. The F ratios can be compared to upper γ percentage points of F distributions to test the hypotheses H_0: all $\beta_i = 0$ and H_0: all $\tau_j = 0$, respectively. Since $F_{0.05}(3, 12) = 3.49$ and $F_{0.05}(4, 12) = 3.26$ we reject H_0: all $\beta_i = 0$ and do not reject H_0: all $\tau_j = 0$ at the 5-percent level of significance. There appears to be an incremental effect on output due to differences in typewriters, but there does not appear to be an incremental effect on output due to typists; that is, a good part of the variability in words per minute is due to the differences in typewriters, not typists.

Comparing Table 4.17 with Table 4.4(b), it is seen that the variation due to differences in typewriters has remained the same ($SS_1 = 270$). However, the residual variation ($SS_R = 166$) is now less, since we have been able to account for part of it by identifying an additional source of variation (differences in typists).

The result is a more sensitive test for typewriter differences. This is indicated by the smaller residual mean square ($MS_R = 13.8$) and hence larger F ratio ($F = 6.52$).

Tukey's Test

We have assumed that the model is additive (no interaction) for the unreplicated two-way classification. If we doubt the validity of this assumption, a test for interaction effects can be performed. To perform this test (due to Tukey[44]), calculate the quantities,

$$SS_A = \frac{[\sum_{i=1}^{I} \sum_{j=1}^{J} (\bar{Y}_{i.} - \bar{Y}_{..})(\bar{Y}_{.j} - \bar{Y}_{..})Y_{ij}]^2}{\sum_{i=1}^{I} (\bar{Y}_{i.} - \bar{Y}_{..})^2 \sum_{j=1}^{J} (\bar{Y}_{.j} - \bar{Y}_{..})^2} \tag{4.31}$$

$$SS_B = \sum_{i=1}^{I} \sum_{j=1}^{J} (Y_{ij} - \bar{Y}_{i.} - \bar{Y}_{.j} + \bar{Y}_{..})^2 \tag{4.32}$$

and

$$SS_R = SS_B - SS_A \tag{4.33}$$

The test statistic for the hypothesis of no interaction effects given the model set out in equations (4.24) and (4.25) is

$$F = \frac{(IJ - I - J)SS_A}{SS_R} \tag{4.34}$$

The hypothesis is rejected at the γ level of significance if F exceeds

$$F_\gamma(1, IJ - I - J)$$

the upper γ percentage point of an F distribution with $\nu_1 = 1$ and $\nu_2 = IJ - I - J$ degrees of freedom.

If the hypothesis of no interactions is rejected, the situation becomes complicated. As Scheffé[8] points out, there is the difficulty of reaching *any* conclusions about the factor main effects if interactions exist, since the

Table 4.16 Typewriter Data

| | | | TYPIST | | | |
	1	2	3	4	5	$\bar{Y}_i$
TYPEWRITER 1	69	62	70	67	62	66
2	67	72	76	69	71	71
3	76	70	71	66	77	72
4	60	64	67	58	66	63
$\bar{Y}_j$	68	67	71	65	69	$\bar{Y}_{..} = 68$

Table 4.17 ANOVA Table for Typewriter Example

Source of Variation	Sum of Squares (SS)	Degrees of Freedom (d.f.)	Mean Squares (MS)	F Ratio
Factor 1 (typewriter)	270	3	90	$F = 6.52$
Factor 2 (typists)	80	4	20	$F = 1.45$
Residual	166	12	13.8	
Total (about mean)	516	19		

residual mean square is no longer an unbiased estimator of σ^2. (Note that the residual mean square appears in the F ratios as well as the simultaneous confidence intervals given in Section 4.12.) In addition, there is the usual practical difficulty of interpreting the factor main effects in the presence of interactions.

To examine the nature of main effects, interaction effects, and the relationship between them, suppose that factor 1 has $I = 2$ levels and factor 2 has $J = 3$ levels and there is no error component. Then, in general, we can write the observation in the (i, j)th cell as

$$Y_{ij} = \mu + \beta_i + \tau_j + (\beta\tau)_{ij}, \qquad i = 1, 2; \quad j = 1, 2, 3 \qquad (4.35)$$

that is, observations may differ from cell to cell due to the factor 1 main effect β_i, the factor 2 main effect τ_j, and the interaction effect $(\beta\tau)_{ij}$, $i = 1, 2$, $j = 1, 2, 3$. On the other hand if there is no interaction between factors 1 and 2, then $(\beta\tau)_{ij} = 0$ for all i and j, and

$$Y_{ij} = \mu + \beta_i + \tau_j, \qquad i = 1, 2; \quad j = 1, 2, 3 \qquad (4.36)$$

Consider the situation in equation (4.36). The observation at the *first* level of factor 1 and the *first* level of factor 2 is given by $Y_{11} = \mu + \beta_1 + \tau_1$. The observation at the *second* level of factor 1 and *first* level of factor 2 is $Y_{21} = \mu + \beta_2 + \tau_1$. Hence the difference in the observations at the first level of factor 2 over the levels of factor 1 is $(\beta_1 - \beta_2)$, the difference in the main effects at the first and second levels of factor 1. It is easily verified that when we move to the second level of factor 2, the difference in the observations over the levels of factor 1 is still $(\beta_1 - \beta_2)$. In fact, this is true for all levels of factor 2. Similarly, if we were to look at the differences in the observations over any two levels of factor 2 for each level of factor 1, the result would be constant. This case is illustrated in the top half of Figure 4.2 where the cell means are plotted for the various combinations of factor levels with $\beta_i \neq 0$, $i = 1, 2$, and $\tau_j \neq 0$, $j = 1, 2, 3$. The result is a series of parallel straight lines.

Thus when there are no interaction effects, if there is a real difference between the first and second levels of factor 1, this difference is the same for all levels of factor 2. Similarly, if there is a real difference between, say, the first and second levels of factor 2, this difference is the same for all levels of factor 1.

The situation is more complex when effects due to the interaction of factors 1 and 2 exist. Using the model in equation (4.35), it is clear that the difference in the observations at the first level of factor 2 $(j = 1)$ over the levels of factor 1 is $(\beta_1 - \beta_2 + (\beta\tau)_{11} - (\beta\tau)_{21})$. If we move to, say, the third level of factor 2, the differences in the observations over the levels of factor 1 is $(\beta_1 - \beta_2 + (\beta\tau)_{13} - (\beta\tau)_{23})$; that is, the difference in the observations is *not* the same for all levels of factor 2. The presence of effects due to the interaction of factors 1 and 2 complicates things. What we can say about the

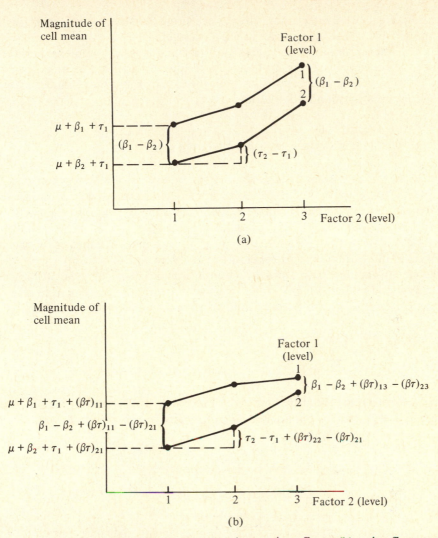

Figure 4.2   (a) Main effects and no interaction effects; (b) main effects and interaction effects

difference between the first and second levels of factor 1 at one level of factor 2 is, in general, *different* from what we can say about this difference at another level of factor 2. A "real" difference between the first and second levels of factor 1 when interactions exist means that there is a difference when the effects are *averaged* over all levels of factor 2. Similar statements can be made for differences between any two levels of factor 2 at given levels of factor 1. The bottom half of Figure 4.2 shows a typical representation of the cell means for the possible combinations of factor levels when main effects and interaction effects are present. The result is a series of straight lines that, in general, are not parallel.

The interpretation of an analysis of variance is much simpler when there are no interaction effects. In fact, it is frequently possible to transform the data to eliminate interactions; that is, the transformed observations can be represented by an additive model.

4.12   MULTIPLE COMPARISONS—TWO-WAY LAYOUT

If the F tests reject the notion of no factor main effects, then it is of interest to compare the effects contributed by different levels of the factor. We can employ an extension of Scheffé's method of multiple comparisons, introduced in Theorem 4.1, for this purpose.

Let μ_i, $i = 1, 2, \ldots, I$, and μ_j, $j = 1, 2, \ldots, J$, denote the means for the ith level of factor 1 and the jth level of factor 2 [see equation (4.28)]. Since the I levels of factor 1 correspond to the rows and the J levels of factor 2 correspond to the columns of the two-way layout, it is convenient to refer to the μ_i's and μ_j's as the "row" and "column" means, respectively.

Theorem 4.2   Let

$$\psi_1 = \sum_{i=1}^{I} c_i \mu_i \qquad \text{and} \qquad \psi_2 = \sum_{j=1}^{J} c_j \mu_j$$

be contrasts in the row and column means, respectively. Then with $100 \times (1 - \gamma)$-percent confidence, *all* contrasts in the means ψ_1 and ψ_2 are bracketed by the bounds,

$$\sum_{i=1}^{I} c_i \bar{Y}_i \pm \sqrt{F_\gamma(I-1, \nu)}\, s \sqrt{\frac{(I-1)}{JK} \sum_{i=1}^{I} c_i^2} \tag{4.37}$$

and

$$\sum_{j=1}^{J} c_j \bar{Y}_j \pm \sqrt{F_\gamma(J-1, \nu)}\, s \sqrt{\frac{(J-1)}{IK} \sum_{j=1}^{J} c_j^2} \tag{4.38}$$

respectively, where s^2 is the residual mean square from the appropriate ANOVA table, $F_\gamma(\nu_1, \nu_2)$ is the upper γ percentage point of an F distribution with ν_1 and ν_2 degrees of freedom, and

Replicated layout		Unreplicated layout	
$\bar{Y}_i = \bar{Y}_{i..}$,	$i = 1, 2, \ldots, I$	$\bar{Y}_i = \bar{Y}_{i.}$,	$i = 1, 2, \ldots, I$
$\bar{Y}_j = \bar{Y}_{.j.}$,	$j = 1, 2, \ldots, J$	$\bar{Y}_j = \bar{Y}_{.j}$,	$j = 1, 2, \ldots, J$
$\nu = IJ(K-1)$		$K = 1$	
		$\nu = (I-1)(J-1)$	

The proof of this theorem can be found in Scheffé [8].

Remark   Using equation (4.28) and the fact that

$$\sum_{i=1}^{I} c_i = \sum_{j=1}^{J} c_j = 0$$

for contrasts, it is easy to show that

$$\sum_{i=1}^{I} c_i \mu_i = \sum_{i=1}^{I} c_i \beta_i \quad \text{and} \quad \sum_{j=1}^{J} c_j \mu_j = \sum_{j=1}^{J} c_j \tau_j$$

Thus simultaneous interval estimates of contrasts in row and column means are interval estimates of the same contrasts in the differential row and column main effects, β_i, $i = 1, 2, \ldots, I$, and τ_j, $j = 1, 2, \ldots, J$.

EXAMPLE 4.10

Consider the packaging data introduced in Example 4.8. In that example we found that there were motivation effects but no training effects or training-motivation interactions. Let μ_1, μ_2, and μ_3 represent the motivation means—that is, the population average outputs associated with the three levels of motivation. The least squares estimates of μ_1, μ_2, and μ_3 are

$$\bar{Y}_{.1.} = 33.375$$
$$\bar{Y}_{.2.} = 51.625$$
$$\bar{Y}_{.3.} = 66.625$$

The differences in the means are simultaneously estimated with 95 percent confidence by

$$\mu_1 - \mu_2: \ \bar{Y}_{.1.} - \bar{Y}_{.2.} \pm \sqrt{F_{0.05}(2, 18)}\ s\ \sqrt{\frac{(3-1)}{2(4)}(1^2 + (-1)^2)}$$

$$= (33.375 - 51.625) \pm \sqrt{3.55}\,(13.9)\ \sqrt{\frac{1}{2}} = -18.25 \pm 18.52$$

$$\mu_1 - \mu_3: \ \bar{Y}_{.1.} - \bar{Y}_{.3.} \pm 18.52 = -33.25 \pm 18.52$$
$$\mu_2 - \mu_3: \ \bar{Y}_{.2.} - \bar{Y}_{.3.} \pm 18.52 = -15 \pm 18.52$$

The results follow from equation (4.38) with $I = 2$, $J = 3$, $K = 4$, and $\nu = IJ(K-1) = 18$.

The interval estimates above also apply to the same contrasts in motivation main effects. Thus the largest effect occurs at motivation level 3, the smallest effect at motivation level 1. The difference in the effects at levels 1 and 3 is real and significant since it is estimated by -33.25 ± 18.52. The effects at motivation levels 2 and 3 are statistically indistinguishable. The effects at motivation levels 1 and 2 are, strictly speaking, indistinguishable, although the 95-percent interval for $\mu_1 - \mu_2$ (and hence $\tau_1 - \tau_2$) just barely includes zero. Clearly with a somewhat smaller confidence coefficient all the differences in effects (means) would be significant.

In summary, there appear to be real differences in the effects of motivation levels on the number of packages produced. Since there were no demonstrated training-motivation interaction effects, our conclusion holds for both types (levels) of training.

4.13 MORE ON DEPARTURES FROM ASSUMPTIONS

The inferential procedures we have presented for the two-way layouts are based on the assumptions of normality, equal variances, independence, and a correct model. For the replicated fixed effects two-way layout these assumptions can be summarized by saying the error terms e_{ijk} are $N(0, \sigma^2)$ random variables, $i = 1, 2, \ldots, I, j = 1, 2, \ldots, J$, and $k = 1, 2, \ldots, K$. For the unreplicated fixed effects two-way layout, we assume the error terms e_{ij} are $N(0, \sigma^2)$ random variables, $i = 1, 2, \ldots, I, j = 1, 2, \ldots, J$, and that the factor 1 and factor 2 effects are additive (no interaction effects). Like the one-factor situation, it is useful to discuss the sensitivity of the inferences to departures from the usual ANOVA assumptions.

In general, provided I and J are reasonably large, the F tests for determining factor main effects and interactions are relatively insensitive to a departure from normality; that is, if the data are not quite normal, we will not err too much by treating them as if they were normal. Once again we have the general fact that tests of means are relatively insensitive (robust) to departures from normality.

In addition, if there is an *equal* number of observations within each cell in the two-way layout, then the F tests for factor main effects and interactions are not affected too much by unequal variances. Note that we will always have an equal number of observations in each cell if the experiment is replicated $K(K \geq 1)$ times.

The problem of correlation (lack of independence) is difficult to deal with. If the observations (or errors) are correlated within a column (within a level of factor 2), then the F test for determining *column* (factor 2) main effects will tend to indicate the presence of these effects when in fact they do not exist; that is, the true probability of a type I error can be much greater than the nominal value γ. The F test for *row* (factor 1) main effects, however, is not affected very much by the correlation within columns. Similar statements can be made for correlation within rows (levels of factor 1).

The conclusion is clear. The F tests for main effects and interactions in the fixed effects two-way layout are relatively insensitive to departures from the normality and equality of variances assumptions provided I and J are fairly large and there is an equal number of observations in each cell. Violation of the independence assumption, however, is more serious and difficult to treat. The comments above apply to Scheffé's multiple comparison methods as well as to the usual F tests.

To check on the assumptions underlying the two-factor models, the data can be plotted or subjected to more sophisticated statistical tests as we pointed out in Section 4.7.

In addition, the assumptions can be checked by examining the residuals, which, if the model is correct, can be regarded as estimates of the error components. The residuals, defined by

$$\hat{e}_{ij} = Y_{ij} - \bar{Y}_{i.} - \bar{Y}_{.j} + \bar{Y}_{..}, \qquad i = 1, 2, \ldots, I; \quad j = 1, 2, \ldots, J \quad (4.39)$$

and

$$\hat{e}_{ijk} = Y_{ijk} - \bar{Y}_{ij.} \qquad i = 1, 2, \ldots, I; \quad j = 1, 2, \ldots, J \quad (4.40)$$

$$k = 1, 2, \ldots, K$$

for the unreplicated and replicated two-way layouts, respectively, can be examined using the plots introduced in Section 4.7 and the methods discussed in Sections 5.6 and 6.6. For example, normality can be checked (roughly) by plotting all the residuals as a frequency diagram or histogram. Equality of variances can be checked for replicated layouts by comparing the spreads of the residuals in each cell. For unreplicated layouts the spreads of the residuals in columns may be compared with one another. Similarly, the spreads of the residuals in the rows may be compared.

If one or more of the assumptions appear to be violated, it is frequently possible to transform the observations to new quantities that (approximately) satisfy the conditions of normality, equal variances, and in some cases independence. Moreover, since factor main effects are more easily interpreted when interaction effects are absent, it is desirable and frequently possible to achieve additivity by transforming the original observations. Happily, a single transformation can often achieve several objectives simultaneously. For example, a transformation to stabilize the variance may at the same time improve the approximation to normality. Procedures for transforming data are given in references [35] and [38].

4.14 FINAL COMMENTS

The models we have presented in this chapter are useful but represent only special cases of a more general "linear model" for characterizing a particular experimental situation. There are many variations that can be introduced depending on the nature and objectives of the experiment. For example, the factor effects may be more appropriately regarded as "random" rather than fixed effects; or it may be appropriate to regard some effects as fixed and other effects as random. A particular model and the appropriate mode of analysis is determined by the circumstances surrounding the experiment; circumstances which, in the ideal situation, are under the control of the investigator. In addition, it is possible to examine the effects of more than two factors in "multifactor experiments." Although the analysis

is more complicated, the availability of general analysis of variance computer programs facilitates matters.

The reader is referred to reference [7] for an elementary account of the
general theory of linear models and the associated analysis of variance.
References [8] and [9] offer more advanced treatments of the same topic.

Supplement 4A

A ROBUST TEST
OF THE EQUALITY
OF SEVERAL VARIANCES

Suppose we have J sample variances $s_1^2, s_2^2, \ldots, s_J^2$ based on relatively large independent samples of sizes $n_1, n_2, \ldots, n_J$ from populations with possibly different means and variances, all unknown. We wish to test the hypothesis $H_0: \sigma_1^2 = \sigma_2^2 = \cdots = \sigma_J^2$ against $H_a:$ not all the σ_i^2 are equal.

Let $Y_{ij}, i = 1, 2, \ldots, n_j, j = 1, 2, \ldots, J$, be the ith observation in the jth sample and let $\bar{Y}_j, j = 1, 2, \ldots, J$ be the sample averages. Define the quantity

$$G = \sum_{j=1}^{J} (n_j - 1) \left[\ln s_j^2 - \frac{\sum_{j=1}^{J}(n_j - 1) \ln s_j^2}{\sum_{j=1}^{J}(n_j - 1)} \right]^2 \Big/ k^2$$

where "ln" denotes the natural logarithm and

$$k^2 = 2 + \left(1 - \frac{1}{\bar{n}}\right) m$$

with

$$\bar{n} = \left(\sum_{j=1}^{J} n_j\right) \Big/ J$$

$$m = \frac{(\sum_{j=1}^{J} n_j)\sum_{j=1}^{J} \sum_{i=1}^{n_j} (Y_{ij} - \bar{Y}_j)^4}{[\sum_{j=1}^{J} \sum_{i=1}^{n_j} (Y_{ij} - \bar{Y}_j)^2]^2} - 3$$

Then if H_0 is true and all the n_j are reasonably large, G has approximately a χ^2 distribution with $\nu = (J - 1)$ degrees of freedom; that is, approximately

$$G \sim \chi^2(J - 1)$$

and thus a formal test of H_0 at the γ level of significance is

Reject H_0 if $G > \chi_\gamma^2(J - 1)$

Do not reject H_0 if $G \leq \chi_\gamma^2(J - 1)$

where $\chi_\gamma^2(J-1)$ is the upper γ percentage point of a χ^2 distribution with $\nu = (J-1)$ degrees of freedom.

Although the test we have presented is a "large sample" test, sampling experiments indicate that it performs fairly well for sample sizes as small as 10 (see [36]). However, following Layard, we recommend that the test *not* be used for sample sizes smaller than 10. The test appears to perform very well for sample sizes in the neighborhood of 25 or more.

The test we have presented does not depend on the *form* of the parent populations, and thus it is insensitive (that is, robust) to departures from normality. (In two-way ANOVA situations with replication, the sample variances $s_1^2, s_2^2, \ldots, s_J^2$ are the sample variances computed for each cell in the layout.)

PROBLEMS

4.1 Define the following terms and give examples when appropriate.
- (a) experimental design
- (b) analysis of variance
- (c) one-factor layout
- (d) two-factor layout
- (e) fixed effect models
- (f) main effects
- (g) interaction effects
- (h) additivity
- (i) multiple comparisons
- (j) residual analysis
- (k) homogeneity of variance
- (l) contrast in means
- (m) robust statistical tests
- (n) analysis of variance table
- (o) factor levels

4.2 Show that the entries in the sum of squares column of the one-factor ANOVA table [Tables 4.4(a) and 4.5] remain the same if a constant, say, c, is subtracted from *all* the observations. (This fact generally makes the computations less tedious.)

4.3 An investigation was conducted to compare three methods for producing a certain commodity. The output in eight successive days from the methods is given below.

	Method	
A	B	C
9	8	16
11	16	19
15	14	21
8	14	19
12	12	22
10	13	22
10	11	18
12	16	23

(a) Stating any assumptions you make, test for differences in the mean outputs of the three methods. (Assume a significance level of $\gamma = 0.05$.)

(b) Compare the means of methods A and B and the mean of C with the average of the means of A and B using Scheffé's multiple comparison method. Interpret the results. Use a 95-percent confidence level.

(c) Plot the residuals for each of the three methods in any way you see fit. Do the assumptions appear to be satisfied?

4.4 A factory manager is contemplating purchasing a new machine to perform a certain operation in a production process. Four manufacturers have submitted their versions of the machine for trial. Each machine is used for five days, and the daily output is recorded. The data follow.

Machine

1	2	3	4
64	41	65	45
39	48	57	51
65	41	56	55
46	49	72	48
63	57	64	47

(a) Are there real differences between machines with respect to output? Be specific and make sure your answer includes (1) the model employed and the underlying assumptions, (2) the ANOVA table and the associated F test, and (3) estimates of the parameters in the model.

(b) Compare all possible pairs of means using the method based on the t distribution [see equation (4.17)] and Scheffé's method of multiple comparison. Comment on the differences.

4.5 Prove the algebraic identity [see equation (4.15)]

$$\sum_{j=1}^{J} \sum_{i=1}^{n} (Y_{ij} - \bar{Y})^2 = \sum_{j=1}^{J} n(\bar{Y}_j - \bar{Y})^2 + \sum_{j=1}^{J} \sum_{i=1}^{n} (Y_{ij} - \bar{Y}_j)^2$$

4.6 The American Car Company is interested in the average number of cars coming off three assembly lines with obvious defects (determined by visual inspection). The number of defects per week for each of the assembly lines is recorded for several weeks. The data follow.

Assembly Line

1	2	3
49	50	61
61	68	44
55	57	53
67	64	47
48	73	58
	66	49
	56	

Given the observations above, test for differences in the mean number of visually defective cars coming off the assembly lines using the F test.

4.7 The makers of "Tasty Tacos" have divided their U.S. market into five regions—Northeast, Mideast, South, Midwest, and West—which have approximately equal populations. They wish to compare their sales performance in the five areas. Data on sales are available for the past 10 months for each region. The data are assumed to be independent samples from normal populations with the same variance but possibly different means. The sales figures are in thousands of tacos.

Northeast	Mideast	South	Midwest	West
22	27	25	37	42
20	29	30	33	38
25	26	32	36	41
23	35	26	40	45
29	31	23	34	43
31	31	27	32	48
26	33	25	29	44
28	28	30	34	45
24	30	33	33	39
24	25	28	38	42

(a) Test for homogeneity of variance using the procedure set out in Supplement 4A.

(b) Write out the ANOVA table, and test for differences in mean sales in the five regions.

(c) Compare average sales in the West and Midwest with average sales for the remaining three regions using Scheffé's multiple comparison procedure. Use a 95-percent confidence level. (Hint: Consider the contrast: $\frac{1}{2}(\mu_4 + \mu_5) - \frac{1}{3}(\mu_1 + \mu_2 + \mu_3)$, where $\mu_1, \mu_2, \ldots, \mu_5$ are the mean sales for the regions in the order in which they are listed above.)

4.8 The following observations represent weight loss (in pounds) of men of similar physique, metabolic activity, and so on, after a certain amount of time on three diets, A, B, and C.

Diet

A	B	C
3	2	7
7	4	10
4	6	8
5	6	9
6	5	4
	3	8
	4	

(a) Stating any assumptions you make, test for differences in mean weight loss between the three diets. Provide point estimates of the mean losses.

(b) If you were in charge of "designing" this experiment, is there anything you would do differently? Explain.

4.9 A transportation analyst wished to compare the "degree of support" for a new rapid transit system among the following three groups of people in the Madison, Wisconsin, area: group A, University of Wisconsin faculty; group B, downtown businessmen; and group C, state government personnel. A questionnaire was designed from which a "rapid transit support index" could be computed for each respondent. The questionnaire was given to randomly selected people from each group. The results follow.

Rapid Transit Support
Index Group

A	B	C
66 4	45 -15	57 - 5
59 -17	53 -7	49 -13
70 0	62 2	61 -1
65 -5	72 12	53 -9
76 6	78 18	72 10
84 14	50 10	80 18

562

Stating any assumptions you make, test for differences between the three populations represented by the three samples using the F test. Specify the null and alternative hypotheses. Show the calculations leading to your conclusion in the form of an ANOVA table. (You should be able to calculate easily all the ANOVA table entries given the following summary measures: $J = 3$, $n = 6$, $\bar{Y}_A = 70$, $\bar{Y}_B = 60$, $\bar{Y}_C = 62$, $s_A^2 = 78.8$, $s_B^2 = 169.2$, and $s_C^2 = 140.0$.)

4.10   Patients with a certain disease were treated with drugs to retard the formation of certain foreign substances in the blood. Three different drugs have, in the past, been used for this purpose with varying degrees of success. An experiment is conducted to compare the efficacy of the three older drugs, say A, B, and C, with a new drug, say D, designed for the same purpose. The patients are randomly assigned to a particular drug group, and the amount of foreign substances in a given volume of blood is measured after a suitable time period. The observations follow.

Drug

A	B	C	D
13	15	14	12
14	16	14	14
17	14	13	13
15	11	16	11
13	15	17	14
16	17	15	13
12	13	18	12

(a)   Write down a model that allows for differential effects of drugs and the accompanying assumptions.

(b)   Test for differences in the drugs using the F test with a 5-percent significance level. Display the ANOVA table.

(c)   List point estimates of the differential drug effects.

(d)   Develop a contrast for comparing drugs A, B, and C with drug D. Make the comparison using Scheffé's multiple comparison method. Assume a 95-percent confidence level.

(e)   Check the residuals for violations of the usual ANOVA assumptions.

4.11   Verify the identity [see equation (4.30)]

$$\sum_{i=1}^{I}\sum_{j=1}^{J}(Y_{ij}-\bar{Y}_{..})^2 = J\sum_{i=1}^{I}(\bar{Y}_{i.}-\bar{Y}_{..})^2 + I\sum_{j=1}^{J}(\bar{Y}_{.j}-\bar{Y}_{..})^2$$

$$+ \sum_{i=1}^{I}\sum_{j=1}^{J}(Y_{ij}-\bar{Y}_{i.}-\bar{Y}_{.j}+\bar{Y}_{..})^2$$

4.12   Consider the typewriter data in Table 4.16.

(a)   Supply point estimates of the model parameters.

(b)   Test for interaction effects using Tukey's test. Let $\gamma = 0.05$.

(c)   Suppose interaction effects are indicated by Tukey's test. How would this affect the interpretation of the F ratios displayed in Table 4.17?

(d)   Explain what can be done to achieve additivity of effects.

4.13   Carefully examine the residuals for the typewriter data in Table 4.16. In particular, plot the residuals as a frequency diagram; compare the groups of residuals for rows (typewriters) with one another, and finally compare the groups of residuals for columns (typists) with one another. Interpret the results.

4.14 Six children are tested for pulse rate before and after watching a violent movie with the following results.

Pulse Rate

Child	Before	After
1	102	112
2	96	108
3	89	94
4	104	112
5	90	102
6	85	96

(a) Using the paired t test, test for differences in the before and after mean pulse rates. Let $\gamma = 0.05$, and use a two-sided test.

(b) Test for differences in the pulse rate means employing a two-factor fixed effects model and the F test. Again, let $\gamma = 0.05$. Are the results consistent with those of part (a)?

(c) Square the observed t value and the corresponding critical point. Compare the results with the observed F ratio and the F critical point, respectively. They should (within rounding error) be the same. Comment on the importance of these results.

(d) Using the t and F distributions, calculate 95-percent confidence intervals for the difference in mean pulse rates. Comment on the results.

4.15 A plant manager is interested in the effect of "time of day" on the output of three men on a car-wax kit assembly line. The number of kits assembled by each man is recorded for three carefully chosen time periods. The results follow.

	Man		
Time	1	2	3
11:00–12:00 A.M.	34	37	36
1:00–2:00 P.M.	39	43	41
4:00–5:00 P.M.	31	36	37

(a) Write out an appropriate model for this situation and list the underlying assumptions.

(b) Display the ANOVA table and test for "time" main effects and "personnel" main effects. Assume a 5-percent significance level.

(c) Provide point estimates of the model parameters.
(d) Use Tukey's test to test for additivity at the 5-percent level.
(e) Compare the time effects pairwise using Scheffé's multiple comparison method. Use a 95-percent confidence level and interpret the results.

4.16 Consider the packaging data in Table 4.11. Let the model for these data be

$$Y_{ijk} = \mu + \beta_i + \tau_j + (\beta\tau)_{ij} + e_{ijk}, \qquad i = 1, 2; \quad j = 1, 2, 3; \quad k = 1, 2, \ldots, 4$$

where e_{ijk} are independent $N(0, \sigma^2)$ random variables.
(a) Supply point estimates of the model parameters.
(b) Examine the residuals using any "informative" plots you can think of. Do the assumptions appear to be satisfied? Explain.

4.17 An engineer is investigating the strength of concrete beams made from four types of cement and employing three curing processes. The breaking strengths are measured and are presented below.

Cement Type

Curing Process	A	B	C	D
1	14	6	22	10
2	15	8	20	7
3	10	7	18	7

(a) List an appropriate model and the underlying assumptions for this situation.
(b) Construct the ANOVA table and test for differences in cement types and curing processes. Assume a 5-percent significance level.
(c) Construct a 95-percent confidence interval for comparing the average breaking strengths of cement types A, C, and D with that of cement type B. Interpret the result.

4.18 Referring back to Problem 4.17, suppose that two additional beams are made for each combination of cement type and curing process and the breaking strengths determined. The complete layout follows.

Cement Type

Curing Process	A	B	C	D
1	14,14,10	6,9,7,	22,25,20	10,11,13
2	15,12,10	8,4,11	20,15,19	7,9,10
3	10,8,10	7,4,7	18,21,22	7,6,6

(a) Specify an appropriate model for this situation and the underlying assumptions.
(b) Construct the ANOVA table and test for main effects and interaction effects. Assume a 5-percent significance level.
(c) Provide point estimates of the model parameters.
(d) Construct a 95-percent confidence interval for comparing the breaking strengths of cement types A, C, and D with that of cement type B. Interpret the result.
(e) Examine the residuals in any way you see fit. Comment on the outcome.
(f) Discuss the advantages, if any, of replicating an experiment.

5

SIMPLE LINEAR REGRESSION AND CORRELATION

5.1 INTRODUCTION

Essentially all scientific investigations are concerned with understanding and explaining observable phenomena. In the search for explanations of these phenomena, one is frequently faced with modeling relationships among several variables. Relationships can often be formulated in terms of a linear[a] model. For example, the one- and two-way ANOVA models considered in Chapter 4 are linear models. ANOVA models are deficient, however, in the sense that they do not indicate exactly *how* the "variables of interest" affect the measured response. Fixed effect models simply indicate the effect on the response (expressed by the parameters μ, β, and τ) at different levels or combinations of levels of the controllable variables included in the experiment. Thus inferences must be interpreted in terms of the particular sets of factor levels (some of which may be categorical) chosen by the investigator and are ordinarily of the following kind: (a) Does a significant factor effect exist? (b) If a factor effect exists, how do the responses at the various factor levels compare with one another? For these ANOVA models, extrapolation of response behavior outside of the set of experimental conditions specifically examined is a dangerous exercise.

In many situations more general models are required indicating "how" and "to what extent" a response variable is related to a set of independent variables. For example, (a) an economist may be interested in the relationship between "demand for housing" and "disposable income," (b) a market researcher may be interested in the relationship between "sales" and the variables "price" and "amount spent on advertising," or (c) a chemical

---

[a] Here the word "linear" refers to linearity in the model *parameters*. A model linking a response variable, Y, to a set of independent variables is linear in the parameters if the partial derivatives of Y with respect to each parameter does not depend on *any* of the parameters in the model.

176

engineer may be interested in the relationship between the "concentration of a reactant" and the variables "reaction time" and "temperature." The list of situations of this kind is virtually endless.

Suppose for a given problem the *true* functional relationship between the response or dependent variable and members of a set of independent variables is known. The investigator is then in a position to understand, predict, and perhaps even control the response. Unfortunately there are very few situations in practice in which the true functional relationship between variables can be determined, particularly in the social sciences. Consequently, one is forced, through the interaction of theory and empirical evidence, to develop models that characterize in some way the *main features* of the behavior of the response variable. Frequently these "approximating" models, if constructed with care, have a reasonably simple structure and yet are powerful enough to aid in the resolution of policy questions and in the further development of underlying theory. In addition, they can provide accurate and reliable forecasts of future response values.

Regression analysis is concerned with modeling the relationships between variables. It is a very useful and commonly employed method of data analysis, particularly with the advent of large-scale computers and readily available regression programs. However, there is a tendency to misuse this technique and therefore, as we shall repeatedly point out, it is of paramount importance to examine critically the results of any regression analysis.

In this chapter we shall be concerned with a first-order relationship between two variables, so-called simple regression. Most of the ideas in regression analysis can be illustrated in this context. The next chapter is concerned with an extension of these ideas to relationships involving more than two variables, or relationships between variables of higher order, that is, multiple regression.

Throughout this chapter we will concentrate on the following two examples.

EXAMPLE 5.1

The marketing department of a Midwest seed supplier has been keeping price and sales figures on various items for the past eight years. The figures for field corn are given below and are plotted in Figure 5.1, where the variable X is the price per bushel in dollars plotted as the abscissa and the variable Y is the sales in thousands of bushels plotted as the ordinate. Sales have been designated as the dependent or response variable Y, and price has been designated as the independent variable X,

X (price per bushel in dollars)	1.25	1.75	2.25	2.00	2.50	2.25	2.75	2.50
Y (sales in 1000s of bushels)	125	100	60	89	75	81	50	55

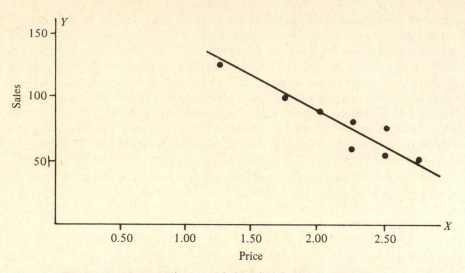

Figure 5.1   Scatter diagram of data for Example 5.1

since it is of interest to determine the dependence of sales on price; that is, we are seeking a relationship of the form $Y = f(X)$. A glance at Figure 5.1 indicates that, in general, an increase in price is associated with a decrease in sales. Moreover it is apparent that the same price (for example, $X = \$2.50$) may be associated with two different sales figures. How can we characterize these features in terms of a mathematical model relating sales to price? It is clear that in lieu of any additional information the true relationship between sales and price is not immediately obvious. In addition, simply connecting the points in the figure in order to get an indication of the relationship leads to problems. For example, how do we explain the two Y values at $X = 2.25$ and at $X = 2.50$? However, if we are content with extracting the general behavior of the response variable Y, the data in Figure 5.1 indicate that the relationship between the variables Y and X is given "approximately" by a function whose graph is a straight line, or $Y = f(X) = a + bX$. One such line is indicated in the figure. On the other hand this function does not give the whole story since it is obvious that the data points in the figure do not lie exactly along a straight line. We shall return to this issue after introducing Example 5.2.

EXAMPLE 5.2

An economist hypothesizes that housing demand and disposable income are related. He postulates the relationship $Y = f(X)$, where Y is the housing demand and X is the disposable income. To investigate the nature of the function $f(X)$, he collects data on aggregate housing demand and aggregate disposable income for a five-year period. The data in appropriate units are given below and plotted in Figure 5.2. Figure 5.2 indicates that, in general, an increase in disposable income is

X (disposable income)	6.8	7.0	7.1	7.2	7.4
Y (housing demand: housing starts)	0.8	1.2	0.9	0.9	1.5

accompanied by an increase in housing demand. The situation pictured in Figure 5.2 is not so clear cut as the one pictured in Figure 5.1, and yet, given only these data, we may, as a first attempt, characterize the overall behavior of the response variable Y by saying that demand increases (approximately) linearly with income. Thus $f(X)$ may be of the form $f(X) = a + bX$. The graph of one function of this type is indicated in the figure. Again, the relationship $f(X) = a + bX$, as it stands, is not entirely adequate.

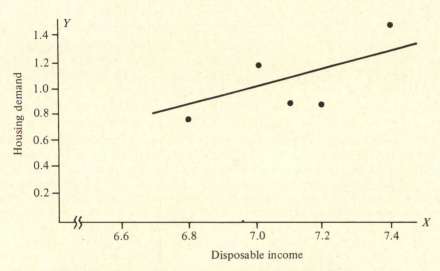

Figure 5.2   Scatter diagram of data for Example 5.2

5.2   THE LEAST SQUARES VALUES

Let us assume that we can represent the overall behavior of the response variables for both examples introduced in the previous section by a function of the form $f(X) = a + bX$, where X denotes the independent variable in each case and a and b are the intercept and slope coefficients. This function will not be particularly useful, however, until we have specified values for the coefficients a and b. Clearly the values for a and b must be such that the function approximates the observable relationship between Y and X; that is, when plotted as a straight line, $f(X)$ should pass through the scatter of points in a manner indicated by the lines pictured in Figures 5.1 and 5.2. Now, the lines in the figures were, in fact, drawn in by eye, using a straightedge. However, this procedure is not satisfactory since different people may draw in different lines and hence reach different conclusions. Moreover, this discrepancy will tend to increase as the scatter of the points in the vertical direction increases. Thus the "best-fitting" straight line cannot be determined in this manner without creating some (potential) controversy. It is evident that we need a *criterion* for determining the "best-fitting" straight line—that is, for determining the values of the coefficients a and b—that will always give the same results for a given set of data. Ideally the criterion we use should be logically appealing—that is, capable of producing a line passing close to all the points in the scatter—and in addition should be capable of producing "best-fitting" functions in situations involving more than one independent variable.

If the "best-fitting" straight line is to pass close to all the data points, then the *deviations* of the individual points from this line must be reasonably small. Therefore we can develop a criterion for determining the appropriate straight line by considering the "distances" of the individual data points from the line.

Consider the function $f(X) = a + bX \geq 0$ for all $X \geq 0$. For a given value of X, say, $X = X_i$, $f(X_i)$ is the height of the line above the horizontal axis at the point X_i. Let us denote this height by $\hat{Y}_i$. Thus $\hat{Y}_i = f(X_i) = a + bX_i$ and hence, for an arbitrary X, we can write $\hat{Y} = a + bX$. One measure[b] of the deviation of the data point (X_i, Y_i) from the line $\hat{Y} = a + bX$ is given by $Y_i - \hat{Y}_i$, the *vertical* distance of the point from the line at X_i. Furthermore, if we denote this vertical distance by e_i then, assuming there are n data points,

$$Y_i = \hat{Y}_i + e_i, \qquad i = 1, 2, \ldots, n \tag{5.1}$$

or equivalently

$$Y_i = a + bX_i + e_i, \qquad i = 1, 2, \ldots, n \tag{5.2}$$

The situation is illustrated for the data of Example 5.2 in Figure 5.3, where some of the $\hat{Y}_i$'s and the e_i's are shown for an arbitrary straight line $\hat{Y} = a + bX$.

The deviations $e_1, e_2, \ldots, e_n$ can be combined to form a criterion measuring the "fit" of a given line to the scatter such that if the value of the criterion is small, a better fit is indicated than if it is large. Intuitively we might feel that the sum of the deviations $\sum_{i=1}^{n} e_i$ would be an appropriate criterion; however, this criterion is unsatisfactory since large positive deviations may be offset by large negative deviations to produce a small total. On the other hand, a criterion of this kind is appealing since it takes into account all of the deviations in a relatively simple manner. It is possible to construct a criterion that overcomes the difficulty associated with the sum of the

---

[b] There are other measures of the deviation of the data point from the line; for example, we might consider the perpendicular distance of the point from the line; however, the vertical distance is convenient to work with mathematically and can be justified on theoretical grounds (see Section 5.5).

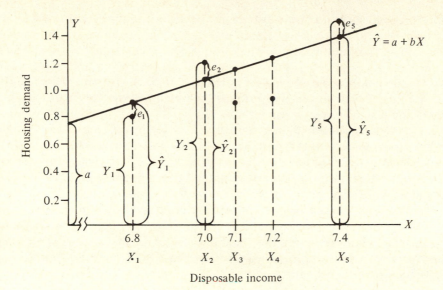

Figure 5.3 Vertical measurements of deviations

deviations by considering a criterion based on the magnitudes of the deviations irrespective of sign. Two simple possibilities are immediately available, namely,

$$\sum_{i=1}^{n} |e_i| = \sum_{i=1}^{n} |Y_i - \hat{Y}_i| \tag{5.3}$$

and

$$\sum_{i=1}^{n} e_i^2 = \sum_{i=1}^{n} (Y_i - \hat{Y}_i)^2 \tag{5.4}$$

The criteria immediately above will produce different "best-fitting" lines, and there are situations in which one criterion may be preferred to the other. However, the "sum of squared errors" criterion given by equation (5.4) is generally used in practice because it is easier to work with mathematically and it emphasizes large errors in the sense that no single deviation from the "best-fitting" line will be "unreasonably" large when compared with the other deviations. In addition, as we shall see in Section 5.5, the criterion $\sum_{i=1}^{n} e_i^2$ can be justified on theoretical grounds.

To find the line that "best" represents the given data, we choose the coefficients a and b such that the sum of the squared deviations of the Y_i's from the corresponding $\hat{Y}_i$'s is minimized. This statement is known as the *least squares principle*, and the resulting values of the coefficients are known as the *least squares values*. We can write the sum of squared errors in terms of the coefficients a, b by substituting for $\hat{Y}_i$ in equation (5.4). Thus

$$S(a, b) = \sum_{i=1}^{n} (Y_i - \hat{Y}_i)^2 = \sum_{i=1}^{n} [Y_i - (a + bX_i)]^2 \tag{5.5}$$

where the notation $S(a, b)$ indicates that the error sum of squares depends on the values of a and b. The problem then is to find the values of a and b that minimize $S(a, b)$.

An analytical solution to this problem is initiated by differentiating $S(a, b)$ with respect to each of the variables a and b, treating the variable not being considered in the differentiation as a constant; that is, the partial derivatives of S with respect to a and b are computed (see Appendix 2). If we set the resulting equations equal to zero and solve them simultaneously, we obtain expressions for a and b that correspond to a point where the function $S(a, b)$ is either a maximum or minimum. It can be shown, however (see Problem 5.11), that these expressions for a and b yield the minimum value of $S(a, b)$ and thus provide us with the least squares values of the coefficients.

Using the chain rule and equation (5.5), the partial derivatives of S with respect to a and b are

$$-2 \sum_{i=1}^{n} (Y_i - a - bX_i)$$

and

$$-2 \sum_{i=1}^{n} (Y_i - a - bX_i)X_i$$

respectively. Setting these partial derivatives equal to zero gives

$$\sum_{i=1}^{n} (Y_i - a - bX_i) = 0$$

$$\sum_{i=1}^{n} (Y_i - a - bX_i)X_i = 0$$

or after some algebraic manipulations

$$na + b \sum_{i=1}^{n} X_i = \sum_{i=1}^{n} Y_i \tag{5.6}$$

$$a \sum_{i=1}^{n} X_i + b \sum_{i=1}^{n} X_i^2 = \sum_{i=1}^{n} X_i Y_i \tag{5.7}$$

Equations (5.6) and (5.7) are known as the "normal equations" and can be solved simultaneously for a and b. In particular if we multiply the first equation through by $\bar{X}$ and subtract it from the second, we get

$$\left(\sum_{i=1}^{n} X_i - n\bar{X} \right) a + \left(\sum_{i=1}^{n} X_i^2 - \bar{X} \sum_{i=1}^{n} X_i \right) b = \sum_{i=1}^{n} X_i Y_i - \bar{X} \sum Y_i$$

However, the coefficient of a is zero (the reader should check this), and thus

$$b = \frac{\sum_{i=1}^{n} X_i Y_i - \bar{X} \sum_{i=1}^{n} Y_i}{\sum_{i=1}^{n} X_i^2 - \bar{X} \sum_{i=1}^{n} X_i} \qquad (5.8)$$

To obtain an expression for a, we can divide equation (5.6) through by n to obtain

$$a + b\bar{X} = \bar{Y} \qquad \text{or} \qquad a = \bar{Y} - b\bar{X} \qquad (5.9)$$

Therefore once b is determined, a can be evaluated.

The expression we have given for b [equation (5.8)] is rather unwieldly and can be written in several equivalent forms (see Problem 5.2). Two of the most commonly employed forms are

$$b = \frac{\sum_{i=1}^{n} (X_i - \bar{X})(Y_i - \bar{Y})}{\sum_{i=1}^{n} (X_i - \bar{X})^2} \qquad (5.10)$$

and

$$b = \frac{\sum_{i=1}^{n} (X_i - \bar{X}) Y_i}{\sum_{i=1}^{n} (X_i - \bar{X})^2} \qquad (5.11)$$

Of the two forms, equation (5.11) requires less computational effort, and we shall therefore adopt it.

EXAMPLE 5.3

Consider the sales price data introduced in Example 5.1. We want to determine the line $\hat{Y} = a + bX$ that best fits the data according to the least squares criterion. Table 5.1 displays the calculations leading to the least squares values a and b.

Using Table 5.1

$$b = \frac{\sum (X_i - \bar{X}) Y_i}{\sum (X_i - \bar{X})^2} = \frac{④}{③} = \frac{-80.22}{1.62} = -49.52$$

$$a = \bar{Y} - b\bar{X} = ② - b① = 79.38 - (-49.52)(2.16) = 186.34$$

and hence the least squares line is given by

$$\hat{Y} = 186.34 - 49.52X \qquad (5.12)$$

EXAMPLE 5.4

The calculations leading to the least squares values a and b for the demand-income data of Example 5.2 are similar in form to those in Table 5.1 and thus will not be reproduced in detail. However, the reader may verify that

$$\bar{X} = 7.10, \qquad\qquad \bar{Y} = 1.06$$

$$\sum_{i=1}^{5} (X_i - \bar{X})^2 = 0.20, \qquad \sum_{i=1}^{5} (X_i - \bar{X}) Y_i = 0.18$$

Table 5.1   Calculations for Least Squares
Values[a] (Example 5.1)

X_i	Y_i	$X_i - \bar{X}$	$(X_i - \bar{X})^2$	$(X_i - \bar{X})Y_i$
1.25	125	−0.91	0.82	−113.28
1.75	100	−0.41	0.17	− 40.63
2.25	60	0.09	0.01	5.63
2.00	89	−0.16	0.02	− 13.91
2.50	75	0.34	0.12	25.78
2.25	81	0.09	0.01	7.59
2.75	50	0.59	0.35	29.69
2.50	55	0.34	0.12	18.91

$\Sigma X_i = 17.25$, $\Sigma Y_i = 635$, ① $\bar{X} = 2.16$, ② $\bar{Y} = 79.38$,
③ $\Sigma(X_i - \bar{X})^2 = 1.62$, ④ $\Sigma(X_i - \bar{X})Y_i = -80.22$.

[a] The entries in the table are the exact values rounded to two
decimal places; therefore some numerical discrepancies exist.

and hence

$$b = \frac{\Sigma(X_i - \bar{X})Y_i}{\Sigma(X_i - \bar{X})^2} = \frac{0.18}{0.20} = 0.90$$

$$a = \bar{Y} - b\bar{X} = 1.06 - (0.90)(7.10) = -5.33$$

The equation defining the least squares line is

$$\hat{Y} = -5.33 + 0.90X \tag{5.13}$$

5.3 FITTED VALUES, RESIDUALS, AND THE ANALYSIS OF VARIANCE

The coefficients a and b of the least squares line are choosen such that the sum of squared deviations,

$$S(a, b) = \sum (Y_i - \hat{Y}_i)^2$$

is as small as possible. The Y coordinates of the points $(X_i, \hat{Y}_i)$ on the least squares line are referred to as the *fitted values* and the vertical deviations $Y_i - \hat{Y}_i$ are called the *residuals*. Thus the minimum value of $S(a, b)$ is the sum of the squared residuals or the "residual sum of squares."

The fitted values, residuals, and residual sum of squares for our two examples are given in Table 5.2. If we start with the expression for the residuals and add and subtract $\bar{Y}$, we can write

$$Y_i - \hat{Y}_i = Y_i - \bar{Y} - (\hat{Y}_i - \bar{Y}) \tag{5.14}$$

Table 5.2 Fitted Values, Residuals, and Residual Sum
of Squares

		(Example 5.1)				(Example 5.2)	
X_i	Y_i	$\hat{Y}_i = 186.34 - 49.52 X_i$	$Y_i - \hat{Y}_i$	X_i	Y_i	$\hat{Y}_i = -5.33 + 0.90 X_i$	$Y_i - \hat{Y}_i$
1.25	125	124.44	0.56	6.8	0.8	0.79	0.01
1.75	100	99.68	0.32	7.0	1.2	0.97	0.23
2.25	60	74.92	-14.92	7.1	0.9	1.06	-0.16
2.00	89	87.30	1.70	7.2	0.9	1.15	-0.25
2.50	75	62.54	12.46	7.4	1.5	1.33	0.17
2.25	81	74.92	6.08			$\sum (Y_i - \hat{Y}_i)^2 = 0.17$	
2.75	50	50.16	-0.16				
2.50	55	62.54	-7.54				
		$\sum (Y_i - \hat{Y}_i)^2 = 475.01$					

Squaring both sides of the equation (5.14) and summing produces[c]

$$\sum (Y_i - \hat{Y}_i)^2 = \sum (Y_i - \bar{Y})^2 - \sum (\hat{Y}_i - \bar{Y})^2 \qquad (5.15)$$

or equivalently

$$\sum (Y_i - \bar{Y})^2 = \sum (\hat{Y}_i - \bar{Y})^2 + \sum (Y_i - \hat{Y}_i)^2 \qquad (5.16)$$

[c]This is not immediately obvious but follows from squaring both sides of equation (5.14) to obtain

$$(Y_i - \hat{Y}_i)^2 = (Y_i - \bar{Y})^2 + (\hat{Y}_i - \bar{Y})^2 - 2(Y_i - \bar{Y})(\hat{Y}_i - \bar{Y})$$

However,

$$-2(Y_i - \bar{Y})(\hat{Y}_i - \bar{Y}) = -2(Y_i - \hat{Y}_i + \hat{Y}_i - \bar{Y})(\hat{Y}_i - \bar{Y}) = -2(Y_i - \hat{Y}_i)(\hat{Y}_i - \bar{Y}) - 2(\hat{Y}_i - \bar{Y})^2$$

Thus

$$(Y_i - \hat{Y}_i)^2 = (Y_i - \bar{Y})^2 - (\hat{Y}_i - \bar{Y})^2 - 2(Y_i - \hat{Y}_i)(\hat{Y}_i - \bar{Y})$$

and

$$\sum (Y_i - \hat{Y}_i)^2 = \sum (Y_i - \bar{Y})^2 - \sum (\hat{Y}_i - \bar{Y})^2 - 2 \sum (Y_i - \hat{Y}_i)(\hat{Y}_i - \bar{Y})$$

Since

$$\hat{Y}_i = a + bX_i = \bar{Y} - b\bar{X} + bX_i = \bar{Y} + b(X_i - \bar{X})$$

the last summation may be written

$$\sum (Y_i - \hat{Y}_i)(\hat{Y}_i - \bar{Y}) = \sum [Y_i - \bar{Y} - b(X_i - \bar{X})][\bar{Y} + b(X_i - \bar{X}) - \bar{Y}]$$
$$= b \sum (Y_i - \bar{Y})(X_i - \bar{X}) - b^2 \sum (X_i - \bar{X})^2$$
$$= b \left[\sum (Y_i - \bar{Y})(X_i - \bar{X}) - b \sum (X_i - \bar{X})^2 \right]$$

Using equation (5.10) and substituting for b within the brackets gives

$$\sum (Y_i - \hat{Y}_i)(\hat{Y}_i - \bar{Y}) = b \left[\sum (Y_i - \bar{Y})(X_i - \bar{X}) - \sum (Y_i - \bar{Y})(X_i - \bar{X}) \right] = 0$$

Therefore

$$\sum (Y_i - \hat{Y}_i)^2 = \sum (Y_i - \bar{Y})^2 - \sum (\hat{Y}_i - \bar{Y})^2$$

which is equation (5.15).

Equation (5.16) expresses the total variation in the Y values about their arithmetic average (the total sum of squares) as the sum of two components. This sum of squares breakup is similar to the sum of squares breakup encountered in the one-factor analysis of variance. The first term on the right in equation (5.16) measures the variation of the fitted values about $\bar{Y}$ and is called the "regression sum of squares." The second term on the right measures the variation of the Y values about the fitted values and consequently is the "residual sum of squares."

We interpret the regression sum of squares, $\Sigma(\hat{Y}_i - \bar{Y})^2$, as that part of the total variation in Y about its arithmetic average explained by the relationship of Y with X (the explained variation) and the residual sum of squares as that part of the total variation in Y about $\bar{Y}$ that is not explained by the relationship of Y with X (the unexplained variation). Equation (5.16) provides us with a means of assessing the relative importance of the independent variable X in accounting for the observed variability in Y. The larger the value for $\Sigma(\hat{Y}_i - \bar{Y})^2$ relative to $\Sigma(Y_i - \bar{Y})^2$, the more pronounced the apparent relationship of Y with X. Equivalently, the smaller the value of $\Sigma(Y_i - \hat{Y}_i)^2$ relative to $\Sigma(Y_i - \bar{Y})^2$, the more pronounced the postulated relationship.

The fraction of the variation in Y about $\bar{Y}$ that is explained (in this case) by the linear relationship of Y with X is known as the *coefficient of determination* and is generally denoted by R^2. Consequently,

$$R^2 = \frac{\Sigma(\hat{Y}_i - \bar{Y})^2}{\Sigma(Y_i - \bar{Y})^2} = 1 - \frac{\Sigma(Y_i - \hat{Y}_i)^2}{\Sigma(Y_i - \bar{Y})^2} \tag{5.17}$$

Clearly $0 \leq R^2 \leq 1$. An R^2 value near unity indicates that almost all of the variability in Y is explained by the linear relationship $Y = a + bX$, and therefore the variable X is useful in explaining the behavior of Y. On the other hand, a value of R^2 near zero indicates that very little of the variability in Y is explained by the linear relationship of Y with X, and hence the equation $Y = a + bX$ is of little use in determining the value of Y. We may interpret R^2 as a measure of the "strength" of the relationship between Y and X. It should be pointed out that the statements we have just made will, in general, hold for the *observed* range of X values, called the "fitting region," and may not be appropriate for other values of the independent variable X.

Before continuing with a discussion of the information contained in equation (5.16) and a tabular display of this information, it is instructive to illustrate further the remarks we have made concerning the explained and unexplained variation in Y.

Suppose changes in the value of X have no effect on the values of the variable Y; that is, the value of X is varied at will, and the overall level of the Y values remains constant. A scatter diagram of the (X_i, Y_i) values would then look something like the one displayed in Figure 5.4. In this case there is

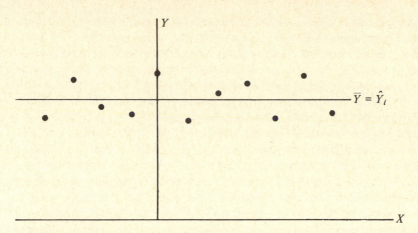

Figure 5.4 Scatter diagram where values of X have no effect on values of Y

no overall change in Y as X changes, and thus if the least squares line were determined, the slope coefficient, b, would be nearly zero. This implies, using equation (5.9), that the intercept coefficient, a, is $\bar{Y}$ and hence that $\hat{Y}_i = \bar{Y}$ for all i. Substituting $\bar{Y}$ for $\hat{Y}_i$ in equation (5.17) gives $R^2 = 0$. None of the variation in Y is explained by X.

On the other hand, suppose as the value of X increases, there is a tendency for the value of Y to increase. In this case a scatter diagram of the (X_i, Y_i) values would look like the one pictured in Figure 5.5. It is clear from

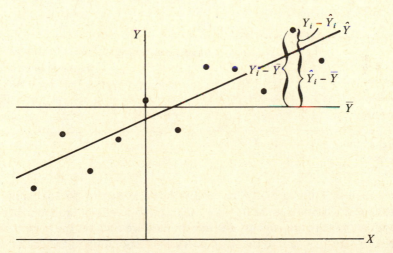

Figure 5.5 Scatter diagram where values of X have some effect on values of Y

the figure that if the least squares line were determined, the slope coefficient would be positive and thus in general $\hat{Y}_i \neq \bar{Y}$. Therefore $\Sigma(\hat{Y}_i - \bar{Y})^2 > 0$, and $R^2 > 0$. Here some of the variability in Y about $\bar{Y}$ is explained by the tendency of Y to increase as X increases. In the extreme case in which all the Y values lie exactly on the least squares line (with $b \neq 0$), the residuals $Y_i - \hat{Y}_i$ are zero for all i, and $R^2 = 1$; that is, all the variation in Y about $\bar{Y}$ is explained by the linear relationship of Y with X.

The reader will note that the coefficient of determination is defined in terms of the pairs of *observations* (X_i, Y_i) and is therefore a data-based quantity. It is a measure of the "strength" of the fitted relationship for a given data set. Clearly R^2 may change if a different data set is generated even though the phenomenon under investigation remains the same.

The sums of squares in equation (5.16) and their associated degrees of freedom are useful for assessing the significance of the relationship between X and Y. The notion of "significance" will become clearer in Section 5.5. At this point we simply take the opportunity to display the analysis of variance table used in simple linear regression analysis to summarize the calculations leading to a formal test of significance of the linear relationship.

The number of degrees of freedom (d.f.) corresponding to the sums of squares given in equation (5.16) are

$$\text{d.f.:} \quad \begin{pmatrix} \text{total sum of} \\ \text{squares about} \\ \text{the mean} \end{pmatrix} = \begin{pmatrix} \text{regression} \\ \text{sum of squares} \end{pmatrix} + \begin{pmatrix} \text{residual} \\ \text{sum of squares} \end{pmatrix} \qquad (5.18)$$
$$n - 1 \quad = \quad 1 \quad + \quad n - 2$$

These numbers represent the number of independent pieces of information involving the Y_i's needed to calculate the associated sum of squares. Using equations (5.16) and (5.18), the ANOVA table has the following form:

Source	Sum of Squares (SS)	Degrees of Freedom (d.f.)	Mean Square (MS)
Regression	$\Sigma(\hat{Y}_i - \bar{Y})^2 = b^2 \Sigma(X_i - \bar{X})^2$	1	$MS_R = \Sigma(\hat{Y}_i - \bar{Y})^2/1$
Residual	$\Sigma(Y_i - \hat{Y}_i)^2$	$n - 2$	$s^2 = \Sigma(Y_i - \hat{Y}_i)^2/(n - 2)$
Total about the mean	$\Sigma(Y_i - \bar{Y})^2$	$n - 1$	

The "source" column in the ANOVA table refers to the source of variation. The "sum of squares" and "degrees of freedom" columns are self-explanatory given our previous discussion, and we note that in both columns the first two entries sum to the third entry corresponding to the "total" row. For example, the residual sum of squares and its associated degrees of freedom may be determined by subtraction if the regression sum of squares and the total sum of squares about the mean and their respective degrees of freedom are known.

The entries in the mean squares column in the ANOVA table are simply the corresponding sum of squares divided by their degrees of freedom. The significance of the mean squares will become apparent later; however, at this point we note the following:

1. Ordinarily only the regression mean square (MS_R) and the residual mean square (s^2) are displayed in the ANOVA table.

2. The residual mean square, s^2, provides an *estimate* based on $n - 2$ degrees of freedom of the variance of Y for a given value of X (see Section 5.4). In addition, if the number of observations is large, s^2 represents a measure of the error with which any observed value of Y could be predicted from a given value of X using the least squares equation. For this reason the positive square root of s^2 is called the *standard error of the estimate.*

3. The ratio MS_R/s^2 can be used to determine the "significance" of the linear relationship between Y and X.

Finally, the reader will note that the coefficient of determination, R^2, is simply the ratio of the regression to the total sum of squares.

It is possible to write the ANOVA table in a more general form making use of the fact that

$$\sum (Y_i - \bar{Y})^2 = \sum Y_i^2 - n\bar{Y}^2$$

Using equation (5.16) and partitioning the degrees of freedom, we have

$$\sum Y_i^2 - n\bar{Y}^2 = \sum (\hat{Y}_i - \bar{Y})^2 + \sum (Y_i - \hat{Y}_i)^2$$

or

$$\sum Y_i^2 = n\bar{Y}^2 + \sum (\hat{Y}_i - \bar{Y})^2 + \sum (Y_i - \hat{Y}_i)^2$$

and (5.19)

$$\text{d.f.:} \quad n \quad = \quad 1 \quad + \quad 1 \quad + \quad (n - 2)$$

The quantity $\sum Y_i^2$ is referred to as the total sum of squares uncorrected for the mean, and $n\bar{Y}^2$ is the sum of squares due to the mean. Employing the breakup in equation (5.19), the general form of the ANOVA table is shown in the table at the top of the following page.

This table, particularly when its format is extended to more complicated problems, is useful primarily for assessing the contribution of additional terms to the equation relating Y with X (or several X's, as the case may be in many relationships) and is not needed for our purposes. In this book we shall concentrate on the form of the ANOVA table as it was initially presented. In more advanced books on regression analysis (for example, reference [2]) both forms of the ANOVA table are employed.

The entries in the ANOVA table associated with our field corn seed

Source	Sum of Squares (SS)	Degrees of Freedom (d.f.)	Mean Squares (MS)
Due to mean	$n\bar{Y}^2$	1	
Regression (given mean)	$\Sigma(\hat{Y}_i - \bar{Y})^2 = b^2\Sigma(X_i - \bar{X})^2$	1	$MS_R = \Sigma(\hat{Y}_i - \bar{Y})^2/1$
Residual	$\Sigma(Y_i - \hat{Y}_i)^2$	$n-2$	$s^2 = \Sigma(Y_i - \hat{Y}_i)^2/(n-2)$
Total (uncorrected)	ΣY_i^2	n	

example can be easily carried out using the information contained in Tables 5.1 and 5.2. From Table 5.2 the residual sum of squares is $\Sigma(Y_i - \hat{Y}_i)^2 = 475.01$. From Table 5.1 $\Sigma(X_i - \bar{X})^2 = 1.62$, and from equation (5.12) $b = -49.52$. Therefore the regression sum of squares is

$$\sum (\hat{Y}_i - \bar{Y})^2 = b^2 \sum (X_i - \bar{X})^2 = (-49.52)^2(1.62) = 3972.61$$

The total sum of squares about the mean (or "corrected for the mean") is then

$$\sum (Y_i - \bar{Y})^2 = \sum (\hat{Y}_i - \bar{Y})^2 + \sum (Y_i - \hat{Y}_i)^2 = 3972.61 + 475.01 = 4447.62$$

The ANOVA table follows.

Table 5.3   The ANOVA Table for the Field Corn Seed Example[a]

Source	Sum of Squares (SS)	Degrees of Freedom (d.f.)	Mean Squares (MS)
Regression	3972.61	1	$MS_R = 3972.61$
Residual	475.01	6	$s^2 = 79.17$
Total (corrected)	4447.62	7	

[a] Note the total (corrected) sum of squares is not quite equal to $\Sigma(Y_i - \bar{Y})^2$ if this quantity is calculated directly due to rounding errors in the calculations of the regression and residual sum of squares. In fact, $\Sigma(Y_i - \bar{Y})^2 = 4453.88$. This example illustrates the need to carry enough decimal places in the calculations to alleviate the rounding error problem.

The coefficient of determination for this example is

$$R^2 = \frac{3972.61}{4447.62} = 0.89$$

which indicates that 89 percent of the variation in sales about its arithmetic average is explained by the linear relationship of sales with price. Most of the movement in sales is explained by the movement of price.

The construction of the ANOVA table and the determination of R^2 for our housing demand-income example is left to the reader (see Problems 5.3 and 5.4).

5.4   THE MATHEMATICAL MODEL AND THE UNDERLYING ASSUMPTIONS

Up to this point we have been concerned only with the overall behavior of the dependent variable Y and have made no attempt to specify the nature of the relationship of Y and X *exactly*. Now it is clear from Figure 5.1 that even though the overall relationship of Y and X can be represented roughly by a function whose graph is a straight line, the points (X_i, Y_i) do not lie *exactly* on a straight line, and in fact the same value of X may be associated with more than one value of Y. To specify the situation completely we proceed in the following way. Suppose, a priori, we regard Y for a given value of X as a *random variable*; that is, we assume that the behavior of Y is described by some probability distribution whose *mean* depends on the value of X and is fixed as soon as X is specified. Thus we are proposing that the conditional expectation of Y, given X, is a function of X. As X varies, the conditional expectation will, in general, change. Suppose that the locus of these conditional means is a straight line. In symbols we are proposing that $E(Y|X) = \alpha + \beta X$. If we represent the deviation of Y from its conditional expectation $\alpha + \beta X$ by e, then e is a random variable[d] and we can write, for any Y_i,

$$Y_i = \alpha + \beta X_i + e_i, \qquad i = 1, 2, \ldots, n \qquad (5.20)$$

It follows that if $E(Y_i|X_i) = \alpha + \beta X_i$, then the expected value of the deviation, or "error," e_i, must be zero. In addition, we assume that the variance of Y is not affected by the value of X and, in fact, remains constant as X changes. Therefore we can write $\text{Var}(Y_i|X_i) = \sigma^2$ for all X_i. Now for a given X, $\alpha + \beta X$ is a fixed constant, and therefore $\text{Var}(e_i) = \sigma^2$, $i = 1, 2, \ldots, n$. Finally, we shall assume that $\text{Cov}(Y_i, Y_j) = 0$ for $i \neq j$. We shall refer to equation (5.20) as the simple linear regression model.

The model for Y_i given by equation (5.20) and the assumptions that follow it are summarized in Table 5.4. Note that the assumptions may be formulated either in terms of the Y_i's or in terms of the e_i's. The formulations are

---

[d] From this point on we will use the symbol e to represent the deviation $[Y - E(Y|X)]$ for an arbitrary X value; that is, $|e|$ represents the magnitude of the distance of the Y value from the value we would *expect* to observe at X, or "error." This deviation represents the effects of extraneous factors on Y and is taken to be a random variable.

Table 5.4   The Mathematical Model and the Underlying
Assumptions

Model: $Y_i = \alpha + \beta X_i + e_i,$ $i = 1, 2, \ldots, n$

Assumptions: 1. $E(Y_i|X_i) = \alpha + \beta X_i$ 1. $E(e_i) = 0$
                      2. $\mathrm{Var}(Y_i|X_i) = \sigma^2$ 2. $\mathrm{Var}(e_i) = \sigma^2$
                      3. $\mathrm{Cov}(Y_i, Y_j) = 0,$ $i \neq j$ 3. $\mathrm{Cov}(e_i, e_j) = 0,$ $i \neq j$

---

equivalent. Usually it is more convenient to specify the distributional
assumptions in terms of the error component.

We have not specified the shape of the probability distribution of the error
component. In practice there is a tendency for the error term to be a sum of
errors from many sources, for example, measurement error, error due to the
effects on Y of variables not specifically included in the relationship $f(X)$,
and so forth. No matter what the probability distribution of the separate
errors may be, as the number of these errors increases, the sum will tend to
be normally distributed because of the central limit effect. Hence we
assume e_i has a $N(0, \sigma^2)$ distribution, or equivalently Y_i has a $N(\alpha +
\beta X_i, \sigma^2)$ distribution. In fact, it is possible to check the normality assumption
by examining the residuals as we shall see in Section 5.6.

The conditional expectation of Y given X as a function of X is known as
the *regression function*. It is now appropriate to relate the regression
function and, in particular, the model displayed in Table 5.4 to the scatter
diagrams given in Figures 5.1 and 5.2 and the least squares lines that were
defined to provide the "best fit" to the two sets of observations.

Model equation (5.20) says that given X_i we can regard the observation Y_i
of the pair (X_i, Y_i) as a realization of a random variable whose probability
distribution is normal with mean $\alpha + \beta X_i$ and variance σ^2. Thus as X_i varies,
the Y values are regarded as independent drawings from normal populations
with the same variance but with means that lie along the straight line
$f(X) = \alpha + \beta X$. The situation is illustrated in Figure 5.6. Consequently, it is
the *expected value of Y* that is exactly related to the independent variable X.
Now, in practice, the function $f(X) = \alpha + \beta X$ is unknown in the sense that
the coefficients (parameters) α and β are unknown. This function is merely a
postulated relationship. All that is available to the investigator are the pairs
of observations (X_i, Y_i) indicated by the dots in Figure 5.6, and from these he
must try to recover the model equation (5.20). In particular, he must try to
determine the values of the parameters α and β. If the model is to be used
for, say, predicting future values of Y corresponding to values of X, we
must have numerical values for α and β.

We have seen that it is possible to fit a straight line mechanically to the
pairs of observations (X_i, Y_i) using the principle of least squares. In fact,
given the assumptions we have made about the behavior of Y (see Table

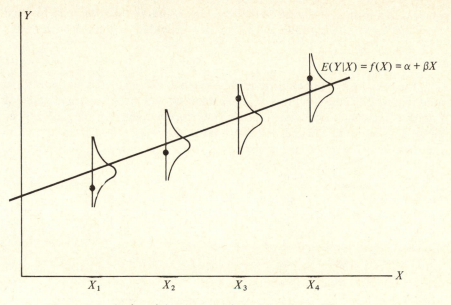

Figure 5.6   A pictorial representation of the simple linear regression model (the dots represent observations)

5.4), the resulting straight line (the least squares line) is our best estimator of the regression line $f(X) = \alpha + \beta X$ in the sense that the least squares intercept and slope coefficients, which we shall now denote by $\hat{\alpha}$ and $\hat{\beta}$, are the best estimators of α and β. In summary, once the data are available, we estimate the regression line $f(X) = \alpha + \beta X$ by the least squares line $\hat{Y} = \hat{\alpha} + \hat{\beta} X$, where

$$\hat{\beta} = \frac{\Sigma(X_i - \bar{X})Y_i}{\Sigma(X_i - \bar{X})^2} \quad \text{and} \quad \hat{\alpha} = \bar{Y} - \hat{\beta}\bar{X}$$

are the values of α and β that minimize

$$S(\alpha, \beta) = \sum_{i=1}^{n} [Y_i - (\alpha + \beta X_i)]^2$$

A justification for this assertion is provided in the next section by the Gauss-Markov theorem.

5.5 INFERENCES ABOUT α, β, AND FUTURE VALUES OF Y

We have suggested that the least squares line is an estimate of the regression line. In this section we will demonstrate the validity of this remark. In addition, we will see that the least squares function may be used to predict or forecast future values of Y—provided it is used with care.

Estimating α and β

Of the two regression coefficients, the slope coefficient, β, is ordinarily the most important since it represents the rate of change of the average value of Y with respect to a unit change in X. Therefore we will consider inferences about β in some detail and merely present the results for α. Some of the theoretical considerations not discussed in this section are covered in the problems at the end of this chapter.

Suppose the values of the independent variable X are fixed. What can we say about the corresponding values of Y? Given the model and the assumptions displayed in Table 5.4 we can say, a priori, that the Y values may be represented by a set of uncorrelated random variables $Y_1, Y_2, \ldots, Y_n$ such that $E(Y_i|X_i) = \alpha + \beta X_i$ and $\text{Var}(Y_i|X_i) = \sigma^2$, $i = 1, 2, \ldots, n$. Now consider the function of $Y_1, Y_2, \ldots, Y_n$,

$$\hat{\beta} = \frac{\Sigma(X_i - \bar{X})Y_i}{\Sigma(X_i - \bar{X})^2} = \Sigma\left[\frac{X_i - \bar{X}}{\Sigma(X_i - \bar{X})^2}\right]Y_i = \Sigma\, w_i Y_i$$

where $w_i = (X_i - \bar{X})/\Sigma(X_i - \bar{X})^2$. We see that the least squares estimator $\hat{\beta}$ can be written as a linear combination of n uncorrelated random variables. Therefore[e]

$$E(\hat{\beta}) = \Sigma\, w_i E(Y_i) = \Sigma\, w_i(\alpha + \beta X_i)$$
$$= \alpha \Sigma\, w_i + \beta \Sigma\, w_i X_i = \beta \tag{5.21}$$

since $\Sigma w_i = 0$, and

$$\Sigma\, w_i X_i = \frac{\Sigma(X_i - \bar{X})X_i}{\Sigma(X_i - \bar{X})^2} = 1$$

(The proofs are left to the reader.) Because $E(\hat{\beta}) = \beta$, $\hat{\beta}$ is called an unbiased estimator of β. In addition

$$\text{Var}(\hat{\beta}) = \sigma_{\hat{\beta}}^2 = \Sigma\, w_i^2 \,\text{Var}(Y_i) = \Sigma\, w_i^2 \sigma^2$$
$$= \sigma^2 \frac{\Sigma(X_i - \bar{X})^2}{[\Sigma(X_i - \bar{X})^2]^2} = \frac{\sigma^2}{\Sigma(X_i - \bar{X})^2} \tag{5.22}$$

and thus the standard deviation of $\hat{\beta}$ is

$$\sigma_{\hat{\beta}} = \frac{\sigma}{\sqrt{\Sigma(X_i - \bar{X})^2}} \tag{5.23}$$

[e] It is convenient to suppress the conditioning factor X_i in the expectations that follow. Therefore we will simply use the notation $E(Y_i)$. The reader should keep in mind, however, that the expectations are, in fact, conditional expectations and are computed for *given* values of X. A similar statement can be made for the variance of Y_i.

The reader will note that the $\text{Var}(\hat{\beta})$ depends on the spread of the X values about their arithmetic average. In particular, as the spread of the X values increases, $\text{Var}(\hat{\beta})$ decreases.

In a similar manner it can be shown (see Problem 5.14) that

$$E(\hat{\alpha}) = E(\bar{Y} - \hat{\beta}\bar{X}) = \alpha \tag{5.24}$$

and

$$\text{Var}(\hat{\alpha}) = \sigma_{\hat{\alpha}}^2 = \left[\frac{1}{n} + \frac{\bar{X}^2}{\Sigma(X_i - \bar{X})^2}\right]\sigma^2 = \left[\frac{\Sigma X_i^2}{n\Sigma(X_i - \bar{X})^2}\right]\sigma^2 \tag{5.25}$$

Thus $\hat{\alpha}$ is an unbiased estimator of α. The major justification for using the least squares coefficients to estimate the corresponding regression parameters is provided by the Gauss-Markov theorem.

Theorem 5.1 (Gauss-Markov theorem) Within the class of unbiased estimators of α and β that are linear combinations of the Y_i's, the least squares estimators $\hat{\alpha}$ and $\hat{\beta}$ have the smallest variance.

The proof of this theorem is given in most mathematical statistics textbooks, for example, reference [9].

The Gauss-Markov theorem is important because it follows from the relatively weak set of assumptions displayed in Table 5.4. In particular it does not require the specification of the *shape* of the distribution of the Y_i's or equivalently the e_i's. For example, the assumption of normality is not required for the theorem to hold. On the other hand, the Gauss-Markov theorem is applicable to only a restricted class of estimators, those which are both linear combinations of the Y_i's and unbiased. Linear functions of the Y_i's are convenient because they are easy to handle; however, in some situations arguments can be advanced for the use of estimators that are *nonlinear* functions of the Y_i's or are biased (or both). The Gauss-Markov theorem has nothing to say about comparing these estimators with those belonging to the class of linear unbiased estimators.

If we focus attention on linear unbiased estimators, the Gauss-Markov theorem provides us with a theoretical justification for employing the principle of least squares. If, in addition, we are willing to assume that the Y_i's (or the e_i's) are normally distributed, then it can be shown that the least squares estimators are the maximum likelihood estimators.

The Gauss-Markov theorem, in its most general form, is applicable to all linear regression models. In addition the statement made about the equivalence of least squares estimators and maximum likelihood estimators is not restricted to regression models whose expectation component is a straight line.

Assume the e_i's, and hence the Y_i's, are normally distributed; then both $\hat{\alpha}$ and $\hat{\beta}$ are linear combinations of normal variables and hence are normally distributed. In particular using equations (5.24), (5.25), (5.21), and (5.22) we see that $\hat{\alpha}$ has the distribution

$$N\left[\alpha, \frac{\Sigma X_i^2}{n\Sigma(X_i - \bar{X})^2}\sigma^2\right]$$

and $\hat{\beta}$ has the distribution

$$N\left[\beta, \frac{\sigma^2}{\Sigma(X_i - \bar{X})^2}\right]$$

or equivalently

$$\frac{\hat{\alpha} - \alpha}{\sqrt{\Sigma X_i^2/[n\Sigma(X_i - \bar{X})^2]}\,\sigma} = \frac{\hat{\alpha} - \alpha}{\sigma_{\hat{\alpha}}} \sim N(0, 1) \tag{5.26}$$

$$\frac{\hat{\beta} - \beta}{\sigma/\sqrt{\Sigma(X_i - \bar{X})^2}} = \frac{\hat{\beta} - \beta}{\sigma_{\hat{\beta}}} \sim N(0, 1) \tag{5.27}$$

The symbol "$\sim$" is read "has distribution."

In practice $\text{Var}(Y|X)$, which we have been denoting by σ^2, is usually unknown. If the postulated model is correct, then σ^2 can be estimated by s^2, the residual mean square as we have noted in Section 5.3. [It is shown in more advanced books that if the model is correct, $E(s^2) = \sigma^2$.] However, if we then substitute s for σ in equations (5.26) and (5.27), the ratios are no longer normally distributed. In fact, in each case the ratios have t distributions with $(n - 2)$ degrees of freedom, the same number of degrees of freedom associated with the residual sum of squares; that is,

$$t = \frac{\hat{\alpha} - \alpha}{\sqrt{\Sigma X_i^2/[n\Sigma(X_i - \bar{X})^2]}\,s} = \frac{\hat{\alpha} - \alpha}{\hat{\sigma}_{\hat{\alpha}}} \sim t(n - 2) \tag{5.28}$$

$$t = \frac{\hat{\beta} - \beta}{s/\sqrt{\Sigma(X_i - \bar{X})^2}} = \frac{\hat{\beta} - \beta}{\hat{\sigma}_{\hat{\beta}}} \sim t(n - 2) \tag{5.29}$$

The denominators in the t ratios are the estimated standard deviations of $\hat{\alpha}$ and $\hat{\beta}$, respectively, and are commonly referred to as the "standard errors" of $\hat{\alpha}$ and $\hat{\beta}$. The distribution theory given by equations (5.28) and (5.29) can be used to construct confidence intervals and tests of hypotheses for α and β.

Confidence Intervals and Hypothesis Tests for β

Given the relationship specified in equation (5.29), it is easily shown that $100(1 - \gamma)$-percent confidence limits for β are provided by

$$\hat{\beta} \pm t_{\gamma/2}(n - 2)\hat{\sigma}_{\hat{\beta}} = \hat{\beta} \pm t_{\gamma/2}(n - 2)\frac{s}{\sqrt{\Sigma(X_i - \bar{X})^2}} \tag{5.30}$$

where $t_{\gamma/2}(n - 2)$ is the upper $\gamma/2$ percentage point of a t distribution with $(n - 2)$ degrees of freedom.

A test of the null hypothesis H_0: $\beta = \beta_0$ versus the alternative H_a: $\beta \neq \beta_0$ will be rejected or not rejected at the γ level of significance depending on whether β_0 falls outside or inside the confidence interval. Equivalently, a test of this hypothesis may be conducted by evaluating the quantity

$$|t| = \left| \frac{\hat{\beta} - \beta_0}{\hat{\sigma}_{\hat{\beta}}} \right| = \left| \frac{\hat{\beta} - \beta_0}{s / \sqrt{\Sigma (X_i - \bar{X})^2}} \right| \tag{5.31}$$

and comparing it with the percentage point $t_{\gamma/2}(n-2)$. The hypothesis test (at the γ significance level) is then

Reject H_0 if $|t| > t_{\gamma/2}(n-2)$
Do not reject H_0 if $|t| \leq t_{\gamma/2}(n-2)$ $\qquad$ (5.32)

Of particular interest is the null hypothesis H_0: $\beta = 0$. If this hypothesis is not rejected, then the data are indicating that the behavior of Y is not significantly influenced (in the statistical sense) by changes in the independent variable X; that is, there is no relationship between Y and X. In this case the test statistic given by equation (5.31) becomes

$$|t| = \left| \frac{\hat{\beta}}{\hat{\sigma}_{\hat{\beta}}} \right| = \left| \frac{\hat{\beta}}{s / \sqrt{\Sigma (X_i - \bar{X})^2}} \right|$$

and this quantity is known as the "t value for testing H_0: $\beta = 0$ versus H_a: $\beta \neq 0$," or simply the "t value."

Testing the Significance of the Regression

If $E(Y_i | X_i) = \alpha + \beta X_i$ and $\text{Var}(Y_i | X_i) = \sigma^2$ hold for all i, then it can be shown that

$$E(MS_R) = \sigma^2 + \beta^2 \sum (X_i - \bar{X})^2 \tag{5.33}$$

$$E(s^2) = \sigma^2 \tag{5.34}$$

where MS_R and s^2 are the mean square entries in the ANOVA table. It is clear that if $\beta = 0$, the regression mean square (MS_R) and the residual mean square (s^2) are both unbiased estimators of the variance σ^2. Therefore *provided* $\beta = 0$, the ratio

$$F = \frac{MS_R}{s^2} \tag{5.35}$$

should be near unity. Of course, in general, this ratio will not be exactly unity even if $\beta = 0$ since it is subject to sampling fluctuations. On the other hand if $\beta \neq 0$, then equation (5.33) indicates that MS_R will tend to be larger than s^2 and hence the ratio $F = MS_R / s^2$ will tend to be greater than unity. Intuitively it follows that a test of the hypothesis H_0: $\beta = 0$ versus H_a: $\beta \neq 0$ can be conducted by examining the magnitude of the F ratio. In fact it can be shown that if the Y_i's are assumed to be normally distributed, then under the *null hypothesis* H_0: $\beta = 0$,

$$F = \frac{MS_R}{s^2} \sim F(1, n-2) \tag{5.36}$$

where the notation $F(1, n-2)$ is used to indicate an F distribution with 1 and $(n-2)$ degrees of freedom. The numbers of degrees of freedom defining the F distribution are simply the numbers of degrees of freedom associated with the regression and residual sum of squares, respectively.

Therefore F values within the bulk of the F distribution with 1 and $(n-2)$ degrees of freedom tend to reinforce H_0, whereas large values of F tend to discredit H_0 and reinforce H_a. In summary, a test of H_0: $\beta = 0$ versus H_a: $\beta \neq 0$ can be conducted by evaluating the F ratio (using the entries in the ANOVA table) and comparing this ratio with the appropriate percentage point of the $F(1, n-2)$ distribution. The appropriate test (at the γ significance level) is then

$$\begin{aligned} &\text{Reject } H_0 \text{ if } F > F_\gamma(1, n-2) \\ &\text{Do not reject } H_0 \text{ if } F \leq F_\gamma(1, n-2) \end{aligned} \tag{5.37}$$

where $F_\gamma(1, n-2)$ denotes the upper γ percentage point of an F distribution with 1 and $(n-2)$ degrees of freedom.

The reader will notice that for the simple linear regression model, the null hypothesis H_0: $\beta = 0$ versus the alternative H_a: $\beta \neq 0$ can be tested in two different ways. In fact, the tests based on the t value and the F ratio are equivalent. Using the relationships between the χ^2, t, and F distributions mentioned in Section 2.4, it follows that

$$t^2(\nu) = F(1, \nu) \tag{5.38}$$

The *square* of a t random variable with ν degrees of freedom is distributed as an F random variable with 1 and ν degrees of freedom. Since any t p.d.f. is symmetric about the origin, there is a *pair* of values of the t random variable that leads to the same value of an F random variable—say, F_0—namely, t_0 and $-t_0$. Thus the following probability statements are equivalent.

$$\Pr[F > F_0] = \Pr[t^2 > t_0^2] = \Pr[t < -t_0 \quad \text{or} \quad t > t_0] = 2\Pr[t > t_0] \tag{5.39}$$

That is, the area under the F p.d.f. to the right of F_0 is precisely twice the area under the t p.d.f. to the right of t_0. The probability statements are indicated schematically in Figure 5.7. If we now let $\nu = n-2$ and $t_0 = t_{\gamma/2}(n-2)$ (this is the value of a t random variable that leaves a proportion $\gamma/2$ of the distribution to the right of it), then using equation (5.39) it follows that

$$\begin{aligned} \gamma &= \Pr[|t| > t_{\gamma/2}(n-2)] = \Pr[t^2 > t_{\gamma/2}^2(n-2)] \\ &= \Pr[F > F_\gamma(1, n-2)] \end{aligned}$$

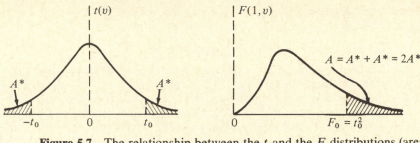

Figure 5.7 The relationship between the t and the F distributions (area = A)

The significance levels associated with the tests in equations (5.32) and (5.37) are the same provided we compare the t value with the $\gamma/2$ percentage point of a t distribution with $(n-2)$ degrees of freedom and the F ratio with the γ percentage point of an F distribution with 1 and $(n-2)$ degrees of freedom. [The reader can easily verify that $F_\gamma(1, n-2) = t^2_{\gamma/2}(n-2)$ using the F and t tables in the Appendix.]

Finally, we point out that the t test and the F test that we have described in this section are equivalent tests only for the simple linear regression model. In more general situations, the usual F ratio calculated from the entries in the ANOVA table is used to test the significance of *all* independent variable terms as a group, whereas the t values are used to examine the significance of *individual* independent variable terms.

EXAMPLE 5.5

To construct a 90-percent confidence interval for β in our field corn seed example we proceed as follows:

$$\hat{\sigma}_{\hat{\beta}}^2 = \frac{s^2}{\Sigma(X_i - \bar{X})^2} = \frac{79.17}{1.62} = 48.87$$

(using results from Tables 5.1 and 5.3) and thus

$$\hat{\sigma}_{\hat{\beta}} = \sqrt{48.87} = 6.99$$

Since $\gamma = 0.10$,

$$t_{\gamma/2}(n-2) = t_{0.05}(6) = 1.943 \qquad \text{(Appendix Table E)}$$

$$\hat{\beta} = -49.52$$

[previously denoted by b, see equation (5.12)]. Therefore 90-percent confidence limits are given by $-49.52 \pm (1.943)(6.99)$ [using equation (5.30)] yielding the interval $(-63.10, -35.94)$; that is, the true value β lies in the interval $(-63.10, -35.94)$ with 90 percent confidence.

A test of the hypothesis $H_0: \beta = 0$ versus $H_a: \beta \neq 0$ at the γ level of significance is provided by comparing the t value

$$|t| = \left| \frac{\hat{\beta}}{\hat{\sigma}_{\hat{\beta}}} \right|$$

with the t percentage point, $t_{\gamma/2}(n-2)$. Using the information above, if $\gamma = 0.10$, $t_{0.05}(6) = 1.943$, $|t| = |-49.52/6.99| = 7.08$ and since

$$|t| = 7.08 > t_{0.05}(6) = 1.943$$

we reject the null hypothesis at the 10-percent level of significance; that is, there appears to be a relationship between sales and price.

Interpreting the 90-percent confidence interval calculated above as a set of plausible values of β would, of course, lead us to the same result since $\beta = 0$ is not included in the interval. Additionally we may test $H_0: \beta = 0$ versus $H_a: \beta \neq 0$ at the γ level of significance by comparing the F ratio

$$F = \frac{MS_R}{s^2}$$

from the ANOVA table with the F percentage point $F_\gamma(1, n-2)$. For the current example (see Table 5.3) with $\gamma = 0.10$,

$$F = \frac{3972.61}{79.17} = 50.18 > F_{0.10}(1, 6) = 3.78$$

(see [8] for an extensive table of F percentage points), and we reject $H_0: \beta = 0$. We note that within rounding error,

$$t^2 = \frac{\hat{\beta}^2}{\hat{\sigma}_{\hat{\beta}}^2} = (7.08)^2 = 50.18 = F$$

and

$$t_{0.05}^2(6) = (1.943)^2 = 3.78 = F_{0.10}(1, 6)$$

so that the t test and the F test are equivalent.

Confidence Intervals and Hypothesis Tests for α

The results for inferences about α parallel those for β and will be quoted without a complete discussion. The reader who wishes to pursue the details is referred to reference [7].

Using the result of equation (5.25) and substituting s^2 for σ^2, we have,

$$\hat{\sigma}_{\hat{\alpha}}^2 = \left[\frac{\Sigma X_i^2}{n \Sigma (X_i - \bar{X})^2} \right] s^2 \tag{5.40}$$

and

$$\hat{\sigma}_{\hat{\alpha}} = \sqrt{\frac{\Sigma X_i^2}{n \Sigma (X_i - \bar{X})^2}}\, s \tag{5.41}$$

Assuming the Y_i's to be normally distributed and using equation (5.28), the ratio $t = (\hat{\alpha} - \alpha)/\hat{\sigma}_{\hat{\alpha}}$ is distributed as a t random variable with $(n-2)$ degrees of freedom. Therefore $100(1-\gamma)$-percent confidence limits are given by

$$\hat{\alpha} \pm t_{\gamma/2}(n-2)\hat{\sigma}_{\hat{\alpha}} = \hat{\alpha} \pm t_{\gamma/2}(n-2)\sqrt{\frac{\Sigma X_i^2}{n \Sigma (X_i - \bar{X})^2}}\, s \tag{5.42}$$

where $t_{\gamma/2}(n-2)$ is the upper $\gamma/2$ percentage point of a t distribution with $(n-2)$ degrees of freedom.

A test of the hypothesis H_0: $\alpha = \alpha_0$ versus the alternative hypothesis H_a: $\alpha \neq \alpha_0$ at the γ level of significance will be rejected or not rejected depending on whether α_0 falls outside or inside the confidence interval. Equivalently, the quantity

$$|t| = \left| \frac{\hat{\alpha} - \alpha_0}{\hat{\sigma}_{\hat{\alpha}}} \right| = \left| \frac{\hat{\alpha} - \alpha_0}{\sqrt{\Sigma X_i^2 / [n \Sigma (X_i - \bar{X})^2]} \, s} \right| \tag{5.43}$$

may be compared with the percentage point $t_{\gamma/2}(n-2)$ using the rule given in equation (5.32). For the particular null hypothesis H_0: $\alpha = 0$, the expression in equation (5.43) reduces to the "t value"

$$|t| = \left| \frac{\hat{\alpha}}{\hat{\sigma}_{\hat{\alpha}}} \right| = \left| \frac{\hat{\alpha}}{\sqrt{\Sigma X_i^2 / [n \Sigma (X_i - \bar{X})^2]} \, s} \right| \tag{5.44}$$

Although the parameter α is ordinarily of less interest than the parameter β, there are situations in which we might wish to make inferences about α. For example, we may be interested in relating the "total cost" to the number of items produced. Suppose total cost is taken to be sum of a "fixed cost" that would be incurred regardless of the number of items produced and an incremental "variable cost" that depends directly on the number of items produced. If a simple linear regression model is postulated to account for the relationship between observed total cost Y and number of units produced X, then α represents the fixed cost and β represents the additional variable cost per unit increase in X. In this context estimation of α may be an important consideration of the analysis.

Predicting Future Values of Y

Suppose that the least squares line has been determined and the data indicate that β is significantly different from zero. Of what use is this information? In addition to simply indicating the nature of the relationship between Y and X, the least squares function may be used to predict the *actual* value of Y, say, Y_0, corresponding to a new setting of the X variable, say, X_0.

A point estimate of Y_0 is given by evaluating the least squares function at the point X_0. Thus

$$\hat{Y}_0 = \hat{\alpha} + \hat{\beta} X_0 = \bar{Y} + \hat{\beta}(X_0 - \bar{X}) \tag{5.45}$$

However it is clear that unless we are extremely lucky, $\hat{Y}_0$ will not coincide with the actual value Y_0 when it becomes available since $\hat{Y}_0$ is constructed from an *estimated* relationship. There are two sources of variability (or uncertainty) associated with our prediction:

1. The variability of the least squares line about the regression line, represented by

$$\text{Var}(\hat{Y}_0) = \text{Var}(\hat{\alpha} + \hat{\beta}X_0) = \text{Var}[\bar{Y} + \hat{\beta}(X_0 - \bar{X})] = \sigma^2_{\hat{Y}_0}$$

2. The inherent variability in Y_0 itself, represented by

$$\text{Var}(Y_0|X_0) = \sigma^2$$

For two linear combinations of uncorrelated random variables of the form

$$W_1 = a_1 Y_1 + a_2 Y_2 + \cdots + a_n Y_n$$

$$W_2 = b_1 Y_1 + b_2 Y_2 + \cdots + b_n Y_n$$

where $\text{Var}(Y_i) = \sigma^2$ for all i,

$$\text{Cov}(W_1, W_2) = (a_1 b_1 + a_2 b_2 + \cdots + a_n b_n)\sigma^2 \qquad (5.46)$$

If we set $a_i = 1/n$, for all i, so that $W_1 = \bar{Y}$ and let $b_i = (X_i - \bar{X})/\Sigma(X_i - \bar{X})^2$ so that $W_2 = \hat{\beta}$; then using equation (5.46), $\text{Cov}(\bar{Y}, \hat{\beta}) = 0$. Consequently,

$$\sigma^2_{\hat{Y}_0} = \text{Var}(\hat{\alpha} + \hat{\beta}X_0) = \text{Var}[\bar{Y} + \hat{\beta}(X_0 - \bar{X})]$$

$$= \text{Var}(\bar{Y}) + (X_0 - \bar{X})^2 \text{Var}(\hat{\beta}) + 2(X_0 - \bar{X}) \text{Cov}(\bar{Y}, \hat{\beta})$$

$$= \text{Var}(\bar{Y}) + (X_0 - \bar{X})^2 \text{Var}(\hat{\beta}) = \left(\frac{1}{n} + \frac{(X_0 - \bar{X})^2}{\Sigma(X_i - \bar{X})^2}\right)\sigma^2 \qquad (5.47)$$

Combining the two sources of variability indicated above, the variability associated with the predicted value of an *individual* observation, say σ^2_{pred}, is

$$\sigma^2_{\text{pred}} = \sigma^2 + \sigma^2_{\hat{Y}_0} = \sigma^2 + \left[\frac{1}{n} + \frac{(X_0 - \bar{X})^2}{\Sigma(X_i - \bar{X})^2}\right]\sigma^2$$

$$= \left[1 + \frac{1}{n} + \frac{(X_0 - \bar{X})^2}{\Sigma(X_i - \bar{X})^2}\right]\sigma^2 \qquad (5.48)$$

This variance can be estimated by inserting s^2 for σ^2 and, hence the estimated standard deviation (or standard error) of the prediction is given by

$$\hat{\sigma}_{\text{pred}} = \sqrt{\left[1 + \frac{1}{n} + \frac{(X_0 - \bar{X})^2}{\Sigma(X_i - \bar{X})^2}\right]}\, s \qquad (5.49)$$

Now it is clear that if the number of observations n is large and the X values are disperse relative to their arithmetic average [so that $\Sigma(X_i - \bar{X})^2$ is large], the quantity under the radical sign will be approximately unity, and hence $\hat{\sigma}_{\text{pred}} \doteq s$, which is why s is called the "standard error of the estimate of Y," or simply the "standard error of the estimate."

If the Y_i's are assumed to be normally distributed, then a 100

$(1 - \gamma)$-percent prediction interval for the value Y_0 is given by[f]

$$\hat{Y}_0 \pm t_{\gamma/2}(n-2)\hat{\sigma}_{\text{pred}} = \hat{Y}_0 \pm t_{\gamma/2}(n-2)\sqrt{\left[1 + \frac{1}{n} + \frac{(X_0 - \bar{X})^2}{\Sigma(X_i - \bar{X})^2}\right]} s \quad (5.50)$$

where $t_{\gamma/2}(n-2)$ is the upper $\gamma/2$ percentage point of a t distribution with $(n-2)$ degrees of freedom.

EXAMPLE 5.6

Suppose the field corn seed supplier decides to charge $X_0 = \$2.20$ per bushel. What can he say about sales?

Point estimate: $\hat{Y}_0 = 186.34 - 49.52(2.20)$ [using equation (5.12)]

$$= 77.40$$

95-percent prediction
   interval: $t_{0.025}(6) = 2.447$

$$\hat{\sigma}_{\text{pred}} = \sqrt{1 + \frac{1}{8} + \frac{(2.20 - 2.16)^2}{1.62}} \sqrt{79.17}$$

$$= 9.44$$

using equation (5.49) and information from Tables 5.1 and 5.3. Therefore a 95-percent prediction interval for Y_0 is $77.40 \pm 2.447(9.44)$, or $[54.30, 100.50]$.

Dangers of Extrapolation

Frequently regression models are suggested by the data themselves (from scatter diagrams) rather than theory, and for convenience the expectation functions are taken to be simple functions, linear in the parameters. If this is the case, then clearly the regression models must be viewed as a useful approximation to reality only *within the range of the available data*, that is, within the observed range of the independent or controllable variable. We have referred to this range as the "fitting region." Moreover, even in cases in which the regression model is suggested by theory and purports to explain the precise mechanism by which the dependent variable and independent variable are related, the model is rarely completely correct, and when we try to verify it experimentally (by collecting data), there is usually some discrepancy which is then accounted for by "experimental error." Or to introduce another possibility, even though a theoretically based regression model is correct (the experimental error is negligible), it may only be correct for a *particular* range of an independent variable, and outside of this range

[f] Once again the reader is asked to accept this statement without proof. It follows from distribution theory that is beyond the scope of this book.

the theory and hence the model may no longer be adequate. (For example, the nature of the relationship between consumption and income may be different for high incomes than for low incomes.)

Therefore extrapolating the least squares function outside of the fitting region can be dangerous. We have already indicated that outside this region the postulated relationship may no longer be valid. This situation is illustrated schematically in Figure 5.8. If we were to use the least squares line to predict values of Y beyond the range of X [represented by the interval (X_{min}, X_{max})] when the true relationship between Y and X is quadratic in X, we would expect to get misleading results.

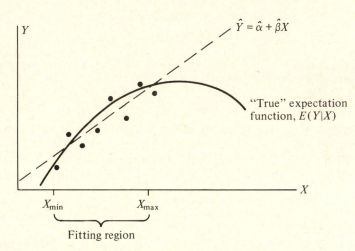

Figure 5.8   Extrapolating a least squares function

In addition to the possibility of an invalid model, there is a statistical risk associated with predictions of future values as we move from the center of the fitting region. This risk is represented by the decrease in precision associated with the prediction of Y_0 as X_0 moves away from $\bar{X}$. From equation (5.49) it is clear that $\hat{\sigma}_{pred}$ increases as the distance $(X_0 - \bar{X})$ increases. Therefore for a fixed confidence level, the width of the prediction interval for Y_0 increases as X_0 moves away from $\bar{X}$ [see equation (5.50)]. This is true even if the assumptions underlying the regression model hold exactly. The situation is illustrated in Figure 5.9 where, for a fixed confidence level, upper and lower prediction limits corresponding to a continuum of values X_0 from X_{min} to X_{max} are plotted. These prediction limits produce the band shown in the figure.

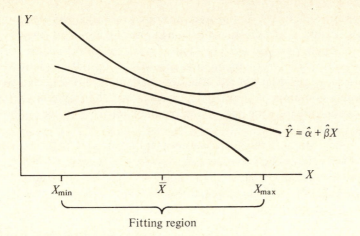

Figure 5.9 Prediction band for future Y values

5.6 THE ANALYSIS OF RESIDUALS

The inferential results of the last section depend on one or more of the assumptions we have spelled out in Table 5.4. If these assumptions are not satisfied, the interpretations we make assuming them to be true may be invalid. The reader will recall that the assumptions in terms of the error e_i are

1. $E(e_i) = 0$, $i = 1, 2, \ldots, n$
2. $\text{Var}(e_i) = \sigma^2$, $i = 1, 2, \ldots, n$ (5.51)
3. $\text{Cov}(e_i, e_j) = 0$, $i \neq j; \ i, j = 1, 2, \ldots, n$

In addition it is generally assumed that the errors are normally distributed. Using the identity

$$Y_i = \hat{Y}_i + (Y_i - \hat{Y}_i), \qquad i = 1, 2, \ldots, n \qquad (5.52)$$

it is clear that if the *regression model* is correct, the fitted value $\hat{Y}_i$ is an estimate of the expected value $E(Y_i|X_i)$ and the residual $(Y_i - \hat{Y}_i)$ is an estimate of the error component e_i. On the other hand, if the regression function is not correct—for example, if $E(Y_i|X_i) = \alpha + \beta_1 X_i + \beta_2 X_i^2$ instead of the postulated $E(Y_i|X_i) = \alpha + \beta X_i$—then $\hat{Y}_i = \hat{\alpha} + \hat{\beta} X_i$ is not an estimate of the "true" expected value.

In summary if the model is correct, we can think of the residuals

$$\hat{e}_i = Y_i - \hat{Y}_i, \qquad i = 1, 2, \ldots, n$$

as representing the values of the error components $e_i, i = 1, 2, \ldots, n$, which have actually occurred. If the assumptions are valid, the behavior of the residuals should reinforce these assumptions (or not deny them).

As we have pointed out, residuals may be analyzed by calculating summary statistics and by examining residual plots. In this chapter we shall concentrate on residual plots. These are particularly informative for regression models containing a single independent variable. In the next chapter we shall concentrate on the use of important summary statistics in residual analysis.

The examination of residuals is important because it is the residuals that are used to criticize the postulated model. As we saw in Chapter 4, Analysis of Variance, residual analysis is a necessary part of any statistical model building endeavor. We will consider this topic once again in the chapters on time series analysis.

Overall Plots

Summarizing the set of assumptions imposed on the error term, we can state that $e_i \sim N(0, \sigma^2)$, $i = 1, 2, \ldots, n$, and that the e_i's are independent. Treating the residuals $\hat{e}_1, \hat{e}_2, \ldots, \hat{e}_n$ as realizations of the corresponding error components and substituting s^2 for σ^2, the residuals should look like random items drawn from an $N(0, s^2)$ population, or equivalently the "standardized" residuals, $\hat{e}_1/s, \hat{e}_2/s, \ldots, \hat{e}_n/s$, should look like drawings from a $N(0, 1)$ population. The standardized residuals can be plotted along a single dimension or plotted on normal probability paper. If normal probability paper is used, a plot of $(i - \frac{1}{2})/n$ versus $\hat{e}_i/s$, $i = 1, 2, \ldots, n$, ordered according to magnitude should be approximately a straight line. Figure 5.10 gives both plots for the standardized residuals from our field corn seed example. The residual plots in Figure 5.10 indicate that there is no reason to doubt the normality of the error component.

Other Plots

Other methods of plotting exist that enable us to pinpoint specific violations of the assumptions. Useful ways of plotting the residuals are

1. In a time sequence (if the chronological order is known).
2. Against the fitted values $\hat{Y}_i$.
3. Against the independent variable values X_i.
4. In any way that is sensible for the particular problem at hand.

In general if the assumptions hold, a plot of the residuals (as the ordinate) versus any of the variables, time, $\hat{Y}$, or X (as the abscissa) should give the

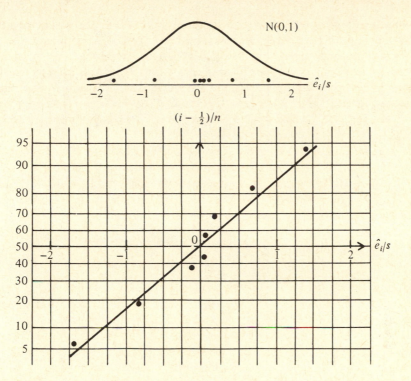

Figure 5.10   Plots of standardized residuals, field corn seed example

impression of a horizontal band of the following form:

However, if one or more of the assumptions are violated, these residual plots frequently display one of the patterns pictured in Figure 5.11. We shall refer to the patterns labeled (a), (b), and (c) in the figure in our discussion of plots (1), (2), and (3) above.

If the residuals are plotted against time and a horizontal band is evident, then this implies that the relationship between Y and X is probably not changing over time; that is, there is no "time effect" that is not accounted for

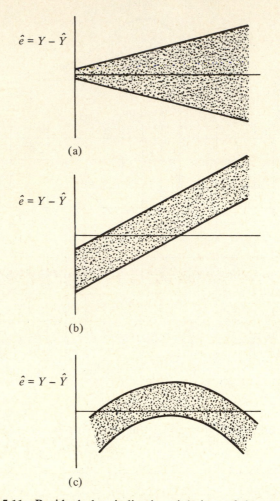

$$\hat{e} = Y - \hat{Y}$$

(a)

$$\hat{e} = Y - \hat{Y}$$

(b)

$$\hat{e} = Y - \hat{Y}$$

(c)

Figure 5.11 Residual plots indicating violations of the regression assumptions

in the model. On the other hand, if we let time be the abscissa in the plots in Figure 5.11, then

1. Pattern (a) implies that the variance σ^2 is not constant,[g] indicating the need for a weighted least squares analysis or a transformation of the observations Y_i (for example, considering $\ln Y_i$ instead of Y_i, see Chapter 7).

---

[g] Nonconstant variance is frequently referred to as "heteroscedasticity."

2. Pattern (b) implies that a first-order term in time should have been included in the regression function.

3. Pattern (c) implies that first- and second-order terms in time should have been included in the regression function.

If the residuals are plotted against the fitted values, then

1. Pattern (a) implies that the variance σ^2 is not constant (heteroscedasticity) indicating the need for weighted least squares or a transformation of the observations Y_i.

2. Pattern (b) implies an error in the analysis (systematic departure from the estimated relationship) or the need for an intercept term α if one has been omitted.

3. Pattern (c) implies that the regression function is inadequate—that is, additional higher-order terms are needed—or that the observations Y_i should be transformed before analysis.

Figure 5.12 is a plot of the residuals versus the fitted values for our field corn seed example. Although it is difficult to argue that the pattern is a horizontal band, none of the violations pictured in Figure 5.11 is strongly indicated. Therefore we regard the evidence in the figure as inconclusive. In fact, the residuals associated with the repeat points can be used to test for "lack of fit" (the need for additional terms in the regression function) as we shall see in Supplement 5A. Using the results given in that supplement along with the

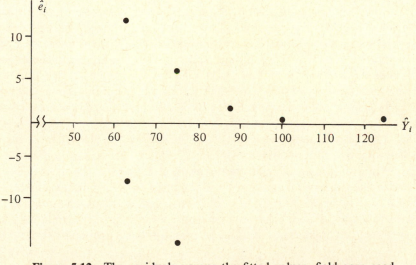

Figure 5.12   The residuals versus the fitted values, field corn seed example

evidence already presented, there is no reason to believe that the assumptions underlying our model linking sales to price are violated.

Finally, if the residuals are plotted against the values of the independent variable,

1.   Pattern (a) implies a nonconstant variance σ^2 (heteroscedasticity) and the need for weighted least squares or a transformation of the observations Y_i.
2.   In general, pattern (b) implies an error in the calculations if a first-order term in X is already in the model—if not, one is needed.
3.   Pattern (c) implies that the regression function is inadequate and that additional higher-order terms in X are needed.

5.7   CORRELATION ANALYSIS

Simple linear regression analysis enables us to study the linear relationship between two variables. Our discussion in this section will concentrate on a *measure* of the linear association between two random variables and the attendant problems of inference. Specifically, we shall discuss correlation, a notion that was introduced in Chapter 2. In the remaining sections of this chapter, we shall consider the relationships between correlation and simple linear regression, introduce additional correlation material needed, in particular, for the time series chapters, and finally present the "rank correlation coefficient," another commonly employed measure of association.

We saw in Chapter 2 that if X and Y are two random variables with a bivariate probability distribution, $\text{Cov}(X, Y)$ is a measure of the linear association between the variables. The covariance is large and positive if, given a large (small) value of X, the probability is high that the value of Y will also be large (small). On the other hand, the covariance is large and negative if, given a large (small) value of X, there is a high probability that the value of Y will be small (large). The same statements can be made about the more easily interpreted correlation coefficient ρ, which is simply the covariance of the standardized variables $(X - \mu_X)/\sigma_X$ and $(Y - \mu_Y)/\sigma_Y$.

The correlation coefficient is the most frequently employed measure of linear association in a bivariate population. This is due to the fact that ρ is invariant under changes of scale and location in X and Y—that is, it does not depend on the orders of magnitudes and the units of the random variables. In addition, for bivariate *normal* random variables, $\rho = 0$ if and only if the random variables X and Y are independent.[h] The bivariate normal distribution is often used (sometimes without justification) as a probability model for

---

[h] Although it is always true that the correlation is zero for independent random variables, the converse is *not* true in general. The converse is true, however, if the joint distribution of X and Y is bivariate normal.

bivariate populations, and therefore in these instances the correlation coefficient is an important measure of relationship.

There are many *sample* measures of association. Some of these measures can be used to make inferences about the underlying population correlation coefficient. Some of these measures are used primarily to describe the association in the sample and they do not have any direct relevance for an underlying population.

Inferences about ρ for a Bivariate Normal Population

Let $(X_i, Y_i), i = 1, 2, \ldots, n$, denote pairs of random variables representing a random sample from a bivariate normal population—that is, a population described by the density function given in equation (2.39). In terms of the random variables $(X_i, Y_i), i = 1, 2, \ldots, n$, the *sample* correlation coefficient is defined as

$$r = \frac{\bar{s}_{XY}}{\bar{s}_X \bar{s}_Y} = \frac{\sum_{i=1}^{n} (X_i - \bar{X})(Y_i - \bar{Y})}{\sqrt{\sum_{i=1}^{n} (X_i - \bar{X})^2 \sum_{i=1}^{n} (Y_i - \bar{Y})^2}} \tag{5.53}$$

where

$$\bar{s}_{XY} = \frac{1}{n-1} \sum_{i=1}^{n} (X_i - \bar{X})(Y_i - \bar{Y})$$

and

$$\bar{s}_X^2 = \frac{1}{n-1} \sum_{i=1}^{n} (X_i - \bar{X})^2$$

$$\bar{s}_Y^2 = \frac{1}{n-1} \sum_{i=1}^{n} (Y_i - \bar{Y})^2$$

are the sample covariance and sample variances, respectively.

The definition of the sample correlation coefficient r is analogous to the definition of the population correlation coefficient ρ. Like the population quantity, $-1 \le r \le 1$.

The sample covariance and sample correlation coefficients are measures of the linear association or dependence in the data. Figure 5.13 illustrates the behavior of $\bar{s}_{XY}$ and r for various sets of observation pairs (X_i, Y_i), $i = 1, 2, \ldots, n$ (realized values of the corresponding random variables). If the pairs (X_i, Y_i) tend to lie along a straight line with a positive slope, $\bar{s}_{XY} > 0$ and hence $r > 0$. If the observations tend to lie along a straight line with a negative slope, $\bar{s}_{XY} < 0$ and hence $r < 0$. Figure 5.13(c) and (d) illustrates cases in which $\bar{s}_{XY}$ (and hence r) is approximately zero.

A value of $\bar{s}_{XY}$ (or r) close to zero indicates that either (a) there is no discernible dependence between the observations X_i and $Y_i, i = 1, 2, \ldots, n$, or (b) there is no *linear* dependence between the observed values of X and the observed values of Y. Thus $\bar{s}_{XY}$, or r, is a measure of the *linear* association, or *linear* dependence, between two sets of observations.

The sample correlation coefficient r can be used to make inferences about

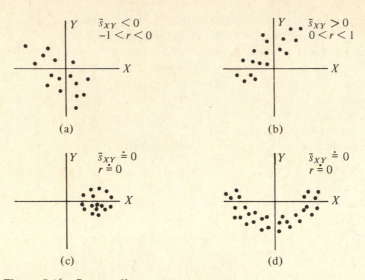

Figure 5.13 Scatter diagrams and the sample covariance

the value of the population correlation coefficient

$$\rho = \frac{\text{Cov}(X, Y)}{\sigma_X \sigma_Y}$$

In order to proceed, however, it is necessary to investigate the behavior of r in repeated samples; that is, we need to know its sampling distribution. If the joint distribution of the random variable pairs (X_i, Y_i) is the bivariate normal distribution with parameters μ_X, μ_Y, σ_X^2, σ_Y^2, and ρ, then the probability distribution of r can be derived theoretically. This probability distribution is, in general, very complicated, but depends only on the quantities n and ρ. The distributions of r for $\rho = 0$ and $\rho = 0.8$ when $n = 8$ are pictured in the top half of Figure 5.14. It is clear from the figure that the shape of the probability distribution of r, say $f(r)$, changes with the value of ρ and when $|\rho|$ is large can be decidedly asymmetrical and nonnormal in appearance. Consequently the precision of r as an estimator of ρ is difficult to determine. It is also true that r is a biased estimator of ρ for $|\rho| \neq 0$; however, for moderate sample sizes the bias is ordinarily negligible.

Although the distribution of r is very complicated, it is possible to define new random variables directly related to r whose distributions are familiar. These distributions can then be used to make inferences about ρ. In particular if $\rho = 0$, the distribution of

$$t = \frac{r\sqrt{(n-2)}}{\sqrt{1-r^2}} \tag{5.54}$$

is the t distribution with $n - 2$ degrees of freedom. This distribution can be

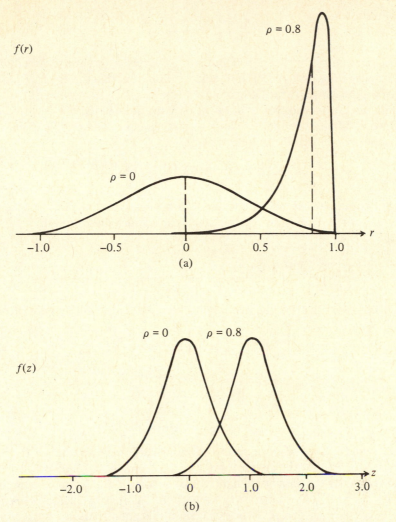

Figure 5.14 The distribution of r (top) and Z (bottom) for $\rho = 0$ and $\rho = 0.8$ when $n = 8$

used to test the "significance" of the observed sample correlation coefficient. Moreover, for any value of ρ the distribution of

$$Z = \frac{1}{2} \ln \frac{1+r}{1-r} \tag{5.55}$$

is *approximately* normal with mean

$$\mu_z = \frac{1}{2} \ln \frac{1+\rho}{1-\rho} \tag{5.56}$$

and variance

$$\sigma_Z^2 = \frac{1}{n-3} \qquad (5.57)$$

The distributions of Z for $\rho = 0$ and $\rho = 0.8$ when $n = 8$ are pictured in the bottom half of Figure 5.14. It is immediately apparent that the distribution of Z is "normal looking" regardless of the value of ρ.

Confidence intervals for ρ may be constructed using the approximate distribution of the random variable Z. Using equations (5.55), (5.56), and (5.57), the standardized random variable

$$Z^* = \frac{Z - \mu_Z}{\sigma_Z} = \frac{\frac{1}{2}\ln\left[(1+r)/(1-r)\right] - \frac{1}{2}\ln\left[(1+\rho)/(1-\rho)\right]}{1/\sqrt{n-3}}$$

has approximately a standard normal distribution, and thus if $z_{\gamma/2}$ denotes the upper $\gamma/2$ percentage point of the standard normal distribution,

$$(1-\gamma) = \Pr[-z_{\gamma/2} \le Z^* \le z_{\gamma/2}]$$

$$= \Pr\left[-z_{\gamma/2} \le \frac{\frac{1}{2}\ln\left[(1+r)/(1-r)\right] - \frac{1}{2}\ln\left[(1+\rho)/(1-\rho)\right]}{1/\sqrt{n-3}} \le z_{\gamma/2}\right]$$

After some manipulation the probability statement above may be written as[i]

where

$$(1-\gamma) = \Pr\left[\frac{e^a - 1}{e^a + 1} \le \rho \le \frac{e^b - 1}{e^b + 1}\right] \qquad (5.58)$$

$$a = \ln\frac{1+r}{1-r} - \frac{2z_{\gamma/2}}{\sqrt{n-3}}$$

$$b = \ln\frac{1+r}{1-r} + \frac{2z_{\gamma/2}}{\sqrt{n-3}} \qquad (5.59)$$

that is, a and b depend only on the random variable r and hence the probability is $(1-\gamma)$ that the random interval $[(e^a - 1)/(e^a + 1), (e^b - 1)/(e^b + 1)]$ contains ρ. After the data are in hand the quantities a and b can be computed, and the interval $[(e^a - 1)/(e^a + 1), (e^b - 1)/(e^b + 1)]$ is a $100(1-\gamma)$-percent confidence interval for ρ. It is clear that the construction of confidence intervals for ρ can be tedious even with the normal approximation. In fact, charts have been developed that allow us to read off confidence intervals for ρ directly given values for r and n. The appropriate chart for constructing 95-percent confidence intervals is given in Figure 5.15.

To illustrate the use of the chart suppose we wish to construct a 95-percent confidence interval for ρ when $n = 15$ pairs of observations yields a sample correlation coefficient of $r = 0.55$. In Figure 5.15 we locate the *pair* of curves labeled $n = 15$ and notice that they contain part of the vertical line through $r = 0.55$. The part of the vertical line contained between the curves represents the 95-percent confidence interval for ρ, and reading across to the vertical axis

---

[i] In the discussion below, e is the base of the natural logarithms. It should not be confused with the error term in the regression model.

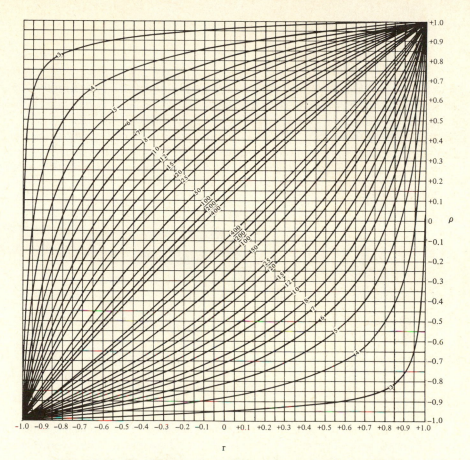

Figure 5.15 The 95-percent confidence bands for the population correlation coefficient ρ, given the sample correlation coefficient r

we may express this interval numerically. For our example the 95-percent confidence interval for ρ is the interval $(0.05, 0.82)$. The considerable length of the confidence interval demonstrates the fact that r can be a very imprecise estimator of ρ for small and moderate sample sizes, and therefore some care must be exercised in drawing conclusions about the parent population.

From the viewpoint of hypothesis testing suppose we specify a value for ρ and then ask if the observed value of r is compatible with this hypothesis. If the hypothesis is $H_0: \rho = 0$, then an exact test of H_0 is provided by calculating the t ratio displayed in equation (5.54) and comparing it with the appropriate percentage point of a t distribution with $n - 2$ degrees of freedom. For example, at the γ level of significance, a test of the hypothesis $H_0: \rho = 0$

versus the alternative H_a: $\rho \neq 0$ is provided by

Reject H_0 if $|t| > t_{\gamma/2}(n-2)$
Do not reject otherwise

where $t_{\gamma/2}(n-2)$ is the upper $\gamma/2$ percentage point of the t distribution with $n-2$ degrees of freedom. The extention to one-sided tests is handled in the usual manner.

On the other hand if H_0: $\rho = \rho_0$ is the hypothesis to be tested versus the alternative H_a: $\rho \neq \rho_0$, then an approximate test is provided by referring the observed value of Z defined in equation (5.55) to a normal distribution with mean

$$\mu_Z = \frac{1}{2} \ln \frac{1 + \rho_0}{1 - \rho_0}$$

and variance $\sigma_Z^2 = 1/(n-3)$. Equivalently we can refer the standardized variate

$$Z^* = \frac{\frac{1}{2} \ln [(1+r)/(1-r)] - \frac{1}{2} \ln [(1+\rho_0)/(1-\rho_0)]}{1/\sqrt{n-3}}$$

to a standard normal distribution. Thus at the γ level of significance we

Reject H_0 if $|Z^*| > z_{\gamma/2}$
Do not reject otherwise

where $z_{\gamma/2}$ is the upper $\gamma/2$ percentage point of the standard normal distribution.

EXAMPLE 5.7

Consider the five pairs of observations representing disposable income and housing demand introduced in Example 5.2. Suppose we regard both disposable income X and housing demand Y as random variables whose joint distribution in the population is given by a bivariate normal density (an assumption that should be checked in practice), and we are interested in the value of the unknown population correlation coefficient ρ. The data are reproduced in the first two columns of Table 5.5. The additional columns of the table summarize the calculations leading to the sample correlation coefficient r. Therefore

$$r = \frac{\Sigma(X_i - \bar{X})(Y_i - \bar{Y})}{\sqrt{\Sigma(X_i - \bar{X})^2 \Sigma(Y_i - \bar{Y})^2}} = \frac{0.18}{\sqrt{(0.20)(0.332)}} = 0.70$$

Use of the chart in Figure 5.15 for $n = 5$, $r = 0.70$ gives a 95-percent confidence interval for ρ of $(-0.39, 0.96)$. The interval estimate is not very precise—that is, the width is large—but this is due to the small sample size.

A test of the hypothesis H_0: $\rho = 0$ (or, equivalently, the random variables X and Y are independent since we are assuming bivariate normality) versus the alternative H_a: $\rho > 0$, say, can be conducted at the 5-percent level of significance by comparing

$$t = \frac{r\sqrt{n-2}}{\sqrt{1-r^2}} = \frac{0.70\sqrt{3}}{\sqrt{1-(0.70)^2}} = 1.698$$

with the upper 0.05 percentage point of a t distribution with three degrees of freedom, $t_{0.05}(3) = 2.353$. Since

Table 5.5

X_i	Y_i	$X_i - \bar{X}$	$Y_i - \bar{Y}$	$(X_i - \bar{X})(Y_i - \bar{Y})$	$(X_i - \bar{X})^2$	$(Y_i - \bar{Y})^2$
6.8	0.8	−0.3	−0.26	0.078	0.9	0.0676
7.0	1.2	−0.1	0.14	−0.014	0.01	0.0196
7.1	0.9	0	−0.16	0	0	0.0256
7.2	0.9	0.1	−0.16	−0.016	0.01	0.0256
7.4	1.5	0.3	0.44	0.132	0.09	0.1936
35.5	5.3			0.180	0.20	0.3320

$$\bar{X} = \frac{35.5}{5} = 7.1, \ \bar{Y} = \frac{5.3}{5} = 1.06.$$

$$t = 1.698 < t_{0.05}(3) = 2.353$$

we do not reject H_0: $\rho = 0$ at the 5-percent level.

To test the hypothesis H_0: $\rho = 0.5$ versus H_a: $\rho \neq 0.5$, we would compute the Z transformation

$$Z = \frac{1}{2} \ln \frac{1+r}{1-r} = \frac{1}{2} \ln \frac{1+0.7}{1-0.7}$$

which can be read directly from Appendix Table H. Entering the table with $r = 0.7$, we find $Z = 0.867$ which may be regarded as an observation from a normal distribution with mean

$$\mu_Z = \frac{1}{2} \ln \frac{1+\rho}{1-\rho} = \frac{1}{2} \ln \frac{1+0.5}{1-0.5} = 0.5493$$

and variance

$$\sigma_Z^2 = \frac{1}{n-3} = \frac{1}{2}$$

Hence $\sigma_Z = 0.7072$. Since the observed value of Z is well within 1.96 standard deviations of the mean μ_Z, we do not reject H_0: $\rho = 0.5$ at the 5-percent level.

The Interpretation of r

If the value of the sample correlation coefficient appears to be significant in the sense that the hypothesis H_0: $\rho = 0$ is untenable, extreme care must be exercised in interpreting r. In particular, a significant value of r merely indicates that the measure of linear association in the population, ρ, is probably nonzero and does not in itself have anything to say about the *effect*, either direct or indirect, one variable has on the other. Both variables may in fact move together because of the effect of other variables not specifically taken into consideration. Thus the correlation observed in the sample may have no meaningful practical interpretation. In situations such as these the correlation is not to be interpreted as causation.

To illustrate these remarks we consider the often-quoted example of the correlation between teachers' salaries and liquor sales. Data on liquor sales and teachers' salaries over a period of years yielded a sample correlation coefficient of $r = 0.98$. This result indicates that there is a strong tendency for teachers' salaries and liquor sales to move together, but it does *not* prove that higher teacher salaries lead to increased liquor sales (that is, that teachers drink) nor does it prove that increased liquor sales yield higher teacher salaries. In this case both variables move together because both are influenced by other variables—for example, growth in the population and growth in per capita national income. The correlation coefficient becomes meaningful only when the effects of these latter variables are held constant (or completely accounted for). This is the objective of *partial* correlation analysis that is discussed in Section 6.5.

The teacher salaries–liquor sales result is an example of spurious correlation in the sense that any inference of a cause and effect relationship is nonsense. On the other hand we do not mean to imply that correlation analysis is useless. Correlation analysis may be used to support a relationship that is suggested by theory, or it may *suggest* a cause and effect relationship that was not previously suspected (for example, cholesterol intake and heart disease).

Finally, ρ is a useful measure of association for bivariate normal populations because a value of $\rho = 0$ implies independence. For populations that are not bivariate normal, however, ρ and hence the estimator r may no longer be a meaningful measure of association. In addition the methodology we have introduced to make inferences about ρ, given the value of r, depends on the assumption of bivariate normality. Thus even in cases in which ρ is a useful measure of association, if the underlying population is far from bivariate normal, the estimator r may be inaccurate and unreliable. The effect of nonnormality on inferences about ρ is the subject of the next section.

The Effect of Nonnormality on Inferences about ρ

We have based our inferences about ρ on the assumption that the random sample variables (X_i, Y_i), $i = 1, 2, \ldots, n$, have a bivariate normal distribution. With this assumption the distribution of the sample correlation coefficient can be derived, and the t and Z transformations given by equations (5.54) and (5.55) are valid. It is important to consider the consequences of applying the normal theory results in situations in which the joint distribution of (X_i, Y_i), $i = 1, 2, \ldots, n$, is *not* bivariate normal. In these cases the correct distribution of r is not the same as that derived under the normality assumption and consequently inferences based on t and Z may be misleading.

The effect of nonnormality on the distribution of r has been studied for many years yielding contradictory conclusions. However, it has now been fairly well established (see reference [34]) that the distribution of r can be very sensitive to nonnormality (even though n is large). In general if $\rho = 0$, the distribution of r is less sensitive to departures from normality than if

$\rho \neq 0$. The conclusion is that the methods of correlation analysis we have introduced should be limited to situations in which the joint distribution of the random variables (X, Y) is bivariate normal or very nearly bivariate normal.

One way to determine whether normal correlation analysis can be safely applied is to check the *marginal* distributions of X and Y. If the marginal distributions of X and Y are both normal or very nearly normal, then the joint distribution of X and Y will very often be *approximately* bivariate normal.[j] The marginal distributions of X and Y can be examined for large samples by seeing whether the individual sets of observations $(X_1, X_2, \ldots, X_n)$ and $(Y_1, Y_2, \ldots, Y_n)$ can be regarded as drawings from normal populations with means and variances $(\bar{X}, \bar{s}_X^2)$ and $(\bar{Y}, \bar{s}_Y^2)$, respectively. If not, it is frequently possible to transform the original observations X_i, $i = 1, 2, \ldots, n$, and Y_i, $i = 1, 2, \ldots, n$, to new observations V_i, $i = 1, 2, \ldots, n$, and W_i, $i = 1, 2, \ldots, n$, say, which can be regarded as drawings from normal populations, and then proceed with the usual methods of analysis. (The transformation of variables is discussed in Chapter 7.)

5.8 THE RELATIONSHIPS BETWEEN REGRESSION AND CORRELATION

The reader will recall the simple linear regression function

$$E(Y|X) = \alpha + \beta X \qquad (5.60)$$

If observations $(X_1, X_2, \ldots, X_n)$ and $(Y_1, Y_2, \ldots, Y_n)$ on X and Y are available, then the least squares estimates of α and β are [see equations (5.10) and (5.9)]

$$\hat{\beta} = \frac{\Sigma(X_i - \bar{X})(Y_i - \bar{Y})}{\Sigma(X_i - \bar{X})^2}$$

and

$$\hat{\alpha} = \bar{Y} - \hat{\beta}\bar{X}$$

It is clear from the definition of the sample correlation coefficient r that

$$\hat{\beta} = \left[\frac{\sqrt{\Sigma(Y_i - \bar{Y})^2}}{\sqrt{\Sigma(X_i - \bar{X})^2}} \right] r = \left[\frac{\sqrt{\Sigma(Y_i - \bar{Y})^2/(n-1)}}{\sqrt{\Sigma(X_i - \bar{X})^2/(n-1)}} \right] r = \frac{\bar{s}_Y}{\bar{s}_X} r \qquad (5.61)$$

where $\bar{s}_X$ and $\bar{s}_Y$ are the sample standard deviations. The least squares estimate of the slope coefficient in the "straight line" regression of Y on X is proportional to the sample correlation coefficient between the pairs of

---

[j] Some care must be exercised here since normal marginal distributions do *not* necessarily imply that the *joint* distribution is bivariate normal; that is, it is possible to define *nonnormal* joint distributions that give rise to normal marginal distributions, but such unusual cases would be hard to check in practice.

observations (X_i, Y_i), $i = 1, 2, \ldots, n$. Note that the constant of proportionality, $\bar{s}_Y / \bar{s}_X$, is always positive. In particular, apart from a scale factor, $\hat{\beta}$ contains all the information about linear association in the sample that r contains.

Suppose we regard *both* X and Y as random variables[k] and suppose further we assume that their joint distribution is bivariate normal. From Section 2.3 the conditional distribution of Y given X is normal with mean

$$E(Y|X) = \mu_Y + \rho \frac{\sigma_Y}{\sigma_X} (X - \mu_X) \tag{5.62}$$

where μ_Y, σ_Y^2 and μ_X, σ_X^2 are the marginal means and variances of X and Y, respectively, and ρ is the population correlation coefficient. After some algebraic manipulation we can write

$$E(Y|X) = \left(\mu_Y - \rho \frac{\sigma_Y}{\sigma_X} \mu_X \right) + \rho \frac{\sigma_Y}{\sigma_X} X \tag{5.63}$$

Comparing equations (5.60) and (5.63) we see that the locus of means of the conditional distributions of Y given X for bivariate normal populations is in the form of a simple linear regression function with

$$\alpha = \mu_Y - \rho \frac{\sigma_Y}{\sigma_X} \mu_X \tag{5.64}$$

and

$$\beta = \rho \frac{\sigma_Y}{\sigma_X} \tag{5.65}$$

Therefore if we take X and Y to be bivariate normal random variables and employ the simple linear regression model of equation (5.60), the parameters α and β are, in fact, the particular functions of the parameters μ_X, σ_X^2, μ_Y, σ_Y^2, and ρ given by equations (5.64) and (5.65). The least squares estimates of α and β are then given by inserting the usual sample estimates of μ_X, σ_X^2, μ_Y, σ_Y^2, and ρ into the relationships (5.64) and (5.65); that is,

$$\hat{\beta} = r \frac{\bar{s}_Y}{\bar{s}_X} \tag{5.66}$$

and

$$\hat{\alpha} = \bar{Y} - \left\{ r \frac{\bar{s}_Y}{\bar{s}_X} \right\} \bar{X} \tag{5.67}$$

It is clear from equation (5.65) that $\beta = 0$ if and only if $\rho = 0$ since $(\sigma_Y / \sigma_X) > 0$. Therefore a test of H_0: $\beta = 0$ is equivalent to a test of H_0: $\rho = 0$, or alternatively H_0: X and Y are independent. However a test of H_0: $\beta = 0$

---

[k] In the regression situation introduced in Section 5.4, we regarded X as a fixed, or controllable, (that is, nonrandom) variable and Y as a random variable. In fact, it makes no difference whether X is regarded as a fixed or random variable as far as the regression methodology is concerned since the regression function purports to describe the behavior of Y given (or conditional on) a value for X.

is provided by comparing the t value [see equation (5.31)],

$$t = \frac{\hat{\beta}}{\hat{\sigma}_{\hat{\beta}}}$$

with the appropriate percentage point of a t distribution with $n - 2$ degrees of freedom. Thus the results from a simple linear regression analysis can be used to make inferences about ρ provided we are willing to assume that the observations (X_i, Y_i), $i = 1, 2, \ldots, n$, represent drawings from a bivariate normal population. If the value of the sample correlation coefficient is of interest, it can be obtained from the ANOVA table associated with the regression analysis by noting that [see equation (5.66)]

$$r^2 = \frac{\hat{\beta}^2 \bar{s}_X^2}{\bar{s}_Y^2} = \frac{\text{regression } SS}{\text{total } SS \text{ about mean}}$$

$$= \text{coefficient of determination}$$

The coefficient of determination (previously denoted by R^2) in a simple linear regression analysis is the *square* of the sample correlation coefficient. The sign of r is the sign of $\hat{\beta}$.

Simple linear regression analysis then yields all the information of a correlation analysis and in addition (a) indicates *how* the variables X and Y are related and (b) is applicable regardless of whether X is taken to be a fixed (controllable) variable or a random variable. That is, correlation may be viewed as a subset of regression analysis.

EXAMPLE 5.8

For our disposable income–housing demand example, a simple linear regression analysis produced the fitted equation (see Table 5.2)

$$\hat{Y} = -5.33 + 0.90X$$

with

$$s^2 = \frac{\Sigma(Y_i - \hat{Y}_i)^2}{n - 2} = \frac{0.17}{3} = 0.0567$$

It can be shown (see Problem 5.4) that the coefficient of determination is

$$r^2 = 0.488$$

which implies, since $\hat{\beta} = 0.90$ is positive, that the sample correlation coefficient is

$$r = +\sqrt{r^2} = 0.70$$

the result obtained in Example 5.7. Moreover,

$$t = \frac{\hat{\beta}}{\hat{\sigma}_{\hat{\beta}}} = \frac{\hat{\beta}}{s/\sqrt{\Sigma(X_i - \bar{X})^2}} = \frac{0.90}{\sqrt{0.0567}/\sqrt{0.20}} = 1.690$$

which is identical (within rounding error) to the value we get for the t statistic

$$t = \frac{r\sqrt{n-2}}{\sqrt{1-r^2}}$$

in Example 5.7. The reader will recall that the latter t statistic was used to test the hypothesis H_0: $\rho = 0$. In fact it can be shown that these results are true in general; that is,

$$\frac{\hat{\beta}}{\hat{\sigma}_{\hat{\beta}}} = \frac{r\sqrt{n-2}}{\sqrt{1-r^2}}$$

(see Problem 5.23). Thus if we are willing to assume that our observations represent drawings from a bivariate normal population, the t value

$$t = \frac{\hat{\beta}}{\hat{\sigma}_{\hat{\beta}}} = 1.690$$

is not inconsistent with the null hypothesis H_0: $\rho = 0$ at any reasonable level of significance—for example, $t_{0.05}(3) = 2.353$—a conclusion we have already reached in Example 5.7.

5.9 AUTOCORRELATION

In this section we will introduce the notion of autocorrelation, or correlation between random variables representing outcomes at different time points. The (sample) autocorrelation coefficient will be useful for checking the independence assumption in regression analysis, which we consider in the next chapter, and it is extremely useful in time series analysis.

Let $y_1, y_2, \ldots, y_n$ denote a *chronological* sequence of observations on a sequence of n random variables at equally spaced time points, $t = 1, 2, \ldots, n$, and let $\bar{y} = \sum_{t=1}^{n} y_t / n$ be the "sample average." Then the sample autocovariance coefficient at lag k is given by

$$c_k = \frac{1}{n} \sum_{t=1}^{n-k} (y_t - \bar{y})(y_{t+k} - \bar{y}), \qquad k = 1, 2, \ldots \qquad (5.68)$$

and the sample autocorrelation[1] coefficient by

$$r_k = \frac{c_k}{c_0} = \frac{1/n \sum_{t=1}^{n-k} (y_t - \bar{y})(y_{t+k} - \bar{y})}{1/n \sum_{t=1}^{n} (y_t - \bar{y})^2} = \frac{\sum_{t=1}^{n-k} (y_t - \bar{y})(y_{t+k} - \bar{y})}{\sum_{t=1}^{n} (y_t - \bar{y})^2}, \qquad (5.69)$$

$$k = 1, 2, \ldots$$

[1] The corresponding population quantities are defined in a supplement to Chapter 10 in which the notion of a stationary stochastic process is introduced. In this section we shall concentrate on the sample autocorrelation coefficient because of its important role in the analysis of residuals and in the identification of appropriate time series models. We are using lower case letters to denote observations in this section to make the presentation consistent with that in the time series chapters which follow.

The sample autocorrelation coefficient is a measure of the linear dependence in the observations, whatever they may be, k time periods apart. In particular, if *adjacent* observations tend to move together, then $r_1 > 0$. If observations two time periods apart tend to move in opposite directions, then $r_2 < 0$, and so forth. Scatter diagrams constructed from the pairs of observations (y_t, y_{t+k}), $t = 1, 2, \ldots, n - k$, provide a pictorial display of the linear dependence between observations separated by k time periods. Figure 5.16 displays several possible scatter diagrams and the resulting sample autocorrelation coefficients.

It is clear from equation (5.68) that as k increases, the number of deviations $(y_t - \bar{y})$ involved in the calculation of c_k decreases. [As the lag k increases, there are fewer pairs of observations (y_t, y_{t+k}), $t = 1, 2, \ldots, n - k$.] As a consequence, c_k is sensitive to sampling fluctuations and can be unstable. For this reason the sample autocovariances (and sample autocorrelations) are ordinarily computed for a selected number of lags K, where K is approximately[m] $n/4$.

A plot of r_k versus k is known as the "sample autocorrelation function" and is an extremely useful device for illustrating the nature of the linear dependence over increasing intervals of time. As we shall see in Chapters 9 and 10, "time series" models can be identified by particular sample autocor-

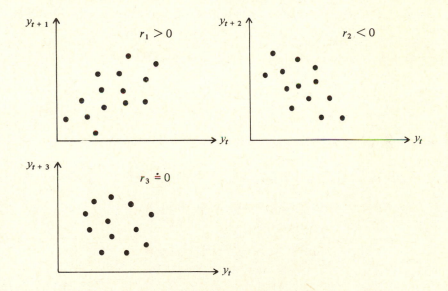

Figure 5.16   Scatter diagrams and the sample autocorrelations

[m] This rule of thumb is merely a guideline, and it will not be appropriate in all situations.

relation patterns. Thus the sample autocorrelation function is an important ingredient of time series analysis.

EXAMPLE 5.9

Suppose the annual rates of return on a portfolio of common stocks over a period of 10 years are as follows:

t (years)	1	2	3	4	5	6	7	8	9	10
y_t (annual rate of return)	0.03	0.05	0.07	0.09	0.08	0.05	0.06	0.03	0.02	0.02

The calculations leading to the sample autocorrelation coefficients r_1 and r_2 are given in Table 5.6.

Table 5.6

t	y_t	y_{t+1}	y_{t+2}	$(y_t - \bar{y})^2$	$(y_t - \bar{y})(y_{t+1} - \bar{y})$	$(y_t - \bar{y})(y_{t+2} - \bar{y})$
1	0.03	0.05	0.07	0.0004	0	−0.0004
2	0.05	0.07	0.09	0	0	0
3	0.07	0.09	0.08	0.0004	0.0008	0.0006
4	0.09	0.08	0.05	0.0016	0.0012	0
5	0.08	0.05	0.06	0.0009	0	0.0003
6	0.05	0.06	0.03	0	0	0
7	0.06	0.03	0.02	0.0001	−0.0002	−0.0003
8	0.03	0.02	0.02	0.0004	0.0006	0.0006
9	0.02	0.02	X	0.0009	0.0009	X
10	0.02	X	X	0.0009	X	X
	0.50			0.0056	0.0033	0.0008

Therefore

$$\bar{y} = \frac{0.50}{10} = 0.05$$

Thus

$$r_1 = \frac{\sum_{t=1}^{9}(y_t - \bar{y})(y_{t+1} - \bar{y})}{\sum_{t=1}^{10}(y_t - \bar{y})^2} = \frac{0.0033}{0.0056} = 0.59$$

and

$$r_2 = \frac{\sum_{t=1}^{8}(y_t - \bar{y})(y_{t+2} - \bar{y})}{\sum_{t=1}^{10}(y_t - \bar{y})^2} = \frac{0.0008}{0.0056} = 0.14$$

If these are the only sample autocorrelations calculated, then a plot of r_k versus k, $k = 1, 2$, produces the sample autocorrelation function displayed in Figure 5.17.

For this example adjacent observations tend to move together ($r_1 = 0.59$), and observations two time periods apart also tend to move together although the "strength" of the linear association is much less ($r_2 = 0.14$).

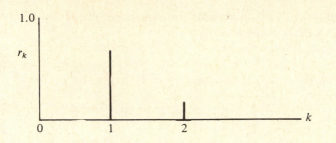

Figure 5.17 The sample autocorrelation function, rate of return data

5.10 RANK CORRELATION—A MEASURE OF MONOTONIC ASSOCIATION

We have seen that if n pairs of observations (X_i, Y_i), $i = 1, 2, \ldots, n$, are drawn from a bivariate population with correlation coefficient ρ, an estimate of ρ is provided by the sample correlation coefficient r. In general, the sampling distribution of r depends on the form of the parent bivariate population and the value of ρ. If the parent population is bivariate normal, the distribution of r can be derived and useful approximations of the kind discussed in Section 5.7 are applicable. However as we have noted, if the parent population is *not* bivariate normal, inferences about ρ based on the normality assumption may be misleading, and in fact there are cases in which ρ itself may not be a particularly good measure of association in the population. Moreover the data may be available as *ordinal* measurements, and we might be interested in an appropriate measure of association in situations of this kind.

One measure of association that can be used for both interval and ordinal data and whose sampling distribution does not depend on the form of the underlying bivariate population is known as Spearman's *rank correlation coefficient*.

Suppose the n observations on the random variable X are ranked according to magnitude, and similarly the n observations on the random variable Y are separately ranked according to magnitude. Then the pairs of observations (X_i, Y_i), $i = 1, 2, \ldots, n$, will be transformed into pairs of *ranks* (r_i, s_i), where

$$r_i = \text{rank}\,(X_i) \quad \text{and} \quad s_i = \text{rank}\,(Y_i) \tag{5.70}$$

and r_i and s_i assume integer values, $1, 2, \ldots, n$, for $i = 1, 2, \ldots, n$. For example, the pairs of observations

$$(X_1, Y_1) = (15.3, 7.2); \quad (X_2, Y_2) = (16.0, 8.9); \quad (X_3, Y_3) = (14.0, 9.3)$$

transform into the pairs of ranks

$$(r_1, s_1) = (2, 1); \quad (r_2, s_2) = (3, 2); \quad (r_3, s_3) = (1, 3)$$

since X_1 has rank 2, X_2 has rank 3, X_3 has rank 1, and Y_1, Y_2, and Y_3 have ranks 1, 2, and 3, respectively.

The ordinary sample correlation coefficient can then be computed for the n pairs of *ranks* (r_i, s_i), $i = 1, 2, \ldots, n$. The statistic is known as Spearman's *rank correlation coefficient* which we will denote by r_s. Thus

$$r_s = \frac{\sum_{i=1}^{n} (r_i - \bar{r})(s_i - \bar{s})}{\sqrt{\sum_{i=1}^{n} (r_i - \bar{r})^2 \sum_{i=1}^{n} (s_i - \bar{s})^2}} \tag{5.71}$$

Now the expression for r_s in equation (5.71) can be greatly simplified. Making use of the facts that

$$\sum_{i=1}^{n} r_i = \sum_{i=1}^{n} s_i = \sum_{i=1}^{n} i = \frac{n(n+1)}{2}, \qquad \bar{r} = \bar{s} = \frac{n+1}{2}$$

and

$$\sum_{i=1}^{n} (r_i - \bar{r})^2 = \sum_{i=1}^{n} (s_i - \bar{s})^2 = \sum_{i=1}^{n} \left(i - \frac{n+1}{2}\right)^2 = \frac{n(n^2 - 1)}{12} \tag{5.72}$$

r_s can be written in the form

$$r_s = \frac{12 \sum_{i=1}^{n} r_i s_i}{n(n^2 - 1)} - \frac{3(n+1)}{n-1} \tag{5.73}$$

A further simplification is provided by considering the *differences* in the rankings

$$d_i = r_i - s_i, \qquad i = 1, 2, \ldots, n \tag{5.74}$$

Since $\bar{r} = \bar{s}$ for all samples

$$d_i = (r_i - \bar{r}) - (s_i - \bar{s})$$

and

$$\sum_{i=1}^{n} d_i^2 = \sum_{i=1}^{n} (r_i - \bar{r})^2 + \sum_{i=1}^{n} (s_i - \bar{s})^2 - 2 \sum_{i=1}^{n} (s_i - \bar{s})(r_i - \bar{r}) \tag{5.75}$$

Equations (5.71), (5.75), and (5.72) then yield

$$r_s = 1 - \frac{6 \sum_{i=1}^{n} d_i^2}{n(n^2 - 1)} \tag{5.76}$$

The expression for r_s given in equation (5.76) is the one most frequently used in practice. Notice that we only need to know the sample size n and the *differences* in the rankings d_i, $i = 1, 2, \ldots, n$, in order to calculate r_s.

The rank correlation coefficient is a measure of the correspondence between *rankings*. It is a *sample* quantity, and an analogous measure for the population is difficult, if not impossible, to define meaningfully. Nevertheless it has some intuitive appeal, and in many situations it can be used as an estimate of ρ or as an estimate of the association between the random variables X and Y in the bivariate population.

Although the population parameter corresponding to r_s is not specifically defined, it is clear that r_s is a "good" measure of association between sample ranks. In particular it can be shown that $-1 \le r_s \le 1$ and

1. If there is perfect agreement in the rankings, that is, $r_i = s_i$, $i = 1, 2, \ldots, n$, then $r_s = 1$.
2. If there is perfect disagreement in the rankings, that is, the Y rankings are the reverse of the X rankings or $s_i = n - r_i + 1$, $i = 1, 2, \ldots n$, then $r_s = -1$.
3. If there is no particular agreement between the sets of rankings, then $r_s \doteq 0$.
4. r_s is invariant (its value remains the same) under all order-preserving transformations of the data. For example, if (r_i, s_i) are the ranks associated with the observation pairs (X_i, Y_i), then the same ranks are associated with the observation pairs $(\log X_i, \log Y_i)$, $i = 1, 2, \ldots, n$.

Interpretation and Inference

Let us assume that the ranks are derived from the values of the underlying random variables X and Y. If X and Y are independent, then the probability distribution of r_s does not depend on the form of the parent population or on any *population* measure of association.

The exact distribution of r_s under the hypothesis of independence can be obtained, but in general this requires an enumeration of all possible values of r_s—a task that can be extremely tedious for anything but small values of n. On the other hand, the exact distribution of r_s is unimodel and symmetric about the origin. In addition under the hypothesis of independence it can be shown that $E(r_s) = 0$ and $\text{Var}(r_s) = 1/(n - 1)$. Although the distribution of r_s approaches normality, it does so rather slowly. However, an interesting analogy exists between the ordinary sample correlation coefficient r and the rank correlation coefficient r_s.

Under the hypothesis of independence

$$t = \frac{r_s \sqrt{n - 2}}{\sqrt{1 - r_s^2}} \tag{5.77}$$

has *approximately* a t distribution with $n - 2$ degrees of freedom provided n is fairly large (for $n \ge 30$, the approximation is excellent; for n in the neighborhood of 20, the approximation is tolerable). The statistic used to test the significance of the rank correlation coefficient is the same as that used to test the significance of the ordinary sample correlation coefficient assuming a bivariate normal population.

We must be careful in interpreting the magnitude of r_s and, consequently,

$$t = \frac{r_s \sqrt{n - 2}}{\sqrt{1 - r_s^2}}$$

If the underlying random variables from which the rankings are derived are continuous, then

1. For r_s significant and positive, we conclude that there appears to be some monotone increasing relationship between the underlying random variables; that is, the variables tend to move together, but we cannot discern the exact form of the association, which may be linear with a positive slope, an increasing exponential, or some other monotone increasing function.
2. For r_s significant and negative, we conclude that there appears to be some monotone decreasing relationship between the underlying variables; again we cannot determine the *form* of the relationship.
3. For $r_s \doteq 0$, we conclude that either the underlying random variables are independent or that the relationship is not monotonic.

A formal test of the hypothesis H_0: X and Y are independent random variables based on the t statistic in equation (5.77) would be carried out by comparing the observed value of t with the appropriate percentage point of a t distribution with $n - 2$ degrees of freedom. We can fail to reject the null hypothesis either because the underlying random variables X and Y are independent or because the relationship is not monotonic. Rejection of the null hypothesis implies a relationship between the underlying variables whose nature depends on the sign of the t statistic or equivalently on the sign of r_s.

When the population correlation coefficient ρ can be defined (as is the case for a bivariate normal population), r_s can be interpreted as an estimator of ρ. In particular, if the population is bivariate normal with $\rho = 0$, the values of r_s and r tend to agree over samples. However, in this case the ordinary sample correlation coefficient r is preferred. On the other hand, there are nonnormal bivariate populations in which r_s is the preferred estimate of the population correlation coefficient. In general we might say that if inference about ρ is of primary interest, r should be used if the parent population is bivariate normal or nearly bivariate normal (it may be worthwhile to transform the data so that this condition is satisfied) and r_s should be used if the parent population is markedly nonnormal.

The Problem of Ties

Our discussion up to this point has assumed that the ranks within each set $\{r_1, r_2, \ldots, r_n\}$ and $\{s_1, s_2, \ldots, s_n\}$ are *distinguishable*. In practice, however, *ties* frequently occur in one or both of the separate sets of rankings. For example, if the sets of ranks are derived from corresponding sets of observations, a pair of sample measurements may be identical (in practice only a limited number of decimal places is retained), and thus separate ranks cannot be assigned. Or, since the theory we have introduced is applicable to any two sets of n pairs of ranks regardless of how they were derived, it is possible that two or more items in either group cannot be distinguished according to the characteristic of interest and therefore must be assigned equal ranks.

In general the problem of ties is simply one of finding a way to assign specific ranks to the tied observations or "identical" items. If the available ranks within each set of tied observations are assigned at random, the theory we have presented is not affected. However, this procedure is not always desirable since we may be unwilling to assign the ranks at random. A more appealing approach is to give each of the tied observations within a set the *average* of the ranks the observations would have received if they were distinguishable. This is the procedure usually employed in practice.

Although the assignment of an average rank or midrank to tied observations has some effect on the behavior of the rank correlation coefficient, the effect will be negligible for n reasonably large provided the ties are not extensive. After assigning the average rank to sets of tied observations, we can proceed in the usual way.

EXAMPLE 5.10

Suppose that 15 stock portfolios are ranked by two measures of "risk adjusted" performance—the Treynor index and Jensen's index. The ranks of the 15 portfolios with respect to each of the performance measures are presented below.

Portfolio	1	2	3	4	5	6	7	8	9	10	11	12	13	14	15
Treynor rank	7	11	3	5	14	10	13	6	12	1	9	15	2	4	8
Jensen rank	6	14	7	5	15	8	11	3	12	4	9	13	1	2	10

We then ask if there is consistency between the two sets of rankings, or do the two measures of risk adjusted performance tend to agree.

The calculations leading to Spearman's rank correlation coefficient are summarized in Table 5.7.

The large and positive value of r_s indicates that the two performance measures produce similar results. There is close agreement between the two sets of rankings. If we compare the t value[n] with, say, $t_{0.005}(13) = 3.012$, the 0.005 percentage point of a t distribution with 13 degrees of freedom, we conclude that r_s is significantly different for zero, and there appears to be a monotonically increasing relationship between the two performance measures.

[n] The reader will recall that we can perform an *approximate* test of independence by treating $t = r_s\sqrt{n-2}/\sqrt{1-r_s^2}$ as a student t random variable with $n-2$ degrees of freedom, provided n is reasonably large. Here $n = 15$ is reasonably large. Moreover, the observed t value falls so far out in the tail of the t distribution with 13 degrees of freedom that we can be almost certain that r_s would be significant if we had used the exact distribution of r_s under the hypothesis of independence.

Table 5.7

i	r_i	s_i	$d_i = r_i - s_i$	d_i^2
1	7	6	1	1
2	11	14	-3	9
3	3	7	-4	16
4	5	5	0	0
5	14	15	-1	1
6	10	8	2	4
7	13	11	2	4
8	6	3	3	9
9	12	12	0	0
10	1	4	-3	9
11	9	9	0	0
12	15	13	2	4
13	2	1	1	1
14	4	2	2	4
15	8	10	-2	4
				66

$$r_s = 1 - \frac{6\sum_{i=1}^{15} d_i^2}{n(n^2 - 1)}$$

$$= 1 - \frac{6(66)}{15(225 - 1)} = 0.88$$

$$t = \frac{r_s \sqrt{n-2}}{\sqrt{1 - r_s^2}} = \frac{(0.88)\sqrt{13}}{\sqrt{1 - (0.88)^2}} = 6.69$$

5.11 FINAL COMMENTS

This chapter contains rather thorough discussions of simple linear regression, correlation, and the relationships between regression and correlation. Both simple linear regression and correlation deal with the "straight line" relationship between two variables. Correlation is concerned with the existence of a relationship between two random variables. Regression, on the other hand, is concerned with the existence of a relationship, as well as with the *nature* of the relationship between the two variables. Regression analysis therefore is, in general, more informative. Moreover, as we have pointed out, the main results in a correlation analysis can be derived from the output of a simple linear regression analysis.

The results of a regression analysis, however, must be interpreted with care—perhaps with even more care than those of a correlation analysis. For example, regression analysis by its very nature implies a cause and effect position. The regression function specifically attempts to describe how values of the "dependent" variable are linked to those of the "independent" variable. But regression functions are formulated by mortals, not handed down from above, and it may be that changes in the independent variable X do not cause changes in the dependent variable Y at all, even though the regression is "significant." Rather, a third variable, not explicitly considered by the investigator in the regression function, may be responsible for the behavior of both X and Y. In this case the observed relationship between Y and X is misleading.

We are particularly vulnerable to the possibility of a faulty cause and effect interpretation if the regression function is estimated from data that is routinely observed; that is, neither the Y values nor the X values are under the control of the investigator. Unfortunately, this is frequently the case in many regression problems arising in business and economics.

In the next chapter we discuss the relationships between a dependent variable, Y, and several independent variables, $X_1, X_2, \ldots, X_p$; that is, we will consider "multiple" regression. Although most of the ideas we have introduced here carry over without modification, the analysis is more complicated. We will use matrix algebra in an attempt to provide a unifying treatment of multiple regression. (The reader is encouraged to review Appendix 3. The matrix algebra approach to simple linear regression is summarized in Supplement 5B).

Supplement 5A

TESTING FOR LACK OF FIT

If there are repeat observations available (several Y values for the same X value), then it is possible to use this information to test the adequacy of the regression function directly. The test is obtained from the following argument.

If the model is correct, s^2 (residual mean square) is an unbiased estimator of σ^2. On the other hand, for inadequate models the residuals represent variance error (due to the variability of the observations about the *true* regression function) *and* "bias" error (due to the fact that the postulated model is incorrect and hence the points on the fitted line are likely to be far from the points on the true regression curve). In fact, if the model is not correct

$$E(s^2) = \sigma^2 + \text{(positive terms due to bias)}$$

To check on the bias terms we compare a separate estimate of σ^2 with s^2 to see if it is inflated. If so, we conclude that lack of fit exists; that is, the regression model is inadequate.

Now with repeat observations a separate estimate of σ^2 can be obtained regardless of whether the model is adequate or inadequate from the "pure error" sum of squares.[a] We let

$Y_{11}, Y_{12}, \ldots, Y_{1n_1}$ be n_1 observations at X_1
$Y_{21}, Y_{22}, \ldots, Y_{2n_2}$ be n_2 observations at X_2
$\vdots$
$Y_{m1}, Y_{m2}, \ldots, Y_{mn_m}$ be n_m observations at X_m

[a] The methodology developed in this section can be applied when there is more than one independent variable in the regression function. The formula for the pure error sum of squares and the associated degrees of freedom is exactly the same. The pure error sum of squares can be subtracted from the residual sum of squares to provide a lack of fit sum of squares and the adequacy of the regression function tested in precisely the same manner as it is for the case of a single independent variable.

then at each setting of X we can calculate the sum of squared deviations

$$SS_i = \sum_{j=1}^{n_i} (Y_{ij} - \bar{Y}_i)^2 \tag{5A.1}$$

where $\bar{Y}_i = (Y_{i1} + \cdots + Y_{in_i})/n_i$ with $(n_i - 1)$ degrees of freedom, $i = 1, 2, \ldots, m$. The pure error sum of squares is defined to be

$$SS_E = SS_1 + SS_2 + \cdots + SS_m = \sum_{i=1}^{m} \sum_{j=1}^{n_i} (Y_{ij} - \bar{Y}_i)^2 \tag{5A.2}$$

with

$$\sum_{i=1}^{m} (n_i - 1) = \sum_{i=1}^{m} n_i - m \qquad \text{d.f.}$$

The pure error mean square

$$s_E^2 = \frac{SS_E}{\Sigma n_i - m} = \frac{\sum_{i=1}^{m} \sum_{j=1}^{n_i} (Y_{ij} - \bar{Y}_i)^2}{\Sigma n_i - m} \tag{5A.3}$$

is an estimate of σ^2 regardless of the true form of the regression model. The pure error sum of squares can be subtracted from the residual sum of squares to give a lack of fit sum of squares, say, SS_L. If we denote the degrees of freedom associated with the residual sum of squares, the pure error sum of squares, and the lack of fit sum of squares, by n_R (temporarily), n_E, and n_L, respectively, then

$$\begin{pmatrix} \text{residual} \\ \text{sum of} \\ \text{squares} \end{pmatrix} = \begin{pmatrix} \text{pure error} \\ \text{sum of} \\ \text{squares, } SS_E \end{pmatrix} + \begin{pmatrix} \text{lack of} \\ \text{fit sum of} \\ \text{squares, } SS_L \end{pmatrix} \tag{5A.4}$$

$$\text{d.f.:} \qquad n_R \qquad = \qquad n_E \qquad + \qquad n_L$$

If the error component satisfies the assumptions of Table 5.4 and the regression function is correct, then the ratio of the lack of fit mean square, say, MS_L, to the pure error mean square, s_E^2, is an F random variable with n_L and n_E degrees of freedom; and a test at the γ level of significance for lack of fit can be made using the rule,

Reject H_0: model adequate: if $F = \dfrac{MS_L}{s_E^2} > F_\gamma(n_L, n_E)$

Do not reject H_0: model adequate   if $F = \dfrac{MS_L}{s_E^2} \le F_\gamma(n_L, n_E)$ (5A.5)

If the F ratio, $F = MS_L/s_E^2$, is significant we must modify the regression model and, if possible, repeat the analysis. If the F ratio is not significant, there appears no reason to doubt the adequacy of the regression model. In this case the pure error sum of squares and the lack of fit sum of squares and their degrees of freedom are recombined, and we use s^2 (the residual mean square) as an estimate of σ^2.

EXAMPLE 5A.1

Let us consider the field corn seed example. There are repeat observations at $X_1 = \$2.25$ and $X_2 = \$2.50$. Therefore using the information in Example 5.1,

$$SS_1 = \sum_{j=1}^{2} (Y_{1j} - \bar{Y}_1)^2 = \frac{(Y_{11} - Y_{12})^2}{2} = \frac{(60 - 81)^2}{2} = 220.5$$

and

$$SS_2 = \sum_{j=1}^{2} (Y_{2j} - \bar{Y}_2)^2 = \frac{(Y_{21} - Y_{22})^2}{2} = \frac{(75 - 55)^2}{2} = 200$$

with 1 degree of freedom associated with each sum of squares. The pure error sum of squares is then

$$SS_E = SS_1 + SS_2 = 220.5 + 200 = 420.5$$

with $n_E = 2$ degrees of freedom. From Table 5.3 the residual sum of squares is 475.01 with 6 degrees of freedom. Therefore the lack of fit sum of squares is [by subtraction, see equation (5A.4)]

$$SS_L = 475.01 - 420.5 = 54.51$$

with

$$n_L = 6 - 2 = 4 \text{ d.f.}$$

Hence

$$s_E^2 = \frac{SS_E}{n_E} = \frac{420.5}{2} = 210.25$$

$$MS_L = \frac{SS_L}{n_L} = \frac{54.51}{4} = 13.63$$

and

$$F = \frac{MS_L}{s_E^2} = \frac{13.63}{210.25} = 0.06$$

This F ratio can be compared with a percentage point of an F distribution with $n_L = 4$ and $n_E = 2$ degrees of freedom. Since the F ratio is not significant at any reasonable significance level [for example, $F_{0.05}(4, 2) = 19.25$] we conclude that the regression model is adequate and repool the pure error and lack of fit sum of squares to obtain the residual sum of squares which, when divided by its degrees of freedom, can be used to estimate σ^2. The results of this example provide the justification for using the residual mean square, s^2, to estimate σ^2 in Sections 5.5 and 5.6.

THE MATRIX APPROACH TO SIMPLE LINEAR REGRESSION

Mathematical Model and Assumptions

In this supplement we present the matrix approach to simple linear regression. It is assumed that the reader is familiar with the material given in Appendix 3.

The simple linear regression model

$$Y_i = \alpha + \beta X_i + e_i, \qquad i = 1, 2, \ldots, n$$

implies the system of equations,

$$\begin{aligned}
Y_1 &= \alpha + \beta X_1 + e_1 \\
Y_2 &= \alpha + \beta X_2 + e_2 \\
&\vdots \\
Y_n &= \alpha + \beta X_n + e_n
\end{aligned} \tag{5B.1}$$

or in matrix notation,

$$\begin{bmatrix} Y_1 \\ Y_2 \\ \vdots \\ Y_n \end{bmatrix} = \begin{bmatrix} 1 & X_1 \\ 1 & X_2 \\ \vdots & \vdots \\ 1 & X_n \end{bmatrix} \begin{bmatrix} \alpha \\ \beta \end{bmatrix} + \begin{bmatrix} e_1 \\ e_2 \\ \vdots \\ e_n \end{bmatrix} \tag{5B.2}$$

[The reader should verify that performing the matrix operations in equation (5B.2) produces the system of equations (5B.1).] With

$$\mathbf{Y} = \begin{bmatrix} Y_1 \\ Y_2 \\ \vdots \\ Y_n \end{bmatrix}, \qquad \mathbf{X} = \begin{bmatrix} 1 & X_1 \\ 1 & X_2 \\ \vdots & \vdots \\ 1 & X_n \end{bmatrix}, \qquad \boldsymbol{\beta} = \begin{bmatrix} \alpha \\ \beta \end{bmatrix}, \qquad \text{and} \qquad \mathbf{e} = \begin{bmatrix} e_1 \\ e_2 \\ \vdots \\ e_n \end{bmatrix}$$

equation (5B.2) can be written compactly as

$$\underset{(n\times1)}{\mathbf{Y}} = \underset{(n\times2)}{X}\ \underset{(2\times1)}{\boldsymbol{\beta}} + \underset{(n\times1)}{\mathbf{e}} \tag{5B.3}$$

where the ordered pairs in the parentheses denote the dimensions of the matrices. [*Any* linear regression model can be expressed in the form of equation (5B.3), provided the elements of the X matrix and the $\boldsymbol{\beta}$ vector are suitably defined.]

In addition the assumptions may be expressed in matrix notation as

$$E(\mathbf{Y}|X) = \underset{(n\times2)}{X}\ \underset{(2\times1)}{\boldsymbol{\beta}}$$

$$\mathrm{Var}(\mathbf{Y}) = \underset{(n\times n)}{I}\ \sigma^2 \tag{5B.4}$$

or, equivalently, as

$$E(\mathbf{e}) = \underset{(n\times1)}{\mathbf{0}}$$

$$\mathrm{Var}(\mathbf{e}) = \underset{(n\times n)}{I}\ \sigma^2 \tag{5B.5}$$

Normal Equations and Least Squares Estimators

The least squares estimates $\hat{\alpha}, \hat{\beta}$ are obtained by minimizing the sum of squared errors

$$S(\alpha, \beta) = \sum_{i=1}^{n} [Y_i - E(Y_i|X_i)]^2 = \sum_{i=1}^{n} (Y_i - \alpha - \beta X_i)^2$$

$$= \sum_{i=1}^{n} e_i^2 \tag{5B.6}$$

Now, it is clear from equation (5B.3) that

$$\mathbf{e} = \mathbf{Y} - X\boldsymbol{\beta}$$

therefore in matrix notation

$$S(\alpha, \beta) = \sum_{i=1}^{n} e_i^2 = \underset{(1\times n)}{\mathbf{e}'}\ \underset{(n\times1)}{\mathbf{e}} = (\mathbf{Y} - X\boldsymbol{\beta})'(\mathbf{Y} - X\boldsymbol{\beta}) \tag{5B.7}$$

If we differentiate $S(\alpha, \beta)$ with respect to α and β—that is, compute the partial derivatives—and set the resulting equations equal to zero, after some algebraic manipulations we get the normal *equations* (in matrix notation)

$$\underset{(2\times n)}{X'}\ \underset{(n\times2)}{X}\ \underset{(2\times1)}{\hat{\boldsymbol{\beta}}} = \underset{(2\times n)}{X'}\ \underset{(n\times1)}{\mathbf{Y}} \tag{5B.8}$$

Thus if $(X'X)^{-1}$ exists, the least squares estimates are given by

$$\underset{(2\times1)}{\hat{\boldsymbol{\beta}}} = \underset{(2\times2)}{(X'X)^{-1}}\ \underset{(2\times n)}{X'}\ \underset{(n\times1)}{\mathbf{Y}} \tag{5B.9}$$

[Again the form of equations (5B.8) and (5B.9) is perfectly general and holds, provided the elements of $\boldsymbol{\beta}$ and X are suitably defined, for any linear regression model. Of course, equation (5B.9) implies the existence of $(X'X)^{-1}$.] It is easily shown that for the simple linear regression case,

$$X'X = \begin{bmatrix} n & \Sigma X_i \\ \Sigma X_i & \Sigma X_i^2 \end{bmatrix}$$

and

$$X'Y = \begin{bmatrix} \Sigma Y_i \\ \Sigma X_i Y_i \end{bmatrix}$$

In addition

$$(X'X)^{-1} = \begin{bmatrix} \dfrac{\Sigma X_i^2}{n\Sigma(X_i - \bar{X})^2} & \dfrac{-\Sigma X_i}{n\Sigma(X_i - \bar{X})^2} \\ \dfrac{-\Sigma X_i}{n\Sigma(X_i - \bar{X})^2} & \dfrac{n}{n\Sigma(X_i - \bar{X})^2} \end{bmatrix}.$$

The reader should verify that multiplying $(X'X)^{-1}$ by $X'Y$ gives, after some algebra, the equations we obtained for the least squares values in the body of the chapter [that is, equations (5.9) and (5.10) with $\hat{\alpha}$ and $\hat{\beta}$ substituted for a and b, respectively]. Of course the advantage of the matrix approach is that if we can remember equation (5B.9) and the procedure for inverting a 2×2 matrix, expressions (5.9) and (5.10) are not needed.

Using matrices the fitted values and the residuals are

$$\underset{(n \times 1)}{\hat{\mathbf{Y}}} = \underset{(n \times 2)}{\mathbf{X}} \ \underset{(2 \times 1)}{\hat{\boldsymbol{\beta}}} \qquad (5B.10)$$

and

$$\underset{(n \times 1)}{\hat{\mathbf{e}}} = \underset{(n \times 1)}{\mathbf{Y}} - \underset{(n \times 1)}{\hat{\mathbf{Y}}} \qquad (5B.11)$$

Analysis of Variance Table

In matrix notation the ANOVA table may be written as

Source	Sum of Squares (SS)	Degrees of Freedom (d.f.)	Mean Squares (MS)	F Ratio
Regression	$\hat{\mathbf{Y}}'\hat{\mathbf{Y}} - n\bar{Y}^2$	1	$MS_R = (\hat{\mathbf{Y}}'\hat{\mathbf{Y}} - n\bar{Y}^2)/1$	$F = MS_R/s^2$
Residual	$(\mathbf{Y} - \hat{\mathbf{Y}})'(\mathbf{Y} - \hat{\mathbf{Y}})$	$n - 2$	$s^2 = \dfrac{(\mathbf{Y} - \hat{\mathbf{Y}})'(\mathbf{Y} - \hat{\mathbf{Y}})}{(n - 2)}$	
Total (about the mean)	$\mathbf{Y}'\mathbf{Y} - n\bar{Y}^2$	$n - 1$		

and the coefficient of determination becomes

$$R^2 = \frac{\hat{\mathbf{Y}}'\hat{\mathbf{Y}} - n\bar{Y}^2}{\mathbf{Y}'\mathbf{Y} - n\bar{Y}^2} = 1 - \frac{(\mathbf{Y} - \hat{\mathbf{Y}})'(\mathbf{Y} - \hat{\mathbf{Y}})}{\mathbf{Y}'\mathbf{Y} - n\bar{Y}^2} \tag{5B.12}$$

The extension to the more general form of the ANOVA table is straightforward.

Variance-Covariance Matrix for $\hat{\boldsymbol{\beta}}$ and $\hat{Y}_0$

If the assumptions given by either equation (5B.4) or equation (5B.5) are correct then

$$\text{Var}(\hat{\boldsymbol{\beta}}) = \begin{bmatrix} \text{Var}(\hat{\alpha}) & \text{Cov}(\hat{\alpha}, \hat{\beta}) \\ \text{Cov}(\hat{\alpha}, \hat{\beta}) & \text{Var}(\hat{\beta}) \end{bmatrix} = (X'X)^{-1}\sigma^2 \tag{5B.13}$$

In general σ^2 is unknown and must be estimated by s^2; therefore the *estimated* variance matrix for $\hat{\boldsymbol{\beta}}$ is given by equation (5B.13) with s^2 in place of σ^2. Hence

$$\text{Vâr}(\hat{\boldsymbol{\beta}}) = \begin{bmatrix} \text{Vâr}(\hat{\alpha}) & \text{Côv}(\hat{\alpha}, \hat{\beta}) \\ \text{Côv}(\hat{\alpha}, \hat{\beta}) & \text{Vâr}(\hat{\beta}) \end{bmatrix} = (X'X)^{-1}s^2 \tag{5B.14}$$

The square root of the diagonal entries in the estimated variance matrix give the standard errors $\hat{\sigma}_{\hat{\alpha}}$ and $\hat{\sigma}_{\hat{\beta}}$, which appear in the expressions for the t values and confidence intervals provided in Section 5.5.

Suppose we want to predict the future value of Y, say $\hat{Y}_0$, at the value $X = X_0$. If we let

$$\mathbf{X}_0 = \begin{bmatrix} 1 \\ X_0 \end{bmatrix}$$

then in matrix notation

$$\hat{Y}_0 = \underset{(1\times1)}{\hat{Y}_0} = \underset{(1\times2)(2\times1)}{\mathbf{X}_0' \, \hat{\boldsymbol{\beta}}} \tag{5B.15}$$

In addition it can be shown that

$$\text{Var(pred)} = \sigma^2_{\text{pred}} = [1 + \mathbf{X}_0'(X'X)^{-1}\mathbf{X}_0]\sigma^2 \tag{5B.16}$$

which is estimated by

$$\text{Vâr(pred)} = \hat{\sigma}^2_{\text{pred}} = [1 + \mathbf{X}_0'(X'X)^{-1}\mathbf{X}_0]s^2 \tag{5B.17}$$

The estimated standard error of the prediction is then

$$\hat{\sigma}_{\text{pred}} = \sqrt{[1 + \mathbf{X}_0'(X'X)^{-1}\mathbf{X}_0]}\,s \tag{5B.18}$$

The latter quantity can be used to construct prediction intervals [see equation (5.50)] for the future value Y_0.

Summary of Important Equations

We conclude this supplement by listing the important matrix expressions associated with the simple linear regression model.

1. Model and assumptions (in terms of the error).

$$\underset{(n\times1)}{\mathbf{Y}} = \underset{(n\times2)}{X}\ \underset{(2\times1)}{\boldsymbol{\beta}} + \underset{(n\times1)}{\mathbf{e}} \qquad E(\mathbf{e}) = \underset{(n\times1)}{\mathbf{0}}$$

$$\text{Var}(\mathbf{e}) = \underset{(n\times n)}{I}\ \sigma^2$$

2. Least squares estimates and the estimated variance matrix.

$$\underset{(2\times1)}{\hat{\boldsymbol{\beta}}} = \underset{(2\times2)}{(X'X)^{-1}}\underset{(2\times n)}{X'}\ \underset{(n\times1)}{\mathbf{Y}}, \qquad \text{[provided } (X'X)^{-1} \text{ exists]}$$

$$\hat{\text{Var}}(\hat{\boldsymbol{\beta}}) = (X'X)^{-1}s^2$$

3. Fitted values, residuals, and sum of squares.

$$\underset{(n\times1)}{\hat{\mathbf{Y}}} = \underset{(n\times2)}{X}\ \underset{(2\times1)}{\hat{\boldsymbol{\beta}}}$$

$$\underset{(n\times1)}{\hat{\mathbf{e}}} = \underset{(n\times1)}{\mathbf{Y}} - \underset{(n\times1)}{\hat{\mathbf{Y}}}$$

$$\mathbf{Y}'\mathbf{Y} - n\bar{Y}^2 = \hat{\mathbf{Y}}'\hat{\mathbf{Y}} - n\bar{Y}^2 + (\mathbf{Y}-\hat{\mathbf{Y}})'(\mathbf{Y}-\hat{\mathbf{Y}})$$

$$\begin{pmatrix} \text{total SS} \\ \text{about the mean} \end{pmatrix} = \begin{pmatrix} \text{regression} \\ \text{SS} \end{pmatrix} + \begin{pmatrix} \text{residual} \\ \text{SS} \end{pmatrix}$$

4. F ratio and the coefficient of determination.

$$F = \frac{MS_R}{s^2} = \frac{\hat{\mathbf{Y}}'\hat{\mathbf{Y}} - n\bar{Y}^2}{(\mathbf{Y}-\hat{\mathbf{Y}})'(\mathbf{Y}-\hat{\mathbf{Y}})/(n-2)}$$

$$R^2 = \frac{\hat{\mathbf{Y}}'\hat{\mathbf{Y}} - n\bar{Y}^2}{\mathbf{Y}'\mathbf{Y} - n\bar{Y}^2} = 1 - \frac{(\mathbf{Y}-\hat{\mathbf{Y}})'(\mathbf{Y}-\hat{\mathbf{Y}})}{\mathbf{Y}'\mathbf{Y} - n\bar{Y}^2}$$

5. Estimate of a future value and its estimated variance.

$$\underset{(1\times1)}{\hat{Y}_0} = \underset{(1\times2)}{\mathbf{X}_0'}\ \underset{(2\times1)}{\hat{\boldsymbol{\beta}}}, \qquad \text{where} \qquad \mathbf{X}_0 = \begin{bmatrix} 1 \\ X_0 \end{bmatrix}$$

$$\hat{\sigma}^2_{\text{pred}} = [1 + \mathbf{X}_0'(X'X)^{-1}\mathbf{X}_0]s^2$$

PROBLEMS

5.1 Define or give an example of each of the following terms.

(a) least squares line
(b) fitted values
(c) residuals
(d) total (corrected) sum of squares
(e) regression sum of squares
(f) residual sum of squares
(g) coefficient of determination
(h) Gauss-Markov theorem
(i) prediction interval
(j) sample correlation coefficient
(k) autocorrelation
(l) spurious correlation
(m) rank correlation coefficient

5.2   Show that the following expressions are algebraically equivalent:

(a) $b = \dfrac{\Sigma X_i Y_i - \bar{X}\Sigma Y_i}{\Sigma X_i^2 - \bar{X}\Sigma X_i}$

(b) $b = \dfrac{\Sigma(X_i - \bar{X})(Y_i - \bar{Y})}{\Sigma(X_i - \bar{X})^2}$

(c) $b = \dfrac{\Sigma(X_i - \bar{X})Y_i}{\Sigma(X_i - \bar{X})^2}$

5.3   Given the data for Example 5.2 displayed in Table 5.2 fill in the entries in the following ANOVA table.

Source	Sum of Squares (SS)	Degrees of Freedom (d.f.)	Mean Squares (MS)
Regression	——	——	——
Residual	0.17	——	$s^2 = $ ——
Total about the mean	——	——	

5.4   Evaluate R^2 for the housing demand–income example and interpret this number. (Hint: Use the entries in the ANOVA table in the preceding problem.)

5.5   Identify regression models that are linear in the parameters from the following list.
   (a) $E(Y|X) = \alpha + \beta X^2$
   (b) $E(Y|X) = \alpha + (\alpha\beta)X = \alpha(1 + \beta)X$
   (c) $E(Y|X) = \beta X$
   (d) $E(Y|X) = \alpha + \beta \ln X$
   (e) $E(Y|X) = \alpha + (1/\beta)X$
   (f) $E(Y|X) = e^{\beta X}$

5.6   In order to determine the profitability of hiring an additional salesman, the sales manager decides to determine the relationship between the number of salesmen and the average sales in thousands of dollars. After looking at a scatter diagram of the data (obtained over several years),

X (number of salesmen)	2	3	4	5	6
Y (sales in $1000's)	20	27	33	38	43

he postulates the model

$$Y_i = \alpha + \beta X_i + e_i, \qquad i = 1, 2, \ldots, 5$$

where the e_i's are uncorrelated normal variables with mean zero and variance σ^2.

(a)   Calculate the least squares estimates $\hat{\alpha}$ and $\hat{\beta}$.

(b)   Construct the ANOVA table and test the hypothesis $H_0: \beta = 0$ versus $H_a: \beta \neq 0$ using the F ratio (assume a 5-percent significance level).

(c)   You are told that the current number of salesmen is $X_0 = 7$. Construct a 95-percent prediction interval for the corresponding value of sales. Explain why it may be dangerous to predict sales for this particular value of X.

5.7   A random sample of $n = 6$ overdue accounts receivable was examined, and the following pairs of observations (X_i, Y_i), where X is the account size in hundreds of dollars and Y is the number of days overdue, were obtained,

X (account size, $100's)	10	5	7	19	11	8
Y (number of days overdue)	15	9	3	25	7	13

The postulated model linking number of days overdue to account size is

$$Y_i = \alpha + \beta X_i + e_i, \qquad i = 1, 2, \ldots, 6$$

with

$$E(e_i) = 0, \operatorname{Var}(e_i) = \sigma^2, \operatorname{Cov}(e_i, e_j) = 0, \qquad i \neq j.$$

(a)   Obtain the least squares estimates $\hat{\alpha}$ and $\hat{\beta}$.

(b)   Compute s^2 and construct a 90-percent confidence interval for β. What additional assumption is necessary to insure the validity of this interval?

(c)   Evaluate R^2 and interpret this number.

(d)   Plot the residuals overall and versus X. Given the evidence above do the assumptions appear to be violated?

(e)   If $X_0 = 12$, what can you say about Y_0, the number of days overdue?

5.8   Suppose a straight line has been fitted by the method of least squares to $n = 9$ pairs of observations (X_i, Y_i) yielding the fitted equation

$$\hat{Y} = -5 + 2X$$

(a)   Write down the underlying model implied by the fitted equation.

(b)   The ANOVA table follows:

Source	Sum of Squares (SS)	Degrees of Freedom (d.f.)	Mean Squares (MS)	F Ratio
Regression	—	—	—	—
Residual	—	—	—	
Total (about mean)	380	—		

Given the following deviations $x_i = X_i - \bar{X}$

$$x_i: -4, -3, -2, -1, 0, 1, 2, 3, 4$$

fill in the blanks of the ANOVA table. Use the F test to test $H_0: \beta = 0$ versus $H_a: \beta \neq 0$ at the 5- and 1-percent significance levels.

5.9   Show that if there is an intercept term in the regression model,
   (a)   The least squares line always passes through the point $(\bar{X}, \bar{Y})$.
   (b) $\sum_{i=1}^{n} (Y_i - \hat{Y}_i) = 0$; that is, the residuals sum to zero.

5.10   Demonstrate algebraically that the square of the t value for testing $H_0: \beta = 0$ is the F ratio; that is, show

$$t^2 = \frac{\hat{\beta}^2}{\hat{\sigma}_{\hat{\beta}}^2} = F = \frac{MS_R}{s^2}$$

5.11   Suppose the postulated regression model is of the form

$$Y_i = \beta X_i + e_i, \qquad i = 1, 2, \ldots, n$$

with the usual assumptions on the error term. (Note the graph of the regression function is a straight line through the origin.)
   (a)   Derive an expression for the least squares estimator $\hat{\beta}$.
   (b)   Verify that $\hat{\beta}$ indeed *minimizes* the error sum of squares by examining the second derivative of the sum of squares function with respect to β.

5.12   It has been proposed that the brightness measured in some unit of color of a commercial product is proportional to the time it is in a certain chemical reaction during the production process, or

$$Y_i = \beta X_i + e_i$$

where Y_i measures brightness, X_i measures time, β is a parameter, and e_i is the random error term satisfying the usual assumptions. The following data on X and Y are available:

Y	3.4	7.0	6.0	9.0	8.0	11.0
X	1.0	1.2	1.4	1.6	1.8	2.0

   (a)   Find the least squares estimate $\hat{\beta}$.
   (b)   Calculate the fitted values $\hat{Y}_i$.
   (c)   Calculate the residuals $\hat{e}_i$.
   (d)   Set up the ANOVA table with sum of squares due to regression, $\sum \hat{\beta}^2 X_i^2$, due to error, $\sum (Y_i - \hat{Y}_i)^2$, and total sum of squares. Does β appear to be different from zero? Explain.

5.13   An alternative representation of the simple linear regression model is

$$Y_i = \alpha_0 + \beta x_i + e_i, \qquad i = 1, 2, \ldots, n$$

where $x_i = X_i - \bar{X}$, $E(e_i) = 0$, $\text{Var}(e_i) = \sigma^2$, and $\text{Cov}(e_i, e_j) = 0$, $i \neq j$.

(a) Obtain expressions for the least squares estimators $\hat{\alpha}_0$ and $\hat{\beta}$. Compare the results with the least squares estimators of α and β in the model

$$Y_i = \alpha + \beta X_i + e_i, \quad i = 1, 2, \ldots, n$$

Note that $\hat{Y} = \hat{\alpha}_0 + \hat{\beta}x = \hat{\alpha} + \hat{\beta}X$. Interpret this result geometrically.

(b) Show that $E(\hat{\alpha}_0) = \alpha_0$ and $E(\hat{\beta}) = \beta$ and derive expressions for $\text{Var}(\hat{\alpha}_0)$ and $\text{Var}(\hat{\beta})$.

5.14 Given the usual simple linear regression model (Table 5.4) show that

$$E(\hat{\alpha}) = \alpha \qquad \text{[equation (5.24)]}$$

and

$$\text{Var}(\hat{\alpha}) = \left[\frac{\Sigma X_i^2}{n \, \Sigma(X_i - \bar{X})^2} \right] \sigma^2 \qquad \text{[equation (5.25)]}$$

5.15 Given the information for Example 5.2 in Table 5.2, plot the residuals
(a) Overall.
(b) Against X_i.
(c) Against $\hat{Y}_i$.
Does the regression model

$$Y_i = \alpha + \beta X_i + e_i, \quad i = 1, 2, \ldots, 5$$

with $E(e_i) = 0$, $\text{Var}(e_i) = \sigma^2$, and $\text{Cov}(e_i, e_j) = 0$, $i \neq j$, appear to be correct? If not, how should it be modified?

5.16 Suppose we are interested in predicting the *mean* value of Y corresponding to the setting $X = X_0$; that is, we want to predict $E(Y_0|X_0) = \mu_{Y_0|X_0} = \alpha + \beta X_0$.
(a) Show that an unbiased estimator of $\mu_{Y_0|X_0}$ is

$$\hat{Y}_0 = \hat{\alpha} + \hat{\beta}X_0$$

(b) A confidence interval for $\mu_{Y_0|X_0}$ can be constructed using an argument similar to the one leading to a prediction interval for the actual value of Y_0 set out in Section 5.5. (Note that the *point* estimators of Y_0 and $\mu_{Y_0|X_0}$ are identical.) The difference is that the variability associated with the predictor comes *only* from the variability of the least squares line about the regression line, which we have denoted by $\sigma_{\hat{Y}_0}^2$ [see equation (5.48)], since there is no inherent variability in $\mu_{Y_0|X_0}$. Therefore the variance associated with the prediction is

$$\sigma_{\text{pred}}^2 = \sigma_{\hat{Y}_0}^2$$

The estimated variance is then $\hat{\sigma}_{\hat{Y}_0}^2$, and a $100(1 - \gamma)$-percent confidence interval for $\mu_{Y_0|X_0}$ is

$$\hat{Y}_0 \pm t_{\gamma/2}(n - 2)\hat{\sigma}_{\hat{Y}_0} = \hat{Y}_0 \pm t_{\gamma/2}(n - 2) \sqrt{\frac{1}{n} + \frac{(X_0 - \bar{X})^2}{\Sigma(X_i - \bar{X})^2}} \, s$$

[compare with (equation 5.50)]

where $t_{\gamma/2}(n-2)$ is the upper $\gamma/2$ percentage point of the t distribution with $(n-2)$ degrees of freedom.

Using the information displayed in Example 5.6 construct a 95-percent confidence interval for the *average* value of field corn seed sales given $X_0 = \$2.20$ per bushel.

5.17 Suppose additional observations on housing demand and disposable income are available and are added to the data of Example 5.2 to give $n = 8$ pairs of observations:

X (disposable income)	6.8	7.0	7.1	7.2	7.4	7.4	6.9	7.1
Y (housing demand)	0.8	1.2	0.9	0.9	1.5	1.2	1.0	1.1

Consider the model

$$Y_i = \alpha + \beta X_i + e_i, \qquad i = 1, 2, \ldots, 8$$

with the usual error assumptions.
(a)   Calculate the least squares estimates $\hat{\alpha}$ and $\hat{\beta}$.
(b)   Write out the ANOVA table.
(c)   Provide an estimate of σ^2 and test for the significance of the regression (at the 5-percent level).

5.18 Given the demand-income data of Example 5.2,
(a)   Obtain the least squares estimates $\hat{\alpha}$ and $\hat{\beta}$ using matrices.
(b)   Display the estimated variance-covariance matrix of the least squares estimates.
(c)   Suppose

$$\mathbf{X}_0 = \begin{bmatrix} 1 \\ 6.9 \end{bmatrix}$$

Obtain a point estimate of Y_0 using matrices.

5.19 A furniture manufacturer feels that it can improve its production scheduling by using the fact that sales of furniture in a given quarter are related to the number of new homes on which construction was started two quarters earlier. The manufacturer decides to study the Los Angeles area. Let Y be quarterly sales in the area in hundreds of thousands of dollars and X be the number of housing starts in thousands, two quarters previously. The data are (read down the columns)

X	Y	X	Y	X	Y
2.0	1.2	5.0	2.5	3.0	2.5
1.5	2.4	4.0	2.3	4.5	2.8
3.0	2.0	3.5	2.0	6.0	3.0
1.1	1.8	2.0	1.8	1.2	1.2

Consider the model

$$Y_i = \alpha + \beta X_i + e_i, \qquad i = 1, 2, \ldots, 12$$

with the usual assumptions on the error term.
(a) Evaluate $X'X, (X'X)^{-1}, X'Y$.
(b) Obtain the least squares estimates $\hat{\beta}$, the fitted values $\hat{Y}$, and the residuals $\hat{e}$.
(c) Set up the ANOVA table and examine the credibility of the hypothesis $H_0: \beta = 0$.
(d) Plot the residuals versus the fitted values. Comment on the appearance of the plot.

5.20 Suppose observation pairs (V_i, W_i) are related to the observation pairs (X_i, Y_i) by

$$V_i = a + bX_i$$
$$i = 1, 2, \ldots, n,$$
$$W_i = c + dY_i$$

where a, b, c, and d are arbitrary constants. Show that the sample correlation coefficient computed from the pairs of observations $(V_i, W_i), i = 1, 2, \ldots, n$, is identical in absolute value to the sample correlation coefficient computed from the pairs of observations $(X_i, Y_i), i = 1, 2, \ldots, n$; that is, the sample correlation coefficient r is invariant under *linear* transformations of the observations. (Note that if b and d are of the same sign, then $r_{xy} = r_{vw}$. Otherwise, $|r_{xy}| = |r_{vw}|$.)

5.21 Data on delinquent accounts receivable was presented in Problem 5.7. The observations $Y_i, i = 1, 2, \ldots, 6$, represent size of delinquent accounts (in 100's), and the observations $X_i, i = 1, 2, \ldots, 6$, represent days overdue.

X	Y
10	15
5	9
7	3
19	25
11	7
8	13

(a) Calculate the sample correlation coefficient and interpret this number.
(b) Test the hypothesis $H_0: \rho = 0$ versus the alternative $H_a: \rho > 0$ at the $\gamma = 0.05$ level of significance. List any assumptions you must make in order to carry out this test. Are these assumptions likely to be satisfied in this instance? Explain.
(c) Given the results in Problem 5.7, (1) verify that the coefficient of determination is the square of the sample correlation coefficient; and (2)

verify that a test of $H_0: \beta = 0$ versus $H_a: \beta > 0$ is equivalent to the test you performed in part (b).

5.22 A new material is being considered for storage battery casings. The casing manufacturer is interested in the relationship between the thickness of the new material and its "resistance to corrosion." Measurements of "resistance to corrosion" and thickness are recorded for 15 pieces of the new material. The results are

X (thickness)	17	32	76	41	63	22	19	51
Y (resistance to corrosion)	0.6	4.7	6.7	5.1	5.5	3.2	2.1	4.9
	54	26	33	12	48	37	72	
	3.8	2.4	2.7	1.9	5.0	4.1	8.3	

Do thickness and resistance to corrosion tend to move together? Compute the sample correlation coefficient and test the hypothesis $H_0: \rho = 0$ versus $H_a: \rho > 0$. (Let $\gamma = 0.05$).

5.23 Let $t = \hat{\beta}/\hat{\sigma}_{\hat{\beta}}$ be the t value associated with the slope coefficient in a simple linear regression model. Show that

$$t = \frac{\hat{\beta}}{\hat{\sigma}_{\hat{\beta}}} = \frac{r\sqrt{n-2}}{\sqrt{1-r^2}}$$

where r is the sample correlation coefficient between the observed values of the dependent and independent variables.

5.24 The end of the week demand deposits for a particular bank over a period of 20 weeks are given below.

Weeks	1	2	3	4	5	6	7	8	9	10
Demand deposits (in appropriate units)	10	13	15	11	9	9	6	4	6	8
Weeks	11	12	13	14	15	16	17	18	19	20
Demand deposits	10	12	13	16	17	15	11	8	6	5

(a) Plot the observations versus weeks. Do successive observations appear to move together?

(b) Let y_t, $t = 1, 2, \ldots, 20$, denote the observed demand deposits. Construct scatter diagrams of the observation pairs (y_t, y_{t+1}), $t = 1, 2, \ldots, 19$, and (y_t, y_{t+2}), $t = 1, 2, \ldots, 18$.

(c) Calculate the sample autocorrelation coefficients r_1 and r_2. Interpret these numbers. Are the values of r_1 and r_2 consistent with the scatter diagrams of part (b)? Explain.

(d) Plot the sample autocorrelations r_k versus the lag k, $k = 1, 2$.

5.25 Two judges are asked to rank 10 pie crusts on the basis of "flakiness." The results are

Pie Crust	Judge Number 1	Judge Number 2
1	6	4
2	4	1
3	3	6
4	1	7
5	2	5
6	7	8
7	9	10
8	8	9
9	10	3
10	5	2

(a) Compute the rank correlation coefficient.
(b) Do the judges tend to agree? Explain.

5.26 The following are two sets of ranks assigned by two corporation presidents to 12 characteristics for "successful" business leadership (ranked in order of importance):

Characteristic	President Number 1	President Number 2
A	11	6
B	1	9
C	2	3
D	9	12
E	8	2
F	6	7
G	7	5
H	5	1
I	4	10
J	3	4
K	10	8
L	12	11

(a) Compute the rank correlation coefficient and interpret this number.
(b) Does the rank correlation coefficient appear to be significantly different from zero? (Use $\gamma = 0.10$.) Interpret the test of significance in terms of the underlying population.

5.27 Consider the data in Problem 5.22. Convert the observations on thickness and resistance to corrosion to ranks.
  (a) Compute the rank correlation coefficient and test the hypothesis H_0: X and Y are independent versus the alternative H_a: X and Y tend to move together (monotone increasing). (Use $\gamma = 0.05$.) Is the result consistent with the result obtained from the test of hypothesis in Problem 5.22. Should it be? Explain.
  (b) Compare the value of the rank correlation coefficient with the value of the ordinary sample correlation coefficient obtained in Problem 5.22. Should they be identical? Comment on any observed difference.

5.28 Discuss correlation and causation.

5.29 Discuss the relationship between correlation and simple linear regression.

5.30 Data are collected on the daily closing price and daily volume in hundreds of shares for the common stock of a certain company listed on the New York Stock Exchange. The data for 15 consecutive trading days are listed below.

t (day)	1	2	3	4	5	6	7	8
X (closing price)	66.5	66.5	66.4	67.5	68.8	66.5	66.3	67.5
Y (volume in 100's)	29.8	34.6	35.0	39.1	44.0	28.1	32.7	30.7

	9	10	11	12	13	14	15
	66.6	66.3	66.3	66.6	67.0	67.5	66.8
	30.3	44.7	33.1	37.1	30.4	30.1	28.1

  (a) Construct a table of the *changes* in closing price (successive differences) and the *changes* in volume.
  (b) Given the data of part (a), compute the sample correlation coefficient between the *changes* in price and *changes* in volume. Do changes in price appear to be related to changes in volume? Explain.

5.31 (a) Given the data in Problem 5.30, part (a), evaluate the lag 1 sample autocorrelation coefficient for both *changes* in daily closing price and *changes* in daily volume.
  (b) Considering the results in part (a), do successive price changes appear to be independent? Do successive volume changes appear to be independent? Explain.
  (c) Would you expect price changes to behave like volume changes a priori? Explain.
  (d) If successive price *changes* were independent, do you think you could predict tomorrow's price from a record of historical prices? If so, what would your best prediction be? Justify your answer.

6

MULTIPLE REGRESSION
AND CORRELATION

6.1 INTRODUCTION

In this chapter we shall extend the ideas introduced in Chapter 5 to situations in which the dependent variable is related to several independent variables. It frequently happens that many factors influence the behavior of a variable of interest, and if these factors can be explicitly accounted for in the regression function, we would expect to obtain, for example, better predictions. Thus we shall be concerned with relationships of the form

$$Y = f(X_1, X_2, \ldots, X_p) + e \qquad (6.1)$$

where e represents the "error component."

In addition, we shall assume that the regression function can be written as

$$f(X_1, \ldots, X_p) = \beta_0 + \beta_1 X_1 + \cdots + \beta_p X_p \qquad (6.2)$$

where the β's represent the parameters. We stress that in spite of the appearance of the X's in equation (6.2), we are *not* restricted to functions that are linear in the independent variables. For example, the regression function

$$f(X_1) = \beta_0 + \beta_1 X_1 + \beta_2 X_1^2 \qquad (6.3)$$

can be written in the form of equation (6.2) by setting $X_2 = X_1^2$, and it therefore falls within the class of regression functions we are considering.[a] We are, however, restricting ourselves to functions that are linear in the parameters $\beta_0, \beta_1, \ldots, \beta_p$. Finally, we point out that $X_1, X_2, \ldots, X_p$ may represent fewer than p *different* independent variables.

---

[a] The function in equation (6.3) is sometimes referred to as a "curvilinear regression function," since its graph is a common curve, in this case, a parabola.

It is convenient and instructive to discuss multiple regression using matrix algebra, and subsequently, this is the approach we will take. The reader is therefore encouraged to review Appendix 3 and Supplement 5B before proceeding.[b]

6.2 THE REGRESSION MODEL

Suppose that there are n observations on the variable Y corresponding to the n sets of values for the independent variables, $X_{i1}, X_{i2}, \ldots, X_{ip}$, $i = 1, 2, \ldots, n$. Using equations (6.1) and (6.2) we can relate each Y value to corresponding X values as follows:

$$
\begin{aligned}
Y_1 &= \beta_0 + \beta_1 X_{11} + \beta_2 X_{12} + \cdots + \beta_p X_{1p} + e_1 \\
Y_2 &= \beta_0 + \beta_1 X_{21} + \beta_2 X_{22} + \cdots + \beta_p X_{2p} + e_2 \\
&\ \vdots \qquad\qquad \vdots \\
Y_n &= \beta_0 + \beta_1 X_{n1} + \beta_2 X_{n2} + \cdots + \beta_p X_{np} + e_n
\end{aligned}
\tag{6.4}
$$

or equivalently

$$
Y_i = \beta_0 + \beta_1 X_{i1} + \beta_2 X_{i2} + \cdots + \beta_p X_{ip} + e_i, \qquad i = 1, 2, \ldots, n
\tag{6.5}
$$

It is possible then to write the set of n equations in equation (6.4) in matrix notation as

$$
\begin{bmatrix} Y_1 \\ Y_2 \\ \vdots \\ Y_n \end{bmatrix} =
\begin{bmatrix}
1 & X_{11} & X_{12} & \ldots & X_{1p} \\
1 & X_{21} & X_{22} & \ldots & X_{2p} \\
\vdots & \vdots & \vdots & & \vdots \\
1 & X_{n1} & X_{n2} & \ldots & X_{np}
\end{bmatrix}
\begin{bmatrix} \beta_0 \\ \beta_1 \\ \vdots \\ \beta_p \end{bmatrix} +
\begin{bmatrix} e_1 \\ e_2 \\ \vdots \\ e_n \end{bmatrix}
$$

or

$$
\underset{(n \times 1)}{Y} = \underset{[n \times (p+1)]}{X} \ \underset{[(p+1) \times 1]}{\beta} + \underset{(n \times 1)}{e}
\tag{6.6}
$$

where Y is the vector of dependent variable values, X is the matrix containing the values of the independent variables, β is the vector of parameters,[c] and e is the vector of error components assumed to be uncorrelated with mean zero and common variance σ^2. If we let X_0 be a column vector of ones and X_i, $i = 1, 2, \ldots, p$, be a column vector of independent variable values, then the X matrix has the form

$$
X = (X_0, X_1, \ldots, X_p)
$$

[b] Most of the material in Sections 6.2–6.4 is summarized for the special case $p = 1$ in Appendix 5B.

[c] For multiple regression functions it is convenient to use one symbol, β, to denote a parameter and then distinguish between different parameters with subscripts.

and we could alternatively express **Y** as

$$\mathbf{Y} = \beta_0 \mathbf{X}_0 + \beta_1 \mathbf{X}_1 + \cdots + \beta_p \mathbf{X}_p + \mathbf{e} \tag{6.7}$$

The $\mathbf{X}_0$ vector is included in the X matrix to allow for the β_0 term in the regression model. If this term is not present in the model, the column of ones is omitted.[d]

Equation (6.6) is the matrix form of the multiple regression model. As in the case of simple regression we assume that the model is correctly specified in the sense that within the appropriate range of the X values, the error component represents random fluctuation and not a systematic deviation due to the omission of an important independent variable. Thus the regression function characterizes the main features of the system being studied and, with the assumptions on the error component, represents the *expected value* of Y given a set of independent variable values. In standard notation, then, we have the relationship

$$E(Y|X_1, \ldots, X_p) = f(X_1, X_2, \ldots, X_p) = \beta_0 + \beta_1 X_1 + \cdots + \beta_p X_p$$

for each setting of the independent variables.

The Least Squares Values

The immediate objective of a regression analysis is to determine completely the function $f(X_1, \ldots, X_p) = E(Y|X_1, \ldots, X_p)$. If the *form* of this function is specified, then the problem reduces to one of specifying the values of the parameters $\beta_0, \beta_1, \ldots, \beta_p$ which, in general, are unknown. The parameter values can be estimated, however, using the principle of least squares. In particular, once the data are available, we choose as the estimates of $\beta_0, \beta_1, \ldots, \beta_p$ those values that minimize

$$S(\beta_0, \beta_1, \ldots, \beta_p) = \sum_{i=1}^{n} [Y_i - E(Y_i|X_{i1}, X_{i2}, \ldots, X_{ip})]^2$$

$$= \sum_{i=1}^{n} (Y_i - \beta_0 - \beta_1 X_{i1} - \cdots - \beta_p X_{ip})^2$$

$$= \sum_{i=1}^{n} e_i^2$$

the sum of squared deviations of the observed Y_i's from the corresponding values given by the regression function.[e] Differentiating $S(\beta_0, \beta_1, \ldots, \beta_p)$

---

[d] We note that the X matrix is frequently called the "design" matrix since in many situations the set of values of the independent variables is known as an experimental plan, or experimental design. The selection of an appropriate design matrix is an important problem; however, in many instances the selection of these values is not under the control of the investigator, and he must then be content with data that are routinely recorded. For this reason we do not discuss the design problem in this book, focusing attention instead on the question how, given the data, to proceed with the analysis.

[e] Since $\sum_{i=1}^{n} e_i^2 = \mathbf{e}'\mathbf{e}$, we can write $S(\beta_0, \beta_1, \ldots, \beta_p)$ in matrix notation as $S(\beta_0, \beta_1, \ldots, \beta_p) = \mathbf{e}'\mathbf{e} = (\mathbf{Y} - X\boldsymbol{\beta})'(\mathbf{Y} - X\boldsymbol{\beta})$. The last equality follows from equation (6.6).

with respect to each parameter produces a set of $p + 1$ equations in $p + 1$ unknowns. If each of the $p + 1$ functions $\partial S(\beta_0, \beta_1, \ldots, \beta_p)/\partial \beta_i$, $i = 0, 1, 2, \ldots, p$, is equated to zero, the result is a system of simultaneous equations that can be solved for the stationary point $(\hat{\beta}_0, \hat{\beta}_1, \ldots, \hat{\beta}_p)$. It can be shown that this point in fact *minimizes* $S(\beta_0, \beta_1, \ldots, \beta_p)$ so that $\hat{\beta}_0, \hat{\beta}_1, \ldots, \hat{\beta}_p$ are the least squares values. The simultaneous equations defining the stationary point can be written in matrix notation as

$$(X'X)\hat{\beta} = X'\mathbf{Y} \tag{6.8}$$

These equations are known as the *normal equations*. Consequently, provided $(X'X)^{-1}$ exists, the least squares values have the representation

$$\hat{\beta} = (X'X)^{-1}X'\mathbf{Y} \tag{6.9}$$

Equation (6.9) is important in regression analysis since it provides an expression for obtaining the least squares values for *any* multiple regression model linear in the parameters. Once we have specified the X matrix and the Y vector, $\hat{\beta} = (\hat{\beta}_0, \hat{\beta}_1, \ldots, \hat{\beta}_p)'$ can be determined by performing the operations indicated on the right-hand side of this equation.

EXAMPLE 6.1

An economist is interested in the relationship between the demand for housing (as measured by housing starts), price, and national disposable income. We shall let Y be the housing demand in appropriate units, X_1 be a variable representing "average price," and X_2 be a variable representing disposable income. The values of the variables (obviously fictitious) for six time periods are given below, and plotted in Figure 6.1.

Period	Y	X_1	X_2
1	0.8	12	6.8
2	0.9	13	7.2
3	1.2	13	7.4
4	0.9	14	7.1
5	1.2	14	7.0
6	1.5	15	7.4

It is clear from Figure 6.1 that Y *tends* to move linearly with X_1 and linearly with X_2. Therefore assuming the effects of X_1 and X_2 to be additive, the economist postulates the regression model

$$Y_i = \beta_0 + \beta_1 X_{i1} + \beta_2 X_{i2} + e_i, \qquad i = 1, 2, \ldots, 6$$

Substituting the observed values for Y, X_1, and X_2, we can write

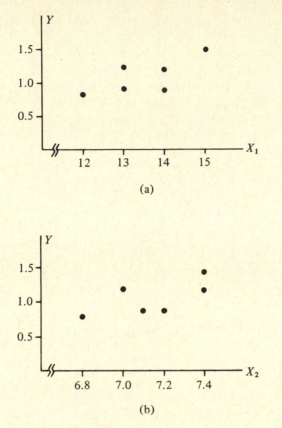

(a)

(b)

Figure 6.1 Plots of housing demand, Y, versus the independent variables, price, X_1, and disposable income, X_2.

$$
\begin{bmatrix} 0.8 \\ 0.9 \\ 1.2 \\ 0.9 \\ 1.2 \\ 1.5 \end{bmatrix} =
\begin{bmatrix} 1 & 12 & 6.8 \\ 1 & 13 & 7.2 \\ 1 & 13 & 7.4 \\ 1 & 14 & 7.1 \\ 1 & 14 & 7.0 \\ 1 & 15 & 7.4 \end{bmatrix}
\begin{bmatrix} \beta_0 \\ \beta_1 \\ \beta_2 \end{bmatrix} +
\begin{bmatrix} e_1 \\ e_2 \\ e_3 \\ e_4 \\ e_5 \\ e_6 \end{bmatrix}
$$

or

$$
\underset{(6\times1)}{\mathbf{Y}} = \underset{(6\times3)}{X}\ \underset{(3\times1)}{\boldsymbol{\beta}} + \underset{(6\times1)}{\mathbf{e}}
$$

We see that **Y** is simply the vector of observed Y values and that the matrix X, apart from a column of ones, contains the observed values of the variables X_1 and X_2 in columns 2 and 3. From equation (6.9), the least squares values $\hat{\boldsymbol{\beta}} = (\hat{\beta}_0, \hat{\beta}_1, \hat{\beta}_2)'$ are given by $\hat{\boldsymbol{\beta}} = (X'X)^{-1}X'\mathbf{Y}$. To illustrate the theory the calculations leading to these values are summarized below. Ordinarily these computations are performed on a computer, and multiple regression programs are available for this purpose.

1. Evaluate $(X'X) = \begin{bmatrix} 1 & 1 & 1 & 1 & 1 & 1 \\ 12 & 13 & 13 & 14 & 14 & 15 \\ 6.8 & 7.2 & 7.4 & 7.1 & 7.0 & 7.4 \end{bmatrix} \begin{bmatrix} 1 & 12 & 6.8 \\ 1 & 13 & 7.2 \\ 1 & 13 & 7.4 \\ 1 & 14 & 7.1 \\ 1 & 14 & 7.0 \\ 1 & 15 & 7.4 \end{bmatrix}$

$= \begin{bmatrix} 6 & 81 & 42.9 \\ 81 & 1099 & 579.8 \\ 42.9 & 579.8 & 307.01 \end{bmatrix}$

2. Evaluate $(X'X)^{-1} = \begin{bmatrix} 188.9832 & 0.8578 & -28.0275 \\ 0.8578 & 0.2523 & -0.5963 \\ -28.0275 & -0.5963 & 5.0459 \end{bmatrix}$

(see Appendix 3).

3. Evaluate $X'\mathbf{Y} = \begin{bmatrix} 1 & 1 & 1 & 1 & 1 & 1 \\ 12 & 13 & 13 & 14 & 14 & 15 \\ 6.8 & 7.2 & 7.4 & 7.1 & 7.0 & 7.4 \end{bmatrix} \begin{bmatrix} 0.8 \\ 0.9 \\ 1.2 \\ 0.9 \\ 1.2 \\ 1.5 \end{bmatrix} = \begin{bmatrix} 6.5 \\ 88.8 \\ 46.69 \end{bmatrix}$

4. Evaluate $\hat{\boldsymbol{\beta}} = (X'X)^{-1}X'\mathbf{Y} = \begin{bmatrix} 188.9832 & 0.8578 & -28.0275 \\ 0.8578 & 0.2523 & -0.5963 \\ -28.0275 & -0.5963 & 5.0459 \end{bmatrix} \begin{bmatrix} 6.5 \\ 88.8 \\ 46.69 \end{bmatrix}$

$= \begin{bmatrix} -4.0419 \\ 0.1367 \\ 0.4587 \end{bmatrix} = \begin{bmatrix} \hat{\beta}_0 \\ \hat{\beta}_1 \\ \hat{\beta}_2 \end{bmatrix}$

The fitted regression function is then

$$\hat{Y} = \hat{\beta}_0 + \hat{\beta}_1 X_1 + \hat{\beta}_2 X_2$$
$$= -4.04 + 0.14 X_1 + 0.46 X_2$$

The Fitted Values and Residuals

The fitted values are always given by evaluating the regression function with the least squares values inserted for the parameters, $\beta_0, \beta_1, \beta_2, \ldots, \beta_p$. Using the notation introduced in Chapter 5, the fitted values are then

$$\hat{Y}_i = \hat{\beta}_0 + \hat{\beta}_1 X_{i1} + \cdots + \hat{\beta}_p X_{ip}, \qquad i = 1, 2, \ldots, n$$

or in matrix notation

$$\hat{\mathbf{Y}} = X\hat{\boldsymbol{\beta}} \tag{6.10}$$

The residuals, $\hat{e}_i = Y_i - \hat{Y}_i$, $i = 1, 2, \ldots, n$, become in matrix notation

$$\hat{\mathbf{e}} = \mathbf{Y} - \hat{\mathbf{Y}} \tag{6.11}$$

The advantage of the matrix approach is that the fitted values and residuals can be easily determined by performing the matrix operations indicated in equations (6.10) and (6.11).

EXAMPLE 6.1 (*Continued*)

Using results from Example 6.1, the fitted values for our housing demand example are

$$\hat{Y} = X\hat{\beta} = \begin{bmatrix} 1 & 12 & 6.8 \\ 1 & 13 & 7.2 \\ 1 & 13 & 7.4 \\ 1 & 14 & 7.1 \\ 1 & 14 & 7.0 \\ 1 & 15 & 7.4 \end{bmatrix} \begin{bmatrix} -4.0419 \\ 0.1367 \\ 0.4587 \end{bmatrix} = \begin{bmatrix} 0.72 \\ 1.04 \\ 1.13 \\ 1.13 \\ 1.08 \\ 1.40 \end{bmatrix}$$

and the vector of residuals is

$$\hat{e} = Y - \hat{Y} = \begin{bmatrix} 0.8 \\ 0.9 \\ 1.2 \\ 0.9 \\ 1.2 \\ 1.5 \end{bmatrix} - \begin{bmatrix} 0.72 \\ 1.04 \\ 1.13 \\ 1.13 \\ 1.08 \\ 1.40 \end{bmatrix} = \begin{bmatrix} 0.08 \\ -0.14 \\ 0.07 \\ -0.23 \\ 0.12 \\ 0.10 \end{bmatrix}$$

6.3 ANALYSIS OF VARIANCE

The reader will recall that in simple linear regression it was possible to partition the total variability in Y into a portion that is due to the apparent movement of Y with X (the explained variation) and a portion that is not due to the movement of Y with X (the unexplained or residual variation). The same thing is true for multiple regression. That part of the total variability in Y, $\sum_{i=1}^{n}(Y_i - \bar{Y})^2$, explained by the estimated relationship of Y with $X_1, X_2, \ldots, X_p$ is given by

$$\sum_{i=1}^{n}(\hat{Y}_i - \bar{Y})^2 = \sum_{i=1}^{n}(\hat{\beta}_0 + \hat{\beta}_1 X_{i1} + \cdots + \hat{\beta}_p X_{ip} - \bar{Y})^2$$

The variability in Y remaining is called the "unexplained" (residual) variation and is given by $\sum_{i=1}^{n}(Y_i - \hat{Y}_i)^2$; that is, we can again write down the sum of squares breakup,

$$\sum_{i=1}^{n}(Y_i - \bar{Y})^2 = \sum_{i=1}^{n}(\hat{Y}_i - \bar{Y})^2 + \sum_{i=1}^{n}(Y_i - \hat{Y}_i)^2$$

$$\begin{pmatrix} \text{total sum} \\ \text{of squares about} \\ \text{the mean} \end{pmatrix} = \begin{pmatrix} \text{regression} \\ \text{sum of squares} \end{pmatrix} + \begin{pmatrix} \text{residual} \\ \text{sum of squares} \end{pmatrix}$$

where the terms on the right-hand side represent the explained and unexplained contributions to the total variability in Y, respectively.

The sum of squares breakup, using matrix notation, and the corresponding degrees of freedom are

$$\text{Sum of squares: } (\mathbf{Y}'\mathbf{Y} - n\bar{Y}^2) = (\hat{\mathbf{Y}}'\hat{\mathbf{Y}} - n\bar{Y}^2) + (\mathbf{Y} - \hat{\mathbf{Y}})'(\mathbf{Y} - \hat{\mathbf{Y}})$$

$$\begin{pmatrix} \text{total sum of} \\ \text{squares about} \\ \text{mean} \end{pmatrix} = \begin{pmatrix} \text{regression} \\ \text{sum of} \\ \text{squares} \end{pmatrix} + \begin{pmatrix} \text{residual} \\ \text{sum of} \\ \text{squares} \end{pmatrix} \quad (6.12)$$

$$\text{d.f.: } (n-1) = p + (n-p-1)$$

The reader should note, in particular, that the number of degrees of freedom associated with the regression sum of squares is the number of independent variable terms in the regression function.

The Analysis of Variance Table

The identities given in equation (6.12) can be displayed in an ANOVA table to facilitate the calculation of statistics useful for making inferences about the nature of the underlying regression function.

The ANOVA table associated with the multiple regression problem is a straightforward generalization of the ANOVA table introduced in Chapter 5, and the interpretation of the table entries is precisely the same. The notable differences are the use of matrix notation and the number of degrees of freedom associated with the residual and regression sum of squares. The general form of the ANOVA table is given in Table 6.1.[f]

Table 6.1   Analysis of Variance Table

Source	Sum of Squares (SS)	Degrees of Freedom (d.f.)	Mean Squares (MS)	F Ratio
Regression	$\hat{\mathbf{Y}}'\hat{\mathbf{Y}} - n\bar{Y}^2$	p	$MS_R = (\hat{\mathbf{Y}}'\hat{\mathbf{Y}} - n\bar{Y}^2)/p$	$F = MS_R/s^2$
Residual	$(\mathbf{Y} - \hat{\mathbf{Y}})'(\mathbf{Y} - \hat{\mathbf{Y}})$	$n - p - 1$	$s^2 = (\mathbf{Y} - \hat{\mathbf{Y}})'(\mathbf{Y} - \hat{\mathbf{Y}})/(n - p - 1)$	
Total (corrected)	$\mathbf{Y}'\mathbf{Y} - n\bar{Y}^2$	$n - 1$		

[f] It is common practice to refer to the total sum of squares about the mean as the total sum of squares corrected for the mean, or simply as "total SS (corrected)." In addition it is convenient to drop the reference to $\bar{Y}$ when referring to the regression sum of squares; however, the reader should keep in mind that the "regression SS" will always refer to the variation of the regression function about $\bar{Y}$ unless we specifically indicate otherwise.

The residual mean square, s^2, provides an estimate of the conditional variance of Y if the regression function is correct (that is, if there are no important terms missing in the regression function that produce systematic behavior in the error component). This quantity is useful in making inferences about the regression parameters $\beta_0, \beta_1, \ldots, \beta_p$ and future Y values.

The F ratio, $F = MS_R/s^2$, can be used to examine the credibility of the hypothesis, $H_0: \beta_1 = \beta_2 = \cdots = \beta_p = 0$. This is equivalent to asserting that the behavior of Y is not influenced by *any* of the independent variables. Of course when $p = 1$, H_0 reduces to the hypothesis $H_0: \beta_1 = 0$ (considered in Chapter 5), and the F ratio can be used to make inferences about the single parameter β_1. In general, however, the F ratio is associated with inferences about the p parameters as a group.

Further comments on the usefulness of the F ratio and s^2 are delayed until the next section in which we discuss inference problems in some detail.

The Coefficient of Determination

The reader will recall that the proportion of variation in Y about $\bar{Y}$ that is explained by the relationship of Y with the independent variables is known as the "coefficient of determination," R^2. In matrix notation

$$R^2 = \frac{\hat{\mathbf{Y}}'\hat{\mathbf{Y}} - n\bar{Y}^2}{\mathbf{Y}'\mathbf{Y} - n\bar{Y}^2} = 1 - \frac{(\mathbf{Y} - \hat{\mathbf{Y}})'(\mathbf{Y} - \hat{\mathbf{Y}})}{\mathbf{Y}'\mathbf{Y} - n\bar{Y}^2} \tag{6.13}$$

It is always possible to increase R^2 by adding independent variables to the regression function (although the increase may be negligible for all practical purposes). Intuitively, this follows from the fact that the additional independent variable terms provide the regression function with "greater flexibility" in explaining the behavior of Y, and hence the fitted regression function will ordinarily pass closer to the observed Y values. Thus the regression sum of squares will increase (or equivalently the residual sum of squares will decrease), whereas the total corrected sum of squares remains the same. However, there is a danger here that can be illustrated by considering the addition of the single term $\beta_{p+1}X_{p+1}$ to the regression function

$$f(X_1, X_2, \ldots, X_p) = \beta_0 + \beta_1 X_1 + \cdots + \beta_p X_p$$

It is likely that R^2 will increase with the addition of $\beta_{p+1}X_{p+1}$ even if $\beta_{p+1} = 0$ since $\hat{\beta}_{p+1}$ will ordinarily be unequal to zero. In addition, the estimated conditional *variance* of Y (given by s^2) may not be reduced by taking X_{p+1} into account although the residual variation (residual sum of squares) will decrease.[g] Therefore additional variables must be selected with some care (see the discussion of variable selection procedures in Chapter 7).

6.4 INFERENCE

There are three problems of inference that are of interest in regression analysis. These are (a) inferences about the *set* of parameters $\beta_1, \beta_2, \ldots, \beta_p$, or equivalently, the "significance" of the regression, (b) inferences about the individual parameter β_i, or equivalently, the "significance" of the $\beta_i X_i$ term in the regression model, and finally, (c) inferences about a future observation, Y_0, corresponding to the setting of the independent variables, $X_{01}, X_{02}, \ldots, X_{0p}$.

We shall consider each of these problems in turn, presenting the main results rather succinctly. The reader is encouraged to work many of the problems at the end of the chapter in order to develop an appreciation for the efficacy of these procedures.

Before proceeding, it is convenient to list the assumptions that provide the basis for the inferential procedures that follow. These assumptions (given in terms of the error) have already been introduced in Chapter 5 (see Table 5.4 and the associated discussion) and are simply reproduced here in matrix notation. We assume[h]

1. $E(\mathbf{e}) = \mathbf{0}$
2. $\text{Var}(\mathbf{e}) = I\sigma^2$ (6.14)
3. $\mathbf{e}' = (e_1, \ldots, e_n)$ is such that each $e_i \sim N(0, \sigma^2)$

In words, assumption 1 implies that the expected value of each e_i is zero, assumption 2 implies the error components are uncorrelated and have the same variance (which does not depend on the values of the independent variables or the values of the dependent variable), and assumption 3 implies that the errors are normally distributed.

It is standard practice to consolidate assumptions 1, 2, and 3 in expression (6.14) and write

$$\mathbf{e} \sim N(\mathbf{0}, I\sigma^2) \qquad\qquad (6.15)$$

which is interpreted as "the components e_i are independently and normally distributed each with mean 0 and variance σ^2."

---

[g] A modified version of R^2, known as the "coefficient of determination corrected for degrees of freedom," is sometimes used to assess the strength of the estimated relationship. This quantity, denoted by $\bar{R}^2$, is given by

$$\bar{R}^2 = 1 - \frac{(\mathbf{Y} - \hat{\mathbf{Y}})'(\mathbf{Y} - \hat{\mathbf{Y}})/(n - p - 1)}{(\mathbf{Y}'\mathbf{Y} - n\bar{Y}^2)/(n - 1)} = 1 - \left(\frac{n - 1}{n - p - 1}\right)(1 - R^2)$$

and represents the proportionate reduction in the estimated *variance* of Y when $X_1, X_2, \ldots, X_p$ are taken into account. For n large and much greater than p, R^2 and $\bar{R}^2$ will be nearly the same.

[h] If the X's are random variables, we also assume that each X_i is uncorrelated with the errors.

As we have pointed out, the regression function represents the expected value of Y for a given setting of the independent variables. To recapitulate,

$$E(Y_i|X_{i1}, X_{i2}, \ldots, X_{ip}) = f(X_{i1}, X_{i2}, \ldots, X_{ip})$$
$$= \beta_0 + \beta_1 X_{i1} + \beta_2 X_{i2} + \cdots + \beta_p X_{ip}$$

or in matrix notation

$$E(\mathbf{Y}|X_1, \ldots, X_p) = X\boldsymbol{\beta} \tag{6.16}$$

Multiple regression analysis is concerned for the most part with identifying and estimating this conditional expectation, in other words, with specifying the form of $f(X_{i1}, \ldots, X_{ip})$ and estimating the parameter vector $\boldsymbol{\beta}$.

The Set of Regression Parameters $\beta_1, \beta_2, \ldots, \beta_p$

In Chapter 5 we saw that if the hypothesis $H_0: \beta = 0$ is true, then the ratio $F = MS_R/s^2$ is distributed as an F random variable. For multiple regression models, the ratio $F = MS_R/s^2$ is distributed as an F random variable provided the hypothesis $H_0: \beta_1 = \beta_2 = \cdots = \beta_p = 0$ is true. In symbols, if $H_0: \beta_1 = \beta_2 = \cdots = \beta_p = 0$ holds,

$$F = \frac{MS_R}{s^2} = \frac{(\hat{\mathbf{Y}}'\hat{\mathbf{Y}} - n\bar{Y}^2)/p}{(\mathbf{Y} - \hat{\mathbf{Y}})'(\mathbf{Y} - \hat{\mathbf{Y}})/(n - p - 1)} \sim F(p, n - p - 1) \tag{6.17}$$

A test of the hypothesis $H_0: \beta_1 = \beta_2 = \cdots = \beta_p = 0$ versus the alternative H_a: at least one $\beta_i \neq 0$, $i = 1, 2, \ldots, p$, is provided by comparing the observed F ratio (obtained from the ANOVA table) with a percentage point of an F distribution with p and $n - p - 1$ degrees of freedom. The appropriate test (at the γ level of significance) is

Reject H_0 if $F > F_\gamma(p, n - p - 1)$
Do not reject H_0 if $F \leq F_\gamma(p, n - p - 1)$

where $F_\gamma(p, n - p - 1)$ is the upper γ percentage point of the F distribution with p and $n - p - 1$ degrees of freedom. We note that a test of $H_0: \beta_1 = \beta_2 = \cdots = \beta_p = 0$ is a test of the "significance of the regression" since clearly if H_0 is *not* rejected, we must conclude that the independent variables as a group do not improve the prediction of Y over that obtained by regression Y on the constant β_0. (Regressing Y on the constant β_0 would produce the least squares estimate $\hat{\beta}_0 = \bar{Y}$.)

Individual Regression Parameters

Since the components of the Y vector are regarded as random variables before the observations are in hand, it follows that the least squares values, given by the components of the vector

$$\hat{\boldsymbol{\beta}} = (X'X)^{-1}X'\mathbf{Y} = A\mathbf{Y} \tag{6.18}$$

are random variables; that is, fluctuations in Y are transformed to fluctuations in $\hat{\boldsymbol{\beta}}$ by the fixed matrix A. Now if assumption 1 in equation (6.14) is satisfied [or equivalently equation (6.16) holds] then using equation (6.18), it can be shown that

$$E(\hat{\boldsymbol{\beta}}) = E(A\mathbf{Y}) = AE(\mathbf{Y}) = AX\boldsymbol{\beta}$$
$$= (X'X)^{-1}(X'X)\boldsymbol{\beta} = \boldsymbol{\beta} \tag{6.19}$$

that is, the least squares values are unbiased estimators of the corresponding regression parameters, and henceforth we shall refer to $\hat{\boldsymbol{\beta}}$ as the least squares estimator.[i]

If assumptions 1 and 2 in expression (6.14) are satisfied,

$$\text{Var}(\mathbf{Y}) = I\sigma^2 \tag{6.20}$$

and consequently[j]

$$\text{Var}(\hat{\boldsymbol{\beta}}) = (X'X)^{-1}\sigma^2 = C\sigma^2 \tag{6.21}$$

Generally σ^2 is unknown. However an unbiased estimator of σ^2, given assumptions 1 and 2, is provided by the residual mean square, $s^2 = (\mathbf{Y} - \hat{\mathbf{Y}})'(\mathbf{Y} - \hat{\mathbf{Y}})/(n - p - 1)$. Therefore the *estimated* variance matrix of the least squares estimator is

$$\hat{\text{Var}}(\hat{\boldsymbol{\beta}}) = (X'X)^{-1}s^2 = Cs^2 \tag{6.22}$$

This estimated variance matrix is a $(p + 1) \times (p + 1)$ symmetric matrix with elements $s^2 C_{ij}$, $i = 0, 1, \ldots, p$, $j = 0, 1, \ldots, p$. Notice that we have labeled the first row and first column of Cs^2 as the zero row and column, respectively. (This is done to keep our notation consistent; for the entries in the first row and first column are associated with the parameter β_0, and the remaining matrix entries are associated with the parameters $\beta_1, \beta_2, \ldots, \beta_p$.) In particular

$$Cs^2 = \begin{bmatrix} s^2C_{00} & s^2C_{01} & s^2C_{02} & \cdots & s^2C_{0p} \\ s^2C_{10} & s^2C_{11} & s^2C_{12} & \cdots & s^2C_{1p} \\ s^2C_{20} & s^2C_{21} & s^2C_{22} & \cdots & s^2C_{2p} \\ \vdots & \vdots & \vdots & & \vdots \\ s^2C_{p0} & s^2C_{p1} & s^2C_{p2} & \cdots & s^2C_{pp} \end{bmatrix}$$

$$= \begin{bmatrix} \hat{\text{Var}}(\hat{\beta}_0) & \hat{\text{Cov}}(\hat{\beta}_0, \hat{\beta}_1) & \hat{\text{Cov}}(\hat{\beta}_0, \hat{\beta}_2) & \cdots & \hat{\text{Cov}}(\hat{\beta}_0, \hat{\beta}_p) \\ \hat{\text{Cov}}(\hat{\beta}_0, \hat{\beta}_1) & \hat{\text{Var}}(\hat{\beta}_1) & \hat{\text{Cov}}(\hat{\beta}_1, \hat{\beta}_2) & \cdots & \hat{\text{Cov}}(\hat{\beta}_1, \hat{\beta}_p) \\ \hat{\text{Cov}}(\hat{\beta}_0, \hat{\beta}_2) & \hat{\text{Cov}}(\hat{\beta}_1, \hat{\beta}_2) & \hat{\text{Var}}(\hat{\beta}_2) & \cdots & \hat{\text{Cov}}(\hat{\beta}_2, \hat{\beta}_p) \\ \vdots & \vdots & \vdots & & \vdots \\ \hat{\text{Cov}}(\hat{\beta}_0, \hat{\beta}_p) & \hat{\text{Cov}}(\hat{\beta}_1, \hat{\beta}_p) & \hat{\text{Cov}}(\hat{\beta}_2, \hat{\beta}_p) & \cdots & \hat{\text{Var}}(\hat{\beta}_p) \end{bmatrix} \tag{6.23}$$

[i] In addition, the Gauss-Markov theorem applies, and the least squares estimators have the smallest variance of any unbiased estimators that are linear functions of the Y_i's.
[j] The proof of equation (6.21) is available in [2].

The diagonal elements of Cs^2 are the estimated variances of the least squares estimators, and the off-diagonal elements are the estimated covariances.

Finally, if we impose the normality assumption, then it can be shown that

$$t = \frac{\hat{\beta}_i - \beta_i}{\sqrt{\hat{\mathrm{Var}}(\hat{\beta}_i)}} = \frac{\hat{\beta}_i - \beta_i}{\hat{\sigma}_{\hat{\beta}_i}} = \frac{\hat{\beta}_i - \beta_i}{s\sqrt{C_{ii}}} \sim t(n - p - 1), \qquad i = 0, 1, \ldots, p \qquad (6.24)$$

The number of degrees of freedom associated with the t random variable defined in expression (6.24) is the number of degrees of freedom associated with the residual sum of squares.

Expression (6.24) can be used to make inferences about individual regression coefficients and hence the significance of individual terms in the regression function. For example, analogous to the results given in Chapter 5, a $100(1 - \gamma)$-percent confidence interval for β_i, $i = 0, 1, \ldots, p$, is provided by

$$\hat{\beta}_i \pm t_{\gamma/2}(n - p - 1)\hat{\sigma}_{\hat{\beta}_i} = \hat{\beta}_i \pm t_{\gamma/2}(n - p - 1)s\sqrt{C_{ii}} \qquad (6.25)$$

where $t_{\gamma/2}(n - p - 1)$ is the upper $\gamma/2$ percentage point of a t random variable with $(n - p - 1)$ degrees of freedom. Since the $100(1 - \gamma)$-percent confidence interval given by equation (6.25) may be regarded as a set of plausible or "acceptable" values of β_i, we can always "test" the hypothesis $H_0: \beta_i = \beta_{i0}$ (at the γ level of significance) by determining if β_{i0} falls in this interval. Alternatively, the hypothesis $H_0: \beta_i = \beta_{i0}$, $i = 0, 1, \ldots, p$, versus the alternative $H_a: \beta_i \neq \beta_{i0}$ may be tested (at the γ level of significance) by comparing the ratio $t = (\hat{\beta}_i - \beta_{i0})/s\sqrt{C_{ii}}$ with the $t_{\gamma/2}(n - p - 1)$ percentage point using the test,

Reject H_0 if $|t| > t_{\gamma/2}(n - p - 1)$
Do not reject H_0 if $|t| \leq t_{\gamma/2}(n - p - 1)$

Specifically, the credibility of the hypothesis $H_0: \beta_i = 0$ versus the alternative $H_a: \beta_i \neq 0$ is examined by comparing the t value

$$t = \frac{\hat{\beta}_i}{\hat{\sigma}_{\hat{\beta}}} = \frac{\hat{\beta}_i}{s\sqrt{C_{ii}}} \qquad (6.26)$$

with the appropriate percentage point of a t distribution with $(n - p - 1)$ degrees of freedom.

We must be careful in interpreting the results of a test of hypotheses involving a single parameter in a multiple regression problem. For $i = 1, 2, \ldots, p$, an examination of the significance of β_i (and hence the significance of the term $\beta_i X_i$) is an examination of the significance of β_i given all other terms in the regression function.

For example, suppose $p = 3$, and the estimated regression function has the form

$$\hat{Y} = \hat{\beta}_0 + \hat{\beta}_1 X_1 + \hat{\beta}_2 X_2 + \hat{\beta}_3 X_3$$

Then rejection of the hypothesis $H_0: \beta_1 = 0$ may be interpreted as indicating that the inclusion of the X_1 variable in the regression function significantly improves the prediction of Y over that obtained by regression Y on X_2 and X_3. On the other hand if $H_0: \beta_2 = 0$ is not rejected, we would conclude that the variable X_2 does *not* significantly improve the prediction of Y over that obtained by regressing Y on X_1 and X_3. Finally, if $H_0: \beta_3 = 0$ is not rejected, we would conclude that the variable X_3 does *not* significantly improve the prediction of Y over that obtained by regressing Y on X_1 and X_2. Thus inferences about individual parameters are conditional on all other terms in the regression function. We *could not conclude* from the information just presented that X_2 and X_3 *together* do not significantly improve the prediction of Y over that obtained by regressing Y on X_1 (unless the variables X_1, X_2, and X_3 happened to be independent of one another; that is, unless $X'X$ is a diagonal matrix). To verify this, we must test the hypothesis[k] $H_0: \beta_2 = \beta_3 = 0$.

EXAMPLE 6.2

The ANOVA table associated with the data of Example 6.1 is given in Table 6.2.
Since $F = 3.421 < F_{0.05}(2, 3) = 9.55$, the hypothesis $H_0: \beta_1 = \beta_2 = 0$ is not rejected at the 5-percent level of significance, and we would conclude that the regression is not

Table 6.2   ANOVA Table

Source	Sum of Squares (SS)	Degrees of Freedom (d.f.)	Mean Squares (MS)	F Ratio
Regression	0.242	2	0.121	$F = \dfrac{0.121}{0.035} = 3.421$
Residual	0.105	3	$s^2 = 0.035$	
Total (corrected)	0.347	5		

significant even though a proportion $R^2 = (0.242)/(0.347) = 0.695$ of the variation in Y (housing demand) is explained by the "relationship" of Y with X_1 (price) and X_2 (disposable income). (We saw that the data in Example 6.1 are artificial, and therefore the conclusions we reach here may not correspond to reality.)
To examine the regression coefficients individually, we would proceed as follows:

$$\text{Vâr}(\hat{\beta}) = (X'X)^{-1} s^2 = \begin{bmatrix} \text{Vâr}(\hat{\beta}_0) & \text{Côv}(\hat{\beta}_0, \hat{\beta}_1) & \text{Côv}(\hat{\beta}_0, \hat{\beta}_2) \\ \text{Côv}(\hat{\beta}_0, \hat{\beta}_1) & \text{Vâr}(\hat{\beta}_1) & \text{Côv}(\hat{\beta}_1, \hat{\beta}_2) \\ \text{Côv}(\hat{\beta}_0, \hat{\beta}_2) & \text{Côv}(\hat{\beta}_1, \hat{\beta}_2) & \text{Vâr}(\hat{\beta}_2) \end{bmatrix}$$

[k] The appropriate procedure for testing the equality of a set of $q(q < p)$ regression parameters can be found in reference [2].

$$= \begin{bmatrix} 188.9832 & 0.8578 & -28.0275 \\ 0.8578 & 0.2523 & -0.5963 \\ -28.0275 & -0.5963 & 5.0459 \end{bmatrix} (0.035)$$

Therefore $\hat{\sigma}_{\hat{\beta}_0} = 2.5862$, $\hat{\sigma}_{\hat{\beta}_1} = 0.0945$, and $\hat{\sigma}_{\hat{\beta}_2} = 0.4226$. The t values associated with $\hat{\beta}_0$, $\hat{\beta}_1$, and $\hat{\beta}_2$ are

$$\hat{\beta}_0: \ t = \frac{-4.0419}{2.5862} = -1.5629$$

$$\hat{\beta}_1: \ t = \frac{0.1367}{0.0945} = 1.4466$$

$$\hat{\beta}_2: \ t = \frac{0.4587}{0.4226} = 1.0855$$

Since $t_{0.025}(3) = 3.182$, we would not reject $H_0: \beta_i = 0$ in favor of $H_a: \beta_i \neq 0$, $i = 0, 1, 2$, at the 5-percent level of significance.

A Future-Value Y_0

We may frequently be interested in predicting the value of Y corresponding to "new" settings of the independent variables, say, $X_{01}, X_{02}, \ldots, X_{0p}$. If we denote the corresponding Y value by Y_0 then the best point estimate of this quantity is the "fitted value,"

$$\hat{Y}_0 = \hat{\beta}_0 + \hat{\beta}_1 X_{01} + \cdots + \hat{\beta}_p X_{0p} \tag{6.27}$$

or in matrix notation

$$\hat{Y}_0 = \mathbf{X}_0' \hat{\boldsymbol{\beta}} \tag{6.28}$$

where $\mathbf{X}_0 = (1, X_{01}, X_{02}, \ldots, X_{0p})'$. To forecast Y_0 we simply evaluate the fitted regression function at the new settings of the X_i's. When we are concerned with the *actual* value of Y_0 [versus, say, the mean value $E(Y_0|X_{01}, X_{02}, \ldots, X_{0p})$], the variance associated with the prediction is

$$\text{Var(pred)} = [1 + \mathbf{X}_0'(X'X)^{-1}\mathbf{X}_0]\sigma^2$$

which can be estimated by

$$\text{Vâr(pred)} = [1 + \mathbf{X}_0'(X'X)^{-1}\mathbf{X}_0]s^2 \tag{6.29}$$

The estimated standard deviation, or standard error, of the prediction is then

$$\hat{\sigma}_{\text{pred}} = s\sqrt{[1 + \mathbf{X}_0'(X'X)^{-1}\mathbf{X}_0]} \tag{6.30}$$

If the usual regression assumptions are satisfied, the ratio

$$t = \frac{(Y_0 - \hat{Y}_0)}{\hat{\sigma}_{\text{pred}}} = \frac{(Y_0 - \hat{Y}_0)}{s\sqrt{[1 + \mathbf{X}_0'(X'X)^{-1}\mathbf{X}_0]}} \sim t(n - p - 1) \tag{6.31}$$

and therefore a $100(1 - \gamma)$-percent prediction interval for Y_0 is given by

$$\hat{Y}_0 \pm t_{\gamma/2}(n - p - 1)s\sqrt{[1 + \mathbf{X}_0'(X'X)^{-1}\mathbf{X}_0]} \tag{6.32}$$

where $t_{\gamma/2}(n - p - 1)$ is the upper $\gamma/2$ percentage point of a t distribution with $n - p - 1$ degrees of freedom.

Ordinarily we would only use the full fitted regression function to predict a future value Y_0 if the hypothesis $H_0: \beta_1 = \beta_2 = \cdots = \beta_p = 0$ is rejected. If the regression is not significant, it does not make much sense to use the full fitted function for prediction. In fact, given the usual error assumptions, when the regression is not significant, the best forecast of Y_0 is $\bar{Y}$.

Confidence intervals for the expected value of Y_0 at given settings of the X's (see Problem 6.4) are similar in form to the interval displayed in equation (6.32). They are not discussed here because in practice it is the actual value of a future observation that is usually of interest, not its expected value.

6.5 MULTIPLE AND PARTIAL CORRELATION

The reader will recall that the coefficient of determination, R^2, in *simple* linear regression analysis involving the variables X and Y is the square of the sample correlation coefficient computed from the observations (X_i, Y_i), $i = 1, 2, \ldots, n$, on these variables. Since the sample correlation coefficient is invariant (remains the same) under a linear transformation of the observations (see Problem 5.20), the sample correlation computed from the observations (X_i, Y_i) is numerically the same as that computed from the observations $(\hat{Y}_i, Y_i)$, where $\hat{Y}_i = \hat{\alpha} + \hat{\beta}X_i$, $i = 1, 2, \ldots, n$, are the fitted values. In symbols, $|r_{XY}| = |r_{\hat{Y}Y}|$.

In multiple regression analysis the positive square root of the coefficient of determination is known as the "sample multiple correlation coefficient." Thus the multiple correlation coefficient, $R = +\sqrt{R^2}$, is constrained to the interval $(0, 1)$ and is a measure of the linear dependence of Y on the *set* of independent variables. The closer the multiple correlation coefficient is to one, the stronger the dependence. Equivalently, the multiple correlation coefficient is a measure of the linear dependence between Y and $\hat{Y}$, the fitted function, since, like the simple regression case, it can be shown that the sample multiple correlation coefficient is identical to the ordinary sample correlation coefficient between the observations and the fitted values.

Both the coefficient of determination and the multiple correlation coefficient measure the "closeness" of the data to the fitted regression function; that is, they contain essentially the same information. Since $0 \le R^2 \le 1$, the multiple correlation coefficient will always be larger than the coefficient of determination; however, the closer R^2 is to one, the smaller the difference between their numerical values. In the extreme case $R = R^2 = 1$, the fitted function passes through all of the data points.

Finally, if we let $\rho_{Y \cdot X_1 \ldots X_p}$ denote the population multiple correlation coefficient defined as a measure of the linear dependence between the random variable, Y, and the regression function,

$$f(X_1, X_2, \ldots, X_p) = \beta_0 + \beta_1 X_1 + \cdots + \beta_p X_p$$

where $X_1, X_2, \ldots, X_p$ are now also regarded as random variables, then a test of $H_0: \rho_{Y \cdot X_1 \ldots X_p} = 0$ is equivalent to a test of $H_0: \beta_1 = \beta_2 = \cdots = \beta_p = 0$ (that is, the regression is not significant), and thus the former hypothesis can be tested using the F statistic from the ANOVA table.

In the simple linear regression case, the least squares value, $\hat{\beta}$, and the t value, $t = \hat{\beta}/\hat{\sigma}_{\hat{\beta}}$, supply information about the correlation between X and Y. In addition, we have noted that in the multiple regression case the t value, $t = \hat{\beta}_i/\hat{\sigma}_{\hat{\beta}_i}$ supplies information about the relationship of Y with the independent variable X_i given the other independent variables $X_1, \ldots, X_{i-1}, X_{i+1}, \ldots, X_p$ in the regression function or given that the effects of the variables $X_1, \ldots, X_{i-1}, X_{i+1}, \ldots, X_p$ on Y have been accounted for. The same information is supplied by the *partial correlation* of Y and X_i which is defined to be the correlation between Y and X_i when the values of all other variables influencing Y and X_i are taken into account. The partial correlation coefficient is not easily defined in terms of the variables $X_1, \ldots, X_p$ and Y; however, we can define the sample partial correlation coefficient[1] in terms of the t value for testing $H_0: \beta_i = 0$, $i = 1, 2, \ldots, p$.

We shall let $r_{YX_i \cdot T}$ denote the sample partial correlation coefficient of Y and X_i, $i = 1, 2, \ldots, p$, where T denotes the full subset of the remaining $p - 1$ variables whose values are held fixed. Then with $t = \hat{\beta}_i/\hat{\sigma}_{\hat{\beta}_i}$,

$$r_{YX_i \cdot T} = \frac{t}{\sqrt{t^2 + (n - p - 1)}}, \qquad i = 1, 2, \ldots, p \qquad (6.33)$$

We point out that (a) $r_{YX_i \cdot T} = 0$ if and only if $t = 0$ (or equivalently $\hat{\beta}_i = 0$) and (b) $r_{YX_i \cdot T}$ has the same sign as t (or equivalently as $\hat{\beta}_i$).

If we define $\rho_{YX_i \cdot T}$ as the population partial correlation, then testing $H_0: \beta_i = 0$ (or that there is no significant contribution due to X_i given the variables in T) is equivalent to testing $H_0: \rho_{YX_i \cdot T} = 0$ (or that there is no significant linear association between the random variables Y and X_i when the variables in T are held fixed). A test of significance of the X_i variable is also a test of significance of the partial correlation of Y and X_i, $i = 1, 2, \ldots, p$. In standard regression analysis computer programs both the t values and the sample partial correlation coefficients are usually printed out.

Although equation (6.33) enables us to calculate the sample partial

---

[1] For three variables Y, X_1, and X_2, the sample partial correlation coefficient of Y and X_1, given that X_2 is held constant, can be expressed in terms of the ordinary correlations between Y and X_1, Y and X_2, and X_1 and X_2. We let $r_{YX_1 \cdot X_2}$ denote the sample partial correlation coefficient and r_{YX_1}, r_{YX_2}, and $r_{X_1X_2}$ denote the ordinary sample correlations, then

$$r_{YX_1 \cdot X_2} = \frac{r_{YX_1} - r_{YX_2} r_{X_1X_2}}{\sqrt{(1 - r^2_{YX_2})(1 - r^2_{X_1X_2})}}$$

The sample partial correlations $r_{YX_2 \cdot X_1}$ and $r_{X_1X_2 \cdot Y}$ are defined analogously.

correlations from the t values, it does not provide much insight into their *nature*. However, some insight can be provided as follows. Consider the regression models

$$Y = \beta_0 + \beta_1 X_1 + \cdots + \beta_{i-1} X_{i-1} + \beta_{i+1} X_{i+1} + \cdots + \beta_p X_p + e_1$$

and

$$X_i = \beta_0^* + \beta_1^* X_1 + \cdots + \beta_{i-1}^* X_{i-1} + \beta_{i+1}^* X_{i+1} + \cdots + \beta_p^* X_p + e_2$$

(The reader will note that both Y and X_i are regressed on the variables in T.) Suppose the models are fitted by the method of least squares, and let $\hat{e}_{1j} = Y_j - \hat{Y}_j$ and $\hat{e}_{2j} = X_{ij} - \hat{X}_{ij}$, $j = 1, 2, \ldots, n$, denote the two sets of residuals. In each case the residuals represent that part of Y and X_i that is not accounted for by the variables in T—that is, the portion of Y and X_i remaining after the effects of the variables in T have been removed. It can be shown that the ordinary sample correlation coefficient computed with these residuals,

$$r_{\hat{e}_1 \hat{e}_2} = \frac{\sum_{j=1}^n (\hat{e}_{1j} - \bar{\hat{e}}_1)(\hat{e}_{2j} - \bar{\hat{e}}_2)}{\sqrt{\sum_{j=1}^n (\hat{e}_{1j} - \bar{\hat{e}}_1)^2 \sum_{j=1}^n (\hat{e}_{2j} - \bar{\hat{e}}_2)^2}}$$

is precisely the sample partial correlation $r_{YX_i \cdot T}$. The partial correlation is the ordinary correlation after the effects of the other variables have been removed.

EXAMPLE 6.3

We shall consider the housing demand problem introduced in Example 6.1. With Y as the housing demand, X_1 as the "average price," and X_2 as the disposable income, the fitted equation linking Y to X_1 and X_2 was

$$\hat{Y} = -4.04 + 0.14 X_1 + 0.46 X_2$$

From Example 6.2 the t values associated with $\hat{\beta}_1$ and $\hat{\beta}_2$ were

$$\hat{\beta}_1 : t = 1.4466$$
$$\hat{\beta}_2 : t = 1.0855$$

Since $n = 6$ and $p = 2$, use of equation (6.33) gives the partial correlations

$$r_{YX_1 \cdot X_2} = \frac{1.4466}{\sqrt{(1.4466)^2 + (6 - 2 - 1)}} = 0.64$$

$$r_{YX_2 \cdot X_1} = \frac{1.0855}{\sqrt{(1.0855)^2 + (6 - 2 - 1)}} = 0.53$$

Although the sample partial correlation coefficients appear to be "large," as we have seen in Example 6.2, the t values were *not* significant at the 5-percent level. Equivalently at the 5-percent level the partial correlations are not significantly different from zero.

6.6   STATISTICS USEFUL IN THE ANALYSIS OF RESIDUALS

We have provided the motivation for analyzing residuals in Chapters 4 and 5. In these chapters several useful plotting techniques designed to test the adequacy of the regression assumptions were presented. These same plots can be used to examine the residuals from a multiple regression analysis. Therefore in this section we shall concentrate on *statistics* that are useful in residual analysis. Attention is focused on the assumption that the errors are uncorrelated since this assumption is frequently violated in practice, particularly when the observations represent routinely recorded economic data collected over time.

Residuals from a regression analysis will tend to look like realizations of an $N(0, \sigma^2)$ random variable. For example, if the regression function contains a β_0 term, it is always true that $\Sigma \hat{e}_i = 0$ and hence $\bar{\hat{e}} = 0$. In addition, there is frequently some linear dependence between the residuals so that they tend to "bunch together," reinforcing normality. In most cases, then, it is difficult to detect a violation of the assumption of normally distributed errors. The independence assumption, however, is another matter.

Consider the assumption $E(e_i e_j) = 0$, $i \neq j$; that is, the errors are uncorrelated. If this assumption is true, we would expect the residuals to exhibit little or no linear dependence.[m] It is possible to construct statistics based on the residuals that would allow us to check this assumption.

The Durbin-Watson Statistic and Residual Autocorrelations

If multiple regression methods are used on data collected over time, the assumption $E(e_i e_j) = 0$, $i \neq j$, is rarely valid. Therefore for time-ordered data it is imperative to examine the correlation in the residuals.

Let $\hat{e}_t$, $t = 1, 2, \ldots, n$, denote the residuals from a multiple regression analysis. (We have introduced the subscript t to emphasize the chronological order of the residuals.) We divide the residuals into two sets as follows:

Set 1	Set 2
$\hat{e}_1$	$\hat{e}_2$
$\hat{e}_2$	$\hat{e}_3$
$\vdots$	$\vdots$
$\hat{e}_{n-1}$	$\hat{e}_n$

The reader will notice that there are $n - 1$ residuals in each set. We define

---

[m] In general, as we have indicated in the preceding paragraph, there will always be some linear dependence between the residuals because of the nature of the least squares estimator $\hat{\beta}$. This contrived dependence will usually be negligible and will ordinarily not be responsible for the "significance" of the statistics designed to measure linear dependence in the residuals.

the lag 1 residual autocovariance coefficient[n]

$$c_1 = \frac{1}{n} \sum_{t=1}^{n-1} \hat{e}_t \hat{e}_{t+1} \tag{6.34}$$

and the corresponding lag 1 residual autocorrelation coefficient

$$r_1 = \frac{c_1}{c_0} = \frac{\sum_{t=1}^{n-1} \hat{e}_t \hat{e}_{t+1}}{\sum_{t=1}^{n} \hat{e}_t^2} \tag{6.35}$$

where

$$c_0 = \frac{1}{n} \sum_{t=1}^{n} \hat{e}_t^2$$

If successive residuals tend to be of the same sign and magnitude (that is, the differences $\hat{e}_t - \hat{e}_{t-1}$, $t = 2, \ldots, n$, are "small"), then r_1 will be positive. On the other hand if successive residuals are of opposite sign (in which case the differences $\hat{e}_t - \hat{e}_{t-1}$, $t = 2, \ldots, n$, are "large"), r_1 will be negative. Therefore a large value of $|r_1|$ indicates that the errors in the regression model are correlated over adjacent time periods. If we denote the correlation[o] between successive error terms e_t and e_{t+1}, $t = 1, \ldots, n-1$, by ρ_1, then a test of the hypothesis $H_0 : \rho_1 = 0$ versus the alternatives $H_a : \rho_1 > 0$ and $H_a : \rho_1 < 0$ can be constructed using the information contained in r_1, since r_1 may be regarded as an estimate of ρ_1.

Durbin and Watson[28] have proposed the test statistic

$$d = \frac{\sum_{t=2}^{n} (\hat{e}_t - \hat{e}_{t-1})^2}{\sum_{t=1}^{n} \hat{e}_t^2} \doteq 2(1 - r_1) \tag{6.36}$$

It is possible to demonstrate that $0 < d < 4$. [Since $-1 \leq r_1 \leq 1$, this result is implied by the approximation on the right-hand side of equation (6.36).] If $r_1 \doteq 0$, $d \doteq 2$. On the other hand, if $r_1 \doteq 1$, $d \doteq 0$, and if $r_1 \doteq -1$, $d \doteq 4$.

---

[n]The rationale for the definitions given in equation (6.34) follows from the fact that $\sum_{t=1}^{n} \hat{e} = 0$ if there is a β_0 term in the regression function; hence for n reasonably large the arithmetic averages of the residuals in sets 1 and 2 will be very nearly zero. Thus the deviations of the residuals from their arithmetic averages in both sets are simply the *residuals themselves*. The divisor $1/n$ can be justified on theoretical grounds. The denominator in equation (6.34) would ordinarily be the square root of the product of the sample variances, but the sample variance for set 1, given $\bar{e} \doteq 0$, can be defined as

$$s_1^2 = \frac{1}{n} \sum_{t=1}^{n-1} \hat{e}^2 \doteq \frac{1}{n} \sum_{t=1}^{n} \hat{e}_t^2$$

Similarly, the sample variance for set 2 can be defined as

$$s_2^2 = \frac{1}{n} \sum_{t=1}^{n-1} \hat{e}_{t+1}^2 \doteq \frac{1}{n} \sum_{t=1}^{n} \hat{e}_t^2$$

and consequently the square root of the sample variances is approximately

$$\sqrt{\left(\frac{1}{n} \sum_{1}^{n} \hat{e}_t^2\right)\left(\frac{1}{n} \sum_{1}^{n} \hat{e}_t^2\right)} = \frac{1}{n} \sum_{t=1}^{n} \hat{e}_t^2 = c_0$$

Finally, the term $1/n$ cancels from both numerator and denominator to yield the algebraic expression for r_1.
[o]A thorough discussion of correlation between time ordered random variables is given in Chapter 10.

Therefore values of d close to two tend to reinforce the null hypothesis $H_0: \rho_1 = 0$, whereas values of d or $d^* = 4 - d$ close to zero tend to reinforce the alternative hypotheses $H_a: \rho_1 > 0$ and $H_a: \rho_1 < 0$, respectively.

A test of the hypotheses

1. $H_0: \rho_1 = 0$
    $H_a: \rho_1 > 0$
2. $H_0: \rho_1 = 0$
    $H_a: \rho_1 < 0$

can be conducted at the 5-percent level of significance using Appendix Table I. Corresponding to each pair (n, p) the table gives two values: d_L and d_U, and the appropriate tests are

Hypothesis 1: $H_0: \rho_1 = 0$ $\qquad$ $\begin{cases} \text{accept } H_0 \text{ if } d \geq d_U \\ \text{do not accept } H_0 \text{ if } d \leq d_L \\ \text{test is inconclusive if } d_L < d < d_U \end{cases}$
$\qquad\qquad\quad H_a: \rho_1 > 0$

Hypothesis 2: $H_0: \rho_1 = 0$ $\qquad$ $\begin{cases} \text{accept } H_0 \text{ if } d^* = 4 - d \geq d_U \\ \text{do not accept } H_0 \text{ if } d^* = 4 - d \leq d_L \\ \text{test is inconclusive if } d_L < d^* < d_U \end{cases}$
$\qquad\qquad\quad H_a: \rho_1 < 0$

If the null hypothesis is rejected, then we conclude that the successive errors, instead of being uncorrelated, are related in some manner. If this is in fact true and if the remaining regression assumptions are satisfied, then carrying out a regression analysis using the methods presented in this chapter will produce unbiased estimates of the regression parameters, but the variances of the least squares estimators may be seriously underestimated (for $\rho_1 > 0$). (This point is discussed in references [7] and [11].) If the usual least squares methods are employed in the presence of autocorrelated errors, parameters may appear to be significantly different from zero when in fact they are not.

The dependency usually assumed for the error term when lag 1 autocorrelation is indicated is given by the model

$$e_t = \rho_1 e_{t-1} + \eta_t \qquad (6.37)$$

where η_t is taken to be normally distributed with mean zero and a constant variance, σ_η^2. If this is the case, then the autocorrelation in the errors can be eliminated by considering the regression model

$$Y_t - \rho_1 Y_{t-1} = \beta_0 (1 - \rho_1) + \beta_1 (X_{t1} - \rho_1 X_{t-11}) + \cdots$$
$$+ \beta_p (X_{tp} - \rho_1 X_{t-1p}) + \eta_t, \qquad t = 2, 3, \ldots, n \qquad (6.38)$$

where t has now replaced the former subscript i. Of course, in practice ρ_1 is unknown, so it is customary to replace ρ_1 by its estimate $r_1 \doteq 1 - (d/2)$, construct a new dependent variable $Y_t^* = Y_t - r_1 Y_{t-1}$, and new independent variables $X_{t1}^* = X_{t1} - r_1 X_{t-11}$, $X_{t2}^* = X_{t2} - r_1 X_{t-2}, \ldots, X_{tp}^* = X_{tp} - r_1 X_{t-1p}$, and apply the usual multiple regression methods with the model

$$Y_t^* = \beta_0^* + \beta_1 X_{t1}^* + \cdots + \beta_p X_{tp}^* + \eta_t, \qquad t = 2, 3, \ldots, n$$

where $\beta_0^* = \beta_0(1 - \rho_1)$ and $\eta_t \sim N(0, \sigma_\eta^2)$. The residuals from this regression are examined and if necessary the whole procedure is repeated.

Although the Durbin-Watson test is useful, it suffers from two main defects: (a) There is a region where the test is inconclusive, and (b) it only uses explicitly the information contained in the lag 1 sample autocorrelation.

A viable alternative to the Durbin-Watson test is to examine simultaneously the correlation in the residuals separated by two time periods, three time periods, and in general k time periods, $k \geq 1$. Analogous to our previous discussion of autocorrelation in Chapter 5, we define the lag k residual autocorrelation coefficient to be

$$r_k = \frac{\sum_{t=1}^{n-k} \hat{e}_t \hat{e}_{t+k}}{\sum_{t=1}^{n} \hat{e}_t^2}, \qquad k = 1, 2, \ldots, K \tag{6.39}$$

The residual autocorrelation r_k is a measure of the linear dependence in the residuals separated by k time periods. The set of residual autocorrelations $r_1, r_2, \ldots, r_K$ provides information about the nature of the correlation in the error component e_t. Values of r_k, $k = 1, 2, \ldots, K$, near zero reinforce the hypothesis of no correlation in the error terms. A specific pattern of nonzero values (within sampling fluctuation) for the r_k's suggests a particular type of dependency among the errors, $e_1, e_2, \ldots, e_n$. This dependency can be exploited to revise the original model so that a straightforward least squares analysis can be justifiably employed to estimate the regression parameters. This procedure is discussed in the context of time series models in Chapter 10. The reader is referred to reference [11] for a complete discussion of the methodology in a regression context.

6.7   FINAL COMMENTS

With the availability of high speed computers and "canned" regression programs, multiple regression analysis is almost never performed by hand. Consequently, readers of this text will probably never have to invert matrices and carry out many of the computations necessary for the inferences we have discussed in the body of this chapter. On the other hand, regression analysis is one of the most misused statistical techniques, and it is our feeling that the results of a regression analysis (ordinarily in the form of computer printout) can only be properly interpreted with some knowledge of the underlying theory. It is necessary to understand where the numbers on the computer printout come from. In this chapter we have tried to present most of the elements of the general theory of regression analysis. Matrix algebra was used in order to present the theoretical results as simply as possible.

The reader is encouraged to work many of the problems at the end of this chapter. Some of them contain real data and displays of typical output from a multiple regression computer program. It is particularly important for this chapter and the time series analysis chapters that follow to solve the problems. Only in this way can the force of the theory and the subsequent morass of equations that follow be appreciated.

PROBLEMS

6.1 Define and give an example of each of the following terms:
   (a) multiple regression
   (b) t value
   (c) prediction interval
   (d) Durbin-Watson statistic
   (e) multiple correlation
   (f) partial correlation

6.2 Show that
   (a) $\hat{\mathbf{Y}} = W\mathbf{Y}$
   (b) $\hat{\mathbf{e}} = (I - W)\mathbf{Y}$
   where $W = X(X'X)^{-1}X'$; that is, both the fitted values and residuals can be written as linear combinations of the observations Y_i, $i = 1, 2, \ldots, n$.

6.3 (a) Show that $\hat{\mathbf{Y}}'\hat{\mathbf{Y}} = \hat{\boldsymbol{\beta}}'X'\mathbf{Y}$. Therefore an alternative expression for the regression sum of squares in the ANOVA table is given by
$$\hat{\mathbf{Y}}'\hat{\mathbf{Y}} - n\bar{Y}^2 = \hat{\boldsymbol{\beta}}'X'\mathbf{Y} - n\bar{Y}^2$$
   (b) Prove
$$F = \left(\frac{R^2}{1 - R^2}\right)\left(\frac{n - p - 1}{p}\right)$$

where $F = MS_R/s^2$ and R^2 is the coefficient of determination. Discuss the significance of this relationship.

6.4 Given the usual regression assumptions, a $100(1 - \gamma)$-percent confidence interval for the expected value of Y at a given setting of the independent variables, $X_{01}, X_{02}, \ldots, X_{0p}$; that is, $E(Y|X_{01}, X_{02}, \ldots, X_{0p})$ is given by
$$\hat{Y}_0 \pm t_{\gamma/2}(n - p - 1)s\sqrt{\mathbf{X}_0'(X'X)^{-1}\mathbf{X}_0}$$

where $\mathbf{X}_0 = (1, X_{01}, X_{02}, \ldots, X_{0p})'$, s is the standard error of the estimate, and $t_{\gamma/2}(n - p - 1)$ is the upper $\gamma/2$ percentage point of a t distribution with $(n - p - 1)$ degrees of freedom. Using the data of Example 6.1 and the results given in Example 6.2, construct a 95-percent confidence interval for $E(Y|X_{01} = 14.5, X_{02} = 7.3)$.

6.5 Using equation (6.32) and the data of Example 6.1, construct a 95-percent confidence interval for the *actual* value of Y at the settings $X_{01} = 14.5$, $X_{02} = 7.3$. Compare the result with that obtained in Problem 6.4, and comment on the difference.

6.6 Suppose newsprint consumption, Y, is related to retail sales (surrogate for advertising space), X, by the regression model

$$Y_i = \beta_0 + \beta_1 X_i + e_i, \qquad i = 1, 2, \ldots, n$$

Data for 10 cities are given below.

Y (in appropriate units)	2.5	2.1	2.2	1.9	3.0	0.3	6.3	1.5	2.8	8.8
X (in appropriate units)	2.1	1.2	0.9	1.1	1.1	0.5	2.5	0.9	1.4	3.6

(a) Specify $\mathbf{Y}$ and X in the matrix form of the regression model,

$$\mathbf{Y} = X\boldsymbol{\beta} + \mathbf{e}$$

(b) Obtain the least squares estimates $\hat{\boldsymbol{\beta}}$.

(c) Set up the ANOVA table and test for the "significance of the regression."

6.7 An economist is interested in the demand for California fresh Bartlett pears.[†] Data are collected on the variables: Y is the weighted-average season's price of California fresh Bartletts on seven auction markets (dollars per box), X_1 is the interstate shipment of all California pears during California Bartlett shipping season (thousands of tons), and X_2 is the Oregon and Washington shipment of all varieties of pears during California's Bartlett season (thousands of tons) for the years 1930–1940. The data are given below.

Year	Y	X_1	X_2
1930	2.31	120	52
1931	2.61	84	23
1932	1.94	66	20
1933	2.30	52	25
1934	2.53	72	17
1935	2.35	53	33
1936	2.33	68	40
1937	2.45	80	36
1938	1.93	89	40
1939	2.53	67	36
1940	2.33	68	44

Consider the model,

$$Y_i = \beta_0 + \beta_1 X_{i1} + \beta_2 X_{i2} + e_i, \qquad i = 1, 2, \ldots, 11$$

where $e_i \sim N(0, \sigma^2)$.

[†] This problem is adapted from data published originally by S. Hoos and S. W. Shear in their paper, "Relation between Auction Prices and Supplies of California Fresh Bartlett Pears," *Hilgardia, 14,* (5), California Experiment Station, 1942.

(a) Obtain the least squares estimates $\hat{\boldsymbol{\beta}}$.

(b) Compute the fitted values $\hat{\mathbf{Y}}$ and the residuals $\hat{\mathbf{e}} = \mathbf{Y} - \hat{\mathbf{Y}}$.

(c) Compute the Durbin-Watson statistic, and test for serial correlation in the error term. Comment on your results.

6.8   A furniture manufacturer feels that production scheduling can be improved by using the fact that sales of furniture in a given quarter can be related to the number of new homes on which construction was started two quarters earlier. The manufacturer decides to study the Los Angeles area. Let Y be the quarterly sales in the area (thousands of dollars) and let X_1 be the number of housing starts (thousands) two quarters previously. The data for a three-year period are given below.

	Quarter	Y	X_1
Year 1	I	190	2.0
	II	240	1.5
	III	200	3.0
	IV	180	1.1
Year 2	I	210	3.4
	II	230	4.0
	III	200	3.5
	IV	180	2.0
Year 3	I	250	3.0
	II	280	4.5
	III	300	6.0
	IV	120	1.2

In addition the manufacturer feels that his business is seasonal and that he should allow for seasonal effects. He decides to consider another independent variable:

$$X_2 = \begin{cases} 1 & \text{if second or third quarter sales} \\ 0 & \text{if first or fourth quarter sales} \end{cases}$$

to account for seasonality after realizing that sales tend to be "high" during the second and third quarters and "low" during the first and fourth quarters.

(a) Fit the regression model

$$Y_i = \beta_0 + \beta_1 X_{i1} + \beta_2 X_{i2} + e_i, \qquad i = 1, 2, \ldots, 12$$

where $e_i \sim N(0, \sigma^2)$. Examine the significance of the regression coefficients β_1, β_2 individually and as a group. Does it appear as if the term $\beta_2 X_{i2}$ is required in the regression model? Comment.

(b) Compute the Durbin-Watson statistic using the residuals in part (a), and test for serial correlation. Comment on the results.

(c) Consider the fitted function, $\hat{Y}_i = \hat{\beta}_0 + \hat{\beta}_1 X_{i1} + \hat{\beta}_2 X_{i2}$. Suppose you wish to predict sales for the third quarter of year 4 and the current value for housing starts (first quarter, year 4) is $X_{01} = 4.2$ What can you say about Y_0?

6.9   Discuss the following:
   (a) F ratio and some but not all β_i are significant.
   (b) F ratio but none of the β_i is significant.
   (c)   Some β_i are significant but not all nor is the F ratio.
   (d)   All β_i are significant but not the F ratio.
   (See reference [26].)

6.10   An investigator is interested in the relationship between demand for a particular beer, Y, and the independent variables, X_1 (price), X_2 (disposable income), and X_3 as demand in previous year. Twenty observations are available. He proposes a model of the form

$$Y = \beta_0 + \beta_1 X_1 + \beta_2 X_2 + \beta_3 X_3 + e$$

where it is assumed $E(e) = 0$, $\mathrm{Var}(e) = I\sigma^2$, and the elements e_i are independent and normally distributed.
   (a)   Consider the matrix formulation of the proposed model: $Y = X\beta + e$. Indicate the dimensions of the vectors and matrices in the model.
   (b)   A regression analysis program is run, and part of the computer printout is given below.

Basic Regression Statistics

Standard Error of Estimate	.0098
Coefficient of Determination	.9810
Corrected Coefficient of Determination	.9774

VARIABLE	TYPE	REGRESSION COEFFICIENT	STANDARD ERROR OF REGRESSION COEFFICIENT	T VALUE WITH 16 DEGREES OF FREEDOM
1 Y	Dependent			
	Constant	1.59428	1.01382	1.57255
2 X1	Independent	−.04797	.14792	−.32431
3 X2	Independent	.05494	.03060	1.79502
4 X3	Independent	.81301	.11594	7.01212

Analysis of Variance Summary Table

SOURCE OF VARIATION	SUM OF SQUARES	DEGREES OF FREEDOM	MEAN SQUARE
Linear Regression	.08012	3	.02671
Residuals from Regression	.00155	16	.00010
Corrected Total	.08167	19	

F Ratio = 275.49 with 3 and 16 Degrees Freedom

(1) Is the regression significant? Explain.
(2) Suppose we are interested in the parameter β_2. Does it appear as if this parameter is significantly different from zero? Interpret the test of the hypothesis: $H_0: \beta_2 = 0$ versus $H_a: \beta_2 \neq 0$.
(3) Write down R^2, the coefficient of determination, and interpret this number. In general do you think it is possible to have R^2 reasonably large and at the same time have *all* the "t values" for testing $H_0: \beta_1 = 0$, $H_0: \beta_2 = 0$, and $H_0: \beta_3 = 0$ insignificant? Discuss this question briefly.
(4) Plot the standardized residuals. Using the Durbin-Watson statistic, test for serial correlation. Do the assumptions underlying the regression analysis appear to be satisfied? Explain.

6.11 In Problem 6.10 we considered the regression model

$$Y_i = \beta_0 + \beta_1 X_{i1} + \beta_2 X_{i2} + \beta_3 X_{i3} + e_i, \qquad i = 1, 2, \ldots, 20$$

OBSER-VATION NUMBER	Y OBSERVED	Y COMPUTED	RESIDUAL	STANDARDIZED RESIDUAL
1	8.853665	8.849766	.003899	.3960
2	8.867850	8.852624	.015226	1.5464
3	8.867850	8.870927	−.003077	−.3125
4	8.860783	8.870927	−.010144	−1.0303
5	8.867850	8.870179	−.002329	−.2365
6	8.867850	8.878651	−.010801	−1.0970
7	8.874868	8.880345	−.005478	−.5563
8	8.895630	8.889859	.005771	.5861
9	8.895630	8.907150	−.011520	−1.1700
10	8.909235	8.910869	−.001633	−.1659
11	8.948976	8.925864	.023112	2.3474
12	8.968269	8.959755	.008514	.8647
13	8.974618	8.978078	−.003460	−.3514
14	8.980927	8.984929	−.004002	−.4065
15	8.987197	8.990059	−.002862	−.2907
16	8.993427	8.998526	−.005098	−.5178
17	8.999619	9.005404	−.005785	−.5875
18	9.011889	9.013221	−.001331	−.1352
19	9.035987	9.023196	.012791	1.2991
20	9.047821	9.049612	−.001791	−.1819

Durbin-Watson Statistic of Correlation   1.6798

linking the demand for a particular beer, Y, with the independent variables: X_1, the price; X_2, the disposable income; and X_3, the demand in the previous year. Refer to the computer printout displayed in Problem 6.10 when discussing the following:

(a)   Compute the multiple correlation coefficient. Interpret this number.

(b)   Compute the sample partial correlations between Y and each of the independent variables from the t values using equation (6.33).

(c)   Interpret the sample partial correlation coefficient associated with the variable, X_2. Is it significant? Explain.

6.12   Consider the linear model

$$E(\mathbf{Y}|\mathbf{X}) = \beta_1 \mathbf{X} + \beta_2 \mathbf{X}^2$$

$$V(\mathbf{Y}|\mathbf{X}) = I\sigma^2, \sigma^2 > 0$$

A designed experiment yields the data,

$$\mathbf{Y} = \begin{bmatrix} 8 \\ 12 \\ -1 \\ 11 \\ 38 \\ 44 \end{bmatrix}, \quad \mathbf{X} = \begin{bmatrix} -1 \\ 1 \\ 0 \\ 1 \\ -2 \\ 2 \end{bmatrix}.$$

(a)   Calculate the least squares estimates $\hat{\beta}_1$ and $\hat{\beta}_2$.

(b)   Fill in the following analysis of variance table:

Source	Sum of Squares (SS)	Degrees of Freedom (d.f.)	Mean Squares (MS)	F Ratio
†Regression	—	—	—	—
Residual	—	—	—	
†Total	—	—		

† Since there is no intercept term (β_0) in the regression model, the total sum of squares is $\mathbf{Y'Y}$, and the regression sum of squares is $\hat{\mathbf{Y}}'\hat{\mathbf{Y}}$; that is, $n\bar{Y}^2$ is *not* subtracted from each of these quantities.

(c)   Using the results in part (b), test for the significance of the regression at the 5-percent level.

(d)   Provide a forecast of Y corresponding to $X_0 = -0.5$.

7

TOPICS IN REGRESSION ANALYSIS

7.1 INTRODUCTION

Chapters 5 and 6 contain what we might call "basic" regression methodology. The linear regression model was presented, the usual inference problems were considered, and residual analysis was discussed. This chapter is devoted to some regression analysis extensions. Specifically, we shall consider the use of "dummy," or categorical, variables, the use of transformations, the multicollinearity problem, procedures for selecting appropriate explanatory variables, and nonlinear estimation.

Dummy variables and transformations are frequently employed in practice. These techniques greatly increase the size of the set of problems amenable to regression analysis. Multicollinearity has received a great deal of attention since the advent of large digital computers and the subsequent tendency to include more independent variables than necessary in the regression function. Multicollinearity is frequently present in routinely recorded or survey data and can often produce results that are misleading because of numerical abnormalities rather than because of any defects in the underlying regression theory. Finally, explanatory variable selection procedures and nonlinear estimation techniques are now so pervasive that they form the basis for many standard regression computer programs. A frequent user of regression analysis is very likely to encounter one (or more) of these programs and as a consequence should understand the procedures on which they are based.

7.2 DUMMY VARIABLES

Sometimes it is necessary in a regression analysis to account for the effects of important "independent variables" that cannot be quantified or, if they can be quantified, cannot be measured for various reasons. Ignoring the effects of these variables can produce misleading or "biased" results.

One way to account for a nonquantifiable variable is to introduce a "dummy" (also called "indicator" or "categorical") independent variable into the regression function in which the values of the dummy variable indicate varying conditions or states of nature. We will illustrate the use of a dummy variable with examples and refer the reader to reference [2] for a more general discussion.

EXAMPLE 7.1

Relia-Ride Taxi Company is interested in the relationship between mileage and the age of cars in its fleet. The 12 fleet cars are the same make and size, and all use the same brand of gasoline. The cars are in good operating condition due to recent maintenance checkups. However, Relia-Ride employs both male and female drivers, and it is felt that some of the variability in mileage may be due to differences in driving technique between the groups of drivers of opposite sex. In fact Relia-Ride feels that, "other things being equal," women tend to get better mileage than men. The cars are randomly assigned to the five female and seven male drivers, and after 200 miles, miles per gallon are computed. The data are given in Table 7.1 with Y being miles per

Table 7.1

Y (miles/gallon)	X_1 (age of car)	X_2 $\begin{pmatrix} 1, \text{ if female driver} \\ 0, \text{ if male driver} \end{pmatrix}$
12.3	3	0
12.0	4	1
13.7	3	1
14.2	2	0
15.5	1	1
11.1	5	0
10.6	4	0
14.0	1	0
16.0	1	1
13.1	2	0
14.8	2	1
10.2	5	0

gallon, X_1 being age of the car in years, and X_2 being a dummy variable introduced to account for the difference in sex of the driver. If the driver is a female, $X_2 = 1$; if the driver is a male, $X_2 = 0$.

The data are plotted in Figure 7.1. Now it is clear from the figure that irrespective of the age of the cars, female drivers tend to get better mileage than their male

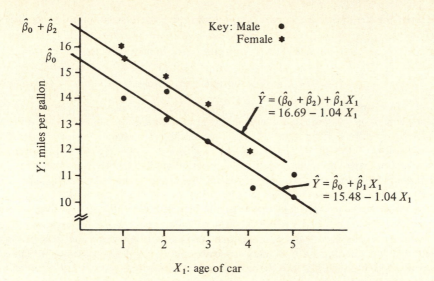

Figure 7.1 Relia-Ride Taxi Company data and the fitted regression function, $\hat{Y} = \hat{\beta}_0 + \hat{\beta}_1 X_1 + \hat{\beta}_2 X_2$

counterparts. Thus if the difference in sex is ignored and miles per gallon are related to the age of the car alone, misleading results will be obtained. Mileage will tend to be underestimated for female drivers and overestimated for male drivers; that is, the fitted regression function would pass somewhere between the lines drawn in Figure 7.1.

The regression model in this situation may be written

$$Y_i = \beta_0 + \beta_1 X_{i1} + \beta_2 X_{i2} + e_i, \qquad i = 1, 2, \ldots, 12 \qquad (7.1)$$

or in matrix notation

$$\underset{(12\times1)}{\mathbf{Y}} = \underset{(12\times3)}{X} \underset{(3\times1)}{\boldsymbol{\beta}} + \underset{(12\times1)}{\mathbf{e}}$$

where

$$X = \begin{bmatrix} 1 & 3 & 0 \\ 1 & 4 & 1 \\ 1 & 3 & 1 \\ 1 & 2 & 0 \\ 1 & 1 & 1 \\ 1 & 5 & 0 \\ 1 & 4 & 0 \\ 1 & 1 & 0 \\ 1 & 1 & 1 \\ 1 & 2 & 0 \\ 1 & 2 & 1 \\ 1 & 5 & 0 \end{bmatrix}$$

It is assumed $\mathbf{e} \sim N(\mathbf{0}, I\sigma^2)$. Since we have assigned values of zero and one to X_2 to account for the difference in sex, the multiple regression model in equation (7.1) is effectively

$$Y_i = (\beta_0 + \beta_2) + \beta_1 X_{i1} + e_i \qquad \text{(for female drivers)}$$

and

$$Y_i = \beta_0 + \beta_1 X_{i1} + e_i \qquad \text{(for male drivers)}$$

We are assuming that the effect of X_1 (age of car) on mileage, as measured by the coefficient β_1, is the same regardless of sex; but for cars of the same age, the mileage of females differs from that of male drivers by a constant amount, β_2. The introduction of the $(0, 1)$ variable X_2 allows us to account for a nonquantifiable source of mileage variability in a single regression model. The reader will notice that it would be possible to fit two regression functions, one to the female data and one to the male data. However, in the former case, for example, we would be fitting a regression model to only five observations, a situation that might produce "nonsignificant" results due to a relatively large s^2 (since there would be few degrees of freedom associated with the residual sum of squares). We can avoid this problem with the introduction of a dummy variable, but we must, of course, assume a constant slope coefficient β_1—an assumption that must be scrutinized in practice.

The estimated relationship for the Relia-Ride example is

$$\hat{Y} = 15.48 - 1.04 X_1 + 1.21 X_2 \qquad (7.2)$$
$$\underset{(0.12)}{} \underset{(0.33)}{}$$

$$R^2 = 0.93, \qquad s^2 = 0.29, \qquad F = 64.55$$

where the quantities in parentheses below the least squares estimates $\hat{\beta}_1$ and $\hat{\beta}_2$ are the estimated standard errors. The fitted regression function is plotted in Figure 7.1. As we would expect from viewing the data, $\hat{\beta}_1$ is negative and $\hat{\beta}_2$, a measure of the shift in the relation accounted for by the sex of the driver, is positive. Of course had we assigned a value of $X_2 = 1$ for male drivers and $X_2 = 0$ for female drivers, $\hat{\beta}_2$ would have been negative.

In the example given above there were two categories for each observation. Sometimes it is necessary to allow for more than two classifications. For example, the year may be divided into quarters to allow for a seasonal effect, the United States may be divided into geographical regions to allow for regional variations, and people may be classified according to political beliefs, age, or some other characteristic to allow for variation between different groups. If k categories are present then $k - 1$ dummy variables must be included in the regression model. Only $k - 1$ dummy variables are used for the same reason that only one dummy variable is included when the classification is dichotomous: The missing category (indicated by zeroes) provides a base. If k dummy variables of the $(0, 1)$ type were included in the regression model, the least squares estimates could not be found in the usual manner since any one of the dummy variables could be written as a linear combination of the remaining dummy variables (multicollinearity in its extreme form, see Section 7.4), and the normal equations would not have a unique solution.

EXAMPLE 7.2

A company is interested in forecasting quarterly earnings from sales. It is assumed initially that earnings move linearly with sales in a given quarter; however, from past experience it is known that earnings tend to be relatively high in the first and fourth quarters and relatively low in the second and third quarters; that is, earnings exhibit "seasonal" fluctuations. To account for the seasonal effect we can introduce the dummy variables X_2, X_3, and X_4, where

$$X_2 = \begin{cases} 1 & \text{if second quarter earnings} \\ 0 & \text{otherwise} \end{cases}$$

$$X_3 = \begin{cases} 1 & \text{if third quarter earnings} \\ 0 & \text{otherwise} \end{cases}$$

and

$$X_4 = \begin{cases} 1 & \text{if fourth quarter earnings} \\ 0 & \text{otherwise} \end{cases}$$

The regression model linking earnings (Y) to sales (X_1) can then be formulated as follows:

$$Y_i = \beta_0 + \beta_1 X_{i1} + \beta_2 X_{i2} + \beta_3 X_{i3} + \beta_4 X_{i4} + e_i, \qquad i = 1, 2, \ldots, n \qquad (7.3)$$

or

$$\underset{(n \times 1)}{\mathbf{Y}} = \underset{(n \times 5)}{\mathbf{X}} \ \underset{(5 \times 1)}{\boldsymbol{\beta}} + \underset{(n \times 1)}{\mathbf{e}}$$

where by assumption $\mathbf{e} \sim N(\mathbf{0}, I\sigma^2)$. Let $n = 12$, and suppose the data are ordered chronologically beginning with the first quarter of the first year. Then the X matrix would look like

$$X = \begin{array}{c} \\ \end{array} \begin{bmatrix} \mathbf{X_0} & \mathbf{X_1} & \mathbf{X_2} & \mathbf{X_3} & \mathbf{X_4} \\ 1 & X_{11} & 0 & 0 & 0 \\ 1 & X_{21} & 1 & 0 & 0 \\ 1 & X_{31} & 0 & 1 & 0 \\ 1 & X_{41} & 0 & 0 & 1 \\ 1 & X_{51} & 0 & 0 & 0 \\ 1 & X_{61} & 1 & 0 & 0 \\ 1 & X_{71} & 0 & 1 & 0 \\ 1 & X_{81} & 0 & 0 & 1 \\ 1 & X_{91} & 0 & 0 & 0 \\ 1 & X_{101} & 1 & 0 & 0 \\ 1 & X_{111} & 0 & 1 & 0 \\ 1 & X_{121} & 0 & 0 & 1 \end{bmatrix} \begin{array}{l} \left.\vphantom{\begin{matrix}1\\1\\1\\1\end{matrix}}\right\} \text{year 1} \\ \left.\vphantom{\begin{matrix}1\\1\\1\\1\end{matrix}}\right\} \text{year 2} \\ \left.\vphantom{\begin{matrix}1\\1\\1\\1\end{matrix}}\right\} \text{year 3} \end{array}$$

For the first quarter, $X_2 = X_3 = X_4 = 0$, and hence the regression model in equation (7.3) reduces to

$$Y_i = \beta_0 + \beta_1 X_{i1} + e_i$$

For the second quarter, $X_2 = 1$, $X_3 = X_4 = 0$, and

$$Y_i = (\beta_0 + \beta_2) + \beta_1 X_{i1} + e_i$$

For the third quarter, $X_3 = 1$, $X_2 = X_4 = 0$, and

$$Y_i = (\beta_0 + \beta_3) + \beta_1 X_{i1} + e_i$$

Finally, for the fourth quarter, $X_4 = 1$, $X_2 = X_3 = 0$, and

$$Y_i = (\beta_0 + \beta_4) + \beta_1 X_{i1} + e_i$$

Thus the coefficients of the three dummy variables measure the shift from the first quarter base. In fact we are essentially fitting lines to the first-quarter observations, the second-quarter observations, the third-quarter observations, and the fourth-quarter observations with the same slope but possibly different intercepts.

7.3   TRANSFORMATIONS

Up to this point we have assumed a specific structure for the multiple regression model. The regression function is taken to be linear in the parameters, and the error terms are assumed to be independent normal random variables with mean zero and constant variance. We are often faced with hypothesized relationships that are not linear in the parameters and situations in which the error terms are, for example, not normally distributed or do not have a constant variance. In instances such as these it is frequently possible (even mandatory) to transform the original variables so that they conform to the usual multiple regression model. The standard regression procedures can then be employed.

Let us consider the (Cobb-Douglas) production function

$$Q = \alpha C^{\beta_1} L^{\beta_2} \tag{7.4}$$

relating output, Q, to the independent variables capital, C, and labor, L, where α, β_1, and β_2 are parameters whose values are to be determined. This function can be transformed to a function linear in the parameters by taking the logarithm[a] of both sides of equation (7.4). For example,

$$\ln Q = \ln \alpha + \beta_1 \ln C + \beta_2 \ln L \tag{7.5}$$

If we add an error term, e, to account for any "unexplained" variation in the independent variable at given values of the dependent variables, then we have the multiple linear regression model

$$Y = \beta_0 + \beta_1 X_1 + \beta_2 X_2 + e \tag{7.6}$$

with $Y = \ln Q$, $X_1 = \ln C$, $X_2 = \ln L$, and $\beta_0 = \ln \alpha$. Observations on Q, C, and L can be converted to their natural logarithms and, assuming $e \sim N(0, \sigma^2)$, the parameters $\beta_0 = \ln \alpha$, β_1, and β_2 can be estimated and inferences made using the standard procedures. [The reader will notice that the regression

---

[a] We will employ natural logarithms for convenience, but logarithms to any base will produce the same result.

model in equation (7.6) implies that if the error term were introduced in the original formulation of the Cobb-Douglas function, it would appear in a multiplicative manner. The only way we can get from equation (7.4) to equation (7.6) is to assume initially that $Q = \alpha C^{\beta_1} L^{\beta_2} u$, where u is an error term such that $\ln u = e \sim N(0, \sigma^2)$. If it is assumed that the error is additive in the original formulation of the Cobb-Douglas function, then we cannot formally take the logarithm of both sides of the equation and arrive at the regression model in equation (7.6). In this case a different approach is necessary (see Section 7.6).]

As we have seen, the logarithmic transformation converts the production function of equation (7.4) into a standard linear regression model. An appropriate transformation is often immediate if we start with a theoretical relationship or physical law. If no theoretical relationship or physical law exists, it is necessary to take an entirely empirical approach to the problem. Inspection of the data may indicate that some of the usual regression assumptions are not satisfied and in addition may lead directly to the choice of a transformation that will correct any defects.

We are not only concerned with finding a transformation to justify assumptions. It is often desirable to be able to express the findings as succinctly as possible. In the sequel we shall suppose that conditions exist whereby it is necessary to transform the data in order[b] (a) to stabilize the error variance, or (b) to produce a normal distribution of errors, or (c) to produce a simple structure for $E(Y|X)$. It is interesting to note that in many cases a single transformation may accomplish more than one of these objectives.

Figure 7.2 contains some plots of some hypothetical Y values against the values of a single independent variable X. In Figures 7.2(a) and (b) we demonstrate the use of a transformation to obtain a simple structure for $E(Y|X)$. We see that if the original Y values are plotted against the X values, the result is, roughly speaking, described by a curve for which the Y values decay rapidly and then begin to level off with increasing values of X. On the other hand, if the reciprocals of the Y values are plotted against the X's, the result is described, apart from some "unexplained" variation, by a function whose graph is a straight line. The implied relationship between Y and X is simpler in Figure 7.2(b) than it is in Figure 7.2(a) provided we are willing to consider the reciprocals of the Y values instead of the original observations.

In fact the data used in Figure 7.2(a) were generated using the relationship

$$Y_i = \frac{1}{\beta_0 + \beta_1 X_i + e_i}, \qquad i = 1, 2, \ldots, 16 \tag{7.7}$$

---

[b] We might also include the "independence of error" assumption in our list, but methods for analyzing data with dependent errors are discussed specifically in Chapters 9, 10 and 11.

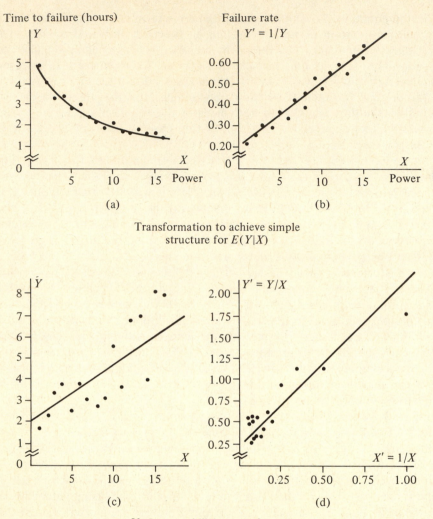

Time to failure (hours)

Failure rate

(a)

(b)

Transformation to achieve simple
structure for $E(Y|X)$

(c)

(d)

Variance stabilizing transformation

Figure 7.2 Scatter plots and possible regression functions

with the e_i's taken to be independent (pseudo) normal deviates with mean
zero and a constant variance. Although this model is not particularly
complex, a simpler form is obtained by considering a reciprocal transformation of the Y values. If we define $Y' = 1/Y$ then the model (7.7) gives

$$Y' = \frac{1}{Y} = \beta_0 + \beta_1 X_i + e_i, \qquad i = 1, 2, \ldots, 16 \qquad (7.8)$$

or, with the standard error term assumptions, the simple linear regression
model.

The point is that if we were confronted with only the data in Figure 7.2(a), the functional form of $E(Y|X)$ may not be immediately apparent. However, if the reciprocal transformation is applied to the Y values, then the relationship between $Y' = 1/Y$ and X is immediately apparent. In this case $E(Y'|X)$ is a function whose graph is a straight line. A simple structure has been obtained for the conditional expectation of the dependent variable by the use of an appropriate transformation.

Figures 7.2(c) and (d) represent a situation in which a transformation can be used to correct a violation of one of the usual error term assumptions, namely, the assumption of constant variance. From Figure 7.2(c) we see that although the relationship between Y and X is roughly linear (as evidenced by the straight line in the figure), the scatter in the Y direction increases as X increases. Equivalently there is greater variability in Y at larger values of X than at smaller values. Situations in which the *variability* of the dependent variable appears to be related to the independent variable(s) are not uncommon in business studies.

Figure 7.2(d) contains a plot of $Y' = Y/X$ against $X' = 1/X$. In this case both the variables have been transformed. The original Y has been divided by the original X to create the new dependent variable Y'. The new independent variable X' is formed by taking the reciprocal of the original independent variable. We see from Figure 7.2(d) that the linearity (apart from error) between the dependent and independent variables has been maintained by the transformations. Moreover the variability in Y' for different X' values appears to be nearly constant. The scatter at small values of X' is about the same as that at larger values of X'. Transformations have stabilized the (conditional) variance of the dependent variable.

The rationale for the variance stabilizing transformation is given by the following argument. Suppose the Y values are generated by the model

$$Y_i = \beta_0 + \beta_1 X_i + e_i, \qquad i = 1, 2, \ldots, 16 \qquad (7.9)$$

where the variance of the error term is proportional to X_i^2. [Since the conditional variance of Y is equal to the variance of the error term, this would produce data like that shown in Figure 7.2(c).] That is, $\text{Var}(e_i) = kX_i^2$, $i = 1, 2, \ldots, 16$, where k is the constant of proportionality. Suppose we divide all the terms in equation (7.9) by X_i. Then

$$\frac{Y_i}{X_i} = \beta_0 \frac{1}{X_i} + \beta_1 + \frac{e_i}{X_i}, \qquad i = 1, 2, \ldots, 16$$

or

$$Y_i' = \beta_0' + \beta_1' X_i' + e_i', \qquad i = 1, 2, \ldots, 16 \qquad (7.10)$$

where $Y_i' = Y_i/X_i$, $X_i' = 1/X_i$, $\beta_0' = \beta_1$, $\beta_1' = \beta_0$, and the new error term $e_i' = e_i/X_i$ (and consequently Y_i') has constant variance k since

$$\text{Var}(e_i') = \frac{1}{X_i^2} \text{Var}(e_i) = k$$

The models in equations (7.9) and (7.10) suggest that if the variability in Y at given values of X increases with the square of these values, a constant variance can be achieved by considering the regression of $Y_i' = Y_i/X_i$ on $X_i' = 1/X_i$. In general if the variance of the dependent variable in a simple linear regression appears to be proportional to some power of an independent variable, say, X^a, $a \neq 0$, then a transformation of the form $Y' = Y/\sqrt{X^a}$ and a corresponding manipulation of the constant term and independent variable will produce a dependent variable with constant variance. These results can be extended in a straightforward manner to the multiple regression model, although in this case it is sometimes easier to use an equivalent variance stabilizing procedure known as "weighted least squares." A discussion of weighted least squares is available in reference [2].

We note that if the usual simple linear regression model were erroneously assumed for the data in Figure 7.2(c), the least squares estimators of β_0 and β_1, although unbiased, would have larger variances than the least squares estimators of the intercept and slope coefficient in a regression model that properly takes the nonconstant error variance into account. In other words given the data in Figure 7.2(c) and equation (7.9) with the usual error term assumptions, the estimates $\hat{\beta}_0$ and $\hat{\beta}_1$ and their estimated variances can be determined by the usual procedures. Alternatively, we can transform the data (thus adjusting for the nonconstant variance) and consider model (7.10). Least squares produces the estimates $\hat{\beta}_0'$ and $\hat{\beta}_1'$ and their estimated variances. As we have seen, however, in the development leading to model (7.10), $\beta_0' = \beta_1$ and $\beta_1' = \beta_0$. As a consequence an estimate of the intercept coefficient in model (7.10) is an estimate of the slope coefficient in model (7.9). Similarly, an estimate of the slope coefficient in model (7.10) is an estimate of the intercept coefficient in model (7.9); that is, β_0 can be estimated by $\hat{\beta}_0$ or by $\hat{\beta}_1'$, and β_1 can be estimated by $\hat{\beta}_1$ or by $\hat{\beta}_0'$. Both sets of estimators of β_0 and β_1 are unbiased; however, the estimators $\hat{\beta}_1'$ and $\hat{\beta}_0'$ have smaller variances than the estimators $\hat{\beta}_0$ and $\hat{\beta}_1$ and therefore are less likely to be far from the true parameters β_0 and β_1 (see Problem 7.5).

Ordinarily when only the data are available for guidance, an appropriate transformation must be selected by trial and error in conjuction with a little common sense. The trial and error portion may be completely subjective, or it may involve statistical procedures specifically designed to select the best way to represent the variables for analysis. One statistical procedure designed for this purpose is briefly discussed in the next paragraph.

A transformation of the dependent variable alone is often sufficient to justify assumptions or to find a metric (units) in which the results may be simply and succinctly expressed. If the Y values are all positive, then a useful transformation of the dependent variable is the power transformation Y^λ, $\lambda \neq 0$. For $\lambda = 0$ we define $Y^\lambda = \log Y$. Consequently, we assume that some number λ exists such that

$$Y^\lambda = \beta_0 + \beta_1 X_1 + \beta_2 X_2 + \cdots + \beta_p X_p + e \tag{7.11}$$

where the error term e is assumed to be independently and normally distributed with zero mean and constant variance. The idea is to use the data to estimate the appropriate transformation, λ, along with the parameters $\beta_0, \beta_1, \ldots, \beta_p$. The details of the procedure are beyond the scope of this book but can be found in reference [16]. Clearly $\lambda = -1$ corresponds to the reciprocal transformation and the linear regression model

$$Y^{-1} = \frac{1}{Y} = \beta_0 + \beta_1 X_1 + \beta_2 X_2 + \cdots + \beta_p X_p + e \qquad (7.12)$$

whereas $\lambda = \frac{1}{2}$ corresponds to the square root transformation and the linear regression model

$$Y^{1/2} = \sqrt{Y} = \beta_0 + \beta_1 X_1 + \beta_2 X_2 + \cdots + \beta_p X_p + e \qquad (7.13)$$

In practice the logarithmic, reciprocal, and square root transformations seem to be the most useful although theoretically λ can be any real number.

7.4 MULTICOLLINEARITY

Linear regression models are commonly employed (a) to predict future response (dependent variable) values, or (b) to depict the effects of important explanatory variables on the response.

Unfortunately in many regression problems, as we have repeatedly pointed out, data are routinely recorded rather than generated from pre-selected settings of the independent variables (the "designed experiment"), and in these cases the independent variables are frequently linearly dependent. For example, in appraisal work, the selling price of homes may be related to independent variables such as age, living space in square feet, lot size, number of bathrooms, number of rooms exclusive of bathrooms, and an index of construction quality. In this example we would expect many of the independent variables to be (nearly) linearly related. Living space, number of bathrooms, and number of rooms should certainly "move together." If one of these variables increases, the others will generally increase.

If this linear dependence in the independent variables is less than perfect, least squares estimates can still be obtained since $(X'X)^{-1}$ exists and hence the normal equations can be solved for $\hat{\beta}$. If the independent variables (or some subset of them) are "nearly" linearly dependent, however, the least squares estimates tend to be unstable and "inflated." More precisely

1. $\hat{\beta}$ is likely to be far from the vector of true values. Indeed, individual coefficients may have the wrong sign.
2. The diagonal elements of $(X'X)^{-1}s^2$ are likely to be large; that is, the estimated variances of $\hat{\beta}_i$, $i = 0, 1, \ldots, p$, are likely to be large and

hence $\hat{\beta}_i$ can be insignificant even though $\beta_i \neq 0$. The sampling fluctuation, as measured by s^2, is small and hence the F ratio can be significant.

3. The calculation of the least squares estimates is sensitive to rounding errors.

In addition, we must be careful in interpreting $\hat{\beta}_i$ as the change in the dependent variable induced by a unit change in X_i, $i = 1, 2, \ldots, p$.

The problems introduced by linear dependence in the X matrix fall under the general heading of problems of "multicollinearity." The multicollinearity problems stem from the fact that the matrix $(X'X)^{-1}$ is involved in the calculation of $\hat{\boldsymbol{\beta}}$ and $\mathrm{Var}(\hat{\boldsymbol{\beta}})$. For example, consider the multiple linear regression model

$$Y_i = \beta_1 X_{i1} + \beta_2 X_{i2} + \cdots + \beta_p X_{ip} + e_i, \qquad i = 1, 2, \ldots, n$$

with the usual assumptions on the error term and the X's scaled[c] so that

$$\sum_{i=1}^{n} X_{ij} = 0 \qquad \text{and} \qquad \sum_{i=1}^{n} X_{ij}^2 = 1, \qquad j = 1, 2, \ldots, p$$

It has been pointed out [see equation (6.21)] that

$$\mathrm{Var}(\hat{\boldsymbol{\beta}}) = C\sigma^2$$

and, in particular,

$$\mathrm{Var}(\hat{\beta}_j) = C_{jj}\sigma^2 \tag{7.14}$$

where $C = (X'X)^{-1}$ and C_{jj} is the jth diagonal element of C. It can be shown that

$$C_{jj} = \frac{1}{1 - R_j^2}, \qquad j = 1, 2, \ldots, p \tag{7.15}$$

where R_j^2 is the coefficient of multiple determination from the regression of the jth *independent* variable on the remaining $p - 1$ independent variables. Thus when there is a "strong" linear relationship between an independent variable and one or more of the remaining independent variables (that is, X_j can be "almost" represented by a linear combination of $X_1, \ldots, X_{j-1}$, $X_{j+1}, \ldots, X_p$), R_j^2 will be near 1 and C_{jj} will be large. This implies that

$$\mathrm{Var}(\hat{\beta}_j) = C_{jj}\sigma^2 = \frac{\sigma^2}{1 - R_j^2}, \qquad j = 1, 2, \ldots, p \tag{7.16}$$

will be large and consequently the probability of obtaining estimates far removed from the true values is high.

What we have just shown is that if multicollinearity is severe, the least

---

[c] We have assumed $\beta_0 = 0$ and $\sum_{i=1}^{n} X_{ij} = 0$, $\sum_{i=1}^{n} X_{ij}^2 = 1$, $j = 1, 2, \ldots, p$, so that the effects of multicollinearity can be demonstrated as simply as possible. The inclusion of a β_0 term in the model and removing the restrictions on the X's does not affect the main points of our argument.

squares estimates can be highly variable. Additional evidence of instability in the least squares estimates in the presence of multicollinearity is provided by examining the calculations leading to the estimators themselves. The C matrix and, in particular, the diagonal elements $C_{jj}, j = 1, 2, \ldots, p$, influences the least squares estimates since [see equation (6.9)]

$$\hat{\beta} = (X'X)^{-1}X'Y = CX'Y$$

If the C_{jj}'s are large, the corresponding $\hat{\beta}_j$'s will tend to be large in absolute value. If one or more of the true parameters is near zero, the inflated variances and magnitudes of the least squares estimators can easily produce estimated coefficients that have the wrong sign.

We illustrate the effect of multicollinearity with some artificial data taken from [42] in the following example.

EXAMPLE 7.3

Consider the regression model

$$Y_i = \beta_1 X_{i1} + \beta_2 X_{i2} + e_i, \qquad i = 1, 2, 3 \tag{7.17}$$

where e_i's are independent $N(0, 1)$ random variables. We shall concern ourselves with two sets of independent variables. Specifically the X matrices are as follows:

$$\text{Set I:} \quad X = \begin{pmatrix} \mathbf{X}_1 & \mathbf{X}_2 \\ 1 & 0 \\ 0 & 1 \\ 0 & 0 \end{pmatrix} \qquad \text{Set II:} \quad X = \begin{pmatrix} \mathbf{X}_1 & \mathbf{X}_2 \\ 0.8 & 0.6 \\ 0.6 & 0.8 \\ 0 & 0 \end{pmatrix}$$

Two sets of Y values, corresponding to the two sets of X's, were generated from model (7.17) by taking $\beta_1 = \beta_2 = 1$ and adding the pseudo $N(0, 1)$ variates $e_1 = -0.31$, $e_2 = 0.22$, and $e_3 = -0.32$. For example, Y_1 for Set I is given by

$$Y_1 = \beta_1 X_{11} + \beta_2 X_{12} + e_1 = 1(1) + 1(0) - 0.31 = 0.69$$

Similarly, for Set II,

$$Y_1 = \beta_1 X_{11} + \beta_2 X_{12} + e_1 = 1(0.8) + 1(0.6) - 0.31 = 1.09$$

The complete Y vectors are

$$\text{Set I:} \quad Y = \begin{pmatrix} 0.69 \\ 1.22 \\ -0.32 \end{pmatrix} \qquad \text{Set II:} \quad Y = \begin{pmatrix} 1.09 \\ 1.62 \\ -0.32 \end{pmatrix}$$

Given the data (X_i, Y_i), $i = 1, 2, 3$, we can now estimate the parameters β_1 and β_2 assuming model (7.17) for each of the two data sets. We point out that the only thing responsible for the differences in the data sets is the X matrix, since the same parameter values and error terms were used in each case.

The least squares estimates of β are easily shown to be

$$\text{Set I:} \quad \hat{\beta} = (X'X)^{-1}X'Y = \begin{pmatrix} 0.69 \\ 1.22 \end{pmatrix}$$

$$\text{Set II:} \quad \hat{\beta} = (X'X)^{-1}X'Y = \begin{pmatrix} -0.37 \\ 2.30 \end{pmatrix}$$

Moreover,

$$\text{Set I:} \quad \text{Vâr}(\hat{\beta}) = (X'X)^{-1}s^2 = \begin{pmatrix} 0.10 & 0 \\ 0 & 0.10 \end{pmatrix}$$

$$\text{Set II:} \quad \text{Vâr}(\hat{\beta}) = (X'X)^{-1}s^2 = \begin{pmatrix} 1.28 & -1.22 \\ -1.22 & 1.28 \end{pmatrix}$$

Keeping in mind that the true parameter values are $\beta_1 = \beta_2 = 1$ we see that the least squares estimates are closer to these values for Set I (no collinearity) than for Set II (severe collinearity). Additionally, $\hat{\beta}_1$ for Set II has the wrong sign. Inspection of the estimated variance-covariance matrices, Vâr($\hat{\beta}$), reveals the effect of multicollinearity on the (estimated) variances of the least squares estimators. For Set II these variances are more than 10 times as large as the corresponding quantities for Set I.

Although Example 7.3 was artificial and carefully constructed to illustrate our points, multicollinearity can be expected to produce similar results in practice. The amount of multicollinearity that can be tolerated before corrective action must be taken is difficult to assess. It depends on the nature of the problem. At the very least the sample correlations between the independent variables should be examined. If one or more are reasonably large in absolute value,[d] problems due to multicollinearity are likely to be forthcoming.

There are several ways to deal with the problems of multicollinearity; however, none of them may be completely satisfactory or feasible. A few of these procedures are outlined below along with some general comments about their usefulness.

Choice of Independent Variables

We should avoid including independent variables in the regression function that "say the same thing"; that is, we should avoid redundant variables. For example, if X_1 is a quantity measured in inches and X_2 is the same quantity measured in feet, these variables should not be in the same regression function. Unfortunately, it is not always clear that redundancy exists in the postulated regression function. However, a little thought before

---

[d] Although no hard and fast rule is available, absolute values of sample correlations larger than 0.5 can be considered large in many cases. Of course the closer the correlations are to ±1, the more unstable the least squares estimators become.

a regression analysis is attempted has been known to produce satisfying results.

Discarding Independent Variables

If redundant variables already in the regression function can be identified, it is possible to eliminate these variables from consideration. This is dangerous since the variables, by their very nature, do not independently affect the response; that is, the total contribution of the variables may be important, not their separate effects. This is particularly true if predicting the behavior of the response or dependent variable is of interest.

Alternative Estimation Procedures

Since the least squares estimates, $\hat{\beta}_i$, $i = 1, 2, \ldots, p$, may be far from the true parameter values $\beta_1, \beta_2, \ldots, \beta_p$, when dependencies among the X's are present, the least squares values cannot be relied on to indicate the "true" effect on Y of changes in the individual X's. However, methods exist that are capable of producing "better" estimators of the β_i's in the sense that the estimates generated by these methods tend to be closer to the true parameter values than the least squares estimates. When multicollinearity exists, these "stable" estimates are likely to present a more correct picture of the nature of the effects of the explanatory variables on the response. The interested reader should consult references [31] and [32] for examples of alternative estimation procedures that produce stable estimates.

7.5 SELECTION OF EXPLANATORY VARIABLES

Whereas the selection of the response (dependent) variable in multiple regression is usually fairly clearcut, the explanatory (independent) variables—particularly in purely empirical studies—must often be chosen from a whole host of possibilities. As a consequence, the choice of explanatory variables involves finding answers to the questions: How many variables should be measured and, if many are measured, how should we proceed to find the most relevant ones?

The questions above are not easily answered. The following considerations,[e] however, will influence the nature and number of explanatory variables measured:

1. Is the study intended to examine a rather specific hypothesis about the phenomenon under investigation, or is the study simply concerned with screening out important variables from a large number of

[e] These were articulated in reference [25].

potential variables, the important variables to be investigated in a subsequent experiment? In the former case we must anticipate the explanations competing with the hypothesis under test and measure only those variables that are relevant with respect to the competing explanations.

2. Can the response variables be observed quickly, so that later parts of the study can be modified, if necessary, in view of the earlier results?

3. The time available, the ease with which the measurements can be made, the availability of "good" official statistics, the amount of money appropriated for the study, and so forth are often crucial in deciding how many variables can be measured.

4. Will the study be used to establish comparisons with earlier, related studies?

Once a given set of explanatory variables has been decided on and measured, the problem of finding the most relevant ones must be tackled. In this regard it is necessary to make the distinction between studies whose objective is primarily prediction and studies whose objective is primarily explanation. For the former, two quite different models, involving different explanatory variables, may fit the data equally well and thus be useful for predicting future outcomes in the absence of major changes in the system. Moreover, it will normally not be desirable to include all "explanatory" variables because, other things being equal, one should strive for simplicity and because the "prediction error" will generally increase if too many variables are included.

When explanation is the objective—that is, a model must be devised that will be compatible with other studies and will predict well under quite different circumstances—it is extremely important to include only those variables that have important effects in a "causal" sense. Here additional information obtained outside the experiment itself (such as which variables are likely to be alternatives to one another and which variables almost certainly must be included) is vitally important. In these situations there is ordinarily only one model that is "correct."

The remainder of this section is devoted to brief discussions of statistical procedures designed to select the "best" subset of explanatory variables from a given set of alternatives. For the same set of data these procedures will frequently point to the same or (arguably) similar subsets. Some of these procedures are illustrated in the case study presented in Chapter 8. The reader should always keep in mind that the procedures are not able to select variables that are not included in the initial set, so it is up to the investigator to be sure all potentially important variables are included at the beginning. The procedures are not specifically designed to guard against multicollinearity. This can still present a problem even after the "best" subset has been selected. Finally, the procedures can be used for both predictive and explanatory studies, although experience has indicated that they are better adapted to prediction (under unchanging conditions) since in this case the particular variables selected are usually not of intrinsic interest.

We suppose there is a continuous response variable Y and a number of explanatory variables $X_1, X_2, \ldots, X_p$ initially under consideration. It is assumed that any preliminary work, for example, transformations, editing to isolate and possibly remove suspected data values, and so forth, has already been done and that a multiple linear regression model of some form is appropriate. The idea is to come up with a subset of variables $X_1, X_2, \ldots, X_k$, $k \leq p$, suitable for model building purposes.

Although significance tests are a useful guide to the importance of different explanatory variables, they must be interpreted with care. Tabulated significance levels of, say, the F distribution, refer to single tests carried out in isolation. In practice the investigator is ordinarily concerned with a sequence of related tests. This makes the usual interpretation of the significance levels incorrect.

Descriptions of the variable selection methods are contained under the headings that follow.

All Possible Regressions

One immediate way to select explanatory variables is to calculate all 2^p (including the one with no X's) possible regressions. Equations clearly inconsistent with the data are rejected, whereas those consistent with the data are retained for further analysis. The determination of consistency can be made on the basis of the residual mean squares (vis à vis the full model), the coefficients of determination R^2, statistics C_k [see equation (7.18)], which look at the total error of prediction, or some combination of these quantities. Among comparable equations, the simplest one will ordinarily be preferred.

The selection based on R^2 can be accomplished by plotting R^2 against the number of explanatory variables k, $k = 1, 2, \ldots, p$. Although the addition of another explanatory variable(s) can never decrease R^2, the increase may be marginal and not worth the additional complexity. The point at which the increase in R^2 ceases to be meaningful can then be used to determine the best subset of explanatory variables.

Figure 7.3 contains a hypothetical plot of the largest possible value of R^2 for each of the possible number, k, of explanatory variables. We see from the figure that once the particular set of $k = 4$ explanatory variables is included in the model, additional variables only provide relatively small increases in R^2. On the other hand, the largest R^2 from a regression with four explanatory variables is substantially bigger than the largest R^2 available with three explanatory variables. Consequently in lieu of any contradictory information the subset of four explanatory variables leading to the R^2 pictured is the "best."

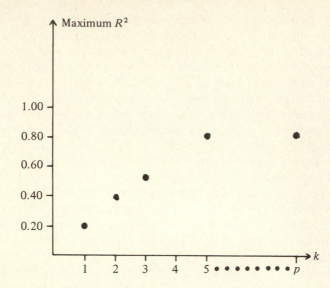

Figure 7.3   Hypothetical plot of maximum R^2 versus the number of explanatory variables

The C_k statistic is defined as

$$C_k = \frac{\text{(residual sum of squares)}}{\hat{\sigma}^2} - (n - 2k), \quad k = 1, 2, \ldots, p + 1 \quad (7.18)$$

where $\hat{\sigma}^2$ is an estimate of σ^2, usually obtained by using the residual mean square from the full model, k is now the *number of regression coefficients* in the current model, and n is the number of observations. The expected value of C_k is roughly k if the model is correct, that is, if no important explanatory variables have been omitted. If important X variables have been omitted, then the least squares estimators are no longer unbiased and $E(C_k) > k$; C_k is, in fact, a measure of the total prediction error. This error involves the variances of the least squares estimators as well as their biases, if any, and the inherent variability in Y.

In a plot of all possible points (k, C_k), those that lie close to the 45-degree line through the origin (assuming identical vertical and horizontal scales) will correspond to equations in which all the important explanatory variables for predicting Y have been included. This plot will provide a basis for deciding which subset of the X's should be included in the final model. We note that it is sometimes better to accept a k coefficient equation that provides a value $C_k > k$ rather than a k' coefficient equation $(k' > k)$ that provides a value $C_{k'} \doteq k'$, provided $C_k < C_{k'}$. The reason is that although the k' coefficient model may yield unbiased estimates, it gives rise to a larger "total prediction error" C_k than the k coefficient model. The latter model is biased in the sense that it gives rise to parameter estimates that are biased by the omission of some important X's. Figure 7.4 contains a typical C_k plot for $p = 4$. We note that there are $p!/(k - 1)!(p - k + 1)!$ C_k's for each k.

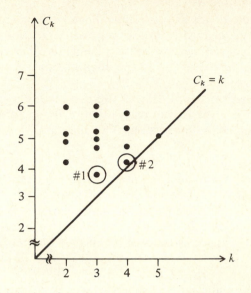

Figure 7.4   A plot of C_k versus the number of explanatory variables ($p = 4$)

The circled dots in Figure 7.4 correspond to explanatory variable subsets that might be included in the final fitted equation. There is a subset of three explanatory variables (indicated by circled dot number 2) that gives an equation yielding an unbiased prediction. On the other hand there is an equation containing two variables (indicated by circled dot number 1) that yields a slightly biased prediction—the least squares parameter estimators are not equal to their expected values—but has a little smaller total prediction error C_k. The final choice can then be made between these two equations.

The major objection to the all possible regressions approach is that it requires a large amount of computational time. With $p = 10$ there are $2^{10} = 1024$ possible regressions, and this involves a tremendous amount of computation even for large-scale digital computers. The all possible regression approach to variable selection is ordinarily not attempted with $p > 10$. The analysis is obviously simplified if some of the regressions can be ruled out a priori. The R^2 and C_k plots can then be constructed for the remaining regressions.

Stepwise Regression

A method of explanatory variable selection that circumvents the computational problem associated with all possible regression is "stepwise" regression. This methodology is available as an option on many standard regression computer programs. The method basically employs a series of F tests to check on the "significance" of explanatory variables added to the regression function. In addition the significance of variables already in the model is reexamined once new variables have been added. If they are not significant, they are deleted from the regression function.

The best way to describe stepwise regression more completely is to list the basic steps (algorithm) involved in the computations.

Step 1 All possible *simple* linear regressions are considered. The X variable that explains the largest proportion of the variation in Y (has the largest correlation with the response) is the first variable to enter the regression function.

Step 2 The next variable to enter is the one (out of those not yet included) that makes the largest significant contribution to the regression sum of squares. The significance of the contribution is determined by an F test. The value of the F statistic that must be exceeded before the contribution of a variable is deemed significant is often called the "F to enter."

Step 3 Once an additional variable has been included in the equation, the individual contributions to the regression sum of squares of the other variables already in the equation are checked for significance using F tests. If the F statistic is less than the one (called the "F to remove") corresponding to a prescribed significance level, the variable is deleted from the regression function.

Step 4 Steps 2 and 3 are repeated until all possible additions are nonsignificant and all possible deletions are significant. At this point the selection stops.

Additional detail on stepwise regression is available in reference [2]. Stepwise regression is an automatic procedure once significance levels have been specified. Although it is a reasonable method in practical situations, it suffers from two major drawbacks. First, there is no guarantee that the final model selected is, in fact, the best model for all purposes. Since the investigator does not see all possible regressions, models that are essentially as good as the final stepwise model and may, in fact, be better for a specific purpose are never revealed. Second, a sequence of F tests is involved in the selection of the final model. As we noted earlier, this means that caution should be used in interpreting the quoted significance levels since these are only (theoretically) appropriate for individual tests.

The practical effect of the stepwise procedure's treating the F tests as isolated individual tests is to include variables in the regression function that may not be significant if the correct significance levels are employed; that is, the quoted significance level understates the true significance level. Unfortunately the computation of exact significance levels for the series of F tests used in stepwise regression is a difficult problem and at present it is not clear how existing programs should be modified.

7.6 NONLINEAR ESTIMATION

Up to this point we have assumed the regression function is *linear in the parameters*. Actually, the principle of least squares can be applied to *any* regression model regardless of how the parameters enter the regression function, provided we are willing to assume that the errors are additive. In situations in which the models are nonlinear in the parameters, the normal equations frequently cannot be solved directly and we must iteratively minimize the sum of squared errors using an algorithm designed to find the minimum of a function of several variables. Many good computer algorithms of this type are currently available. In general these procedures require initial guesses of the parameter values as starting points. They then proceed iteratively to the values that minimize the error sum of squares for a given data set. Since a thorough understanding of the "nonlinear" estimation procedure requires mathematical arguments beyond the scope of this book, we shall merely present the main ideas in the context of two examples. The first example illustrates the principle involved, and the second example contains a simplified version of the procedure actually employed to obtain the least squares estimates. For those wishing to learn more about the subject, an excellent treatment of nonlinear estimation is available in reference [2].

EXAMPLE 7.4

The Cobb-Douglas production function was introduced in connection with transformation of variables. If we assume that the error term is additive, rather than multiplicative, then the Cobb-Douglas regression model can be written

$$Q_i = \alpha C_i^{\beta_1} L_i^{\beta_2} + e_i, \qquad i = 1, 2, \ldots, n \qquad (7.19)$$

where the variables Q, C, and L represent measures of quantity, capital, and labor, respectively, and the e_i's are assumed to be uncorrelated normal variables, each with mean zero and variance σ^2. The regression function $E(Q_i|C_i, L_i) = \alpha C_i^{\beta_1} L_i^{\beta_2}$ is not linear in the parameters; however, by definition, the least squares estimates of α, β_1, and β_2 are given by those values that minimize the error sum of squares

$$S(\alpha, \beta_1, \beta_2) = \sum_{i=1}^{n} (Q_i - \alpha C_i^{\beta_1} L_i^{\beta_2})^2 = \sum_{i=1}^{n} e_i^2 \qquad (7.20)$$

for a given set of data. In this case a nonlinear estimation procedure would determine the least squares values $\hat{\alpha}$, $\hat{\beta}_1$, and $\hat{\beta}_2$ beginning with initial guesses, say, α^*, β_1^*, and β_2^*, and a specification of the function to be minimized, $S(\alpha, \beta_1, \beta_2)$.

Let us return temporarily to the usual multiple linear regression model

$$Y_i = \beta_0 + \beta_1 X_{i1} + \cdots + \beta_p X_{ip} + e_i, \qquad i = 1, 2, \ldots, n \qquad (7.21)$$

and in particular the regression function

$$E(Y_i | X\text{'s}) = \beta_0 + \beta_1 X_{i1} + \cdots + \beta_p X_{ip}$$

Given the data, the least squares estimates, $\hat{\beta}_0, \hat{\beta}_1, \ldots, \hat{\beta}_p$, are those values of the β's that minimize the error sum of squares. The minimum error sum of squares (residual sum of squares) is

$$S(\hat{\beta}_0, \hat{\beta}_1, \ldots, \hat{\beta}_p) = \sum_{i=1}^{n} (Y_i - \hat{\beta}_0 - \hat{\beta}_1 X_{i1} - \cdots - \hat{\beta}_p X_{ip})^2 \qquad (7.22)$$

where the least squares estimates are given, in matrix notation, by $\hat{\boldsymbol{\beta}} = (X'X)^{-1}X'Y$. It is also possible to compute the least squares estimates using a two-step procedure. A description follows.

Suppose that initial *guesses* of the $\hat{\beta}$'s, say, $\beta_0{}^*, \beta_1{}^*, \ldots, \beta_p{}^*$, are available. These guesses can be anything as long as they do not coincide with the least squares values. Then we can write the *minimum* error sum of squares as

$$S(\hat{\beta}_0, \hat{\beta}_1, \ldots, \hat{\beta}_p) = \sum_{i=1}^{n} (Y_i - \beta_0{}^* - \beta_1{}^* X_{i1} - \cdots - \beta_p{}^* X_{ip} - (\hat{\beta}_0 - \beta_0{}^*)$$

$$- (\hat{\beta}_1 - \beta_1{}^*) X_{i1} - \cdots - (\hat{\beta}_p - \beta_p{}^*) X_{ip})^2 \qquad (7.23)$$

Let us denote $\beta_0{}^* + \beta_1{}^* X_{i1} + \cdots + \beta_p{}^* X_{ip}$ by $Y_i{}^*$, $i = 1, 2, \ldots, n$; $Y_i - Y_i{}^*$ by R_i, $i = 1, 2, \ldots, n$; and $(\hat{\beta}_j - \beta_j{}^*)$ by δ_j, $j = 0, 1, \ldots, p$. Equation (7.23) becomes

$$S(\hat{\beta}_0, \hat{\beta}_1, \ldots, \hat{\beta}_p) = \sum_{i=1}^{n} (R_i - \delta_0 - \delta_1 X_{i1} - \cdots - \delta_p X_{ip})^2 \qquad (7.24)$$

Our objective is to obtain the least squares values. Since the guesses $\beta_0{}^*, \beta_1{}^*, \ldots, \beta_p{}^*$ are known, the Y^*'s and hence the R's can always be computed. Moreover, since $\delta_i = \hat{\beta}_j - \beta_j{}^*$, $j = 0, 1, \ldots, p$, and the least squares estimates yield the minimum error sum of squares, $\partial S / \partial \hat{\beta}_j = \partial S / \partial \delta_j = 0$, $j = 0, 1, \ldots, p$; that is, the minimum error sum of squares is given by inserting the $\hat{\beta}$'s in equation (7.22) or by using the δ's in equation (7.24).

Since equation (7.24) is in the same form as equation (7.22) with the R's playing the role of the Y's and the δ's playing the role of the $\hat{\beta}$'s, the δ's must be given, in matrix notation, by

$$\boldsymbol{\delta} = (X'X)^{-1}X'\mathbf{R} \qquad (7.25)$$

Once the δ's are calculated, the least squares estimates are given by $\hat{\beta}_j = \beta_j{}^* + \delta_j$, $j = 0, 1, \ldots, p$, or, in matrix notation, by

$$\hat{\boldsymbol{\beta}} = \boldsymbol{\beta}^* + \boldsymbol{\delta} \qquad (7.26)$$

The δ's represent the "corrections" needed to turn our initial guesses into the least squares values.

To summarize, the least squares estimates for linear multiple regression models can always be calculated by a two-step procedure.

Step 1:   Use any initial guesses $\boldsymbol{\beta}^*$ to calculate $\mathbf{R} = \mathbf{Y} - \mathbf{Y}^*$ and use equation (7.25) to calculate $\boldsymbol{\delta}$.

Step 2:   Use equation (7.26) to obtain the least squares estimates $\hat{\boldsymbol{\beta}}$.

In practice, of course, we would never get the least squares estimates in this manner because they can be obtained directly from $\hat{\boldsymbol{\beta}} = (X'X)^{-1}X'\mathbf{Y}$. We introduce the two-step procedure because it is of fundamental importance to the nonlinear estimation methodology.

Before proceeding, we make two more important points. First,

$$Y^* = \beta_0^* + \beta_1^* X_1 + \cdots + \beta_p^* X_p$$

represents a guess of the regression function derived by using the original model and the guesses $\beta_0^*, \beta_1^*, \ldots, \beta_p^*$. Second, the independent variables $X_1, \ldots, X_p$ are the partial derivatives of the regression function with respect to each of the parameters. For example,

$$\frac{\partial E(Y_i|X\text{'s})}{\partial \beta_0} = 1, \qquad i = 1, 2, \ldots, n$$

$$\frac{\partial E(Y_i|X\text{'s})}{\partial \beta_1} = X_{i1}, \qquad i = 1, 2, \ldots, n$$

$$\vdots$$

$$\frac{\partial E(Y_i|X\text{'s})}{\partial \beta_p} = X_{ip}, \qquad i = 1, 2, \ldots, n$$

(7.27)

We illustrate the nonlinear estimation algorithim with the next example.

EXAMPLE 7.5

Consider the model

$$Y_i = \beta_0 + \beta_1 e^{-\beta_2 T_i} + e_i, \qquad i = 1, 2, \ldots, n \tag{7.28}$$

(the error terms $e_1, \ldots, e_n$ are assumed to have the usual characteristics) relating the yields on certain bonds, Y, to years to maturity, T. The regression function $E(Y_i|T_i) = \beta_0 + \beta_1 e^{-\beta_2 T_i}$ is nonlinear in the parameters.

The least squares estimates of the parameters in model (7.28) give the minimum error sum of squares

$$S(\hat{\beta}_0, \hat{\beta}_1, \hat{\beta}_2) = \sum_{i=1}^{n} (Y_i - \hat{\beta}_0 - \hat{\beta}_1 e^{-\beta_2 T_i})^2$$

Since the parameters enter the error sum of squares in a "nonlinear" fashion, the least squares estimates *cannot* be obtained in the usual way. However, suppose we have a guess of the regression function, say, $Y^* = \beta_0^* + \beta_1^* e^{-\beta_2^* T}$, calculated by using guesses β_0^*, β_1^*, and β_2^* of the true parameters. Once the Y^*'s are available, suppose we calculate the corrections, δ_0, δ_1, and δ_2, and the "least squares" values using the two-step procedure [equations (7.25) and (7.26)] outlined above. It turns out that one pass through this procedure will *not* produce the least squares estimates. The least squares estimates will only be obtained at this point if the model is linear in the parameters. It can be shown, however, that one pass through steps 1 and 2 will improve our guesses in the sense that the new parameter estimates are now closer to the least squares values than the previous guesses. This suggests that we repeat the two-step procedure several times, obtaining successively better estimates, until there is no appreciable change in our current estimates and the error (residual) sum of squares. This is precisely what is done by nonlinear estimation routines.

The efficiency of the nonlinear estimation algorithm (the two-step procedure) depends on the choice of starting parameter values (the initial guesses); the closer these are to the least squares estimates, the faster the least squares estimates are obtained. Also note that we effectively treat the nonlinear regression function as we would a function that is linear in the parameters. Thus in nonlinear situations the columns of the X matrix are provided by the partial derivatives of this regression function with respect to each of the parameters.

Suppose the following data are available:

(Years to maturity) T_i	1	2	5	10	15	18	23	25
(Bond yield) Y_i	0.067	0.072	0.076	0.079	0.081	0.077	0.082	0.078

We propose to fit the model (7.28) to this data.

Let the vector of initial parameter guesses be

$$\beta^* = \begin{pmatrix} \beta_0^* \\ \beta_1^* \\ \beta_2^* \end{pmatrix} = \begin{pmatrix} 0.08 \\ -0.02 \\ 0.20 \end{pmatrix}$$

The partial derivatives of the regression function are

$$X_{i0} = \frac{\partial E(Y_i|T_i)}{\partial \beta_0} = 1, \qquad i = 1, 2, \ldots, 8$$

$$X_{i1} = \frac{\partial E(Y_i|T_i)}{\partial \beta_1} = e^{-\beta_2 T_i}, \qquad i = 1, 2, \ldots, 8$$

$$X_{i2} = \frac{\partial E(Y_i|T_i)}{\partial \beta_2} = -\beta_1 T_i e^{-\beta_2 T_i}, \qquad i = 1, 2, \ldots, 8$$

where it is seen that, in general, the X's involve the unknown parameters and consequently they are not readily evaluated. However, they can be evaluated using the initial guesses β_0^*, β_1^*, and β_2^*. As a rule the X's are calculated using the latest estimates of the parameters. The X matrix in our example is then

$$\mathbf{X}_0 = 1 \quad \mathbf{X}_1 = e^{-0.20T} \quad \mathbf{X}_2 = 0.02Te^{-0.20T}$$

$$X = \begin{bmatrix} 1 & 0.8187 & 0.0164 \\ 1 & 0.6703 & 0.0268 \\ 1 & 0.3679 & 0.0368 \\ 1 & 0.1353 & 0.0271 \\ 1 & 0.0498 & 0.0149 \\ 1 & 0.0273 & 0.0098 \\ 1 & 0.0101 & 0.0046 \\ 1 & 0.0067 & 0.0034 \end{bmatrix}$$

Since $Y^* = \beta_0^* + \beta_1^* e^{-\beta_2^* T} = 0.08 - 0.02e^{-0.20T}$, the vectors $\mathbf{Y}^*$ and $\mathbf{R} = \mathbf{Y} - \mathbf{Y}^*$ are

$$\mathbf{Y}^* = \begin{bmatrix} 0.0636 \\ 0.0666 \\ 0.0726 \\ 0.0773 \\ 0.0790 \\ 0.0795 \\ 0.0798 \\ 0.0799 \end{bmatrix} \quad \mathbf{R} = \begin{bmatrix} 0.0034 \\ 0.0054 \\ 0.0034 \\ 0.0017 \\ 0.0020 \\ -0.0025 \\ 0.0022 \\ -0.0019 \end{bmatrix}$$

At this point the error sum of squares

$$S(\beta_0^*, \beta_1^*, \beta_2^*) = \sum_{i=1}^{8} (Y_i - \beta_0^* - \beta_1^* e^{-\beta_2^* T_i})^2$$

corresponding to the initial guesses of the parameters is obtained by calculating the sum of squares of the elements in $\mathbf{R}$. This gives

$$S(\beta_0^*, \beta_1^*, \beta_2^*) = 0.00007387 \tag{7.29}$$

Matrix multiplication and inversion give

$$X'X = \begin{bmatrix} 8 & 2.0861 & 0.1398 \\ 2.0861 & 1.2766 & 0.0497 \\ 0.1398 & 0.0497 & 0.0034 \end{bmatrix}$$

$$X'\mathbf{R} = \begin{bmatrix} 0.0137 \\ 0.0079 \\ 0.0004 \end{bmatrix}$$

and

$$(X'X)^{-1} = \begin{bmatrix} 0.4447 & -0.0333 & -17.7860 \\ -0.0333 & 1.8216 & -25.1986 \\ -17.7860 & -25.1986 & 1393.8145 \end{bmatrix}$$

Consequently, equations (7.25) and (7.26) yield

$$\delta = (X'X)^{-1}X'\mathbf{R} = \begin{bmatrix} -0.0013 \\ 0.0039 \\ 0.1148 \end{bmatrix}$$

and

$$\hat{\boldsymbol{\beta}}^{(1)} = \boldsymbol{\beta}^* + \boldsymbol{\delta} = \begin{bmatrix} 0.0787 \\ -0.0161 \\ 0.3148 \end{bmatrix}$$

where we have denoted the new estimates by $\hat{\boldsymbol{\beta}}^{(1)}$ instead of $\hat{\boldsymbol{\beta}}$ since they are not yet the least squares values. The error sum of squares corresponding to the estimates $\hat{\boldsymbol{\beta}}^{(1)}$ is

$$S(\hat{\beta}_0^{(1)}, \hat{\beta}_1^{(1)}, \hat{\beta}_2^{(1)}) = \sum_{i=1}^{8} (Y_i - \hat{\beta}_0^{(1)} - \hat{\beta}_1^{(1)} e^{-\hat{\beta}_2^{(1)} T_i})^2$$

$$= 0.00002549 \qquad (7.30)$$

A comparison of equations (7.29) and (7.30) shows that the estimates $\hat{\boldsymbol{\beta}}^{(1)}$ have produced a smaller error sum of squares than the estimates $\boldsymbol{\beta}^*$. This indicates that the current estimates are closer (as a group) to the least squares estimates than the initial guesses.

The whole procedure can now be repeated with $\hat{\boldsymbol{\beta}}^{(1)}$ used in place of $\boldsymbol{\beta}^*$. The iteration continues until the corrections are essentially zero and the error sum of squares has stabilized.

7.7 FINAL COMMENTS

This chapter has been devoted to a discussion of several important topics in regression analysis, namely, dummy variables, transformations, multicollinearity, the selection of explanatory variables, and nonlinear estimation. We have relied primarily on examples, keeping the mathematics to a minimum. Hopefully this approach will promote understanding. More detail is available in the references cited in the various sections. The "case study" chapter which follows relies heavily on the material in Chapters 6 and 7. We shall have occasion to refer to some of the topics introduced in the regression chapters in the sequence of time series chapters which follow our case study.

PROBLEMS

7.1 Define, discuss, or give an example of each of the following terms:
    (a) dummy variable
    (b) seasonal data
    (c) transformation of data
    (d) multicollinearity
    (e) stepwise regression
    (f) nonlinear estimation

7.2 An economist is interested in the relationship between household consumption, say, Y, and household disposable income, say, X. Data are collected for several families and plotted as a scatter diagram. It is evident that the data fall into two essentially distinct clusters corresponding to households (families) headed by a woman and households headed by a man. Generally speaking, families headed by a woman have lower incomes than families headed by a man. In addition, for a given level of income, families headed by females tend to consume more.

(a) Given the information above, indicate the general appearance of the scatter diagram (that is, draw a picture with Y as the dependent variable and X as the independent variable).

(b) Suppose that, regardless of the sex of the head of the household, consumption increases linearly with income. Specify a possible model, linking consumption and income, that also allows for the difference between households headed by females and those headed by males. Define all terms, and specifically list any assumptions you make.

(c) Discuss the difficulties that might arise if consumption was regressed on income and no allowance was made for the sex of the head of the household.

7.3 The Penmery department stores hired a graduate student in business to study the factors affecting its sales. A large number of factors were considered, leading to the conclusion that perhaps the most influential factor is the advertising expenditure. Also, $1\frac{1}{2}$ years ago, the company introduced a new charge account system, and it was thought that this might have an important influence on the sales. A tentative model

$$Y = \beta_0 + \beta_1 X_1 + \beta_2 X_2 + e$$

was then proposed, where Y = sales, X_1 = advertising expenditure, and $X_2 = 0$ before and $X_2 = 1$ after the charge account system was introduced. The data on quarterly sales and advertising expenditure (in millions of dollars) for the last three years are given below in chronological order.

Advertising	4	4	5	7	5	8	7	8	6	8	11	9
Sales	8	10	13	15	9	15	18	17	15	21	26	20

(a) Write down the matrix of independent variables corresponding to the proposed model.

(b) What is the meaning of the two parameters (β_1, β_2)?

(c) Is β_0 of interest? Why, or why not?

(d) The fitted equation is

$$\hat{Y} = 1.48 + 1.85 X_1 + 2.89 X_2$$
$$\quad\quad\quad\;\; (0.30) \quad\; (1.24)$$

where the numbers in parentheses beneath $\hat{\beta}_1$ and $\hat{\beta}_2$ are their standard errors. In addition, $R^2 = 0.92$, $s^2 = 2.62$, and $F = 53.75$.

1. What would your reaction be to the proposition that both advertising and the charge account have no effect on sales?

2. If interest centers on the parameter β_2 only, what inference can you make?

3. Suppose someone suggests that each dollar spent on advertising would produce approximately $2 worth of sales and that the charge account has contributed, approximately, to a $3-million increase in sales, what would be your reaction? Explain.

7.4 Consider the following regression models. Indicate how they might be converted to linear regression models with the usual error term assumptions.

(a)
$$Y_i = \beta_0 e^{\beta_1 X_{i1}} e_i, \qquad i = 1, 2, \ldots, n$$
e_i's are independent errors.

(b)
$$\sqrt{Y_i} = \frac{1}{\beta_1 X_{i1} + \beta_2 X_{i2} + e_i}, \qquad i = 1, 2, \ldots, n$$
e_i's are independent $N(0, \sigma^2)$ errors.

(c)
$$Y_i = \beta_0 + \beta_1 X_{i1} + \beta_2 X_{i1}^2 + e_i, \qquad i = 1, 2, \ldots, n$$
e_i's are independent $N(0, \sigma^2 X_{i1})$ errors.

(d)
$$Y_i = \beta_0 + \beta_1(1/X_{i1}) + \beta_2 X_{i2} + \beta_3 X_{i1} X_{i2} + e_i, \qquad i = 1, 2, \ldots, n$$
e_i's are independent $N(0, \sigma^2 X_{i2}^2)$ errors.

7.5 The data plotted in Figures 7.2(c) and (d) are given below.

Figure 7.2(c):

X_i	1	2	3	4	5	6	7	8	9	10	11	12	13	14	15	16
Y_i	1.8	2.3	3.4	3.8	2.6	3.8	3.1	2.8	3.2	5.6	3.7	6.8	7.0	4.0	8.2	8.0

Figure 7.2(d):

$X'_i = 1/X_i$	1.00	0.50	0.33	0.25	0.20	0.17	0.14	0.13
$Y'_i = Y_i/X_i$	1.80	1.15	1.13	0.95	0.52	0.63	0.44	0.35

	0.11	0.10	0.09	0.08	0.08	0.07	0.07	0.06
	0.35	0.56	0.34	0.57	0.54	0.29	0.55	0.50

(a) Using the data from Figure 7.2(c), obtain the least squares estimates of β_0, β_1 in the model

$$Y_i = \beta_0 + \beta_1 X_i + e_i, \qquad i = 1, 2, \ldots, 16 \qquad \text{[see equation (7.9)]}$$

and their estimated variances.

(b) Using the data from Figure 7.2(d), obtain the least squares estimates of β'_0 and β'_1 in the model

$$Y'_i = \beta'_0 + \beta'_1 X'_i + e'_i \qquad \text{[see equation (7.10)]}$$

and their estimated variances.

(c) Estimates of the intercept coefficient β_0 in the original model are provided by $\hat{\beta}_0$ and $\hat{\beta}'_1$. Compare these estimates and their estimated variances. Comment on the result. Similarly, estimates of the slope coefficient β_1 in the original model are provided by $\hat{\beta}_1$ and $\hat{\beta}'_0$. Compare these estimates and their estimated variances. Comment on the results. ($\hat{\beta}_0$ and $\hat{\beta}'_1$ will not agree exactly because of rounding error. Similarly for $\hat{\beta}_1$ and $\hat{\beta}'_0$.)

7.6 It is observed that consumption of textiles per capita in the United States in the years 1923–1931 was subject to an increasing trend, after which the development became more irregular during the later part of the 1930s. The relevant data are shown below. (See reference [11].)

(a) Given that classical demand theory indicates real income and relative price are the variables that determine the consumption of various commodities, to what extent does a statistical analysis of the data below show that these variables account for the variation of textile consumption over time?

Year	Volume of Textile Consumption Per Capita[a]	Real Income Per Capita[a]	Relative Price of Textiles[a]
1923	99.2	96.7	101.0
1924	99.0	98.1	100.1
1925	100.0	100.0	100.0
1926	111.6	104.9	90.6
1927	122.2	104.9	86.5
1928	117.6	109.5	89.7
1929	121.1	110.8	90.6
1930	136.0	112.3	82.8
1931	154.2	109.3	70.1
1932	153.6	105.3	65.4
1933	158.5	101.7	61.3
1934	140.6	95.4	62.5
1935	136.2	96.4	63.6
1936	168.0	97.6	52.6
1937	154.3	102.4	59.7
1938	149.0	101.6	59.5
1939	165.5	103.8	61.3

[a] Index base 1925 = 100.

(b) Can we estimate income and price elasticities from these data? If yes, what are your estimates and their associated 95-percent confidence intervals? If not, why not?

(c) Can we test the hypothesis that, for example, income elasticity is equal to one against the alternative hypothesis that it is larger than one? If yes, what would be your conclusion, and if no, why not?

(d) Carefully examine the residuals of your model against the usual assumptions of linear models, and suggest appropriate modifications to your analysis if any of the assumptions are violated.

Hint: Consider the multiplicative model

$$C_t = \alpha_0 I_t^{\beta_1} P_t^{\beta_2}$$

where C_t is the per capita textile consumption in year t, I_t is the per capita real income in year t, and P_t is the relative price of textiles in year t. Then apply some suitable transformation in order to convert the above multiplicative model to a convenient linear model known to you. Proceed with the linear model and the transformed data set.

7.7 Review Example 7.3. Consider the following two sets of values for two independent variables X_1 and X_2.

$$\text{Set I: } X = \begin{pmatrix} X_1 & X_2 \\ 1 & 0 \\ 0 & 1 \\ 0 & 0 \end{pmatrix} \quad \text{Set II: } X = \begin{pmatrix} X_1 & X_2 \\ 0.8 & 0.6 \\ 0.6 & 0.8 \\ 0 & 0 \end{pmatrix}$$

You are given two observation vectors,

$$\text{Set I: } \mathbf{Y} = \begin{pmatrix} 0.65 \\ 2.39 \\ -0.14 \end{pmatrix} \quad \text{Set II: } \mathbf{Y} = \begin{pmatrix} 1.05 \\ 2.79 \\ -0.14 \end{pmatrix}$$

As in Example 7.3 the Y's were generated from the multiple regression model,

$$Y_i = \beta_1 X_{i1} + \beta_2 X_{i2} + e_i, \qquad i = 1, 2, 3$$

Set I X's were used to generate Set I Y's, and Set II X's were used to generate Set II Y's. Both cases used the same (pseudo) $N(0, 1)$ deviates for error terms and the same true parameter values $\beta_1 = \beta_2 = 1$. Calculate the least squares estimates for both data sets. Comment on the influence of multicollinearity on the results.

7.8 Give examples of situations in which stepwise regression may be a useful tool for screening out unimportant explanatory variables. Also provide examples of situations in which an automatic procedure like stepwise regression may not be beneficial.

7.9 You are given the nonlinear regression model

$$Y_i = \beta_0 + \beta_1 e^{-\beta_2 T_i} + e_i, \qquad i = 1, 2, \ldots, 16$$

where the e_i's are independent $N(0, \sigma^2)$ variables and the data

T_i	1	2	5	10	15	18	23	25
Y_i	0.067	0.072	0.076	0.079	0.081	0.077	0.082	0.078

With the initial guesses $\beta_0^* = 0.07$, $\beta_1^* = -0.01$, and $\beta_2^* = 0.10$, calculate new estimates $\hat{\beta}_0^{(1)}$, $\hat{\beta}_1^{(1)}$, and $\hat{\beta}_2^{(1)}$ using one iteration of the nonlinear estimation algorithm described in Section 7.6. (It will be helpful to review Example 7.5.)

8

REGRESSION ANALYSIS:
A CASE STUDY

8.1 INTRODUCTION

In this chapter we shall demonstrate selected regression modeling techniques using real data. In our experiments a large data set is treated as a population. A small random sample is drawn from the population and used to build a number of plausible regression models. These models are then tested for predictive power by assessing their ability to predict 10 additional observations randomly drawn from the entire data set (that is, the population).

The large data set[a] consists of observations on variables associated with 500 U.S. cities in the year 1960. These data were originally collected for the purpose of studying newsprint consumption in the United States. The specific variables considered in our case study are listed below.

Dependent variable
    Y: newsprint consumption

Independent variables
    X_1:   number of newspapers in the city
    X_2:   proportion of the city population under age 18
    X_3:   median school years completed (city residents)
    X_4:   proportion of city population employed in white collar occupations
    X_5:   logarithm of the number of families in the city
    X_6:   logarithm of total retail sales

[a] We gratefully acknowledge the kindness of Professor Gil Churchill, who made these data available to us.

308

Our objective is to build a model relating newsprint consumption to one or more of the independent variables. Since there is no "theory" linking newsprint consumption to a specific set of independent variables, our approach must necessarily be an empirical one. We start with the set of "candidate" independent variables (X_1 through X_6)—that is, variables that are likely to influence the demand for newsprint—use these independent variables to build linear regression models, and then check the models for adequacy of fit. Finally the performance of the models is investigated by observing how well they predict "new" observations.

8.2 THE DATA

A random sample of $n = 39$ cities was chosen for analysis. The sample data are listed in Table 8.1. Preliminary dot diagrams indicated that the variables X_5 (number of families), X_6 (total retail sales), and Y (newsprint consumption) are highly skewed. Transformations of these variables were sought in order to make the dot diagrams of the transformed observations symmetric. This was done so that the sample correlation coefficients for pairs of variables may be interpreted unambiguously. One or two large "outlying" observations can greatly influence the magnitude of the sample correlation coefficient. It is particularly important that the dependent variable be approximately symmetric because confidence intervals and tests of significance are based on the assumption that the Y's are normally distributed and the normal distribution is symmetric. Figure 8.1 contains the dot diagrams of Y and $Y^{-1/2}$. The reader will note that the $Y^{-1/2}$ values are roughly symmetric about 0.03 and are "normal looking." The variable $Y^{-1/2}$ will be used as the dependent variable in some of the subsequent regressions.

Figure 8.2 displays an interesting feature of the data that will have important implications in the sequel. The figure is a scatter plot of $Y^{-1/2}$ versus X_5. There appears to be a curvilinear relationship between these variables. This means that we may need to include polynomial terms of degree greater than one in the regression equation. The same remarks hold for a plot of $Y^{-1/2}$ versus X_6.

Table 8.2 displays the correlation matrix of all the variables considered. The correlation coefficient between X_5 and X_6 is 0.842, indicating a high degree of linear relationship between these two variables (that is, multicollinearity). The correlation between $Y^{-1/2}$ and X_5 is -0.684, indicating a fairly high degree of linear relationship, but Figure 8.2 suggests that there may also be an important nonlinear aspect of the relationship. We note that there is a fairly high degree of correlation between Y and X_5, between X_3 and X_5, and between X_4 and X_5.

Table 8.1   Data from a Random Sample of $n = 39$ Cities

X_1	X_2	X_3	X_4	X_5	X_6	Y	$Y^{-1/2}$
1	398	111	462	9.05951	11.25390	961	0.03225
1	328	107	373	8.83491	10.81384	469	0.04617
1	287	96	356	8.98293	10.92602	3511	0.01687
2	310	125	587	10.28045	12.40269	9256	0.01039
1	470	70	414	9.19786	11.01949	556	0.04240
1	370	106	346	9.42351	11.38852	1252	0.02826
1	359	110	463	8.84289	11.36694	902	0.03329
1	337	100	390	9.52981	11.15439	1399	0.02673
1	339	116	458	8.91570	11.21228	1877	0.02308
2	290	109	513	10.38887	12.43562	13907	0.00847
1	351	119	484	8.81105	10.83897	921	0.03295
1	380	119	493	10.40186	12.22769	6959	0.01198
2	336	105	460	9.38546	11.39104	2260	0.02103
2	284	117	506	11.70504	13.66537	73993	0.00367
1	311	121	486	8.91233	10.99804	494	0.04499
2	299	96	372	9.64949	11.61746	7255	0.01174
1	397	117	422	8.84347	10.65124	530	0.04343
1	284	125	547	8.90910	10.70743	488	0.04526
1	377	121	510	8.91233	11.28648	1253	0.02825
1	355	108	413	9.02737	10.70398	878	0.03374
1	406	110	373	9.10597	11.42671	637	0.03962
2	281	108	563	8.80192	10.79859	625	0.04000
1	352	102	413	9.37500	11.37393	3291	0.01743
1	307	91	341	9.84755	11.65175	2470	0.02012
1	377	96	379	9.05380	11.16008	916	0.03304
2	351	120	467	9.76646	11.49217	785	0.03569
2	336	111	474	9.94280	11.99071	4376	0.01511
1	353	99	423	9.21243	10.95816	291	0.05862
2	390	124	594	9.30764	11.65288	1235	0.02845
1	448	123	501	9.08783	11.28756	525	0.04364
2	305	100	436	11.93080	13.79489	99448	0.00317
1	332	100	355	10.64077	12.41086	2921	0.01850
1	188	127	719	9.05298	12.28072	1159	0.02937
1	337	102	352	8.84072	11.00208	1138	0.02964
2	437	124	531	8.99590	11.32535	1097	0.03019
1	362	121	496	8.86220	11.51983	979	0.03196
1	378	113	376	9.29798	10.95411	1899	0.02294
1	280	89	373	9.60211	11.86669	5022	0.01411
1	368	100	429	9.29862	9.45883	81781	0.00349

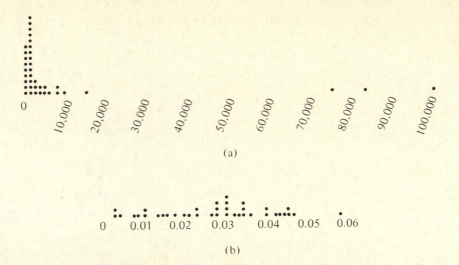

Figure 8.1 (a) Dot diagram of Y; (b) dot diagram of $Y^{-1/2}$

Table 8.2 The Correlation Matrix for the Data in Table 8.1

	X_1	X_2	X_3	X_4	X_5	X_6	Y	$Y^{-1/2}$
X_1	1.000							
X_2	−0.190	1.000						
X_3	0.179	−0.067	1.000					
X_4	0.369	−0.247	−0.104	1.000				
X_5	0.493	−0.262	0.609	0.558	1.000			
X_6	0.485	−0.343	0.100	0.228	0.842	1.000		
Y	0.301	−0.178	−0.107	0.015	0.672	0.388	1.000	
$Y^{-1/2}$	−0.391	0.326	0.128	0.039	−0.684	−0.526	−0.588	1.000

8.3 REGRESSION MODELS INCORPORATING ALL INDEPENDENT VARIABLES

Initially we will consider models in which the dependent variable is regressed on all the candidate independent variables. This allows us to look at all the individual regression coefficients and to make some preliminary judgments about the importance of the various candidate variables.

In all the regressions in this chapter the independent variables have been "standardized"; that is, for any independent variable X_i, the "standardized" variable X' is defined by

$$X' = \frac{X - \bar{X}}{S}$$

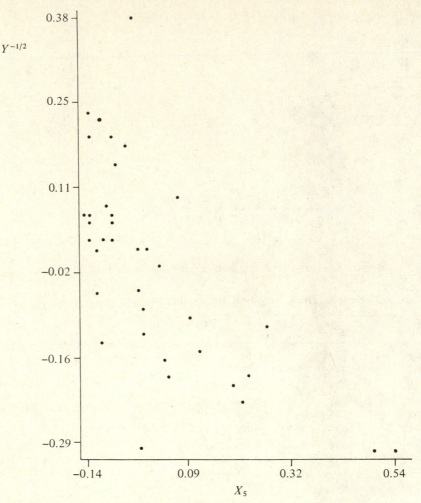

Figure 8.2   Scatter plot of $Y^{-1/2}$ versus X_5

where

$$\bar{X} = \frac{1}{n} \sum_{i=1}^{n} X_i$$

$$S = \sqrt{\Sigma_{i=1}^{n}(X_i - \bar{X})^2}$$

Table 8.3 gives the values of $\bar{X}$ and S for each of the independent variables used in this chapter. When the independent variables are standardized, the estimate of β_0 in the regression function is given by $\hat{\beta}_0 = \bar{Y}$, the sample mean of the dependent variable. Moreover, standardization removes what might be called "nonessential ill conditioning" and results in an equation that is

Table 8.3 Values of $\bar{X}$ and S Used for "Standardization"

Variable	$\bar{X}$	S
X_1	1.28205	2.81024
X_2	344.87179	327.67111
X_3	109.17949	75.77429
X_4	452.56410	505.71493
X_5	9.43686	4.58592
X_6	11.44788	4.86901
X_5^2	89.59361	92.96146
X_6^2	131.66192	115.22400
$X_5 X_6$	108.51430	100.53274

easier to interpret and use. Nonessential ill conditioning refers to the numerical problems that can arise due to the (frequently arbitrary) origins on which the independent variables are expressed. Standardization is appropriate whenever there is a constant term, β_0, in the model.

The fitted regression equation relating Y to all six of the standardized candidate independent variables is

$$\hat{Y} = 8710.67 - \underset{(19,921.83)}{11,516.95 X_1'} - \underset{(17,077.60)}{6,407.93 X_2'} - \underset{(22,213.29)}{8,273.95 X_3'}$$

$$+ \underset{(24,890.23)}{36,935.33 X_4'} + \underset{(33,385.29)}{190,185.74 X_5'} - \underset{(32,620.59)}{110,083.96 X_6'} \quad (8.1)$$

where the estimated standard deviations of the coefficients are given in parentheses beneath the coefficients. The summary statistics are

Residual standard deviation = 15,613.48 (32 d.f.)
$F = 8.058$ (6 and 32 d.f.)
$R^2 = 0.6017$

The F statistic quoted is that used in the overall test of the significance of the regression. A comparison of $F = 8.058$ with, say, the upper 1-percent point of an F distribution with 6 and 32 degrees of freedom indicates that the regression is significant; R^2 reveals that about 60 percent of the variation in Y (about $\bar{Y}$) is explained by movement in the independent variables.

Figure 8.3 shows two residual plots: (a) the standardized residuals versus the predicted values and (b) the normal probability plot. A normal probability plot of a random sample of normal random variables should appear as a straight line. Thus gross deviations from a straight line are indicative of nonnormal residuals and hence the normality assumption in the regression model is suspect. Needless to say, neither of the residual plots in Figure 8.3 is satisfactory.

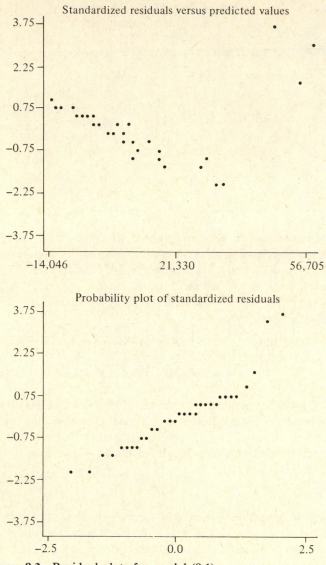

Figure 8.3   Residual plots for model (8.1)

The regression analysis using model (8.1) can be summarized as follows:

1. The residual standard deviation (15,613.48) is very large indicating that unreliable predictions may be expected from this model.
2. Many of the fitted values, $\hat{Y}$, are *negative*, a very unpleasant state of affairs for a model attempting to predict a positive-valued variable.
3. Based on the ratios of coefficients to their estimated standard

deviations, X_5 and X_6 are the most "significant" variables in the regression, with X_4 a distant third. (Recall that inferences about individual correlated independent variables are conditional on all other X's in the regression function. At this point we are simply making rough judgments.)

The fitted regression equation relating $Y^{-1/2}$ to all six "standardized" candidate independent variables is

$$\hat{Y}^{-1/2} = 0.02718 - \underset{(0.01299)}{0.00921} X_1' + \underset{(0.01114)}{0.01639} X_2' + \underset{(0.01449)}{0.00145} X_3'$$

$$+ \underset{(0.01624)}{0.00505} X_4' - \underset{(0.02178)}{0.06322} X_5' + \underset{(0.02128)}{0.01840} X_6' \qquad (8.2)$$

with

Residual standard deviation $= 0.01019$ (32 d.f.)
$F = 5.715$ (6 and 32 d.f.)
$R^2 = 0.5173$

The residual plots corresponding to equation (8.2) appear in Figure 8.4.
   The regression of $Y^{-1/2}$ on $X_1', X_2', \ldots, X_6'$ reveals

1. Satisfactory residual plots; no overwhelming evidence of violation of the usual regression assumptions is indicated.
2. The variable X_5 is most "significant" on the basis of the ratio of the regression coefficient to its estimated standard deviation. The next highest ratio belongs to X_2. We point out that X_5 turned out to be a "significant" variable in both models (8.1) and (8.2).
3. Predicted values of Y can be generated from model (8.2) by squaring the reciprocals of the predicted values of $Y^{-1/2}$; that is,

$$\hat{Y} = (\hat{Y}^{-1/2})^{-2} \qquad (8.3)$$

All of these predicted values will be *positive*.

8.4 VARIABLE SELECTION

As mentioned in the chapter introduction, our goal is to develop a regression model with as few parameters as necessary to maintain adequate predictive performance. On the basis of the analysis done thus far we might be led to discard variables X_1, X_2, X_3, and X_4 when Y is the dependent variable, and to discard X_1, X_3, X_4, X_6, and possibly X_2 when $Y^{-1/2}$ is the dependent variable.

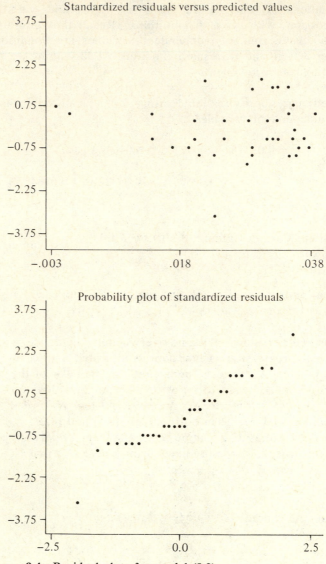

Figure 8.4   Residual plots for model (8.2)

One approach to variable selection is to use stepwise regression (see Section 7.5). Table 8.4 shows the results of four stepwise regressions that were performed using all six candidate independent variables. The results in Table 8.4 tend to confirm our conclusions based on the analysis of models (8.1) and (8.2).

Table 8.4 Results of Four Stepwise Regressions

Dependent Variable	Significance Levels F to enter	F to remove	Independent Variables Retained
Y	0.05	0.10	X_5, X_6
Y	0.25	0.25	X_4, X_5, X_6
$Y^{-1/2}$	0.05	0.10	X_5
$Y^{-1/2}$	0.25	0.25	X_2, X_5

8.5 FITS OF SELECTED MODELS

In this section we will present the results of fitting a number of models to our data. The first model involves Y as the dependent variable and the independent variables X_5, X_6, X_5^2, X_6^2, and X_5X_6. Thus we are fitting a full second-degree polynomial regression function in the variables X_5 and X_6. The reader will recall that the scatter diagrams in Figure 8.2 suggested a possible curvilinear relationship between $Y^{-1/2}$ and X_5. The first model guards against the possibility of curvilinear relationships between the original dependent variable, Y, and the (apparently) important independent variables, X_5 and X_6. The fitted equation is

$$\hat{Y} = 8710.67 - 99{,}612.17X_5' - 1{,}614{,}140.40X_6' + 781{,}521.45(X_5^2)'$$
$$\underset{(310{,}058.86)}{} \quad \underset{(429{,}323.30)}{} \quad \underset{(781{,}718.09)}{}$$

$$+ 2{,}195{,}729.20(X_6^2)' - 1{,}159{,}802.90(X_5X_6)' \qquad (8.4)$$
$$\underset{(203{,}339.38)}{} \quad \underset{(844{,}606.76)}{}$$

with

Residual standard deviation = 5079.60 (33 d.f.)
$F = 145.224$ (5 and 33 d.f.)
$R^2 = 0.9565$

Remarks

1. We note the dramatic drop in the residual standard deviation from model (8.1) to model (8.4). The addition of polynomial terms is clearly warranted in this case.
2. The residual plots (Figure 8.5) are not very satisfactory. The three largest values of Y are causing difficulties in the modeling.
3. Many of the predicted values $\hat{Y}$ are still negative for this model.
4. The P value for testing for the "significance" of X_5X_6 given the other independent variable terms is 0.179. Thus we might be justified in

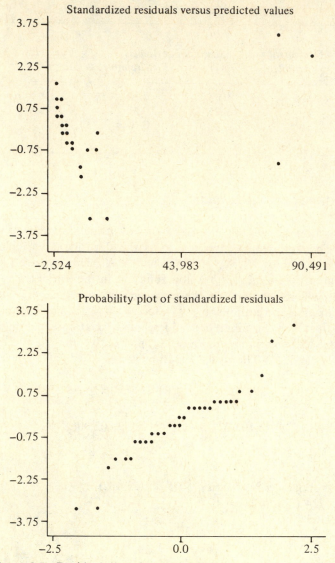

Figure 8.5 Residual plots for model (8.4)

eliminating this term from the model. On the other hand we might argue that the term should be left in the equation so that a full second-degree polynomial is fit to the data. A full second-degree polynomial may be more useful for interpreting the nature of the surface describing the response Y.

5. Notice that the R^2 for model (8.4) is 0.9565, whereas for model (8.1), $R^2 = 0.6017$.

For purposes of comparison a second-degree regression function *without* the cross-product term X_5X_6 was fit to the data. The results are given below.

$$\hat{Y} = 8710.67 + 252,296.76X_5' - 2,144,651.90X_6' - 259,827.78(X_5^2)'$$
$$\quad\quad\quad\quad (176,790.22) \quad\quad\quad (189,674.17) \quad\quad\quad (192,181.84)$$

$$+ 2,210,310.90(X_6^2)' \tag{8.5}$$
$$\quad (205,689.63)$$

with

Residual standard deviation = 5145.34 (d.f.)
$F = 176.462$ (4 and 34 d.f.)
$R^2 = 0.9540$

Remarks

1. The signs of the coefficients of X_5' and $(X_5^2)'$ change as we move from model (8.4) to model (8.5).
2. The residual plots (Figure 8.6) are not very satisfactory.
3. Many of the predicted values of Y derived from equation (8.5) are negative.

Table 8.5 summarizes the fits of three models using $Y^{-1/2}$ as the dependent variable. These models incorporate the standardized variables X_2' and X_5' since these variables appear to be important from our initial analysis. All of the models produce fairly satisfactory residual plots. Figure 8.7 shows the residual plots for the model

$$\hat{Y}^{-1/2} = 0.02718 - 0.37157X_5' + 0.31519(X_5^2)' \tag{8.6}$$

Table 8.5 Summary of Three Regression Models

Dependent Variable	INDEPENDENT VARIABLES[a] X_5'	$(X_5^2)'$	X_2'	Residual Standard Deviation	R^2
$Y^{-1/2}$	− 0.05675 (0.00994)			0.0099	0.4681
$Y^{-1/2}$	− 0.37157 (0.20311)	0.31519 (0.20311)		0.0098	0.4985
$Y^{-1/2}$	− 0.38571 (0.20061)	0.33305 (0.20075)	0.01412 (0.00999)	0.0096	0.5284

[a] The body of Table 8.5 contains the least squares coefficients for the independent variables included in the model. Estimated standard deviations of the coefficients appear in parentheses.

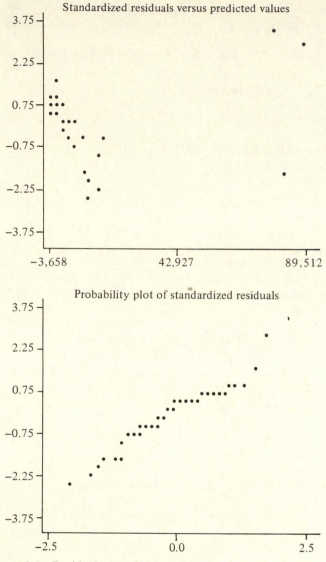

Figure 8.6   Residual plots for model (8.5)

As noted previously, all predictions of Y based on the models in Table 8.5 will be *positive*.

The entries in Table 8.5 suggest that a model with one or two independent variables may be adequate. Once a (relatively) small number of independent variables is decided on, we can examine all possible regressions with that number of independent variables. Figure 8.8 is a variation on the R^2 plot discussed in Chapter 7. The figure contains the R^2 values for all the models

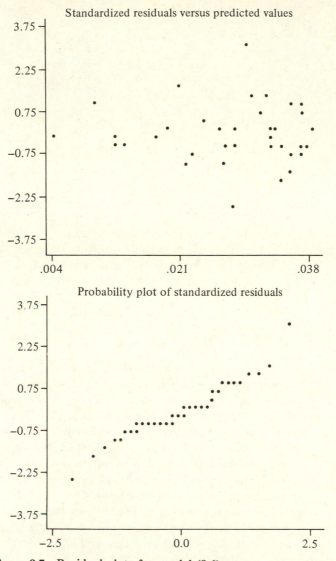

Figure 8.7   Residual plots for model (8.6)

using one or two of the basic candidate independent variables and the model using X_5' and $(X_5')^2$. This plot is informative because it provides a visual ranking of the importance of the individual variables and certain combinations of the variables. To compare the simple models with the more complicated models, the reader will recall that R^2 from the model using X_1', X_2', X_3', X_4', X_5', and X_6' was 0.5173, and the R^2 from the model using X_5, $(X_5')^2$, and X_2 was 0.5284.

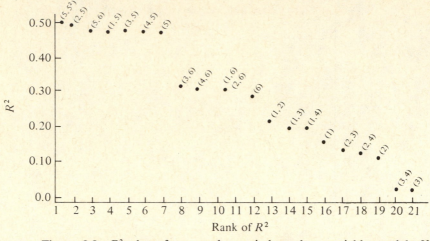

Figure 8.8 R^2 plot of one and two independent variable models [for example, $(1,5)$ stands for the model with X_1' and X_5' as independent variables; the point for (4) is equal to 0 and is not shown]

8.6 ANALYSIS OF PREDICTIONS

As pointed out in Section 8.1, we have the "population" of values for Y and the X's. Consequently we can draw new random samples from the population and compare various models on the basis of their ability to predict newsprint consumption for these new samples. This provides us with a means of validating our models and is an extremely useful technique in practice. We will soon demonstrate that "good fit," as measured by R^2, to the data does not necessarily guarantee good predictions of new observations. Thus it is important that model validation using data not involved in the model building be carried out whenever possible.

In order to test the predictive power of the various models, 10 new cities were drawn at random from the population. The observations are given in Table 8.6. For each model and each new city, predictions of Y (newsprint consumption) were obtained.

If a model used $Y^{-1/2}$ as the dependent variable, then $Y^{-1/2}$ was converted to a prediction of Y using equation (8.3). An assessment of a model's predictive power over the 10 cities is based on the sum of squared prediction errors, namely,

$$\sum_{i=1}^{10} (Y_i - \hat{Y}_i)^2 \tag{8.7}$$

Table 8.7 shows the assessments of seven different models based on three different functions of $\sum_{i=1}^{10}(Y_i - \hat{Y}_i)^2$.

Table 8.6 Ten Randomly Drawn Items from the Population to Be Used for Testing Model Predictions

i	X_{i1}	X_{i2}	X_{i3}	X_{i4}	X_{i5}	X_{i6}	Y_i
1	1	335	87	329	10.291	12.001	5985
2	2	336	105	460	9.385	11.391	2260
3	1	364	109	469	9.613	11.424	2727
4	1	358	115	460	9.695	11.160	7993
5	1	283	98	453	8.952	11.141	630
6	1	432	123	507	9.494	11.482	916
7	2	323	123	586	9.111	11.548	2500
8	1	303	96	397	9.108	11.179	1784
9	1	329	97	429	9.117	11.278	1639
10	1	346	100	394	9.094	11.033	656

A study of Table 8.7 suggests: (a) That the worst predictor is, in a certain sense, the most natural model, namely, the model involving Y as a linear function of all the candidate independent variables. (b) The two models involving Y and polynomial terms in X_5 and X_6 produced by far the highest values of R^2 (0.9565 and 0.9540); yet these models are relatively poor predictors. (c) It appears that any model using $Y^{-1/2}$ as a dependent variable will be a better predictor than any model using Y as a dependent variable. This result illustrates the value of transforming the dependent variable so that its distribution is roughly symmetric before applying the usual regression methodology. (d) The model using $Y^{-1/2}$ as the dependent variable and X'_5 and $(X'_5)^2$ as the independent variables turned out to be the best predictor for the sample of new cities considered. Thus we have the rather pleasing result that one of the simplest models is a better predictor than any of the more complicated models.

8.7 FINAL COMMENTS

In this chapter we have used real data to demonstrate a number of features of regression analysis. Our emphasis has been on prediction. Specifically we have pointed out (a) the value of plots in identifying and evaluating the adequacy of models, (b) the value of transforming both independent and dependent variables, (c) the value of considering polynomial terms of degree greater than one (including cross-product terms), and (d) the very useful insights that can be gained by using new data to validate models.

Table 8.7   Assessment of Predictive Power of Various Regression Models

Dependent Variable (before standardization)	Independent Variables	Sum of Squared Prediction Error $\times 10^{-6}$ $\sum_{i=1}^{10}(Y_i - \hat{Y}_i)^2/10^6$	Root Mean Square Prediction Error $\sqrt{\sum_{i=1}^{10}(Y_i - \hat{Y}_i)^2/10}$	Relative Root Mean Square Prediction Error $[\sqrt{\sum_{i=1}^{10}(Y_i - \hat{Y}_i)^2/10}]/\bar{Y}^a$
Y	$X_1, X_2, X_3, X_4, X_5, X_6$	1181.01	10,867.41	1.25
Y	X_5, X_6, X_5^2, X_6^2	251.10	5,011.02	0.58
Y	$X_5, X_6, X_5^2, X_6^2, X_5 X_6$	101.38	3,184.04	0.37
$Y^{-1/2}$	X_5	53.28	2,308.25	0.26
$Y^{-1/2}$	$X_1, X_2, X_3, X_4, X_5, X_6$	49.72	2,229.87	0.26
$Y^{-1/2}$	X_5, X_5^2, X_2	39.92	1,997.88	0.23
$Y^{-1/2}$	X_5, X_5^2	23.65	1,537.93	0.18

[a] $\bar{Y} = 8710.67$.

PROBLEMS

8.1 Verify that the model in equation (8.1) yields negative predicted values for some items in the sample.

8.2 Plot $Y^{-1/2}$ versus X_6 and X_5 versus X_6. Interpret these plots.

8.3 Fit the model

$$Y^{-1/2} = \beta_0 + \beta_1 X_2' + \beta_2 X_5' + \beta_3 (X_2^2)' + \beta_4 (X_5^2)' + \beta_5 (X_2 X_5)'$$

to the data in Table 8.1. Use the model to predict the items in Table 8.6. What do you conclude?

8.4 Selected items from the population of 500 cities are listed below. Reproduce the case study in this chapter using some or all of these observations. (Alternatively, each class member can select his or her own sample and compare the results of a regression analysis with those of his or her classmates. The variability in conclusions is often rather sobering.)

X_1	X_2	X_3	X_4	X_5	X_6	Y
1	359	90	331	15941	97543	1743
2	312	122	554	22171	181416	11970
1	369	98	436	8140	64332	745
2	364	98	397	11046	76830	1395
1	373	92	400	6909	44325	358
2	356	106	409	27102	202215	8487
1	396	121	591	6695	57443	550
2	354	123	588	11037	95121	3239
1	360	121	530	8052	69811	1745
1	377	121	510	7423	79737	1253
1	345	108	404	75531	421600	26409
1	339	110	423	7320	50802	792
1	330	99	369	29503	192756	8607
2	321	97	420	125921	799748	66998
3	326	96	329	220538	1278144	133856
2	337	104	396	65473	470991	30064
1	378	113	376	10916	57189	1899
1	355	120	461	8007	56206	1198
1	366	99	394	18829	107848	1986
1	348	108	397	8086	63089	785
2	358	109	423	13042	100699	4481
1	376	103	324	16972	89443	2200
1	325	109	416	12741	121261	2611
1	361	108	420	9661	62508	1366
1	338	103	383	8271	50612	926

X_1	X_2	X_3	X_4	X_5	X_6	Y
1	349	106	419	11352	86539	1932
1	340	109	408	11337	78395	1122
1	319	94	453	9181	68694	1513
1	341	107	396	8320	61068	1126
2	342	106	416	21146	135030	2886
1	308	99	395	8455	78081	2072
2	332	104	426	81348	496643	29984
1	344	109	362	15644	105692	3009
1	332	100	355	41805	245455	2921
1	345	100	391	10029	65051	1737
2	335	116	496	10547	64415	1400
2	400	120	494	16368	94182	1360
2	333	108	502	10204	54595	900
1	269	123	566	7356	51175	309
2	350	122	541	70426	429340	20526
1	310	125	565	12594	176538	3208
2	282	115	552	11023	133897	2850
2	289	100	415	28900	234559	10475
1	317	101	392	18342	94474	2257
1	323	105	403	19515	86472	1710
1	349	89	307	15520	91608	2711
1	287	96	356	7966	55605	3511
2	352	109	430	34890	202627	7362
2	294	108	512	20091	173996	12463
2	294	102	367	8666	47903	1338
1	323	92	344	13910	93445	3675
2	299	96	372	15514	110200	7255
1	325	99	333	8030	56332	1313
1	306	91	374	12039	76404	2770
1	326	99	376	11557	78461	1351
1	282	94	359	8550	80175	1613
1	282	120	452	15198	177189	5383
1	419	105	422	13684	93948	1165
1	292	109	446	7014	63255	250
2	331	105	441	120464	1015750	71169
2	357	87	374	16799	146250	4617
2	387	98	404	30054	191214	4081
2	348	94	403	17304	184380	3767
1	387	115	488	6691	80671	546
1	355	92	394	8413	66705	610
2	378	105	415	36693	190503	6619
1	404	104	395	7423	50403	498
1	417	101	433	5520	44225	335
4	379	121	477	64261	507522	19096
2	323	123	586	9052	103600	2500

X_1	X_2	X_3	X_4	X_5	X_6	Y
2	437	124	531	8070	82897	1097
1	382	122	491	7061	59506	613
1	340	102	388	11374	71118	1940
1	341	112	434	16348	122028	2293
1	320	98	498	10104	77745	1260
1	313	114	514	9437	86450	2557
2	285	125	594	17052	137543	4275
4	311	100	417	909204	5630939	373616
1	343	98	442	11095	95031	2127
2	336	111	474	20802	161250	4376
1	373	87	348	20281	120688	1885
1	275	105	431	11635	94032	1287
1	326	110	447	7052	44329	761
1	312	117	448	9429	67757	1218
1	330	109	457	16898	150626	2571
1	350	100	375	7114	82687	1683
1	325	114	462	11601	105009	1925
1	353	106	397	7825	46282	702
2	318	105	441	26315	220128	8182
1	317	102	445	11512	81530	1983
2	345	110	455	33569	262212	7525
1	327	114	448	13723	71882	1694
2	306	111	550	22138	177692	5739
1	358	117	453	14371	128507	2617
1	343	101	322	12909	106541	2000
1	238	126	555	6792	63979	696
1	352	113	436	10470	79043	1211
2	344	102	429	36840	219556	8367
2	354	116	479	41354	314611	13989
1	392	101	316	43435	258525	6458
1	367	108	350	29433	177403	6064
2	340	108	441	120624	916535	59666
1	370	106	346	12376	88303	1252
1	356	118	464	10921	96389	2870
2	347	106	376	9367	80653	1390
1	346	100	394	8906	61866	656
2	342	105	387	17339	122103	2881
1	353	99	423	10021	57421	291
1	341	108	403	11616	89526	1304
1	353	114	466	34507	229514	13593
2	311	107	433	19264	176910	3828
1	273	133	609	5888	42342	215
1	377	96	379	8551	70269	916
2	354	101	421	86029	567165	30171
1	369	110	489	7823	54982	580

X_1	X_2	X_3	X_4	X_5	X_6	Y
1	390	102	428	7987	64661	1007
1	369	109	449	8163	56723	569
1	375	98	378	15029	78490	1119
1	387	118	541	18783	178005	1837
1	399	112	477	48981	283197	10540
1	380	119	493	32921	204371	6959
1	399	102	408	6652	43413	308
1	323	107	455	13379	84086	981
2	390	124	594	11022	115023	1235
1	412	120	431	8373	97623	260
1	379	118	483	110878	804942	32987
2	371	121	488	53076	368955	8818
1	360	120	445	6843	43087	470
1	353	112	492	14337	117209	1975
1	283	98	453	7725	68966	630
2	312	121	515	28007	233105	15297
1	365	105	431	11048	73166	1514
1	300	121	503	14950	57806	586
1	253	121	568	15972	123481	1382
1	412	123	519	26461	289389	1367
1	369	120	513	14534	212523	3924
1	249	129	613	26761	174763	3000
1	188	127	719	8544	215501	1159
1	303	122	508	25862	180968	467
1	448	123	501	8847	79823	525
1	375	121	439	9803	84532	780
1	346	122	504	8790	127643	897
1	339	116	458	7448	74034	1877
1	351	117	514	34026	374227	13715
1	402	125	552	14255	112821	953
1	451	123	487	20537	146707	2500
1	251	123	578	34550	236305	1986
1	444	121	446	17911	176649	4045
1	193	111	460	8824	102845	2208
1	253	122	571	18880	181612	1100
1	273	121	508	90928	558118	18258
4	305	121	503	636522	4463965	310512
1	340	120	547	9807	142382	2090
1	311	121	486	7423	59757	494
1	291	114	453	97193	694345	36246
1	376	119	421	12239	85477	1455
1	353	122	510	6907	58838	230
1	406	110	373	9009	91740	637
1	359	133	691	13217	153157	2720
1	256	124	530	30200	340361	5652

X_1	X_2	X_3	X_4	X_5	X_6	Y
1	374	121	477	17354	172932	2718
1	350	123	514	6721	46288	250
1	379	119	413	12057	113476	3326
1	347	122	489	12450	97894	1152
1	371	111	398	18935	121535	2717
1	369	124	568	21810	192072	5775
2	309	121	546	49913	463654	27797
1	362	121	496	7060	100693	979

9

TIME SERIES ANALYSIS: AN INTRODUCTION

9.1 INTRODUCTION

Frequently data become available in the form of a time-ordered sequence called a "time series." Examples of time series of interest in business and economics include daily closing stock prices, quarterly earnings, monthly sales, and annual gross national product. Analysis of data of this kind generally has two immediate objectives: description of the generating mechanism and the forecasting of future values. The body of statistical methodology available for accomplishing these objectives is referred to as "time series analysis."

From a statistical point of view description is accomplished when a mathematical model can be formulated which characterizes the behavior of the series;[a] that is, time series models attempt to describe the mechanism generating the observations.

The derivation of a suitable time series model for a particular set of data is not immediate and can frequently be a very frustrating task. As is the case with most statistical model building procedures, a suitable model must often be obtained iteratively. Any single iteration of the search for a time series model can be conveniently broken down into the three distinguishable stages: identification, estimation, and diagnostic checking.[b] These stages are concerned with the selection of a tentative model, estimation of the model parameters, and testing for adequacy of fit, respectively. The iterative nature of model building is apparent since the last stage, diagnostic checking, may reveal an inappropriate model, and hence the original model will be modified

---

[a] In this book we shall concentrate on the parametric modeling of time series. Alternative descriptions of a given time series are available. For example, "spectral analysis" attempts to describe the series in terms of frequency components. The reader is referred to references [1] and [5] for discussions of spectral analysis.

[b] This formulation of the model building problem is due to Box and Jenkins [1].

or discarded altogether, thus leading to a new iteration in the model building process. In this chapter and the next, methods will be presented for tackling the problems presented at each of the three model building stages.

Once a suitable time series model has been obtained, forecasting future observations is relatively easy; for, as we shall see, the forecasts follow directly from the time series model. This implies that most of the forecasting effort must be directed toward obtaining a suitable model. Forecasting is a by-product of the more formidable task of description, although in many cases it is the forecasts that are of primary interest.

It should be noted that the validity of forecasting procedures developed from models that have been suggested by and fitted to historical data is necessarily predicated on the assumption that the future tends to behave like the past. Rarely in the social sciences is this assumption warranted for an *extended* time horizon, that is, over the *long term*. Hence accurate long-term forecasting in the dynamic social science disciplines is a very difficult, if not impossible, task, and successes may be followed by agonizing failures.

In this chapter we shall illustrate our approach to time series analysis with a rather simple example. We intend to rely heavily on the reader's intuition, leaving most of the mathematical details to Chapter 10.

9.2 EXAMPLE OF A TIME SERIES

One of the distinguishing features of time series analysis, as opposed to other statistical techniques, is that the observations are not assumed to vary independently. In fact it is precisely the nature of the *dependence* that is of interest. Table 9.1 is a list of hypothetical yields from, say, 70 consecutive batches of a chemical product. The observations (accumulated yield over batch time) have been plotted in Figure 9.1. Although the observations are discrete, they have been joined by straight lines in the figure to dramatize some of the characteristics of the series.[c] First, it is evident that the series appears to vary about a fixed level. Time series that exhibit this characteristic are said to be *stationary in the mean*. Second, a pronounced "up and down" effect is apparent; that is, a "high" yield tends to be followed by a "low" yield, and vice versa. This suggests that the observations are not independent and that, in fact, observations one time period apart are negatively correlated. The implications of these characteristics for model building are discussed in detail in later sections of this chapter. We point them out initially to emphasize the fact that a visual examination of the data often reveals prominent characteristics of the series that can provide important clues in the search for an appropriate model. We note, however,

[c] Discrete time series are often displayed in this way. The reader should keep in mind that all time series in this book are discrete even though they may be represented by continuous curves.

Table 9.1 Series of 70 Consecutive Yields from a Batch Chemical Process

t:	1–15	16–30	31–45	46–60	61–70
	40	59	53	56	48
	54	51	43	49	44
	48	55	66	52	49
	52	48	48	33	44
	41	51	52	52	49
	52	50	42	59	69
	38	52	44	34	40
	56	44	56	57	54
	48	65	44	39	58
	45	40	58	60	49
	66	65	41	40	
	17	41	54	52	
	62	64	51	44	
	50	53	56	65	
	38	48	38	43	

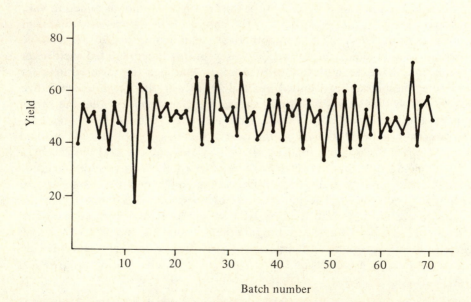

Figure 9.1 Hypothetical yields from a chemical process

that "eyeballing" the series must be done with an open mind, and we must always be wary of reading something into the data that is not there.

9.3 MODEL BUILDING

In this section we shall consider each of the components of our model building strategy individually. The objective is to obtain a suitable mathematical representation of the chemical yield series.

Identification

Identification refers to an informal analysis of available data and prior beliefs about the system under study which leads to a tentative model. We have already implied that a time series model will reflect the nature of the dependence that exists among the observations. Hence an important statistical tool at the identification stage is the sample autocorrelation function.

Let the N observations comprising the time series be denoted by $y_1, y_2, \ldots, y_N$. The reader will recall that a numerical measure of the linear relationship between observations k time periods apart is provided by the kth sample autocorrelation coefficient (see equation 5.69)

$$r_k = \frac{(1/N)\sum_{t=1}^{N-k}(y_t - \bar{y})(y_{t+k} - \bar{y})}{(1/N)\sum_{t=1}^{N}(y_t - \bar{y})^2} = \frac{\sum_{t=1}^{N-k}(y_t - \bar{y})(y_{t+k} - \bar{y})}{\sum_{t=1}^{N}(y_t - \bar{y})^2}, \qquad k = 1, 2, \ldots$$

(9.1)

where $\bar{y} = (1/N)\sum_{t=1}^{N}y_t$ is the "sample average."

Plots of the pairs of observations (y_t, y_{t+1}) and (y_t, y_{t+2}) are presented as scatter diagrams in Figure 9.2. A glance at this figure indicates that adjacent observations tend to move together in a linear fashion with large values of y_t associated with small values of y_{t+1} and small values of y_t associated with large values of y_{t+1}. On the other hand observations two time periods apart tend to move together in a linear fashion with large values of y_t associated with large values of y_{t+2} and small values of y_t associated with small values of y_{t+2}. Hence we would expect r_1 to be negative and r_2 to be positive. The sample autocorrelations have been computed for several lags and are plotted in Figure 9.3(a). The oscillating behavior exhibited by the sample autocorrelation function for this particular series is significant and can be used to identify an appropriate time series model. For instance, the sample autocorrelations suggest that the relationship between pairs of observations at adjacent time points can be described, approximately, by a straight line with a negative slope. Regarding each observation at a given time point t as a realization of a random variable Y_t (that is, the observation y_t is just one of a population of values that might have occurred at time t) and recalling that the series appears to vary about some overall level, say, μ, different from

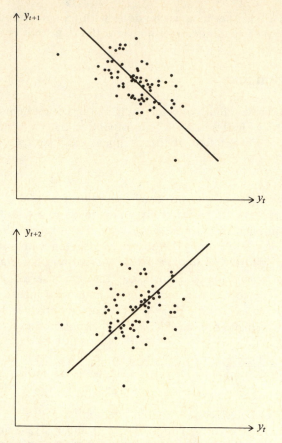

Figure 9.2 Scatter diagrams of (y_t, y_{t+1}) and (y_t, y_{t+2}) for the chemical process yield data

zero, we might postulate the "regression" model

$$(Y_t - \mu) = \phi(Y_{t-1} - \mu) + e_t, \qquad t = \cdots, -1, 0, 1, \ldots \qquad (9.2)$$

where $\phi < 0$ and e_t is an error term, to describe the relationship between adjacent observations. Considering the observations y_t, y_{t-1} as realizations of the random variables Y_t and Y_{t-1}, we are postulating that the deviation of an observation from the overall level at time t is a constant multiple of the deviation at time $t-1$, although opposite in sign, plus a term, e_t, which accounts for the effects of additional sources of variation impinging on the system at time t. In addition we assume the error term, or random "shock," is such that

$$E(e_t) = 0, \qquad \text{Var}(e_t) = \sigma_e^2 \text{ for all } t$$
$$E(e_t e_{t'}) = 0, \qquad \text{for } t \neq t' \qquad (9.3)$$
$$E(e_t Y_{t'}) = 0, \qquad \text{for } t' < t$$

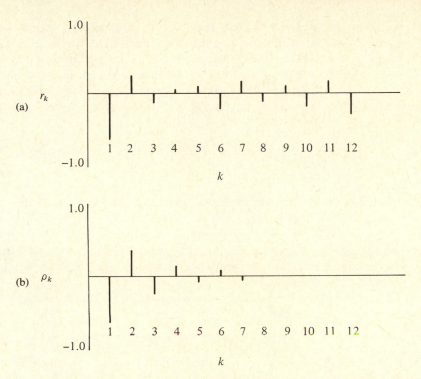

Figure 9.3 (a) Estimated autocorrelation function for the chemical yield data; (b) autocorrelation function for a first-order autoregressive process with $\phi = -0.6$

The reader will note the close similarity of the model given by equation (9.2) to the simple linear regression model, the difference being that the "independent" variable in equation (9.2) is the *dependent* variable lagged one time period; that is, we are regressing the process variable, Y, on itself. This model is called an *autoregressive* (AR) model and, in particular, a *first-order* autoregressive model.

As it stands the first-order autoregressive model is not completely specified. In order for this model to represent variables which vary about a fixed level, that is, variables which are stationary in the mean, it is neessary to require that $|\phi| < 1$.

Given the complete specification of the first-order autoregressive model, the question now arises as to whether a *particular* first-order autoregressive model (that is, a first-order autoregressive model with given values for the parameters ϕ and μ) adequately represents the chemical yield data. Specifically, is it possible for realizations of the random variables Y_t, Y_{t-1}, related to each other by equation (9.2) with $-1 < \phi < 0$, to produce a sample autocorrelation function like the one pictured in Figure 9.3(a)?

Consider a time-ordered process that can be described by a first-order autoregressive model. Actually *all* of the random variables $Y_t, Y_{t-1}, Y_{t-2}, \ldots$ are related to one another since Y_t depends on Y_{t-1}, which in turn depends on Y_{t-2}, and so forth. Now, a measure of the linear association between two *random variables* describing the same time-ordered phenomenon but separated by k time periods is given by the (theoretical) autocorrelation coefficient,

$$\rho_k = \frac{\text{Cov}(Y_t, Y_{t+k})}{\sqrt{\text{Var}(Y_t)\,\text{Var}(Y_{t+k})}} = \frac{E(Y_t - \mu)(Y_{t+k} - \mu)}{\sigma_Y^2}, \qquad k = \pm 1, \pm 2, \ldots$$

(9.4)

for any time t.

In the definition of the autocorrelation coefficient, we have made the tacit assumption that the random variables at different time points vary about the same level, μ, and have the same variance, σ_Y^2. This assumption is common in time series analysis, and it is discussed in more detail in Supplement 10A. We shall assume it holds for any sequence of time-ordered random variables introduced in this chapter.

In particular for random variables following the first-order autoregressive model it can be shown that

$$\rho_k = \rho_{-k} = \phi^k, \qquad k = 1, 2, \ldots$$

(9.5)

The autocorrelation function is symmetric about $k = 0$ and, since $|\phi| < 1$, tends to die out as k increases. In addition, if $\phi > 0$, ρ_k will always be positive, whereas if $\phi < 0$, ρ_k will oscillate from negative to positive values. The function ρ_k is plotted in Figure 9.3(b) for the special case, $\phi = -0.6$. We note the close similarity between the autocorrelation function of Figure 9.3(b) and the *sample* autocorrelation function produced by pairs of observations from the chemical yield series of Figure 9.3(a).

The identification of a time series model is frequently accomplished by matching *sample* autocorrelation patterns with the autocorrelation patterns produced by members of a *known* class of time series models. This procedure is warranted since the sample autocorrelations r_k (under suitable conditions) can be interpreted as estimates of the corresponding values ρ_k. Obviously, this procedure can be highly subjective and thus the initial selection of a model is regarded as tentative.

The striking similarity between the theoretical autocorrelation function of Figure 9.3(b) and the sample (or estimated) autocorrelation function of Figure 9.3(a) suggests that the observations in Figure 9.1 may have been generated by a process that can be represented by the autoregressive model given by equation (9.2). Therefore from the autocorrelation evidence and considerations of the observed behavior of the series itself, we will assume that a tentative time series model for the chemical yield series is given by equation (9.2) with $\phi < 0$. This tentative selection of a time series model completes the identification stage of the investigation. The next problem is that of determining values for the parameters μ and ϕ. This is the subject of the next section.

Estimation

As in regression analysis, to determine the "best" values (estimates) of the unknown parameters in the tentatively selected model we will rely on the principle of least squares. In the simple linear regression model

$$Y_i = \alpha + \beta X_i + e_i, \qquad i = 1, \ldots, N$$

the least squares estimates of the coefficients α and β were those values that minimized the sum of squares

$$S(\alpha, \beta) = \sum_{i=1}^{N} [y_i - E(Y_i|X_i)]^2 = \sum_{i=1}^{N} (y_i - \alpha - \beta X_i)^2$$

where $y_1, \ldots, y_N$ are observations on the random variables $Y_1, Y_2, \ldots, Y_N$ at the corresponding values of the independent variable X. When we fit time series models, we proceed in exactly the same manner. In particular, using equation (9.2) and changing the subscript from i to t, to conform with our earlier notation, the conditional expectation of Y_t given $Y_{t-1} = y_{t-1}$ is

$$\begin{aligned}
E(Y_t|Y_{t-1} = y_{t-1}) &= E[\mu + \phi(y_{t-1} - \mu) + e_t|y_{t-1}] \\
&= \mu + \phi(y_{t-1} - \mu) + E(e_t|y_{t-1}) \\
&= \mu + \phi(y_{t-1} - \mu)
\end{aligned} \qquad (9.6)$$

since $E(e_t|y_{t-1}) = E(e_t) = 0$ by assumption. Therefore given the observations $y_1, y_2, \ldots, y_N$ and employing the principle of least squares, we seek the values of μ and ϕ that minimize the sum of squares

$$S(\mu, \phi) = \sum_{t=1}^{N} [y_t - E(Y_t|y_{t-1})]^2 = \sum_{t=1}^{N} [y_t - \mu - \phi(y_{t-1} - \mu)]^2 \qquad (9.7)$$

The observant reader will note that minimizing $S(\mu, \phi)$ is not quite as straightforward as it might appear to be. Since the sum on the right-hand side of equation (9.7) begins at $t = 1$, $S(\mu, \phi)$ depends on the *unobserved* quantity y_0. A natural way out of this difficulty is to begin the summation at $t = 2$ and hence find those values of μ and ϕ that minimize

$$S^*(\mu, \phi) = \sum_{t=2}^{N} [y_t - \mu - \phi(y_{t-1} - \mu)]^2 \qquad (9.8)$$

This procedure is justified if N is fairly large, for then $S^*(\mu, \phi)$ will be nearly equal to $S(\mu, \phi)$.

The least squares estimates can be obtained in the usual manner by differentiating $S^*(\mu, \phi)$ partially with respect to μ and ϕ and then solving the equations

$$\frac{\partial S^*(\mu, \phi)}{\partial \mu} = 0 \qquad \text{and} \qquad \frac{\partial S^*(\mu, \phi)}{\partial \phi} = 0$$

simultaneously. Denoting the least squares estimates by $\hat{\mu}$ and $\hat{\phi}$ it can be shown that[d]

$$\hat{\mu} \doteq \bar{y}$$

$$\hat{\phi} \doteq \frac{\sum_{t=2}^{N}(y_t - \bar{y})(y_{t-1} - \bar{y})}{\sum_{t=2}^{N}(y_{t-1} - \bar{y})^2} = \frac{\sum_{t=1}^{N-1}(y_t - \bar{y})(y_{t+1} - \bar{y})}{\sum_{t=1}^{N-1}(y_t - \bar{y})^2} \qquad (9.9)$$

Comparing the expression for $\hat{\phi}$ with the expression for the first sample autocorrelation r_1, given by equation (9.1) with $k = 1$, it is clear that $\hat{\phi} \doteq r_1$; that is, an *approximate* least squares estimate of ϕ is provided by the first sample autocorrelation coefficient.

The chemical yield data gives (rounded to the first decimal place)

$$\hat{\mu} \doteq \bar{y} = 49.7 \qquad \text{and} \qquad \hat{\phi} \doteq -0.59$$

Hence continuing the regression analogy, the estimated or "fitted values," $\hat{y}_t$, of the series are given by evaluating the *observational* equation

$$y_t - \mu = \phi(y_{t-1} - \mu)$$

with the least squares estimates $\hat{\mu}$ and $\hat{\phi}$ inserted for μ and ϕ. Thus transposing $\hat{\mu}$,

$$\hat{y}_t = \hat{\mu} + \hat{\phi}(y_{t-1} - \hat{\mu})$$
$$= 49.7 - 0.59(y_{t-1} - 49.7)$$

Since $\hat{y}_1$ depends on y_0, which is unknown, we ignore this quantity and begin the calculation of the fitted values with $\hat{y}_2$. For example,

$$\hat{y}_2 = 49.7 - 0.59(y_1 - 49.7)$$
$$= 49.7 - 0.59(40 - 49.7) = 55.4$$

The fitted values $\hat{y}_2, \hat{y}_3, \ldots, \hat{y}_{70}$ for the chemical yield series are plotted in Figure 9.4 along with the original observations. The figure indicates that the fitted model

$$\hat{y}_t - 49.7 = -0.59(y_{t-1} - 49.7)$$

does a reasonable job of describing the behavior of the series.

It may be that the process generating the observed series is more complex than first imagined and hence that a different time series model is called for. To see if this is, in fact, the case, we must examine the adequacy of the present model.

[d] Actually

$$\hat{\mu} = \frac{y_N - \hat{\phi}y_1}{(N-1)(1-\hat{\phi})} + \frac{\sum_{t=2}^{N-1} y_t}{N-1}$$

which is not much different from $\bar{y}$ regardless of the value of $\hat{\phi}$ provided N is reasonably large and $|\hat{\phi}| < 1$. For the chemical yield data $\bar{y} = 49.7$, $\hat{\mu} = 49.8$. In addition,

$$\hat{\phi} = \frac{\sum_{t=2}^{N}(y_t - \hat{\mu})(y_{t-1} - \hat{\mu})}{\sum_{t=2}^{N}(y_{t-1} - \hat{\mu})^2}$$

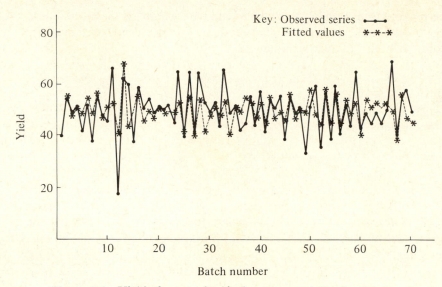

Figure 9.4 Yields from a chemical process and fitted values

Diagnostic Checking

Diagnostic checking is concerned both with testing the time series model for "goodness of fit" and with indicating possible modifications if inadequacies exist.

We have already considered the topic of diagnostic checking in connection with the examination of residuals in the regression and analysis of variance chapters. Any statistical model can be viewed as a device for transforming observations into uncorrelated residuals—that is, as a device for extracting all the "explainable" information in the data. These notions hold true for time series models as well. Consequently, most tests of model adequacy in time series analysis are concerned with detecting "nonrandomness" in the residuals.

The reader will recall that the residuals are defined to be the difference between the observations and the corresponding fitted values. Therefore denoting the residual by $\hat{e}_t$,

$$\hat{e}_t = y_t - \hat{y}_t, \qquad t = 2, \ldots, N \tag{9.10}$$

Now assuming that the model is correct, $\hat{y}_t$ can be viewed as an estimate of

$$E(Y_t | Y_{t-1} = y_{t-1}) = \mu + \phi(y_{t-1} - \mu)$$

and hence $\hat{e}_t$ can be viewed as an estimate of the error component $e_t = Y_t - E(Y_t | Y_{t-1})$. By assumption the e_t's are uncorrelated at any lag with mean zero and common variance, and therefore if the model is correct and the parameter estimates are close to the true values of the parameters, the $\hat{e}_t$'s should behave in a similar manner. In particular, within sampling variation both the average value of

$$\bar{\hat{e}} = \frac{1}{N-1} \sum_{t=2}^{N} \hat{e}_t,$$

and the residual autocorrelations $r_k(\hat{e})$, $k = 1, 2, \ldots$, should be zero.
We define[e]

$$r_k(\hat{e}) = \frac{\sum_{t=1}^{N-k} \hat{e}_t \hat{e}_{t+k}}{\sum_{t=1}^{N} \hat{e}_t^2}, \qquad k = 1, 2, \ldots, K \tag{9.11}$$

to be the autocorrelation coefficients of the residuals. These autocorrelations are ordinarily computed for a selected number of lags, K. If K is relatively small compared with N, it can be shown that for the first-order autoregressive model, the *estimated* means and variances of the residual autocorrelations are

$$E[r_k(\hat{e})] \doteq 0, \qquad \text{for all } k > 0$$

$$\text{Vâr}[r_1(\hat{e})] \doteq \frac{|\hat{\phi}|^2}{N}$$

$$\text{Vâr}[r_2(\hat{e})] \doteq \frac{1 - \hat{\phi}^2 + \hat{\phi}^4}{N}$$

$$\text{Vâr}[r_k(\hat{e})] \doteq \frac{1}{N}, \qquad k = 3, \ldots, K \tag{9.12}$$

Therefore to determine whether or not the residuals are indeed random, that is, whether the population autcorrelation coefficient, $\rho_k(e) = 0$ for a *given* $k > 0$, the autocorrelation coefficient, $r_k(\hat{e})$, can be compared with its estimated standard deviation, $\hat{\sigma}_{r_k(\hat{e})} = +\sqrt{\text{Vâr}[r_k(\hat{e})]}$.

The first 15 residual autocorrelations for our chemical yield example are plotted in Figure 9.5 along with their two standard deviation limits. The standard deviations were obtained by evaluating the expressions in equations (9.12) with $\hat{\phi} = -0.59$ and $N = 70$ and then taking the positive square roots to give

$$\hat{\sigma}_{r_1(\hat{e})} = 0.07$$
$$\hat{\sigma}_{r_2(\hat{e})} = 0.11$$

and

$$\hat{\sigma}_{r_k(\hat{e})} = 0.12, \qquad k = 3, \ldots, 15$$

Hence an approximate test of the hypothesis H_0: $\rho_k(e) = 0$, for a *given* lag k

---

[e] The reader will note that the definition of the residual autocorrelation coefficients implies that there are N residuals available; this will frequently not be the case (we have seen for the chemical process yield example that there are only $N - 1$ residuals). In practice the N appearing in the definition of the residual autocorrelation function will represent the number of residuals.

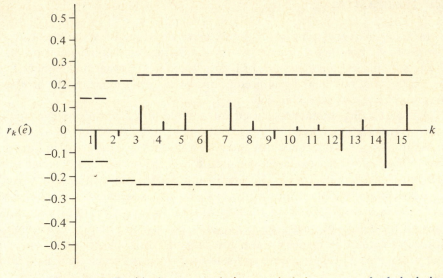

Figure 9.5 Residual autocorrelations and their two standard deviation limits (broken line)

(at the 5-percent level) is provided by the test,

Do not reject H_0 if $|r_k(\hat{e})| \leq 2\hat{\sigma}_{r_k(\hat{e})}$
Reject H_0 otherwise

Using this test and the results displayed in Figure 9.5, there is no reason to reject the null hypothesis, H_0: $\rho_k(e) = 0$, for any lag k. The residuals produced by fitting a first-order autoregressive model to the chemical yield data appear to be uncorrelated. On the basis of this evidence together with the fact that $\bar{\hat{e}} = 5.4 \times 10^{-6}$, there is no reason to doubt the adequacy of our model.

We conclude this section by pointing out that we have by no means covered all of the ramifications of diagnostic checking. For example, there are other analytical tests of model adequacy. (Some of these additional tests are discussed in Chapter 10.) In addition there frequently comes a time when analytical methods have been exhausted and informed judgment must be used. Two competing models may describe a given time series equally well in the sense that neither one can be rejected as inadequate on the basis of an examination of residuals or analytical tests of goodness of fit. It is then up to the investigator to "discriminate" between these two models on the basis of considerations external to the statistical tests used in diagnostic checking. For example, one model may be preferred on the basis of its implications for forecasting.

9.4   FORECASTING

In many instances time series models are developed for the purpose of forecasting, or predicting, *future* observations since vital policy decisions are frequently based on perceptions of future outcomes. Corporate planners, for example, must rely heavily on predictions of sales, earnings, costs, production capacity, and so forth.

All forecasting procedures based on a model fitted to an observed time series necessarily involve the extrapolation of previous (historical) data. As in regression analysis, some care must be taken when extrapolating beyond the fitting region. For example, the nature of a time series may change and hence the fitted model may no longer be appropriate. In addition, for a fixed forecast origin, the precision associated with the forecasts often decreases as the forecasts are projected further into the future. Although forecasters should be aware of these dangers, they do not imply that meaningful forecasts cannot be made. On the contrary, if future observations tend to behave like past ones, accurate and reliable forecasts can be obtained. Also previous forecasts may be "updated" once new information is in hand.

The Next Observation

We shall let y_N denote the last value of the time series and assume we wish to forecast the future values y_{N+l}, $l = 1, 2, \ldots, q$. Then the time period $t = N$ is called the "forecast origin," and l is called the "lead time." Now we consider the problem of forecasting the next value of the chemical process yield series. We have seen that this series is well represented by a first-order autoregressive model; therefore the observation at time $N + 1$ will be a realization of the random variable Y_{N+1} where

$$Y_{N+1} - \mu = \phi(Y_N - \mu) + e_{N+1} \qquad (9.13)$$

At the forecast origin N, the observed value y_N of the random variable Y_N is available. Assuming the parameters ϕ and μ are known, the only unknown quantity in the right-hand side of equation (9.13) is the value of e_{N+1}. Thus we could use the relationship in equation (9.13) to forecast y_{N+1}, the future value of Y_{N+1}, provided we had a value for e_{N+1}. A natural way to proceed is to replace e_{N+1} by its expected value, which at time N is zero. Since all other quantities in the model are constant at time N—the random variable Y_N having been observed—this procedure is equivalent to forecasting the value of Y_{N+1} by computing its expected value at the forecast origin. This expectation is really a conditional expectation since it is the expected value of Y_{N+1} given all the information up to and including time N. Using the notation $E_N(Y)$ to denote the conditional expectation at time N and denoting the forecast of the next observation by $\hat{y}_{N+1}$ we have

$$\hat{y}_{N+1} = E_N(Y_{N+1}) = \mu + \phi(y_N - \mu) \tag{9.14}$$

It can be shown (see reference [1]) that this method of forecasting produces forecasts that are optimal according to the commonly employed mean squared error criterion.

The error actually incurred in forecasting the next observation will be the difference between y_{N+1}, the observed value of Y_{N+1}, and the forecast $\hat{y}_{N+1}$. At time N, however, the observed value of Y_{N+1} is unavailable, and hence this difference is the random variable,

$$Y_{N+1} - \hat{y}_{N+1} = e_{N+1}$$

The accuracy of the forecast can then be judged by the magnitude of the forecast error, and a measure of the precision associated with the forecast is provided by its variance, $\text{Var}(e_{N+1}) = \sigma_e^2$. The variance of the one-step ahead forecast error can be estimated by the residual mean square[f]

$$\hat{\sigma}_e^2 = \frac{1}{N-3} \sum_{t=2}^{N} \hat{e}_t^2 \tag{9.15}$$

The forecast, $\hat{y}_{N+1}$, and the estimated variance of the forecast error, $\hat{\sigma}_e^2$, can then be used to construct limits within which the next observation is likely to fall. (For a discussion of the construction of these limits, the reader is referred to Section 10.8.)

Using equation (9.14) with $N = 70$ and ϕ and μ equal to their estimated values $\hat{\phi} = -0.59$ and $\hat{\mu} = 49.7$, respectively, we have for the chemical yield series

$$\hat{y}_{N+1} = 49.7 - 0.59(y_{70} - 49.7)$$
$$= 49.7 - 0.59(49 - 49.7) = 50.1$$

In addition it can be shown that $\hat{\sigma}_e^2 = 55.7$ and hence a rough indication of the precision associated with the forecast can be obtained by comparing the magnitude of the forecast with the estimated standard deviation of the forecast error, $\hat{\sigma}_e = 7.5$. If the forecast error is normally distributed, then with probability approximately 0.95, the error, when it occurs, will be within two standard deviations of zero. Since the next observation is equal to the forecast (a constant) plus the forecast error, the probability is approximately

---

[f] The

$$\text{residual mean square} = \frac{\text{residual sum of squares}}{\text{degrees of freedom}}$$

The number of degrees of freedom associated with the residual sum of squares $\sum_{t=2}^{N} \hat{e}_t^2$ is equal to the number of residuals minus the number of parameters that must be estimated in the model. For our example,

$$\text{d.f.} = N - 1 - 2 = N - 3$$

0.95 that the next observation will fall in the interval

$$\hat{y}_{N+1} \pm 2\hat{\sigma}_e$$

For the chemical yield series 95-percent probability limits for the next observation are

$$50.1 \pm 2(7.5) = 50.1 \pm 15$$

More than One Step Ahead

The procedure for forecasting two time periods ahead is a straightforward extension of the method for forecasting the first future observation. Specifically, the forecast is obtained by (a) writing down the model for Y_{N+2}, and (b) evaluating the conditional expectation of Y_{N+2} at the forecast origin $t = N$. Using equation (9.2) with $t = N + 2$,

$$Y_{N+2} - \mu = \phi(Y_{N+1} - \mu) + e_{N+2} \tag{9.16}$$

The reader will notice that Y_{N+2} depends on two quantities, Y_{N+1} and e_{N+2}, whose values at the forecast origin are unknown. Taking the expectation of Y_{N+2} at time N gives the forecast

$$\hat{y}_{N+2} = E_N(Y_{N+2}) = \mu + \phi[E_N(Y_{N+1}) - \mu]$$

$$= \mu + \phi(\hat{y}_{N+1} - \mu) \tag{9.17}$$

since

$$E_N(e_{N+2}) = 0$$

by assumption and

$$E_N(Y_{N+1}) = \hat{y}_{N+1}$$

which is the forecast of the first future observation, by definition. Therefore given values for μ and ϕ, $\hat{y}_{N+2}$ can be calculated by suitably modifying the forecast of the next observation, $\hat{y}_{N+1}$.

Alternatively, it is possible to express $\hat{y}_{N+2}$ entirely in terms of μ, ϕ and the available observations. Substituting the expression for $\hat{y}_{N+1}$ [given by equation (9.14)] into equation (9.17) gives

$$\hat{y}_{N+2} = \mu + \phi[\mu + \phi(y_N - \mu) - \mu]$$

$$= \mu + \phi^2(y_N - \mu) \tag{9.18}$$

so that $\hat{y}_{N+2}$ may be calculated directly from the last observation, y_N.

It is possible to construct "error limits" for the observation two time periods ahead by considering the forecast error random variable $Y_{N+2} - \hat{y}_{N+2}$. Using equations (9.16) and (9.17),

$$Y_{N+2} - \hat{y}_{N+2} = \phi(Y_{N+1} - \hat{y}_{N+1}) + e_{N+2}$$

$$= \phi e_{N+1} + e_{N+2} \tag{9.19}$$

and therefore

$$\text{Var}(Y_{N+2} - \hat{y}_{N+2}) = \phi^2 \text{Var}(e_{N+1}) + \text{Var}(e_{N+2})$$
$$= \phi^2 \sigma_e^2 + \sigma_e^2 = (1 + \phi^2)\sigma_e^2 \qquad (9.20)$$

since e_{N+1}, e_{N+2} are assumed to be statistically independent. The estimated variance of the forecast error

$$\text{Vâr}(Y_{N+2} - \hat{y}_{N+2}) = (1 + \hat{\phi}^2)\hat{\sigma}_e^2$$

where $\hat{\phi}$ is the least squares estimate of ϕ and $\hat{\sigma}_e^2$ is the residual mean square, can then be used to construct approximate 95-percent probability limits for y_{N+2} of the form

$$\hat{y}_{N+2} \pm 2\sqrt{(1 + \hat{\phi}^2)}\hat{\sigma}_e$$

For the chemical yield series with μ and ϕ replaced by their least squares estimates, equation (9.17) gives

$$\hat{y}_{72} = 49.7 - 0.59(50.1 - 49.7) = 49.5$$

and approximate 95-percent probability limits are

$$\hat{y}_{72} \pm 2\sqrt{1 + \hat{\phi}^2}\hat{\sigma}_e = 49.5 \pm 2\sqrt{1 + (-0.59)^2}(7.5)$$
$$= 49.5 \pm 17.4.$$

Employing the same procedure that was used for obtaining the forecasts of y_{N+1} and y_{N+2}, it is easy to show (see Problem 9.9) that the forecast of the observation l time periods ahead ($l = 1, 2, \ldots$) is given by the recursive relationship

$$\hat{y}_{N+l} = \mu + \phi(\hat{y}_{N+l-1} - \mu), \qquad l = 1, 2, \ldots \qquad (9.21)$$

or alternatively by

$$\hat{y}_{N+l} = \mu + \phi^l(y_N - \mu), \qquad l = 1, 2, \ldots \qquad (9.22)$$

In addition the forecast error can be expressed as

$$Y_{N+l} - \hat{y}_{N+l} = e_{N+l} + \phi e_{N+l-1} + \phi^2 e_{N+l-2} + \cdots + \phi^{l-1} e_{N+1} \qquad (9.23)$$

with variance (see Problem 9.10)

$$\text{Var}(Y_{N+l} - \hat{y}_{N+l}) = (1 + \phi^2 + \phi^4 + \cdots + \phi^{2(l-1)})\sigma_e^2 \qquad (9.24)$$

Approximate 95-percent probability limits are then given by

$$\hat{y}_{N+l} \pm 2\sqrt{\text{Vâr}(Y_{N+l} - \hat{y}_{N+l})} = \hat{y}_{N+l} \pm 2\sqrt{1 + \hat{\phi}^2 + \cdots + \hat{\phi}^{2(l-1)}}\hat{\sigma}_e \quad (9.25)$$

The nature of the forecasts derived from a first-order autoregressive model is apparent from an examination of equations (9.22) and (9.24). Since $|\phi| < 1$, it is clear that as the forecast lead time increases (that is, as $l \to \infty$),

$\hat{y}_{N+l} \to \mu$, the process mean, and

$$\text{Var}(Y_{N+l} - \hat{y}_{N+l}) \to \frac{\sigma_e^2}{1 - \phi^2} = \sigma_Y^2$$

the process variance (see Problem 9.5). Thus although initial forecasts may fluctuate about the process level, μ, in a manner determined by ϕ and the last observation, y_N, as the lead time increases, the information contained in the last observation becomes less relevant, and the best we can do is to predict future values using the overall level (or an estimate of it). In addition, for long lead times, a measure of uncertainty associated with the forecast is provided by the process variance.

Forecasts of the next 10 observations for the chemical yield process along with their approximate 95-percent probability limits are plotted in Figure 9.6. The forecasts and probability limits were calculated using equations (9.21) and (9.25), respectively, with the unknown parameters μ, ϕ, and σ_e set equal to their estimated values $\hat{\mu} = 49.7$, $\hat{\phi} = -0.59$, and $\hat{\sigma}_e = 7.5$.

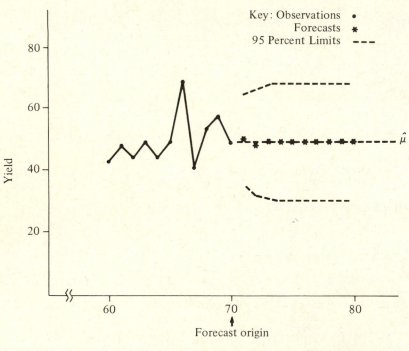

Figure 9.6 Forecasts and 95-percent probability limits, chemical process yield series

Updating Forecasts

One desirable property of any forecasting procedure is that it should provide a convenient means of modifying or updating previous forecasts as new observations become available.

To illustrate updating we consider the following situation. Suppose the forecasts, $\hat{y}_{N+1}$, $\hat{y}_{N+2}$, of the next two observations have been calculated at the forecast origin $t = N$. Now suppose that the next time period has elapsed and that the observation y_{N+1} is available. It seems reasonable to use this new information to "update" the forecast $\hat{y}_{N+2}$, which at time $t = N + 1$ is a forecast of the next observation, albeit an "old" one. To update $\hat{y}_{N+2}$, however, we must consider the problem of forecasting the *next* observation at forecast origin $t = N + 1$.

It is now necessary to introduce some new notation to avoid confusing subscripts. Let $\hat{y}_N(2)$ denote the forecast of the observation two time periods ahead made at forecast origin $t = N$ (that is, $\hat{y}_N(2) = \hat{y}_{N+2}$) and let $\hat{y}_{N+1}(1)$ denote the forecast of the next observation made at forecast origin $t = N + 1$. Clearly, $\hat{y}_N(2)$ and $\hat{y}_{N+1}(1)$ are forecasts of the same future observation made at different forecast origins. Now, using the procedure introduced previously

$$\hat{y}_{N+1}(1) = E_{N+1}(Y_{(N+1)+1}) = \mu + \phi[E_{N+1}(Y_{N+1}) - \mu]$$

$$= \mu + \phi(y_{N+1} - \mu) \qquad (9.26)$$

since at time $N + 1$, the observed value of Y_{N+1}, y_{N+1} is in hand. The reader will recall that

$$\hat{y}_N(2) = E_N(Y_{N+2}) = \mu + \phi[E_N(Y_{N+1}) - \mu]$$

$$= \mu + \phi[\hat{y}_N(1) - \mu] \qquad (9.27)$$

Using equations (9.26) and (9.27)

$$\hat{y}_{N+1}(1) - \hat{y}_N(2) = \phi[y_{N+1} - \hat{y}_N(1)]$$

or

$$\hat{y}_{N+1}(1) = \hat{y}_N(2) + \phi[y_{N+1} - \hat{y}_N(1)] \qquad (9.28)$$

Equation (9.28) now provides us with a way to update the forecast $\hat{y}_N(2)$ when the new observation y_{N+1} becomes available. Specifically, the updated forecast, $\hat{y}_{N+1}(1)$, is equal to the previous forecast, $\hat{y}_N(2)$, plus a correction factor, $\phi[y_{N+1} - \hat{y}_N(1)]$, which is recognized as a proportion (since $|\phi| < 1$) of the error made in forecasting the observation y_{N+1} at time N.

To illustrate the method, we consider the forecasts of the next two observations made at the end of the series (forecast origin $t = N = 70$) for the chemical process yield data. The reader will recall that

$$\hat{y}_{N+1} = \hat{y}_N(1) = \hat{y}_{70}(1) = 50.1$$

and

$$\hat{y}_{N+2} = \hat{y}_N(2) = \hat{y}_{70}(2) = 49.5$$

Now suppose the observation $y_{N+1} = y_{71} = 58$ is available. We can make use of this new observation to update the forecast $\hat{y}_{70}(2)$ of what is now the next observation. Using equation (9.28) with $\phi = \hat{\phi} = -0.59$, the updated forecast is given by

$$\hat{y}_{N+1}(1) = \hat{y}_{71}(1) = 49.5 + (-0.59)(58 - 50.1)$$
$$= 44.8.$$

A general expression can be developed (see Problem 9.12) for updating all of the previous forecasts given the new observation y_{N+1}. Essentially the updated forecast for any lead time, given y_{N+1}, is the previous forecast plus a proportion of the forecast error, $y_{N+1} - \hat{y}_N(1)$, where the proportion decreases in magnitude as the lead time increases. Specifically,

$$\hat{y}_{N+1}(l-1) = \hat{y}_N(l) + \phi^{l-1}[y_{N+1} - \hat{y}_N(1)], \qquad l = 2, 3, 4, \ldots \qquad (9.29)$$

The original forecasts and the updated forecasts for the chemical yield series for $l = 2, 3, 4, \ldots, 10$ are given in Table 9.2, assuming $y_{71} = 58$ and $\phi = \hat{\phi} = -0.59$.

Table 9.2 Original and Updated Forecasts for the Chemical Yield Series

l	ORIGINAL FORECASTS $\hat{y}_{70}(l)$	CORRECTION FACTOR $\phi^{l-1}[y_{71} - \hat{y}_{70}(1)]$	UPDATED FORECASTS $\hat{y}_{71}(l-1)$
1	50.1	X	X
2	49.5	-4.7	44.8
3	49.8	2.8	52.6
4	49.7	-1.6	48.1
5	49.7	1.0	50.7
6	49.7	-0.6	49.1
7	49.7	0.3	50.0
8	49.7	-0.2	49.5
9	49.7	0.1	49.8
10	49.7	-0.1	49.6

9.5 FINAL COMMENTS

We have tried in this chapter to cover the various facets of time series modeling and forecasting in the context of a single example. The methods that have been introduced, however, are general and can be applied to any time series problem. As we shall see in subsequent chapters the class of time series models can be extended considerably to handle other kinds of stationary series as well as series that do not appear to vary about a fixed level. In addition it is possible to model and forecast series that appear to possess seasonal components. In the course of embarking on these more difficult cases, more theoretical structure must be introduced. However, the reader should keep in mind that beneath all the mathematical detail, the basic ideas, which have been introduced in this chapter, remain the same.

PROBLEMS

9.1 Given the chemical process yield data in Table 9.1, plot the pairs of observations (y_t, y_{t-3}). What does the resulting scatter diagram indicate about the correlation between variables three time periods apart? Is the pattern of scatter consistent with the sign of the sample autocorrelation coefficient r_3 in Figure 9.3(a)?

9.2 Given the time series

$$64 \quad 55 \quad 41 \quad 59 \quad 48 \quad 71 \quad 35 \quad 57 \quad 40 \quad 58$$

calculate r_1 and r_2.
(Note: In practice reliable autocorrelation estimates are only obtained from series consisting of approximately 50 observations or more.)

9.3 The following sample autocorrelations were computed for a time series consisting of $N = 50$ observations:

k:	1	2	3	4	5	6	7	8	9	10
r_k:	0.78	0.67	0.49	0.40	0.29	0.23	0.19	0.13	0.11	0.07

On the basis of the behavior of the sample autocorrelations, identify an appropriate first-order autoregressive model by supplying an initial estimate of the parameter ϕ.

9.4 The following sample autocorrelations were computed for a time series consisting of $N = 60$ observations:

k:	1	2	3	4	5	6	7	8	9	10
r_k:	-0.70	0.51	-0.32	0.22	-0.11	0.08	-0.04	0.03	-0.01	-0.01

Identify an appropriate first-order autoregressive model by supplying an initial estimate of the parameter ϕ. Justify your choice.

9.5 Suppose the process variable Y follows the first-order autoregressive model

$$Y_t - \mu = \phi(Y_{t-1} - \mu) + e_t, \qquad t = \ldots, -1, 0, 1, \ldots, -1 < \phi < 1$$

Show that

$$\text{Var}(Y_t) = E(Y_t - \mu)^2 = \frac{\sigma_e^2}{1 - \phi^2} \qquad \text{for all } t$$

where $\sigma_e^2 = \text{Var}(e_t)$.

{Hint: Recall from the assumptions underlying the first-order autoregressive model that $E(e_t) = 0$, all t, and $E(Y_t e_{t'}) = 0$, $t < t'$. Then from the definition of Y_t we have

$$
\begin{aligned}
\text{Var}(Y_t) = E(Y_t - \mu)^2 &= E[\phi(Y_{t-1} - \mu) + e_t]^2 \\
&= \phi^2 E(Y_{t-1} - \mu)^2 + 2\phi E[(Y_{t-1} - \mu)e_t] + E(e_t^2) \\
&= \phi^2 \text{Var}(Y_{t-1}) + \text{Var}(e_t)
\end{aligned}
$$

Assuming the variables Y_t, Y_{t-1} have the same variance, the required result follows.}

9.6   Assume the following 10 observations:

$$8, -5, 5, -13, 6, -2, 3, -5, 7, 0$$

were generated by a first-order autoregressive process that can be described by a model of the form (note that $\mu = 0$),

$$Y_t = \phi Y_{t-1} + e_t$$

Obtain the least squares estimate of ϕ.

9.7   Suppose you are fitting the model

$$(Y_t - \mu) = \phi(Y_{t-1} - \mu) + e_t$$

to a time series whose first 10 values are

$$60, 54, 63, 49, 52, 51, 57, 46, 58, 50$$

Obtain the (approximate) least squares estimates of ϕ and μ.

9.8   A first-order autoregressive model has been fitted to a time series of $N = 50$ observations giving $\hat{\mu} = 15$ and $\hat{\phi} = 0.6$. The first 12 residual autocorrelations are

k:	1	2	3	4	5	6	7
$r_k(\hat{e})$:	0.13	-0.07	-0.12	0.19	-0.01	0.30	0.18

	8	9	10	11	12
	-0.16	-0.05	0.01	0.08	-0.21

Is there any reason to doubt the adequacy of the fitted model? Explain. Assume that an estimated standard deviation of $r_k(\hat{e})$ is about 0.15 for $k = 1, 2, \ldots, 12$.

9.9   Consider the problem of forecasting future observations at time $t = N$ from a first-order autoregressive process. Use an argument similar to the one employed in Section 9.4 to show that

$$\hat{y}_{N+1} = \mu + \phi(y_N - \mu)$$

and derive the general result

$$\hat{y}_{N+l} = \mu + \phi^l (y_N - \mu)$$

9.10   Show that the forecast error, $Y_{N+l} - \hat{y}_{N+l}$, for a first-order autoregressive process may be written as

$$Y_{N+l} - \hat{y}_{N+l} = e_{N+l} + \phi e_{N+l-1} + \cdots + \phi^{l-1} e_{N+1}$$

and that $\text{Var}(Y_{N+l} - \hat{y}_{N+l})$ is given by

$$\text{Var}(Y_{N+l} - \hat{y}_{N+l}) = (1 + \phi^2 + \cdots + \phi^{2(l-1)})\sigma_e^2$$

where $\sigma_e^2 = \text{Var}(e_t)$.
[Hint: In general

$$Y_{N+l} - \hat{y}_{N+l} = \phi(Y_{N+l-1} - \hat{y}_{N+l-1}) + e_{N+l}, \qquad l = 2, 3, \ldots$$

where

$$Y_{N+l-1} - \hat{y}_{N+l-1} = \phi(Y_{N+l-2} - \hat{y}_{N+l-2}) + e_{N+l-1}$$

and so forth. Recalling that $Y_{N+1} - \hat{y}_{N+1} = e_{N+1}$, appropriate substitution gives the required result.]

9.11   The model

$$(Y_t - \mu) = \phi(Y_{t-1} - \mu) + e_t$$

has been fitted to a time series giving $\hat{\phi} = 0.7$, $\hat{\mu} = 1.0$, and $\hat{\sigma}_e^2 = 0.0009$. The last five values of the series are

$$1.10, 1.15, 1.07, 1.02, 0.99$$

Using the time corresponding to the last observation as the forecast origin, calculate the forecasts and approximate 95-percent probability limits for the next five observations.

9.12   Show that a general expression for updating forecasts made from first-order autoregressive models at time $t = N$ is given by

$$\hat{y}_{N+1}(l - 1) = \hat{y}_N(l) + \phi^{l-1}[y_{N+1} - \hat{y}_N(1)]$$

where y_{N+1} is the observation available at time $t = N + 1$ and $\hat{y}_N(1)$ is its forecasted value at time $t = N$.
{Hint: In general

$$\hat{y}_{N+1}(l - 1) = \mu + \phi[\hat{y}_{N+1}(l - 2) - \mu]$$
$$\hat{y}_N(l) = \mu + \phi[\hat{y}_N(l - 1) - \mu]$$

so that

$$\hat{y}_{N+1}(l - 1) - \hat{y}_N(l) = \phi[\hat{y}_{N+1}(l - 2) - \hat{y}_N(l - 1)]$$
$$\hat{y}_{N+1}(l - 2) - \hat{y}_N(l - 1) = \phi[\hat{y}_{N+1}(l - 3) - \hat{y}_N(l - 2)]$$

and so forth. Recalling that $\hat{y}_{N+1}(1) - \hat{y}_N(2) = \phi[y_{N+1} - \hat{y}_N(1)]$, appropriate substitution gives the required result.}

9.13   Using the result in Problem 9.12, update the forecasts calculated in Problem

9.11 assuming the observation following 0.99 in Problem 9.11 is now available and is 0.91.

9.14   Generate 50 random normal deviates using the method in Section 2.4. These numbers are to serve as errors, $e_1, e_2, \ldots, e_{50}$, in some time series you are to construct. Each series will be generated from the first-order autoregressive model

$$Y_t = \mu + \phi(Y_{t-1} - \mu) + e_t, \qquad t = 1, 2, \ldots, 50$$

Given values of Y_0, μ, and ϕ, and your random normal deviates, a series can be constructed by forming a worksheet as outlined below.

(1) t	(2) e_t	(3) $(Y_{t-1} - \mu)$	(4) $= \phi(3)$ $\phi(Y_{t-1} - \mu)$	(5) $= \mu + (4) + (2)$ $Y_t = \mu + \phi(Y_{t-1} - \mu) + e_t$
0				Y_0
1	e_1	$Y_0 - \mu$	$\phi(Y_0 - \mu)$	$\mu + \phi(Y_0 - \mu) + e_1$
2	e_2	$Y_1 - \mu$	$\phi(Y_1 - \mu)$	$\mu + \phi(Y_1 - \mu) + e_2$
$\vdots$	$\vdots$	$\vdots$	$\vdots$	$\vdots$
50	e_{50}	$(Y_{49} - \mu)$	$\phi(Y_{49} - \mu)$	$\mu + \phi(Y_{49} - \mu) + e_{50}$

(A computer would be very helpful in carrying out such calculations.)

(a)   Construct a time series using the method outlined above for the following sets of parameter values:

Series	y_0	μ	ϕ
1	0	0	0.90
2	75	78	0.01
3	75	72	0.01
4	75	72	0.30
5	75	72	-0.30
6	75	72	0.70
7	75	72	-0.70

(b)   Plot each of the series in part (a).
(c)   Compute and plot the theoretical and sample autocorrelation functions for each of the series in (a). (The sample autocorrelation functions should be computed only if a computer is available.)

9.15   Collect 51 consecutive closing Dow Jones industrial averages. Denote these by $x_1, x_2, \ldots, x_{51}$. Compute the differences $y_t = x_{t+1} - x_t$, $t = 1, 2, \ldots, 50$. Plot the y_t series, and compare it to series 1 in Problem 9.14. Point out any similarities and differences you observe. Calculate the sample autocorrelation function of your y_t series.

10

TIME SERIES ANALYSIS: MODEL BUILDING AND FORECASTING†

10.1 INTRODUCTION

The material in Chapter 9 was designed to acquaint the reader with some of the rudiments of time series analysis. In this chapter we will enlarge on some of these rudiments and illustrate the methodology with three real-world examples.

Much of the theoretical rationale for the models and procedures introduced in this chapter is available in Supplement 10A in which "stationary stochastic processes and their properties" are discussed. The reader wishing to gain a deeper insight into the theoretical underpinnings of time series analysis is encouraged to read this supplement. We shall rely mainly on intuition to justify the practices advocated in later sections.

We will present some relatively simple time series models that have been successfully used to describe the behavior of series arising in a variety of fields. Generalizations of these simple models are given in Supplement 10B. The model building strategy of identification, estimation, and diagnostic checking is reconsidered as a prerequisite for the development of suitable time series models for the three actually occurring series given in Section 10.2. The problem of forecasting future values is discussed in some detail.

We remind the reader that we will be concerned with the analysis of a chronological sequence of *dependent* observations that become available at equally spaced time points. That is a *discrete time series*. Our objectives are description and projection (or forecasting), and in this sense the goals of time series analysis are similar to those of regression analysis. In fact, many of the techniques introduced in connection with regression analysis reappear in time series analysis. For example, we will continue to use the least squares criterion to determine estimates of model parameters, and we will still be concerned with the analysis of residuals.

†The influence of G.C. Tiao on the presentation in this chapter is gratefully acknowledged.

In general, ordinary regression models with time as the independent variable are not suitable for describing time series for two reasons. First, the observations making up the time series are usually dependent. (The reader will recall that one of the assumptions underlying the regression model is that the errors, and hence the observations, are independently distributed.) Second, forecasting future values necessarily entails an extrapolation of historical data, and as we indicated in Chapter 6 the use of regression models to predict values of the dependent variable outside of the fitting region can be extremely dangerous. Hence regression models can lead to inaccurate forecasts with misleading error allowances.

Our treatment of time series analysis parallels that of Box and Jenkins [1].

10.2   EXAMPLES OF TIME SERIES

It is instructive to discuss the methodology of time series analysis with reference to series that actually occur. Examples of time series are given in Figures 10.1, 10.2, and 10.3. Figure 10.1 shows the sales (demand) for refrigerators with left-handed doors for 52 consecutive months. Figure 10.2 is a plot of the interest rates on three-month Treasury bills for the period January 1956 to January 1969, and Figure 10.3 depicts a U.S. quarterly unemployment rate for the period March 1948 to March 1963. In addition we shall occasionally refer to the chemical process yield time series depicted in

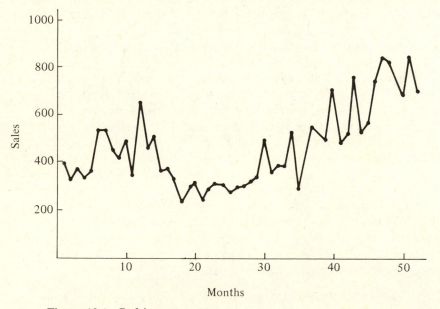

Figure 10.1   Refrigerator sales

Figure 10.2   Interest rates on three-month Treasury bills (January 1956–
January 1969)

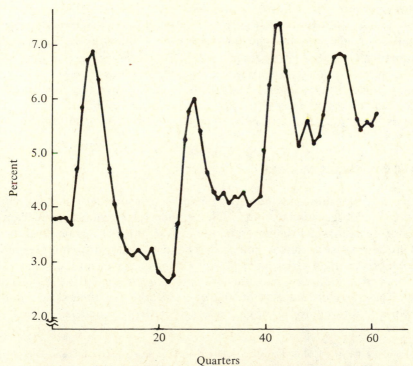

Figure 10.3   U.S. unemployment rates [1948(1)–1963(1)]

Figure 9.1. Both the refrigerator sales series and the interest rates series tend to drift upward over the latter half of the series. These series do not appear to vary about a fixed or average level. Series that do not vary about a fixed level are said to be *nonstationary in the mean*. Typically nonstationarity of this kind is indicated if the series appears to wander away from a horizontal line. The time series models we will study in later sections are especially suited for representing series which are stationary (that is, vary about a fixed level) or nonstationary in the mean.

We note at this point that, in general, nonstationary series are those series whose statistical properties change over time.[a] Nonstationary series can occur when the underlying physical mechanism generating the series changes—that is, when the principal movements of the series are due to a shift in the influence of factors which occur periodically. A sudden shift in U.S. economic policy may have a severe effect on the interest rate of three-month Treasury bills, and, in fact, a contraction of the money supply by the Federal Reserve System would tend to drive the interest rate upward. Of course, the influence of certain factors on a given time series may produce a stabilizing effect as well. This phenomenon might explain the behavior of the unemployment rate series. Whenever the unemployment rate increases beyond "acceptable" levels, government controls are usually implemented to bring it back to a satisfactory position. On the other hand when unemployment is low, another aspect of the economy is frequently unsatisfactory (for example, the inflation rate), and attempts to control this "diametrically opposed" aspect generally lead to an increase in the unemployment rate. Although the preceding argument is a gross oversimplification, it is approximately true, and hence we might expect, a priori, the unemployment rate series to be stationary in the mean. As we shall see in Section 10.5, this appears to be the case, at least for the time period considered.

10.3 TIME SERIES MODELS

The idea of using a mathematical model incorporating random components to link observational reality with the underlying generating mechanism has already been well established in previous sections of this book. Here we will introduce mathematical models that have proved useful in representing both stationary time series and time series nonstationary in the mean. (The concepts of stationarity and nonstationarity are discussed in Supplement 10A.) These models are part of an important class of time series models known as "autoregressive-moving average" (ARMA) models.

[a] In this book the only kind of nonstationarity we consider is nonstationarity in the mean (a common occurrence in practice); however, we note that other statistical properties of the series may change over time (for example, the variability) producing a different kind of nonstationarity.

First we shall consider the simple model

$$Y_t = \mu + e_t \tag{10.1}$$

where $e_t \sim N(0, \sigma_e^2)$ for all t and $E(e_t e_{t'}) = 0$, $t \neq t'$. This model implies that the process varies about a fixed level μ (the process mean) in accordance with a normal probability law, and in addition the random variables Y_t and Y_{t+k} are uncorrelated for all values of k; that is,

$$\gamma_k = E(Y_t - \mu)(Y_{t+k} - \mu) = E(e_t e_{t+k}) = 0, \qquad \text{for all } k \neq 0$$

(see Supplement 10A). This stochastic process is simply a chronological sequence of *independent* normal random variables. Thus what occurs at time t is not influenced by what occurred in the past nor will it influence the occurrence of any future event. The model given by equation (10.1) is referred to in time series analysis as a "white noise" model since the e_t can be interpreted as a random variable representing the "shock," or "noise," that enters the system at time t causing the process to deviate from its overall level μ. The white noise model provides a convenient starting point for introducing a class of time series models known as "moving average" models.

First-Order Moving Average Model

An immediate generalization of expression (10.1) can be obtained by allowing the deviation of Y_t from the mean to depend not only on the shock which enters the system at time t but also on the shock, e_{t-1}, which enters the system during the previous time period; that is, we can write

$$Y_t = \mu + e_t - \theta_1 e_{t-1} \tag{10.2}$$

or equivalently with $\dot{Y}_t = Y_t - \mu$,

$$\dot{Y}_t = e_t - \theta_1 e_{t-1} \tag{10.3}$$

where θ_1 is a constant such that $|\theta_1| < 1$ and represents the proportionate influence of the previous shock, e_{t-1}, on the current deviation, $\dot{Y}_t$. We note that when $\theta_1 = 0$, equation (10.2) reduces to equation (10.1). The model given by equation (10.3) is referred to as a "first-order moving average" (MA) process or an MA(1) process—the number in parentheses denoting the order.

For an MA(1) process adjacent random variables are correlated, whereas variables separated by more than one time period are uncorrelated. In fact, it can be shown (see Problems 10.3 and 10.4) that

$$\rho_1 = \frac{-\theta_1}{1 + \theta_1^2}$$

$$\rho_k = 0, \qquad k \geq 2 \tag{10.4}$$

where ρ_1 is the lag 1 autocorrelation coefficient and ρ_k measures the correlation between random variables of lags, k, greater than one. The sequence of autocorrelation coefficients for a first-order moving average process has one of the shapes pictured in Figure 10.4(a). We notice that since $|\theta_1| < 1$, $|\rho_1| < 0.5$. The process defined by equation (10.3), then, has an autocorrelation function that "cuts off" (that is, is zero) after lag $k = 1$. In addition, we notice that the ρ_1 depends solely on θ_1, and if we restrict ourselves to values of θ_1 such that $|\theta_1| < 1$, then there is a *unique* θ_1 corresponding to a given value of ρ_1.[b] Table 10.1 gives the value of θ_1

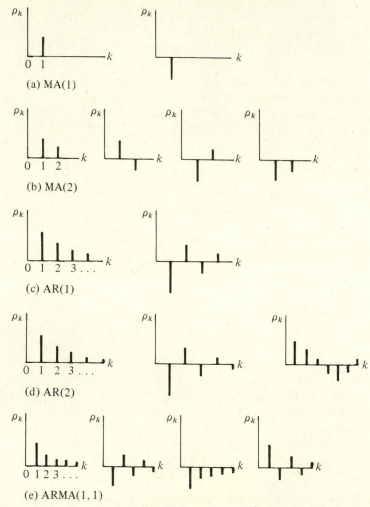

(a) MA(1)

(b) MA(2)

(c) AR(1)

(d) AR(2)

(e) ARMA(1, 1)

Figure 10.4   Shapes of ρ_k $(k \geq 0)$ for some stationary stochastic processes

[b] The reader will note that the equation $\rho_1 = -\theta_1/(1 + \theta_1^2)$ or, equivalently, $\theta_1^2\rho_1 + \theta_1 + \rho_1 = 0$ is quadratic in θ_1 and therefore has two roots, say, $\theta_1{}^*$ and $\theta_2{}^*$, given by $(-1 + \sqrt{1 - 4\rho_1^2})/2\rho_1$ and $(-1 - \sqrt{1 - 4\rho_1^2})/2\rho_1$, respectively. Since the product of the roots is unity, if $\theta_1{}^*$ is a solution, then $\theta_2{}^* = 1/\theta_1{}^*$. Hence only one of the roots satisfies $|\theta_1| < 1$.

Table 10.1 The Relationship between ρ_1 and θ_1 for a First-order Moving Average Process

θ_1	ρ_1	θ_1	ρ_1
0.00	0.000	0.00	0.000
0.05	−0.050	−0.05	0.050
0.10	−0.099	−0.10	0.099
0.15	−0.147	−0.15	0.147
0.20	−0.192	−0.20	0.192
0.25	−0.235	−0.25	0.235
0.30	−0.275	−0.30	0.275
0.35	−0.315	−0.35	0.315
0.40	−0.349	−0.40	0.349
0.45	−0.374	−0.45	0.374
0.50	−0.400	−0.50	0.400
0.55	−0.422	−0.55	0.422
0.60	−0.441	−0.60	0.441
0.65	−0.457	−0.65	0.457
0.70	−0.468	−0.70	0.468
0.75	−0.480	−0.75	0.480
0.80	−0.488	−0.80	0.488
0.85	−0.493	−0.85	0.493
0.90	−0.497	−0.90	0.497
0.95	−0.499	−0.95	0.499
1.00	−0.500	−1.00	0.500

corresponding to a particular ρ_1 for a range of values $-0.5 < \rho_1 < 0.5$. As we shall see, this table can be used to obtain a preliminary estimate of the model parameter, θ_1.

Second-Order Moving Average Model

The second-order moving average model is obtained by letting the deviation from the process mean, $\dot{Y}_t$, depend not only on the current shock e_t and the previous shock e_{t-1}, but also on the shock entering the system two time periods previously, e_{t-2}. Thus we can write the MA(2) model as

$$\dot{Y}_t = e_t - \theta_1 e_{t-1} - \theta_2 e_{t-2} \tag{10.5}$$

where the parameters θ_1, θ_2 are constrained to lie in the triangular region in the θ_1, θ_2 plane defined by the inequalities

$$\theta_2 + \theta_1 < 1$$
$$\theta_2 - \theta_1 < 1 \tag{10.6}$$
$$-1 < \theta_2 < 1$$

The parameters θ_1, θ_2 are constrained so that the model in equation (10.5) represents processes for which the current observation does not depend overwhelmingly on observations in the *remote* past (a sensible restriction for most practical applications). This condition has been called the "invertibility" condition by Box and Jenkins and leads to a unique autocorrelation structure.

It can be shown (see Problems 10.3 and 10.4) that for the second-order moving average process the autocorrelation function is given by

$$\rho_1 = -\theta_1 \frac{1 - \theta_2}{1 + \theta_1^2 + \theta_2^2}$$

$$\rho_2 = \frac{-\theta_2}{1 + \theta_1^2 + \theta_2^2}$$

$$\rho_k = 0, \qquad k \geq 3 \tag{10.7}$$

We see again that the autocorrelation function "cuts off" after a lag equal to the order of the model. Appendix Table K allows us to relate the values of ρ_1 and ρ_2 to the allowable values of θ_1 and θ_2. This chart can be used to obtain preliminary estimates of θ_1 and θ_2. The possible shapes for the autocorrelation function are displayed in Figure 10.4(b).

In practical applications it is rarely necessary to consider moving average models with more than two moving average parameters θ_1 and θ_2.[c] For this reason we have relegated a discussion of more general moving average models to Supplement 10B. In fact, all of the time series models introduced in the text are simply important special cases of the more general time series models considered in Supplement 10B.

First-Order Autoregressive Model

We have already introduced a first-order autoregressive (AR) model [that is, an AR(1) model] in Chapter 9. Moreover we have shown that an AR(1) model provides an adequate description of the stochastic behavior of the yields from a hypothetical chemical batch process. To recapitulate, an AR(1) process is defined by the recursive relationship

$$\dot{Y}_t = \phi_1 \dot{Y}_{t-1} + e_t, \qquad |\phi_1| < 1 \tag{10.8}$$

where the shock, or noise, term e_t has the statistical properties given in equation (9.3) and $\dot{Y}_t = Y_t - \mu$. The current deviation $\dot{Y}_t$ is composed of two parts—a fraction ϕ of the previous deviation $\dot{Y}_{t-1}$ plus an extraneous shock e_t. We can interpret equation (10.8) as a regression model in which the dependent variable is regressed on itself lagged one time period (hence *auto*regressed). The order simply refers to the number of independent variables (number of lagged $\dot{Y}$'s). In general a pth-order autoregressive

---

[c] Provided the time series does not exhibit a recurring seasonal pattern (see Chapter 11).

model, or AR(p) model, is one in which $\dot{Y}_t$ is regressed on $\dot{Y}_{t-1}, \dot{Y}_{t-2}, \ldots, \dot{Y}_{t-p}$.

An AR(1) model has an alternative representation as an infinite moving average model (see Problem 10.7) in which the coefficients $\theta_1 = -\phi_1$, $\theta_2 = -\phi_1^2, \ldots$ decay geometrically, since $|\phi_1| < 1$; that is, the relationship in equation (10.8) can be expressed as

$$\dot{Y}_t = e_t + \phi_1 e_{t-1} + \phi_1^2 e_{t-2} + \cdots \tag{10.9}$$

We notice that the current deviation depends on the shocks entering the system in all previous time periods.

We have seen that for the particular infinite moving average process given by equation (10.9) a considerable economy of parameters is achieved by employing the AR(1) representation given by equation (10.8). This economy of parameterization is referred to as the *principle of parsimony*. Thus the AR(1) model is a parsimonious representation of a particular infinite moving average model.

The autocorrelation function for the AR(1) process has been described in the previous chapter [see equation (9.5)]. The possible shapes of ρ_k, $k \geq 0$, are displayed in Figure 10.4(c).

If ϕ_1 is near 1—that is, when $\dot{Y}_t$ is very much like $\dot{Y}_{t-1}$—and hence the realized time series is rather smooth, ρ_k decays slowly to zero, indicating a process with a large amount of inertia. If ϕ_1 is near zero, $\dot{Y}_t$ is very nearly e_t (that is, random), and ρ_k dies away rapidly to zero. Finally, for ϕ_1 near -1, $\dot{Y}_t$ is likely to change sign from $\dot{Y}_{t-1}$, and thus the process has a tendency to alternate between high and low values, inducing a corresponding effect in the autocorrelation function.

The first-order autoregressive model is important in applications since with one adjustable parameter (apart from the process mean μ) it is capable of representing a wide variety of stationary processes, from smooth processes to processes that tend to oscillate rapidly.

Second-Order Autoregressive Model

An immediate extension of the AR(1) model is provided by the second-order autoregressive model, or AR(2) model,

$$\dot{Y}_t = \phi_1 \dot{Y}_{t-1} + \phi_2 \dot{Y}_{t-2} + e_t \tag{10.10}$$

where to ensure stationarity (the reader is referred to reference [1] for a discussion of this point) the coefficients must satisfy

$$\phi_2 + \phi_1 < 1$$
$$\phi_2 - \phi_1 < 1 \tag{10.11}$$
$$-1 < \phi_2 < 1$$

Thus the coefficients ϕ_1, ϕ_2 lie within the same triangular region as do the coefficients θ_1, θ_2 for the MA(2) process; however, the justification is different. For the AR(2) process the restriction is necessary to ensure stationarity. For the MA(2) process the restriction is necessary to ensure invertibility, since *any* moving average process of *finite* order is necessarily stationary. As a general rule, for $-1 < \phi_2 < 0$ the AR(2) process tends to exhibit sinusoidal behavior, regardless of the value of ϕ_1. This is particularly true for ϕ_2 close to -1. On the other hand if $0 < \phi_2 < 1$, the process exhibits different kinds of behavior depending on the sign of ϕ_1. For $\phi_1 > 0$ a particular realization of the AR(2) process will tend to consist of successive "runs" of observations of the same magnitude. Equivalently, if deviations from the mean are considered, the observed series will tend to consist of runs of observations of the same sign. For $\phi_1 < 0$, the process tends to oscillate, and the time series will exhibit marked "up and down" behavior. The situation is illustrated in Figure 10.5 in which three particular realizations of an AR(2) process (with $\mu = 0$) are plotted corresponding to the following pairs of values for (ϕ_1, ϕ_2): $(1.6, -0.7), (0.50, 0.40), (-0.50, 0.40)$.

The autocorrelation coefficients for an AR(2) process satisfy the recursive relationship

$$\rho_k = \phi_1 \rho_{k-1} + \phi_2 \rho_{k-2}, \qquad k \geq 1 \tag{10.12}$$

or equivalently

$$\rho_k - \phi_1 \rho_{k-1} - \phi_2 \rho_{k-2} = 0 \tag{10.13}$$

In particular with $k = 1$

$$\rho_1 - \phi_1 \rho_0 - \phi_2 \rho_{-1} = \rho_1 - \phi_1 \rho_0 - \phi_2 \rho_1 = 0$$

or

$$\rho_1 = \frac{\phi_1}{1 - \phi_2} \tag{10.14}$$

since $\rho_0 = 1$. Setting $k = 2$ in equation (10.13) and using equation (10.14) gives

$$\rho_2 - \phi_1 \rho_1 - \phi_2 \rho_0 = 0$$

or

$$\rho_2 = \phi_2 + \frac{\phi_1^2}{1 - \phi_2} \tag{10.15}$$

Subsequent values of ρ_k can be generated recursively using equation (10.12) along with equations (10.14) and (10.15).

It will be noted that equations (10.14) and (10.15) represent a pair of simultaneous equations in two unknowns, ϕ_1, ϕ_2. If ϕ_1, ϕ_2 are constrained to the triangular region defined in expressions (10.11), then ρ_1 and ρ_2 correspond to a unique pair of values ϕ_1, ϕ_2. A chart relating ρ_1 and ρ_2 to ϕ_1 and ϕ_2 for an AR(2) process is available in appendix tables at the end of this book (Appendix Table J). This chart can be used to obtain initial estimates of ϕ_1 and ϕ_2. The possible shapes for the autocorrelation functions are pictured in Figure 10.4(d).

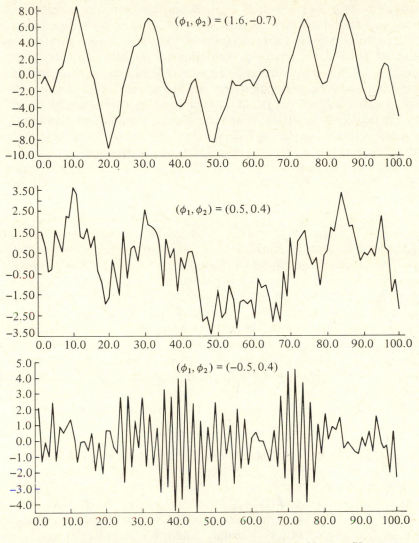

Figure 10.5 Realizations of the process: $Y_t = \phi_1 Y_{t-1} + \phi_2 Y_{t-2} + e_t$

Equation (10.12) implies that, apart from noise, the autocorrelations ρ_k bear *exactly* the same relationship to one another as do the process variables $\dot{Y}_t, \dot{Y}_{t-1}, \dot{Y}_{t-2}, \ldots$ [compare with equation (10.10) after setting $e_t = 0$]. Therefore we would expect the remarks made about the characteristic behavior of an AR(2) process for different values of ϕ_1, ϕ_2 to hold for the autocorrelations as well. Figure 10.4(d) indicates that this is precisely the case.

Once again experience has indicated that a small number of autoregressive parameters is usually sufficient for describing the behavior of a given time series. This is particularly true if we combine autoregressive and moving average terms into a single model as we shall do in the next section.

We conclude this section by remarking that all stationary autoregressive models of finite order have representations as invertible moving average models of infinite order. Similarly, all invertible moving average models of finite order have representations as stationary autoregressive models of infinite order (see Problem 10.6). A particular representation is preferred primarily on the basis of parsimony (economy of parameters). Intuitively we might feel that an even greater degree of economy is available if the two kinds of time series models are combined. This is frequently the case and leads us into a discussion of "mixed" autoregressive-moving average models.

Mixed Autoregressive-Moving Average Model

We have noted that the AR(1) process has a representation as a particular infinite moving average process and Problem 10.6 indicates that an MA(1) process has a representation as a particular infinite autoregressive process. In certain cases a great deal of economy can be achieved (in terms of the number of model parameters) by writing moving average processes as autoregressive processes and vice versa. Further economy can frequently be achieved by combining autoregressive and moving average formulations. The simplest example is the autoregressive-moving average (ARMA) model of order $(1, 1)$, or the ARMA$(1, 1)$ model, in which the numbers in the parentheses indicate the order of the autoregressive and moving average terms, respectively. Thus the ARMA$(1, 1)$ model is given by

$$\dot{Y}_t - \phi_1 \dot{Y}_{t-1} = e_t - \theta_1 e_{t-1} \tag{10.16}$$

with $\dot{Y}_t = Y_t - \mu$, $|\phi_1| < 1$, $|\theta_1| < 1$, and the e_t's have the usual properties associated with the "error" component.

Notice that for $\theta_1 = 0$, the ARMA$(1, 1)$ model reduces to the AR(1) model. Similarly for $\phi_1 = 0$, the ARMA$(1, 1)$ model reduces to the MA(1) model. Specifically, we can regard the AR(1) and MA(1) models as special cases of the more general ARMA$(1, 1)$ model. Therefore if $|\theta_1|$ is close to zero, ARMA$(1, 1)$ processes exhibit behavior very similar to the behavior of first-order autoregressive processes. On the other hand with $|\phi_1|$ close to zero, an ARMA$(1, 1)$ process exhibits the characteristics of a first-order moving average process.

Since an ARMA$(1, 1)$ process is essentially a composite of an AR(1) process with an MA(1) process, we might expect the autocorrelation function associated with this process to have the characteristics of the autocorrelation function associated with the first-order autoregressive and moving average processes, respectively; that is, we would expect ρ_k to have a spike at $k = 1$ whose magnitude is in part determined by the moving average parameter θ_1 and to exhibit behavior reminiscent of an AR(1) process at higher lags. In fact, the autocorrelation function is defined by

$$\rho_k = \frac{(\phi_1 - \theta_1)(1 - \phi_1\theta_1)}{(1 - 2\phi_1\theta_1 + \theta_1^2)} \phi_1^{k-1}, \qquad k \geq 1 \qquad (10.17)$$

We notice that except for ρ_1, ρ_k has the form $\rho_k = C\phi_1^{k-1}$, where C is a constant depending on ϕ_1 and θ_1, which can be either positive or negative. Therefore ρ_k dies out with increasing lag. A chart relating ρ_1 and ρ_2 to θ_1 and ϕ_1 is provided in the appendix tables at the end of the book (see Appendix Table L). Figure 10.4(e) indicates the possible shapes for this autocorrelation function.

Extensions of the ARMA(1, 1) model are considered briefly in Supplement 10B.

Differencing to Produce Stationarity

Many time series arising in the "real world" behave as if they have no fixed mean level even though parts of the series display a certain kind of homogeneity. For example, we have already mentioned that the refrigerator sales series and the three-month Treasury bill interest rates series do not appear to vary about a fixed level. On the contrary both series appear to be steadily increasing over time. Even so, the *local* behavior for each series is similar in the sense that for short time intervals, apart from a difference in level or possibly level and "trend," one part of the series looks very much like any other part. The situation is illustrated in Figure 10.6 in which two typical series nonstationary in the mean are plotted. Although the level of the series in Figure 10.6(a) and the level and trend of the series in Figure 10.6(b) are constantly changing, the behavior of a given series within the blocks outlined in the figure is very similar. Box and Jenkins have termed this kind of behavior *homogeneous* nonstationary behavior and the resulting time series a *homogeneous* nonstationary series. Models useful for representing such behavior can be obtained by supposing a suitable *difference* of the process is stationary.

The refrigerator sales series is displayed in Figure 10.7 along with its first-difference series. Although the original series tends to increase over time, the first-difference series appears to vary about zero. This indicates that the nonstationary process generating the original series may be transformed into a process stationary in the mean by differencing. Hence to describe this process we are led to consider models for which the *first difference* can be represented by an ARMA model.

In some cases it will be necessary to difference more than once in order to achieve stationarity. For example, the kind of behavior illustrated in Figure 10.6(b) can be described by a model in which the second difference of the original process is taken to be stationary.

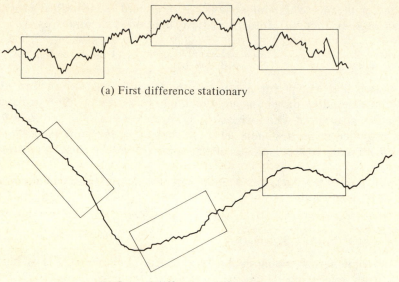

(a) First difference stationary

(b) Second difference stationary

Figure 10.6   Examples of homogeneous nonstationary time series

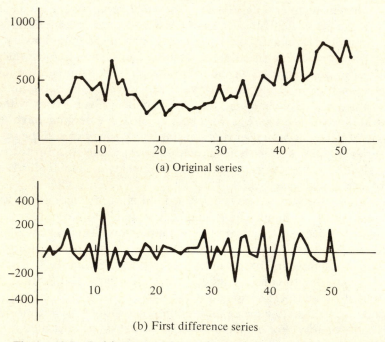

(a) Original series

(b) First difference series

Figure 10.7   Refrigerator sales series and the associated first-difference series

In general, models for nonstationary time series can be obtained by supposing that the dth difference ($d > 0$) of the original process is stationary and can be represented by an ARMA model. Such a model is referred to as an autoregressive *integrated* moving average (ARIMA) model. Thus all of the time series models we have introduced are special cases of an ARIMA model obtained by selecting an appropriate value for d and the "right" combination of autoregressive and moving average parameters (see Supplement 10B).

To summarize, we have discussed models capable of representing many commonly occurring stochastic processes. It has been suggested that the original process or a suitable difference of the original process may be represented by an autoregressive-moving average model. These models are extremely flexible; that is, they can describe a wide range of both stationary and nonstationary behavior, and as we shall see they can provide easily computed and reliable forecasts.

Effect of Nonstationarity on the Estimated Autocorrelation Function

Nonstationary stochastic processes tend to generate series whose estimated autocorrelation functions fail to die out rapidly; that is, the estimated autocorrelations for nonstationary processes tend to persist for a large number of lags. Persistently large values of r_k, then, indicate that the time series is nonstationary and that *at least* one difference is needed. In practice the behavior of r_k for the original series and the associated differenced series is routinely inspected. The amount of differencing needed to achieve stationarity is identified as that value of d for which the sample autocorrelations of the dth differenced series[d] die out rapidly and, within sampling fluctuations, have a recognizable pattern.

To illustrate our remarks, Figure 10.8 depicts the sample autocorrelation function of the original series, y_t, $t = 1, 2, \ldots, N$, and the differenced series, $w_t = y_t - y_{t-1}$, $t = 2, 3, \ldots, N$, for the refrigerator sales and the Treasury bill interest rates data. The reader will notice that in both cases values of r_k for the original series tend to be persistently large. However, the sample autocorrelations for the differenced series, in each case, tend to die out quickly. It appears as if both series are realizations of stochastic processes whose first difference is stationary.

We note at this point that for the refrigerator sales example, the differenced series produces an estimated autocorrelation function with a large spike at lag $k = 1$ and considerably smaller spikes at other lags that behave rather sporadically. For the Treasury bill differenced series, the estimated autocorrelation function has a damped sinusoidal appearance. The

---

[d] For example, if $d = 2$, the second differenced series w_t is generated by setting $w_t = y_t - 2y_{t-1} + y_{t-2}$ for $t = 3, 4, \ldots, N$ (see Supplement 10B).

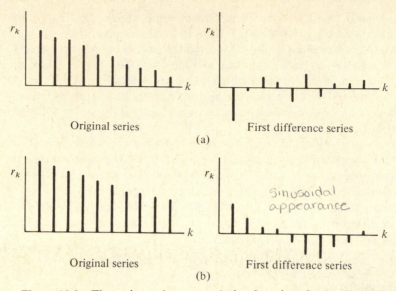

Figure 10.8   The estimated autocorrelation functions for (a) the refrigerator
sales series and (b) the Treasury bill interest rate series

implications of these findings will be apparent in Section 10.5 when
appropriate models for these series are considered in more detail. The
observant reader making use of the *theoretical* autocorrelation patterns for
ARMA processes displayed in Figure 10.4, may already have a tentative
model in mind for each of these series.

The Random Walk Model

For stationary series the overall behavior of the observations stays
roughly the same over time. This is largely accounted for in the ARMA
models by the parameter μ, the process mean, which represents the
constant, long-run level about which the stochastic process fluctuates.
However, in many applications of time series analysis, the use of a
stationary model is unrealistic.

For example, time series representing certain facets of economic activity
would not be expected to remain stationary over time. Rather, we would
expect such time series as stock prices, income, sales, and consumption to
"develop." As an illustration we consider the plot of 100 observations in
Figure 10.9. This series looks, for example, very much like plots of daily
closing stock prices. In fact the observations $y_1, y_2, \ldots, y_{100}$ are nothing
more than the cumulative sums

$$y_t = \sum_{j=0}^{t-1} e_{t-j}, \qquad t = 1, 2, \ldots, 100$$

Table 10.2   One Hundred Random Standard Normal Deviates (read down the columns)

Column 1	Column 2	Column 3	Column 4
0.697	− 0.747	2.945	0.446
3.521	0.790	0.881	− 2.127
0.321	0.145	0.971	− 0.656
0.595	0.034	1.033	1.041
0.769	0.234	− 0.511	− 0.899
− 0.136	− 0.736	0.181	− 1.114
− 0.345	− 1.206	− 0.486	− 0.515
0.761	− 0.491	− 0.256	− 0.451
− 1.229	− 0.109	0.065	1.410
− 0.561	0.574	1.147	− 1.045
1.598	− 0.509	− 0.199	1.378
− 0.725	0.394	− 0.508	0.499
1.231	1.810	− 0.992	0.665
1.046	0.060	0.969	0.754
0.360	− 0.491	0.983	0.298
0.424	− 1.186	− 1.096	1.456
1.377	− 0.762	0.250	− 0.106
− 0.873	− 1.541	1.265	− 1.579
0.542	0.993	− 0.927	0.532
0.882	− 1.407	− 0.227	− 0.899
− 1.210	− 0.504	− 0.577	0.410
0.891	− 0.463	− 0.291	1.375
− 0.649	0.833	− 2.828	− 1.851
− 0.219	0.926	0.247	1.974
0.084	0.571	− 0.584	− 0.934

formed from the observed values of 100 random standard normal deviates that are given in Table 10.2.

The level of the time series in Figure 10.9 appears to be changing over time with no predictable pattern of movement. This series appears to have no fixed mean level and is therefore nonstationary. Since the observations were formed by taking cumulative sums, it is clear that

$$y_t - y_{t-1} = e_t$$

or equivalently

$$y_t = y_{t-1} + e_t, \qquad t = 2, 3, \ldots, 100$$

If we regard

$$y_{t-1} = \sum_{j=0}^{t-2} e_{t-1-j}$$

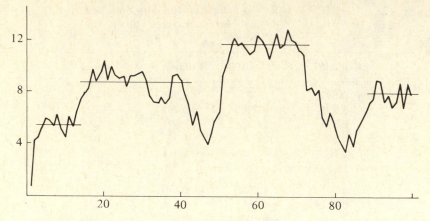

Figure 10.9   A time series formed by summing independent normal deviates

as the "position," or "level," of the series at time t—that is, the value y_t would have if the shock $e_t = 0$—then our series may be characterized as a series in which successive changes in level are determined by chance, in this case by a "random drawing" from an $N(0, 1)$ population. A time series (or more precisely, a stochastic process) that evolves in this way is sometimes referred to as a "random walk." There are various kinds of random walks, and their behavior has been extensively studied. For our purposes a random walk is any stochastic process that can be represented by the model (hereafter called the "random walk model"),

$$Y_t - Y_{t-1} = e_t$$

or

$$W_t = e_t, \qquad \text{with} \quad W_t = Y_t - Y_{t-1} \tag{10.18}$$

with the usual assumptions imposed on the error term; that is, a random walk is a stochastic process for which the successive increments $W_t = Y_t - Y_{t-1}$, $t = \ldots, -1, 0, 1, \ldots$, are independent and identically distributed $N(0, \sigma_e^2)$ random variables.

It is clear from equation (10.18) and the nature of the e_t's that the *differenced* process, $W_t = Y_t - Y_{t-1}$, $t = \ldots, -1, 0, 1, \ldots$, is stationary and that $\rho_k = 0$, $k > 0$. Therefore a random walk model may be appropriate if the *sample* autocorrelations, r_k, $k = 1, 2, \ldots$, associated with a particular time series fail to die out quickly for the original series but are uniformly small—that is, close to zero in absolute value—for the first difference series.

Random walk behavior occurs frequently in the real world. For example, random walk models have been used in various forms to describe the amount of water behind a dam, the amount of capital in an insurance company, the escape of comets from the solar system, common stock prices, and the movement of small particles suspended in a fluid.

The IMA(1, 1) Model

An important generalization of the random walk model is provided by the model

$$W_t = Y_t - Y_{t-1} = e_t - \theta_1 e_{t-1}, \qquad |\theta_1| < 1 \qquad (10.19)$$

that is, equation (10.19) represents a model for a nonstationary process that can be transformed to a stationary first-order moving average process by taking first differences. We shall refer to equation (10.19) as an *integrated moving average* model of order $(1, 1)$, or an IMA$(1, 1)$ model—the numbers in the parentheses denoting the amount of differencing and the number of moving average parameters, respectively. From the form of equation (10.19) it is evident that the differenced process W_t, $t = \ldots, -1, 0, 1$, is stationary and that *for this process*

$$\rho_1 = \frac{-\theta_1}{1 + \theta_1^2}$$

$$\rho_k = 0, \qquad k > 1$$

Therefore an IMA$(1, 1)$ model may be appropriate if the sample autocorrelations, r_k, associated with a particular time series fail to die out rapidly for the original series but exhibit first-order moving average behavior for the differenced series; that is, an IMA$(1, 1)$ model is suggested if, for the *differenced series*, r_1 is relatively large in absolute value whereas the remaining r_k, $k > 1$, are uniformly close to zero.

A realization of an IMA$(1, 1)$ process generated from equation (10.19) with $\theta_1 = 0.5$ and $\sigma_e^2 = 4$ is given in Figure 10.10(a). We note that the level tends to change over time, indicated by the horizontal lines in the figure. A plot of the first-difference series is given in Figure 10.10(b). We notice that this series appears to vary about a fixed level; that is, it appears to be stationary. The IMA$(1, 1)$ model is important because it describes the stochastic process for which the much-used technique of simple exponential smoothing provides the optimal forecast of future values.[e] For this reason it is of interest to examine the nature of the IMA$(1, 1)$ model.

Using equation (10.19) we can write

$$\begin{aligned}
Y_t - Y_{t-1} &= e_t - \theta_1 e_{t-1} \\
Y_{t-1} - Y_{t-2} &= e_{t-1} - \theta_1 e_{t-2} \\
Y_{t-2} - Y_{t-3} &= e_{t-2} - \theta_1 e_{t-3}
\end{aligned} \qquad (10.20)$$

$$\cdot$$
$$\cdot$$
$$\cdot$$

After summing the quantities on both sides of the equal signs, it is evident that everything on the left-hand side cancels except Y_t and the terms on the

---

[e] The reader not familiar with exponential smoothing should not be concerned at this point. This topic will be covered more thoroughly in Section 10.8 in which the whole problem of forecasting is considered.

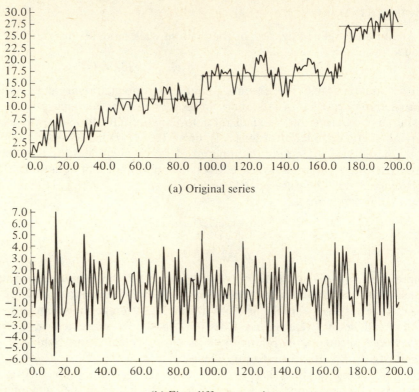

(a) Original series

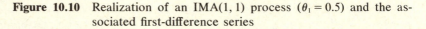

(b) First difference series

Figure 10.10   Realization of an IMA(1, 1) process ($\theta_1 = 0.5$) and the associated first-difference series

right-hand side can be combined to give

$$Y_t = e_t + (1 - \theta_1)e_{t-1} + (1 - \theta_1)e_{t-2} + \cdots$$

or

$$Y_t = e_t + (1 - \theta_1) \sum_{j=1}^{\infty} e_{t-j}, \qquad \text{for all } t \qquad (10.21)$$

where

$$L_t = (1 - \theta_1) \sum_{j=1}^{\infty} e_{t-j}$$

can be interpreted as the level of the process at time t. Hence by "integrating" (that is, summing) we have obtained an expression for the original process variable Y_t. Equation (10.21) is referred to as the "integrated" form of the model and expresses Y_t entirely in terms of the current and previous shocks. In a similar manner we can multiply both sides of the equations in expression (10.20) by 1, θ_1, θ_1^2, θ_1^3, . . . , respectively, and sum to obtain (see Problem 10.11)

$$Y_t = (1 - \theta_1)\{Y_{t-1} + \theta_1 Y_{t-2} + \theta_1^2 Y_{t-3} + \cdots\} + e_t$$

or

$$Y_t = (1 - \theta_1) \sum_{j=1}^{\infty} \theta_1^{j-1} Y_{t-j} + e_t \qquad (10.22)$$

where

$$L_t = (1 - \theta_1) \sum_{j=1}^{\infty} \theta_1^{j-1} Y_{t-j}$$

represents the level of the process at time t in terms of previous Y's. More precisely, Y_t can be expressed in terms of the current shock e_t and an *exponentially weighted sum* of previous Y's, where the weights $(1 - \theta_1)\theta_1^{j-1}$ decay exponentially, alternating in sign if $-1 < \theta_1 < 0$.

10.4 A MODEL-BUILDING STRATEGY

As we mentioned in Chapter 9, the process of determining a suitable time series model can be conveniently broken down into the three components: (a) *identification*, (b) *estimation*, and (c) *diagnostic checking*. A useful model-building strategy is to iterate on the three distinguishable component problems until a suitable model is determined. In this section we shall elaborate on the comments made in Section 9.3 in order to display more completely the techniques we shall use to derive models for the time series pictured in Figures 10.1–10.3.

Identification

The selection of a tentative time series model is frequently accomplished by "matching" estimated autocorrelations with the theoretical autocorrelations produced by particular members of the class of time series models. Since theoretical autocorrelations are only independent of time for stationary processes, it is often necessary to difference the original series until it can be assumed to be a realization of a stationary process. Consequently, estimated autocorrelations of the original series and of its first and second differences are routinely inspected. (In practice differencing beyond second order is rarely necessary.) In general if large values of r_k persist, nonstationarity is suspected, and differencing is required.

We have already seen an example of the identification procedure in connection with the chemical process yield data presented in Chapter 9. The reader will recall that matching the estimated autocorrelations produced by the series to those of an AR(1) process led to the time series model,

$$\dot{Y}_t - \phi_1 \dot{Y}_{t-1} = e_t$$

with $-1 < \phi_1 < 0$.

The calculation of sample autocorrelations is ordinarily carried out on an electronic computer. It should be pointed out that these autocorrelation estimates can be unstable for short series and hence are ordinarily calculated only for series consisting of (approximately) 50 or more observations.

Parameter Estimation

The objective at the estimation stage is to determine the "best" estimates of the unknown parameters in the tentatively selected model. In regression analysis the best estimates of the parameters β were found by choosing those values that minimized the error sum of squares

$$S(\beta) = [\mathbf{Y} - E(\mathbf{Y}|X\text{'s})]'[\mathbf{Y} - E(\mathbf{Y}|X\text{'s})]$$

$$= \sum [y_i - E(Y_i|X\text{'s})]^2$$

The principle of least squares can be employed in time series analysis as well and leads to selecting those parameter values that minimize

$$S(\mu, \phi, \theta) = \sum [y_t - E(Y_t|\text{previous } y\text{'s})]^2$$

or if the series is nonstationary,

$$S(\phi, \theta) = \sum [w_t - E(W_t|\text{previous } w\text{'s})]^2$$

where ϕ represents the autoregressive parameters, θ the moving average parameters, μ is the process mean, and w_t is the differenced series. We have seen what the sum of squares function looks like for an AR(1) process. The following examples illustrate the nature of the function for other processes.

EXAMPLE 10.1 MA(1) Model: $Y_t - \mu = e_t - \theta_1 e_{t-1}$

At time t, e_{t-1} (the error associated with the previous time period) has been realized. Thus at time t,

$$E(Y_t|\text{previous } y\text{'s}) = E(\mu + e_t - \theta_1 e_{t-1})$$

$$= \mu - \theta_1 e_{t-1}, \qquad t = 1, 2, \ldots, N$$

Therefore

$$y_t - E(Y_t|\text{previous } y\text{'s}) = y_t - (\mu - \theta_1 e_{t-1}), \qquad t = 1, 2, \ldots, N$$

and hence given a value for e_0, μ, and θ_1,

$$S(\mu, \theta_1) = \sum_{t=1}^{N} [y_t - E(Y_t|\text{previous } y\text{'s})]^2 = \sum_{t=1}^{N} (y_t - \mu + \theta_1 e_{t-1})^2$$

can be computed.

EXAMPLE 10.2   IMA(1, 1) model: $W_t = Y_t - Y_{t-1} = e_t - \theta_1 e_{t-1}$

This example is similar to Example 10.1. We notice, however, that $\mu = 0$ and that the observations w_t are formed by differencing the original series, $y_1, \ldots, y_N$. In this case the sum of squares function will depend on only one parameter, θ_1. Using an argument similar to the one presented in the previous example we have at time t,

$$E(W_t | \text{previous } w\text{'s}) = E(e_t - \theta_1 e_{t-1}) = -\theta_1 e_{t-1}, \qquad t = 2, \ldots, N$$

and hence given the original observations $y_1, \ldots, y_N$ and values for e_1 and θ_1,

$$S(\theta_1) = \sum_{t=2}^{N} [w_t - E(W_t | \text{previous } w\text{'s})]^2 = \sum_{t=2}^{N} (w_t + \theta_1 e_{t-1})^2$$

can be computed.

The reader will notice that in each of the examples the sum of squares function is a function of the unknown model parameters. It is also apparent that the calculation of this function requires a set of "initial values" to get the process started. This will be true in general for the time series models considered in this book. The problem of initial values is considered in Section 10.6.

In many cases the sum of squares function is not a quadratic function of the parameter values, so that the search for least squares estimates is not as straightforward as it was for linear regression models. In general a special computer algorithm must be used to find the least squares values. A number of good "nonlinear" estimation computer programs are currently available for this purpose. For problems involving one or two parameters, contours of the sum of squares function can be plotted, and the least squares values selected "by inspection."

Diagnostic Checking

Diagnostic checking is concerned both with testing the adequacy of the model and with indicating possible modifications if inadequacies exist.

One method of testing the adequacy of the model is to fit a model that is more complicated (that is, contains more parameters) than the given model and see if a better fit, as measured by the minimum value of the sum of squares function, is obtained. This "overparametization" can be an effective technique; however a word of caution is in order. If the parameters in an overparameterized model are being estimated by a nonlinear least squares procedure, computing difficulties (convergence problems) can arise.

More effective tests of model adequacy are based on detecting departures from randomness among the residuals, $\hat{e}_1, \ldots, \hat{e}_N$ (that is, the e_t's calculated with the least squares estimates of the parameters inserted). If the model is correct and if the model is fitted using the true parameter values, the resulting e_t's should be independent and identically distributed normal variates. Consequently, on the assumption of model adequacy, the $\hat{e}_t$'s should have very similar properties. Any other behavior would indicate model inadequacy.

A simple test of model fit is to examine the residual autocorrelations. If these autocorrelations are effectively zero when compared with standard deviation limits, the hypothesis of model adequacy remains tenable.

Recall the residual autocorrelations are given by

$$r_k(\hat{e}) = \frac{\sum_{t=1}^{N-k} \hat{e}_t \hat{e}_{t+k}}{\sum_{t=1}^{N} \hat{e}_t^2}, \qquad k = 1, 2, \ldots, K$$

If K is small compared to N, where N is the number of residuals, it can be shown that for a correctly specified model, the $r_k(\hat{e})$'s are approximately normally distributed about zero with a particular variance-covariance matrix from which the standard deviations can be determined. Except for small lags, one standard deviation is approximately $1/\sqrt{N}$.

Finally, a useful overall test of the adequacy of the time series model is provided by referring $X^2 = N\sum_{k=1}^{K} r_k^2(\hat{e})$ to a χ^2 distribution with $K - r$ degrees of freedom, where r is the total number of parameters of any kind in the postulated model. Too large a value for X^2 can be viewed as evidence *against* model adequacy.

Analysis of residuals and tests of goodness of fit either provide no reason to doubt the model, suggest ways in which the original model should be modified, or indicate the need for further data and hence lead to a new cycle in the process of model building. We will see examples of this kind of analysis shortly.

10.5 MODEL IDENTIFICATION

We indicated in Chapter 9 and Section 10.4 that time series models can be tentatively selected by matching known theoretical autocorrelation patterns with the sample autocorrelation pattern produced by the data.[f] Therefore to derive tentative models for the time series presented in Section 10.2, we must examine the behavior of their sample autocorrelation coefficients.

Table 10.3 gives the sample autocorrelations and their estimated standard deviations[g] as a function of the lag, k, for the refrigerator sales series, the Treasury bill interest rates series, and the unemployment rates series displayed in Figures 10.1, 10.2, and 10.3. In order to identify models for the time

---

[f] It should be pointed out that other identification methods are available. These methods (which may involve an examination of partial autocorrelations, "inverse" autocorrelations, and spectral densities) will not be discussed in this book. The interested reader can find a detailed account of these methods in references [1], [5], and [23]. These alternative identification procedures should not be viewed as substitutes for the autocorrelation methods; rather, they should be viewed as *complementary* procedures.

[g] The estimated standard deviations associated with the sample autocorrelations are computed using the theorem in Supplement 10A.

Table 10.3   Sample Autocorrelations

Refrigerator sales[a]

Original series y_t

k	1	2	3	4	5	6	7	8	9	10	11	12
r_k	0.70	0.64	0.63	0.54	0.41	0.41	0.31	0.27	0.23	0.15	0.08	0.02
$\sqrt{\text{Vâr}(r_k)}$	0.14	0.20	0.23	0.26	0.28	0.30	0.31	0.31	0.32	0.32	0.32	0.32

First-difference series $w_t = y_t - y_{t-1}$

k	1	2	3	4	5	6	7	8	9	10	11	12
r_k	-0.48	-0.04	0.12	0.05	-0.24	0.22	-0.07	0.01	0.01	0.03	0.02	-0.17
$\sqrt{\text{Vâr}(r_k)}$	0.14	0.17	0.17	0.17	0.17	0.18	0.18	0.18	0.18	0.18	0.18	0.18

Treasury bill interest rates[a]

Original series y_t

k	1	2	3	4	5	6	7	8	9	10	11	12
r_k	0.95	0.89	0.82	0.76	0.69	0.63	0.58	0.53	0.50	0.47	0.45	0.43
$\sqrt{\text{Vâr}(r_k)}$	0.08	0.13	0.17	0.19	0.21	0.22	0.23	0.24	0.25	0.26	0.26	0.27

First-difference series $w_t = y_t - y_{t-1}$

k	1	2	3	4	5	6	7	8	9	10	11	12
r_k	0.39	0.20	0.07	0.02	-0.10	-0.23	-0.30	-0.16	-0.07	0.01	-0.01	0.08
$\sqrt{\text{Vâr}(r_k)}$	0.08	0.09	0.09	0.09	0.09	0.10	0.10	0.10	0.11	0.11	0.11	0.11

Unemployment rates

Original series y_t

k	1	2	3	4	5	6	7	8	9	10	11	12
r_k	0.90	0.69	0.45	0.24	0.12	0.06	0.05	0.05	0.07	0.08	0.08	0.07
$\sqrt{\text{Vâr}(r_k)}$	0.13	0.21	0.24	0.26	0.26	0.26	0.26	0.26	0.26	0.26	0.26	0.26

[a] The autocorrelation functions are pictured in Figure 10.8 for lags $k = 1, 2, \ldots, 10$.

series being considered, we must ask, which process within the class of ARIMA processes is likely to produce the sample autocorrelation patterns exhibited in Table 10.3? We will answer this question for each series in turn.

Refrigerator Sales Series

We shall consider the sample autocorrelations associated with the refrigerator sales series. The behavior of the sample autocorrelations for the original series and the first-difference series can be summarized as follows.

Original Series   The sample autocorrelations are fairly large and fail to die out *quickly* although the rate of decay is moderately fast. Here the situation is not clear cut, and we might be led to consider a stationary process whose autocorrelation function exhibits exponential decay. However, the fact that sample autocorrelation values of about the same size (0.70, 0.64, 0.63, 0.54) tend to persist for several lags suggests that the original series may be nonstationary. Therefore the autocorrelations associated with the *first-difference* series should be examined. Note that nonstationarity is also indicated in Figure 10.1, where a plot of the series shows that the series tends to drift upward over time.

First-Difference Series   The sample autocorrelation at lag $k = 1$ is large (in absolute value) when compared with its standard error. The autocorrelations at other lags ($k > 1$) are small when compared with their standard errors and do not exhibit any consistent pattern of behavior. Therefore we conclude that the autocorrelation at lag $k = 1$ is probably "real" and that the nonzero values of r_k at other lags simply represent sampling fluctuation.

The behavior of the sample autocorrelations for the refrigerator sales series suggests that a possible model for this series is one for which the *first difference* process has a nonzero autocorrelation at lag $k = 1$ (that is, $\rho_1 \neq 0$) and no correlation at the remaining lags (that is, $\rho_k = 0$, $k > 1$). From Figure 10.4(a) we know that MA(1) processes have this kind of autocorrelation function, and therefore a tentative model for the refrigerator sales series is the model that represents the *differenced* process as a first-order moving average, that is the IMA(1, 1) model

$$W_t = Y_t - Y_{t-1} = e_t - \theta_1 e_{t-1}$$

Moreover, since $r_1 = -0.48$ for the differenced series, we can obtain an *initial* estimate of θ_1 from Table 10.1 by treating r_1 as if it were ρ_1 (r_1 is an estimate of ρ_1). Entering the table with $\rho_1 = -0.48$ and reading the value of θ_1 corresponding to this value of ρ_1, we find that $\theta_1 = 0.75$. Since this is really an *estimate* of θ_1 (ρ_1 is *estimated* to be -0.48), we write $\hat{\theta}_1 = 0.75$.

Treasury Bill Interest Rates

The sample autocorrelations associated with the Treasury bill interest rates series are characterized below.

Original Series   The sample autocorrelations are large and do not die out rapidly. The Treasury bill interest rates series is clearly nonstationary, and the sample autocorrelations for the differenced series should be inspected.

First-Difference Series   The sample autocorrelations die out rapidly and exhibit sinusoidal behavior. The first-difference series appears to be stationary (that is, a realization of a stationary stochastic process).

Comparing the behavior of the sample autocorrelations for the first-difference series with the theoretical autocorrelation functions displayed in Figure 10.4(d) it is clear that a tentative model for the Treasury bill interest rates series is the model

$$W_t - \phi_1 W_{t-1} - \phi_2 W_{t-2} = e_t \qquad (10.23)$$

where $W_t = Y_t - Y_{t-1}$; that is, it appears as if this series can be regarded as a realization of a nonstationary stochastic process whose associated first difference process is stationary and can be represented by a second-order autoregressive model. Initial estimates of the parameters ϕ_1, ϕ_2 can be obtained from Appendix Table J by treating the first two sample autocorrelations r_1 and r_2 for the first-difference series *as if* they were ρ_1 and ρ_2. Use of the chart gives $\hat{\phi}_1 = 0.40$ and $\hat{\phi}_2 = 0.10$.

Unemployment Rates Series

Finally, we examine the behavior of the sample autocorrelations for the unemployment rates series to identify an appropriate time series model.

Original Series   The sample autocorrelations die out quickly and, in particular, decay exponentially. In conclusion, the original series is stationary and may be a realization of a stochastic process whose theoretical autocorrelation function exhibits exponential decay.

Figure 10.4 indicates that AR(1), AR(2), and ARMA(1, 1) processes can all produce autocorrelation functions that decay exponentially. Hence we must regard each of the models

$$\dot{Y}_t - \phi_1 \dot{Y}_{t-1} = e_t$$
$$\dot{Y}_t - \phi_1 \dot{Y}_{t-1} - \phi_2 \dot{Y}_{t-2} = e_t \qquad (10.24)$$

and

$$\dot{Y}_t - \phi_1 \dot{Y}_{t-1} = e_t - \theta_1 e_{t-1}$$

where $\dot{Y}_t = Y_t - \mu$ as possible descriptions of the mechanism generating the unemployment rate series—at least initially. In practice each of these models could be fitted, the residuals examined, and the best model selected based on some criterion such as the smallest residual variance (assuming one or more models is not ruled out by diagnostic checking procedures). It turns out that the "best" model for the unemployment rates series is the AR(2) model. We ask the reader to accept this statement without verification. In later sections of this chapter we will demonstrate that the AR(2) model provides an adequate description of the unemployment rates series, although we will not show that the other models do not.

Again initial estimates of ϕ_1 and ϕ_2 in the AR(2) model may be obtained from Appendix Table J by treating r_1 and r_2 as if they were ρ_1 and ρ_2. This procedure gives $\hat{\phi}_1 = 1.50$ and $\hat{\phi}_2 = -0.60$.

Results of Identification Procedure

The results of the identification procedure are summarized in Table 10.4, where the tentative time series models for the refrigerator sales series, the Treasury bill interest rates series, and the unemployment rates series are displayed.

Table 10.4   Summary of Tentative Models

Series	Model	Initial Parameter Estimates
Refrigerator sales	$W_t = e_t - \theta_1 e_{t-1},$ where $W_t = Y_t - Y_{t-1}$	$\hat{\theta}_1 = 0.75$
Treasury bill interest rates	$W_t - \phi_1 W_{t-1} - \phi_2 W_{t-2} = e_t$ where $W_t = Y_t - Y_{t-1}$	$\hat{\phi}_1 = 0.40, \hat{\phi}_2 = 0.10$
Unemployment rates	$\dot{Y}_t - \phi_1 \dot{Y}_{t-1} - \phi_2 \dot{Y}_{t-2} = e_t,$ where $\dot{Y}_t = Y_t - \mu$	$\hat{\phi}_1 = 1.50, \hat{\phi}_2 = -0.60$ $\hat{\mu} = 5.0$

We conclude this section by stressing that the selection of an appropriate model should be influenced by the principle of parsimony. If two models appear to be equally viable candidates, the one chosen, at least initially, should be that with the fewest parameters. We should seek an adequate representation, but one that is as simple as possible. This will prevent difficulties at the estimation stage as well as simplify the subsequent inference.

10.6 PARAMETER ESTIMATION

In order to use a time series model for, say, forecasting purposes, values for the unknown parameters must be specified. In practice this generally means *estimating* the values of the parameters from the available data since only rarely will the value of a process parameter be known.

It has been suggested that the "best" estimates of the parameter values in a time series model are the least squares estimates; that is, if we denote the parameters in a general ARIMA process (see Supplement 10B) by $\phi = (\phi_1, \ldots, \phi_p)'$, $\theta = (\theta_1, \ldots, \theta_q)'$, and μ, we should take as the parameter estimates those values that minimize the sum of squared deviations

$$S(\mu, \phi, \theta) = \sum_{t=1}^{N} [y_t - E(Y_t | \text{previous } y\text{'s})]^2 \qquad (10.25)$$

where $y_1, y_2, \ldots, y_N$ represent the observed series and $E(Y_t | \text{previous } y\text{'s})$ is the expected value of the process variable Y_t taken at time t.[h] It turns out that the least squares estimators have desirable statistical properties, provided N is large, and moreover are easily obtained using standard computer algorithms. It has been pointed out in Section 10.4 that $E(Y_t | \text{previous } y\text{'s})$ will not, in general, be a linear function of the *parameters* so that a "nonlinear" estimation algorithm may be required to obtain the least squares estimates. Of course, if $S(\mu, \phi, \theta)$ is a function of only one or two parameters, it may be plotted over a grid of possible parameter values, and the least squares estimates determined by inspection. We shall, in fact, use this approach to get the least squares estimate of the moving average parameter θ_1 in the model proposed for the refrigerator sales series.

Given a nonlinear estimation algorithm (a procedure for finding the minimum of a function of several variables), the least squares estimates are usually easily derived, particularly if reasonable initial estimates are available. The primary responsibility of the user of these algorithms is the specification of the time series model so that $E(Y_t | \text{previous } y\text{'s})$ and hence the function to be minimized, $S(\mu, \phi, \theta)$, can be evaluated. We have already given expressions for the sum of squares function for several models in Chapter 9 and Section 10.4. The examples below give the sum of squares functions associated with the time series models of Table 10.4.

EXAMPLE 10.3

Let us consider the proposed unemployment rate model

$$\dot{Y}_t - \phi_1 \dot{Y}_{t-1} - \phi_2 \dot{Y}_{t-2} = e_t$$

where $\dot{Y}_t = Y_t - \mu$. Solving this equation for Y_t gives

$$Y_t = \mu + \phi_1(Y_{t-1} - \mu) + \phi_2(Y_{t-2} - \mu) + e_t$$

Now at any time t, the random variables Y_{t-1} and Y_{t-2} will have been observed. If we denote these observed values by y_{t-1} and y_{t-2}, then the expected value of Y_t given the

[h] If simple differencing is necessary to achieve stationarity, the y_t's are replaced by the *differenced* series $w_t = y_t - y_{t-1}$, $t = 2, 3, \ldots, N$, and Y_t's are replaced by the *differenced* process variable $W_t = Y_t - Y_{t-1}$, $t = 2, 3, \ldots, N$.

previous y's is

$$E(Y_t|\text{previous } y\text{'s}) = E(Y_t|y_{t-1}, y_{t-2}) = \mu + \phi_1(y_{t-1} - \mu) + \phi_2(y_{t-2} - \mu)$$
(10.26)

since $E(e_t) = 0$ by assumption. We note that equation (10.26) is simply the expected value of Y_t taken at time $t - 1$. Thus the sum of squares function is

$$S(\mu, \boldsymbol{\phi}) = \sum_{t=1}^{N} [y_t - \mu - \phi_1(y_{t-1} - \mu) - \phi_2(y_{t-2} - \mu)]^2$$
(10.27)

where $\boldsymbol{\phi} = (\phi_1, \phi_2)'$. This function can be evaluated given the data $y_1, y_2, \ldots, y_N$ and values for μ, ϕ_1, ϕ_2, y_0, and y_{-1}.

EXAMPLE 10.4

The tentative model for the refrigerator sales series is

$$W_t = e_t - \theta_1 e_{t-1}$$

where $W_t = Y_t - Y_{t-1}$. As indicated in the footnote at the beginning of Section 10.6 and illustrated in Example 10.2, if differencing is necessary to achieve stationarity, the y_t's in the sum of squares function are replaced by the differences $w_t = y_t - y_{t-1}$, $t = 2, \ldots, N$, and $E(Y_t|\text{previous } y\text{'s})$ is replaced by $E(W_t|\text{previous } w\text{'s})$. From Example 10.2 we have

$$S(\theta_1) = \sum_{t=2}^{N} [w_t - (-\theta_1 e_{t-1})]^2 = \sum_{t=2}^{N} (w_t + \theta_1 e_{t-1})^2$$
(10.28)

where e_{t-1} is the *observed* error in the previous time period. Since $e_{t-1} = w_{t-1} + \theta_1 e_{t-2}$ from the model, the e_{t-1}'s can be generated recursively. Thus given the data and a value for θ_1 and e_1, $S(\theta_1)$ can be computed.

EXAMPLE 10.5

The tentative model for the Treasury bill interest rate series is

$$W_t - \phi_1 W_{t-1} - \phi_2 W_{t-2} = e_t$$

where $W_t = Y_t - Y_{t-1}$. At time $t - 1$, W_{t-1} and W_{t-2} will have been observed. If we denote the observed values by w_{t-1} and w_{t-2}, then, since $E(e_t) = 0$,

$$E(W_t|\text{previous } w\text{'s}) = E(W_t|w_{t-1}, w_{t-2}) = \phi_1 w_{t-1} + \phi_2 w_{t-2}$$
(10.29)

and

$$S(\boldsymbol{\phi}) = \sum_{t=1}^{N} (w_t - \phi_1 w_{t-1} - \phi_2 w_{t-2})^2$$
(10.30)

where $\boldsymbol{\phi} = (\phi_1, \phi_2)'$. Given the data, values for ϕ_1 and ϕ_2 and values for w_0 and w_{-1}, $S(\boldsymbol{\phi})$ can be computed. The reader will note that the Treasury bill interest rate model is *linear* in the parameters ϕ_1 and ϕ_2, so that given w_0 and w_{-1}, the least squares estimates could be obtained in the usual way (see Chapter 6).

Problem of Starting Values

We have seen in Examples 10.3, 10.4, and 10.5 that the evaluation of the sum of squares function requires the data, values for the parameters, and "starting values" in the form of previous y's, previous w's, or previous e's. For example, the sum of squares function given by equation (10.27) requires the starting values y_0 and y_{-1}.

Perhaps the easiest solution to the problem of starting values is the one suggested in Chapter 9 in connection with the AR(1) model; that is, simply ignore the terms in the sum of squares that involve unknown quantities like y_0, y_{-1}. For example if we ignore the first two terms in the sum of squares functions in equations (10.27) and (10.30), we have the approximations

$$S^*(\mu, \boldsymbol{\phi}) = \sum_{t=3}^{N} [y_t - \mu - \phi_1(y_{t-1} - \mu) - \phi_2(y_{t-2} - \mu)]^2 \qquad (10.31)$$

and

$$S^*(\boldsymbol{\phi}) = \sum_{t=3}^{N} (w_t - \phi_1 w_{t-1} - \phi_2 w_{t-2})^2 \qquad (10.32)$$

respectively. We then find the values of the parameters that minimize the sum of squares functions in equations (10.31) and (10.32). This procedure is justified if N, the number of observations, is fairly large (for example, $N \geq 50$).

The sum of squares function in equation (10.28) presents a somewhat different problem. Here the starting value e_1 is required. The solution is to replace e_1 by its expected value at time $t = 0$; that is, at time $t = 0$, $E(e_1) = 0$; so at time $t = 1$, we *assume* the value $e_1 = 0$ has occurred. At time $t = 1$, this represents as good a guess for e_1 as any. Thus we approximate the sum of squares function in equation (10.28) with

$$S^*(\theta_1) = \sum_{t=2}^{N} (w_t + \theta_1 e_{t-1})^2 \qquad (10.33)$$

where $e_1 = 0$ and subsequent e_t's are computed recursively using the expression

$$e_t = w_t + \theta_1 e_{t-1}, \qquad t = 2, \ldots, N \qquad (10.34)$$

obtained from the model.

The solutions to the starting values problem suggested above are easily generalized to more complex models.

Initial Parameter Values

If the time series model is nonlinear in the parameters and the least squares estimates cannot be conveniently determined by inspection, that is, by plotting the sum of squares function, then a "nonlinear" estimation algorithm must be employed. Since these algorithms iterate toward the least squares values, they require initial values for the parameters. If the tentative time series model is a parsimonious representation of the process, then initial values of the autoregressive and moving average parameters calculated from the sample autocorrelations—like the ones displayed in Table 10.4—will be satisfactory. Moreover an initial value for μ is supplied by $\bar{y}$, the arithmetic average of the observed series. If it is not convenient to derive initial estimates from the sample autocorrelations, most nonlinear estimation algorithms will "converge" to the least squares estimates with starting values of $\hat{\mu} = \bar{y}$ and the remaining parameters set equal to a small nonzero value like 0.1.

Least Squares Estimates for Models in Table 10.4

The least squares estimates of the parameters[i] for the models of Table 10.4 obtained using a nonlinear estimation algorithm are given in Table 10.5. Since the proposed model for refrigerator sales is particularly simple it is instructive to demonstrate the evaluation of the sum of squares function for a particular value of θ_1. Moreover, since the model contains only one parameter, the sum of squares function is easily plotted and the least squares estimate can be determined by inspection.

The refrigerator sales series and the associated first-difference series are listed in Table 10.6. We can generate the terms in the sum of squares expression (10.33) as follows. We note that $S^*(\theta_1)$ can be written

$$S^*(\theta_1) = \sum_{t=2}^{N} e_t^2$$

since from equation (10.34), $w_t + \theta_1 e_{t-1} = e_t$. Therefore to evaluate the sum of squares function, it is only necessary to generate the errors, $e_2, e_3, \ldots, e_N$.

Let us suppose that $\theta_1 = 0.50$. Then using the differenced series given in Table 10.6 and setting $e_1 = 0$, we have

$$e_1 = 0$$
$$e_2 = w_2 + \theta_1 e_1 = -67 + 0.5(0) = -67$$
$$e_3 = w_3 + \theta_1 e_2 = 48 + 0.5(-67) = 14.5$$
$$e_4 = w_4 + \theta_1 e_3 = -45 + 0.5(14.5) = -37.75$$

$$\cdot$$
$$\cdot$$
$$\cdot$$

$$e_{52} = w_{52} + \theta_1 e_{51} = -162 + 0.5(135.25) = -94.375$$

and

$$S^*(0.50) = \sum_{t=2}^{52} e_t^2 = (-67)^2 + (14.5)^2 + (-37.75)^2 + \cdots + (-94.375)^2 = 586,937.3$$

[i] The least squares estimates can be compared with the initial estimates given in Table 10.4.

Table 10.5 Fitted Time Series Models

Series	Model	Least Squares Estimates	Fitted Model
Refrigerator sales	$W_t = e_t - \theta_1 e_{t-1}$	$\hat{\theta}_1 = 0.59$	$W_t = e_t - 0.59 e_{t-1}$
Treasury bill interest rates	$W_t - \phi_1 W_{t-1} - \phi_2 W_{t-2} = e_t$	$\hat{\phi}_1 = 0.38, \hat{\phi}_2 = 0.07$	$W_t - 0.38 W_{t-1} - 0.07 W_{t-2} = e_t$
Unemployment rates	$\dot{Y}_t - \phi_1 \dot{Y}_{t-1} - \phi_2 \dot{Y}_{t-2} = e_t$	$\hat{\phi}_1 = 1.56, \hat{\phi}_2 = -0.72$ $\hat{\mu} = 4.97$	$(Y_t - 4.97) - 1.56(Y_{t-1} - 4.97)$ $+ 0.72(Y_{t-2} - 4.97) = e_t$

Table 10.6   Refrigerator Sales Series and the Associated First-Difference Series (in appropriate units)

t	y_t	t	y_t	t	w_t	t	w_t
1	390	27	298	2	− 67	27	14
2	323	28	318	3	48	28	20
3	371	29	340	4	− 45	29	22
4	326	30	497	5	32	30	157
5	358	31	349	6	180	31	− 148
6	538	32	380	7	− 5	32	31
7	533	33	379	8	− 75	33	− 1
8	458	34	526	9	− 44	34	147
9	414	35	272	10	75	35	− 304
10	489	36	401	11	− 183	36	129
11	306	37	553	12	348	37	152
12	654	38	527	13	− 196	38	− 26
13	458	39	485	14	49	39	− 42
14	507	40	722	15	− 145	40	237
15	362	41	474	16	5	41	− 248
16	367	42	510	17	− 61	42	36
17	306	43	760	18	− 83	43	250
18	223	44	515	19	58	44	− 245
19	281	45	560	20	36	45	45
20	317	46	751	21	− 79	46	191
21	238	47	842	22	48	47	91
22	286	48	818	23	20	48	− 24
23	306	49	746	24	1	49	− 72
24	307	50	672	25	− 32	50	− 74
25	275	51	854	26	9	51	182
26	284	52	692			52	− 162

$S^*(\theta_1)$ was evaluated for other values of θ_1 in the neighborhood of the *initial* estimate $\hat{\theta}_1 = 0.75$. The sum of squares function is plotted in Figure 10.11, and the least squares estimate of θ_1 can be easily determined from the figure by inspection. For example, it is seen that the minimum value of $S^*(\theta_1)$, to the degree of accuracy allowed by "eyeballing" the figure, occurs at $\theta_1 = 0.60$. The least squares estimate is therefore $\hat{\theta}_1 = 0.60$, a result that is consistent with the least squares value given in Table 10.5 obtained by the more sophisticated nonlinear estimation algorithm.

The Fitted Values

The fitted values, $\hat{y}_t$, of the series are defined to be the "best" estimates of the conditional expectation, $E(Y_t | \text{previous } y\text{'s})$, where "best" is used in the least squares sense; that is, the fitted values are generated by evaluating the

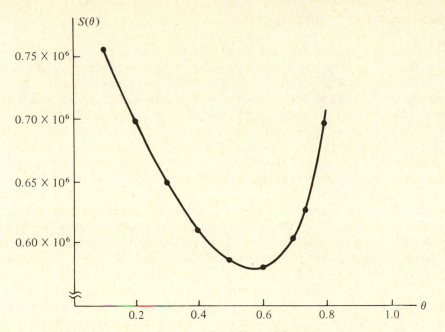

Figure 10.11 Sum of squares function, refrigerator sales example

expression $E(Y_t|\text{previous } y\text{'s})$ with the least squares estimates substituted for the unknown "true" parameter values.[j] To illustrate these remarks we can consider the unemployment rate series. We have already seen that

$$E(Y_t|\text{previous } y\text{'s}) = \mu + \phi_1(y_{t-1} - \mu) + \phi_2(y_{t-2} - \mu)$$

In addition the least squares estimates are given by $\hat{\mu} = 4.97$, $\hat{\phi}_1 = 1.56$, and $\hat{\phi}_2 = -0.72$. Since observations y_0 and y_{-1} are not available we begin the calculation of the fitted values with y_3. The first three observations in the unemployment rate series are $y_1 = 3.82$, $y_2 = 3.82$, and $y_3 = 3.83$; thus

$$\hat{y}_3 = \hat{\mu} + \hat{\phi}_1(y_2 - \hat{\mu}) + \hat{\phi}_2(y_1 - \hat{\mu})$$
$$= 4.97 + 1.56(3.82 - 4.97) - 0.72(3.82 - 4.97) = 4.00$$
$$\hat{y}_4 = 4.97 + 1.56(3.83 - 4.97) - 0.72(3.82 - 4.97) = 4.01$$

.
.
.

The rest of the fitted values are generated in a similar manner.

The calculation of fitted values is important because it is the differences between the observations and the corresponding fitted values that are examined to determine the "adequacy of fit" of the current time series model.

---

[j] Similarly the fitted values, $\hat{w}_t$, are generated by evaluating the conditional expectation $E(W_t|\text{previous } w\text{'s})$ with the least squares estimates inserted for the "true" parameter values.

10.7 DIAGNOSTIC CHECKING

It has been pointed out that diagnostic checking is concerned with the problem of determining if a given time series model is a satisfactory representation of the data. In addition diagnostic checks can suggest modifications if the model is not satisfactory. Almost all diagnostic checking procedures are based on the examination of the residuals, $\hat{e}_t = y_t - \hat{y}_t$, where $\hat{y}_t$ is the fitted value, or some function of the residuals. The rationale is that if the model is satisfactory, the residuals, which are estimates of the error component e_t, should be uncorrelated at any lag and should be approximately normally distributed with mean zero and a variance estimated by the residual mean square.

The normality assumption can be checked by plotting the residuals as suggested in Section 5.6. In addition we can compute summary measures like the average, the variance, and skewness and kurtosis statistics to see if these quantities are consistent with the hypothesis of normality. In practice the normality assumption is not as crucial as the assumptions $E(e_t) = 0$ and $E(e_t e_{t'}) = 0$, $t \neq t'$, although we will assume normality when we discuss forecasting and the associated inferences about future values. Therefore we will concentrate on the assumptions of zero mean and no correlation at any lag.

The assumption that the expected value of e_t is zero can be tested by comparing the value of the arithmetic average of the residuals with its estimated standard deviation.

To test for correlation we examine the autocorrelation function of the residuals. The residual autocorrelations, $r_k(\hat{e})$ [see equation (9.11)], for the three examples we have been considering are given in Table 10.7 and plotted in Figure 10.12. The estimated standard deviations of $r_k(\hat{e})$ presented in Table 10.7 are calculated using the theoretical results of Box and Pierce [21]. These authors have shown that *if the model is correct* and N is the number of residuals, then these autocorrelations are approximately normally distributed about zero, and for moderate and large lags the estimated standard deviation of $r_k(\hat{e})$ is approximately $1/\sqrt{N}$. For small lags, however, the estimated standard deviation of $r_k(\hat{e})$ is less than this value. In particular for the time series models we are considering we have the following results.

Refrigerator sales series [IMA(1, 1) model]:

$$\text{Vâr}[r_1(\hat{e})] \doteq \frac{\theta_1^2}{N}$$

$$\text{Vâr}[r_2(\hat{e})] \doteq \frac{1 - \theta_1^2 + \hat{\theta}_1^4}{N}$$

$$\text{Vâr}[r_k(\hat{e})] \doteq \frac{1}{N}, \qquad k \geq 3 \tag{10.35}$$

Table 10.7 Residual Autocorrelations

Refrigerator sales

k	1	2	3	4	5	6	7	8	9	10	11	12
$r_k(\hat{e})$	-0.11	-0.02	0.13	0.03	-0.16	0.20	0.04	0.05	0.04	0.05	-0.04	-0.21
$\sqrt{\hat{\text{Var}}[r_k(\hat{e})]}$	0.08	0.12	0.14	0.14	0.14	0.14	0.14	0.14	0.14	0.14	0.14	0.14

$X^2 = N\sum_{k=1}^{12} r_k^2(\hat{e}) = 51(0.15) = 7.65$ can be referred to a χ^2 distribution with $12 - 1 = 11$ degrees of freedom.

Treasury bill interest rates

k	1	2	3	4	5	6	7	8	9	10	11	12
$r_k(\hat{e})$	-0.00	0.01	-0.02	0.04	-0.03	-0.13	-0.23	-0.05	-0.01	0.05	-0.04	0.10
$\sqrt{\hat{\text{Var}}[r_k(\hat{e})]}$	0.01	0.03	0.08	0.08	0.08	0.08	0.08	0.08	0.08	0.08	0.08	0.08

$X^2 = N\sum_{k=1}^{12} r_k^2(\hat{e}) = 154(0.09) = 13.86$ can be referred to a χ^2 distribution with $12 - 2 = 10$ degrees of freedom.

Unemployment rates

k	1	2	3	4	5	6	7	8	9	10	11	12
$r_k(\hat{e})$	0.16	0.04	-0.11	-0.13	0.11	0.15	0.20	-0.05	0.12	0.08	0.16	-0.06
$\sqrt{\hat{\text{Var}}[r_k(\hat{e})]}$	0.09	0.11	0.13	0.13	0.13	0.13	0.13	0.13	0.13	0.13	0.13	0.13

$X^2 = N\sum_{k=1}^{12} r_k^2(\hat{e}) = 59(0.18) = 10.62$ can be referred to a χ^2 distribution with $12 - 3 = 9$ degrees of freedom.

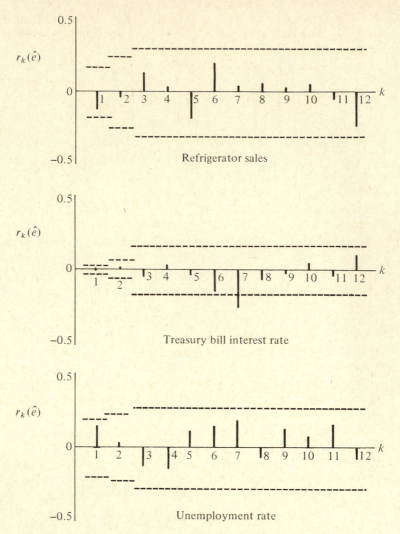

Figure 10.12   Residual autocorrelation functions and estimated two standard deviation limits

Treasury bill interest rate series [AR(2) model for first differences] and unemployment rate series [AR(2) model]:

$$\text{Vâr}[r_1(\hat{e})] \doteq \frac{\phi_2^2}{N}$$

$$\text{Vâr}[r_2(\hat{e})] \doteq \frac{\phi_2^2 + \phi_1^2(1 + \phi_2)^2}{N}$$

$$\text{Vâr}[r_k(\hat{e})] \doteq \frac{1}{N}, \qquad k \geq 3 \tag{10.36}$$

The estimated standard deviations in Table 10.7 were calculated using equations (10.35) and (10.36) with the least squares estimates inserted for the unknown parameter values.

It is well known that residual autocorrelation coefficients are themselves correlated. Therefore it is not enough to examine the $r_k(\hat{e})$'s individually. An overall test of model adequacy is provided by referring the quantity

$$X^2 = N \sum_{k=1}^{K} r_k^2(\hat{e}) \tag{10.37}$$

to a χ^2 distribution with $\nu = K - r$ degrees of freedom, where r is the total number of parameters (of any kind) in the postulated model.[k] As before, N is the number of residuals, and K is the number of residual autocorrelations being considered. (As a rule of thumb, K should be taken at least as large as $\sqrt{N}$.) Too large a value for $N\sum_{k=1}^{K} r_k^2(\hat{e})$ is evidence *against* model adequacy; that is, the residuals are not behaving like uncorrelated identically distributed normal variables.

To illustrate the procedure we shall consider the X^2 value given in Table 10.7 for the refrigerator sales series. From the table,

$$X^2 = 51 \sum_{k=1}^{12} r_k^2(\hat{e}) = 7.65$$

This value can be referred to a χ^2 distribution with $12 - 1 = 11$ degrees of freedom. Since the mean of the χ^2 distribution is equal to the number of degrees of freedom, our observed value falls to the left of the mean, that is, well within the bulk of the distribution. Thus there is no reason on the basis of this test to believe that the model is misspecified. Similar statements can be made for the treasury bill interest rate series and the unemployment rate series.

Table 10.8 gives the arithmetic averages of the residuals, $\bar{\bar{e}}$, and their estimated standard deviations, $\hat{\sigma}_e/\sqrt{N}$, where $\hat{\sigma}_e^2$ is the residual mean square

Table 10.8 Residual Means, Residual Mean Squares, and Estimated Standard Deviation of Residual Means

Series	$\bar{\bar{e}}$	$\hat{\sigma}_e^2$ (residual mean square)	$\hat{\sigma}_e/\sqrt{N}$
Refrigerator sales	16.82	11.58×10^3	15.07
Treasury bill interest rate	0.015	0.045	0.017
Unemployment rate	-2.2×10^{-6}	0.146	0.049

[k] For a more detailed discussion of this test the reader is referred to reference [21].

given by

$$\hat{\sigma}_e^2 = \frac{\sum_{t=1}^{N}(y_t - \hat{y}_t)^2}{N-r} = \frac{\sum_{t=1}^{N}\hat{e}_t^2}{N-r} \tag{10.38}$$

and N is the *number of residuals* and r is the *total* number of parameters in the model. The residual mean square is an *estimate* of $\sigma_e^2 = \text{Var}(e_t)$ and hence we use the notation $\hat{\sigma}_e^2$. It is seen from the table that the $\bar{e}$'s are small relative to their estimated standard deviations for all of our examples and thus there is no reason to doubt the assumption $E(e_t) = 0$.

An Example of Model Misspecification

We have suggested that if the model is *not* adequate, an examination of the residuals may reveal the nature of the inadequacy. In the previous examples the models appeared to be adequate. However, suppose we had before us the hypothetical residual autocorrelations and estimated standard deviations given in Table 10.9.

Table 10.9   Hypothetical Residual Autocorrelations

k	1	2	3	4	5	6
$r_k(\hat{e})$	0.62	0.35	0.17	0.08	0.03	0.01
$\sqrt{\hat{\text{Var}}[r_k(\hat{e})]}$	0.09	0.12	0.15	0.15	0.15	0.15

	7	8	9	10	11	12
	0.01	−0.02	0.03	−0.01	−0.03	0.05
	0.15	0.15	0.15	0.15	0.15	0.15

In addition suppose the time series model currently being considered is the first-order moving average model

$$\dot{Y}_t = e_t - \theta_1 e_{t-1} \tag{10.39}$$

where $\dot{Y}_t = Y_t - \mu$. We note that the residual autocorrelation function has a large value at lag 1 and then dies out exponentially, reminiscent of an AR(1) process. This suggests that the original model is not correct and, in particular, indicates that the errors e_t are not uncorrelated. Instead the error term e_t itself appears to be an AR(1) process and thus can be modeled as

$$e_t - \phi_1 e_{t-1} = a_t \tag{10.40}$$

where now the errors a_t are assumed to be uncorrelated normal random variables with mean zero. Therefore the original model must be modified as follows.

Using equation (10.39),

(a) $\dot{Y}_t = e_t - \theta_1 e_{t-1},$ for all t

thus

(b) $\dot{Y}_{t-1} = e_{t-1} - \theta_1 e_{t-2}$

and hence

(c) $\phi_1 \dot{Y}_{t-1} = \phi_1 e_{t-1} - \theta_1 \phi_1 e_{t-2}$

We now subtract $\phi_1 \dot{Y}_{t-1}$ from *both* sides of equation (a) to get

(d) $\dot{Y}_t - \phi_1 \dot{Y}_{t-1} = e_t - \theta_1 e_{t-1} - \phi_1 \dot{Y}_{t-1}$

By using (c), however, the right-hand side of (d) can be written entirely in terms of the e_t's. Thus

$$\dot{Y}_t - \phi_1 \dot{Y}_{t-1} = e_t - \theta_1 e_{t-1} - \phi_1 e_{t-1} + \theta_1 \phi_1 e_{t-2}$$

or

(e) $\dot{Y}_t - \phi_1 \dot{Y}_{t-1} = (e_t - \phi_1 e_{t-1}) - \theta_1 (e_{t-1} - \phi_1 e_{t-2})$

The terms in parentheses on the right-hand side of (e) are, using equation (10.40), just a_t and a_{t-1}, respectively. Therefore the residual autocorrelations suggest that the correct time series model is an ARMA(1, 1) model of the form

$$\dot{Y}_t - \phi_1 \dot{Y}_{t-1} = a_t - \theta_1 a_{t-1}$$

where now the a_t's denote uncorrelated normal random variables with mean zero and *not* the MA(1) model that was originally considered.

Overfitting

Overfitting is concerned with testing the adequacy of the time series model by introducing more parameters into the model than may be indicated by the sample autocorrelation function and seeing whether or not a *significant* reduction in the residual sum of squares is obtained. This appears, at first glance, to be a sensible idea since the sample autocorrelation function provides only a rough indication of the appropriate time series model, and moreover we know that there are several processes that are capable of producing similar sample autocorrelations (recall Figure 10.4). For example, an AR(1) model may be suggested by an inspection of the sample autocorrelation function, and in the interest of parsimony this model may be fitted, and an examination of the residual autocorrelations may reveal no apparent inadequacy. However, we know AR(2) processes can be expected to produce sample autocorrelation functions similar to those produced by AR(1) processes in certain cases, and it may be that the addition of a second autoregressive parameter will lead to a significant decrease in the residual sum of squares. That is, a significant amount of the residual variation from the first model may be explained by the addition of a second-order autoregressive term in the model.

Extreme care is called for in overparameterizing a model, and there should be a good reason for adding more parameters to the original model. For example, adding parameters simultaneously to *both* sides of an ARMA model may lead to estimation difficulties if a nonlinear estimation algorithm is being used—in particular, the algorithm may not "converge" to the least squares values. Also, even if least squares values are obtained and a subsequent analysis indicates that the additions are not needed, we shall not have proved that the original model is correct. We may have merely created a new model that is essentially no different from the old one in the sense that the parameters being added simultaneously to each side "cancel" one another out.

10.8   FORECASTING

Like regression analysis one of the major objectives in time series analysis is to predict future observations since, very often, vital policy decisions are based on the "best guess" of future outcomes. This is particularly true in business and economics. Corporate planners must concern themselves with predicting future sales, earnings, costs, expenditures, and other relevant quantities. Economists involved in formulating and investigating the ramifications of a particular monetary or fiscal policy must concern themselves with forecasting quantities like the unemployment rate, interest rates, and gross national product (GNP).

Once an adequate time series model is in hand it can be used to make inferences about future values. A word of warning, however, is in order. It was pointed out in Chapter 9 that forecasting procedures based on a model fitted to an observed time series necessarily involves the extrapolation of historical data. To the extent that the future behaves like the past and the model correctly describes the stochastic process of interest, good forecasts can be obtained. However, it should always be kept in mind that the nature of the stochastic process may change in time and the current time series model may no longer be appropriate. If this happens, forecasts computed from the inadequate model may be misleading. This is particularly true for forecasts with a long lead time. Therefore we recommend a periodic monitoring of the time series model, for the model as well as the forecasts should be updated as new information (data) becomes available.

At this point then we assume that identification, estimation, and diagnostic checking have been completed and an adequate time series model has been obtained. Initially we assume the values of the model parameters are known exactly. Later, of course, we will insert the least squares estimates of the parameter values.

We let y_N denote the last value of the time series, and suppose we are interested in forecasting the value that will be observed l time periods ($l > 0$)

in the future, y_{N+l}; that is, at forecast origin $t = N$ we are interested in forecasting the future value y_{N+l}. We denote the forecast of y_{N+l} by $\hat{y}_N(l)$, where the subscript denotes the forecast origin and the number in parentheses denotes the lead time. It can be shown (see reference [1]) that the "best" forecast of y_{N+l} is given by the *expected value* of Y_{N+l} *at time N*, where "best" is defined as that forecast which minimizes the expected (mean) squared error, $E_N[Y_{N+l} - \hat{y}_N(l)]^2$.

The expected value of Y_{N+l} at the forecast origin N is really a conditional expectation since, in general, it will depend on information already available. Thus using the notation introduced in Chapter 9 for this expectation, we have

$$\hat{y}_N(l) = E_N(Y_{N+l}) \qquad (10.41)$$

Equation (10.41) says that to obtain the forecast $\hat{y}_N(l)$ of y_{N+l}, we write down the expression for Y_{N+l} (using the time series model) and evaluate the expected value at time N. This procedure has already been illustrated for the case of an AR(1) model in Chapter 9. The following examples illustrate the procedure for other models representing both stationary and nonstationary processes.

EXAMPLE 10.6

Consider the AR(2) model

$$\dot{Y}_t - \phi_1 \dot{Y}_{t-1} - \phi_2 \dot{Y}_{t-2} = e_t$$

where $\dot{Y}_t = Y_t - \mu$. Setting $t = N + l$ in the equation above and solving for Y_{N+l} gives

$$Y_{N+l} = \mu + \phi_1(Y_{N+l-1} - \mu) + \phi_2(Y_{N+l-2} - \mu) + e_{N+l} \qquad (10.42)$$

In particular for $l = 1$,

$$Y_{N+1} = \mu + \phi_1(Y_N - \mu) + \phi_2(Y_{N-1} - \mu) + e_{N+1} \qquad (10.43)$$

Now at the forecast origin, N, the random variables Y_{N-1} and Y_N have been observed, the observed values being the last two observations, y_{N-1}, y_N, in the time series. Therefore, given values for the model parameters, the only unknown quantity on the right-hand side of equation (10.43) is e_{N+1}, the shock entering the system during the next time period. Thus the expected value of Y_{N+1} at time N, given all the information up to and including this time point, is

$$E_N(Y_{N+1}) = E_N\{\mu + \phi_1(y_N - \mu) + \phi_2(y_{N-1} - \mu) + e_{N+1}\}$$
$$= \mu + \phi_1(y_N - \mu) + \phi_2(y_{N-1} - \mu) + E_N(e_{N+1})$$

By assumption

$$E_N(e_{N+1}) = 0$$

so the best forecast of the next observation y_{N+1} is given by

$$\hat{y}_N(1) = E_N(Y_{N+1}) = \mu + \phi_1(y_N - \mu) + \phi_2(y_{N-1} - \mu) \tag{10.44}$$

Of course, in practice μ, ϕ_1, and ϕ_2 are replaced by their least squares estimates. Setting $l = 2$ in equation (10.42) gives

$$Y_{N+2} = \mu + \phi_1(Y_{N+1} - \mu) + \phi_2(Y_N - \mu) + e_{N+2}$$

Since Y_N has been observed,

$$\hat{y}_N(2) = E_N(Y_{N+2}) = \mu + \phi_1[E_N(Y_{N+1}) - \mu] + \phi_2(y_N - \mu) + E_N(e_{N+2})$$

$$= \mu + \phi_1[\hat{y}_N(1) - \mu] + \phi_2(y_N - \mu) \tag{10.45}$$

Therefore given values for the parameters, the last observation in the time series, y_N, and the *forecast*, $\hat{y}_N(1)$, of the *next* observation, the forecast of the future value, y_{N+2}, can be computed. Continuing in this way the forecast of any future value may be obtained. The reader should have no trouble seeing, for example, that

$$\hat{y}_N(3) = E_N(Y_{N+3}) = \mu + \phi_1[E_N(Y_{N+2}) - \mu] + \phi_2[E_N(Y_{N+1}) - \mu] + E_N(e_{N+3})$$

$$= \mu + \phi_1[\hat{y}_N(2) - \mu] + \phi_2[\hat{y}_N(1) - \mu]$$

and, in general,

$$\hat{y}_N(l) = \mu + \phi_1[\hat{y}_N(l-1) - \mu] + \phi_2[\hat{y}_N(l-2) - \mu], \qquad l \geq 3 \tag{10.46}$$

The reader will note that the forecasts in this case are generated recursively and thus can be obtained very easily and quickly, particularly if an electronic computer is employed.

To illustrate the forecasting methodology more completely we will calculate the forecasts of the next three observations in the unemployment rate series pictured in Figure 10.3 assuming we are at the end of the series. The reader will recall that this series consisted of $N = 61$ observations and appeared to be adequately described by an AR(2) model. The last two observations in this series are $y_{60} = 5.53$ and $y_{61} = 5.77$. Using equation (10.44) with the least squares estimates given in Table 10.5 substituted for the true parameter values, the forecast of the next observation is

$$\hat{y}_{61}(1) = \mu + \phi_1(y_{61} - \mu) + \phi_2(y_{60} - \mu)$$

$$= 4.97 + 1.56(5.77 - 4.97) - 0.72(5.53 - 4.97) = 5.82$$

Similarly, using equations (10.45) and (10.46),

$$\hat{y}_{61}(2) = \mu + \phi_1[\hat{y}_{61}(1) - \mu] + \phi_2(y_{61} - \mu)$$

$$= 4.97 + 1.56(5.82 - 4.97) - 0.72(5.77 - 4.97) = 5.72$$

and

$$\hat{y}_{61}(3) = \mu + \phi_1[\hat{y}_{61}(2) - \mu] + \phi_2[\hat{y}_{61}(1) - \mu]$$

$$= 4.97 + 1.56(5.72 - 4.97) - 0.72(5.82 - 4.97) = 5.54$$

The three forecasts above may be compared with the actual unemployment rates for the last three quarters of 1963, which were 5.77, 5.53, and 5.67 percent, respectively.

EXAMPLE 10.7

Consider the MA(1) model

$$\dot{Y}_t = e_t - \theta_1 e_{t-1}$$

where $\dot{Y}_t = Y_t - \mu$. Setting $t = N + l$ and solving for Y_{N+l} gives

$$Y_{N+l} = \mu + e_{N+l} - \theta_1 e_{N+l-1} \tag{10.47}$$

Now for any $l > 1$, the expected values, at time N, of the errors e_{N+l-1} and e_{N+l} are zero by assumption. Therefore

$$\hat{y}_N(l) = E_N(Y_{N+l}) = \mu, \qquad \text{for } l > 1 \tag{10.48}$$

that is, the best forecast of any future value beyond the first one is simply the process mean. In practice μ is replaced by its least squares estimate $\hat{\mu}$.

For $l = 1$ the situation is somewhat different. In this case equation (10.47) becomes

$$Y_{N+1} = \mu + e_{N+1} - \theta_1 e_N$$

and at time N, the error e_N will have been observed. Since the true values of the model parameters are not available, the magnitude of the observed error cannot be determined exactly. However, an *estimate* of the observed value of e_N is given by the residual $\hat{e}_N$. Substituting $\hat{e}_N$ for e_N and taking the conditional expectation at time N gives the forecast of the next observation

$$\hat{y}_N(1) = E_N(Y_{N+1}) = \mu - \theta_1 \hat{e}_N \tag{10.49}$$

Again in practice least squares estimates are substituted for the true parameter values.

EXAMPLE 10.8

The IMA(1, 1) model is given by

$$Y_t - Y_{t-1} = e_t - \theta_1 e_{t-1}$$

Setting $t = N + l$, solving for Y_{N+l}, and taking the expectation of Y_{N+l} at time N gives

$$\hat{y}_N(l) = E_N(Y_{N+l}) = E_N(Y_{N+l-1}) + E_N(e_{N+l}) - \theta_1 E_N(e_{N+l-1}) \tag{10.50}$$

which for $l > 1$ reduces to

$$\hat{y}_N(l) = \hat{y}_N(l-1) \tag{10.51}$$

since

$$E_N(e_{N+l-1}) = E_N(e_{N+l}) = 0$$

that is, successive forecasts are equal. Equivalently, the forecasts lie along a horizontal line whose intercept is given by $\hat{y}_N(1)$.

For $l = 1$,

$$\hat{y}_N(1) = y_N - \theta_1 e_N$$

where e_N represents the *observed* error entering the system at time N. The quantities θ_1 and e_N can be estimated by the least squares value $\hat{\theta}_1$ and the residual $\hat{e}_N$, respectively.

The procedure for obtaining minimum mean square error forecasts for any ARIMA process can be summarized as follows.

1.   Using the fitted mathematical model, solve for Y_{N+l}, where N denotes the forecast origin and l denotes the lead time.
2.   Obtain the expression for the expectation of Y_{N+l} conditional on all the information available up to and including the forecast origin using the relationships

$$E_N(Y_{N+j}) = \begin{cases} \hat{y}_N(j), & j > 0 \\ y_{N+j}, & j \leq 0 \end{cases} \tag{10.52}$$

and

$$E_N(e_{N+j}) = \begin{cases} 0, & j > 0 \\ e_{N+j}, & j \leq 0 \end{cases} \tag{10.53}$$

where e_{N+j} represents the *observed* error for $j \leq 0$.
3.   With the least squares estimates inserted for the true parameter values and the observed errors, e_{N+j}, $j \leq 0$, estimated by the corresponding residuals, if necessary, evaluate the forecast, $\hat{y}_N(1)$, of the next observation, y_{N+1}. Generate the subsequent forecasts recursively using the expression for the conditional expectation derived in step 2.

Probability Limits

Estimates of future values supplied by the forecasts $\hat{y}_N(l)$, $l = 1, 2, \ldots$, are of little use unless some indication of their variability is provided. This variability is given by the variance of the forecast error, where, in general, the forecast error, $e_N(l)$, associated with a prediction at time N of an observation l time periods in the future is defined by

$$e_N(l) = Y_{N+l} - \hat{y}_N(l) \tag{10.54}$$

We note that at the forecast origin N, the error $e_N(l)$ is a random variable since the value of Y_{N+l} has not yet been observed.

In order to evaluate the variance of $e_N(l)$, it is necessary to express the forecast error *entirely* in terms of the errors that enter the system at times $N + 1, N + 2, \ldots, N + l$. It has been suggested in Section 10.3 that it is always possible to express the process variable Y_t appearing in an ARIMA model entirely in terms of the "past" errors, $e_t, e_{t-1}, \ldots$. (If an autoregressive term is present, this will require an infinite number of e's.) In particular it is always possible to write[1] (see Supplement 10C)

$$Y_t = e_t + \psi_1 e_{t-1} + \psi_2 e_{t-2} + \psi_3 e_{t-3} + \cdots \qquad (10.55)$$

where the coefficients $\psi_1, \psi_2, \ldots$ are functions of the autoregressive and moving average parameters. As an example, for the IMA(1, 1) process we have shown [see equation (10.21)], that

$$Y_t = e_t + (1 - \theta_1)e_{t-1} + (1 - \theta_1)e_{t-2} + \cdots$$

and thus

$$\psi_j = (1 - \theta_1), \qquad \text{for } j = 1, 2, 3, \ldots$$

A general method for determining the ψ coefficients for any ARIMA model is discussed in Supplement 10C.

If we set $t = N + l$ in equation (10.55) then

$$Y_{N+l} = e_{N+l} + \psi_1 e_{N+l-1} + \psi_2 e_{N+l-2} + \psi_3 e_{N+l-3} + \cdots + \psi_{l-1} e_{N+1} + \psi_l e_N + \cdots$$
$$(10.56)$$

Now at time N, the errors $e_N, e_{N-1}, e_{N-2}, \ldots$ have been observed, and the yet to be observed errors $e_{N+l}, e_{N+l-1}, e_{N+l-2}, \ldots$ have an expected value of zero. Therefore it is possible to write the forecast $\hat{y}_N(l)$ in terms of the observed errors. In particular

$$\hat{y}_N(l) = E_N(Y_{N+l}) = \psi_l e_N + \psi_{l+1} e_{N-1} + \psi_{l+2} e_{N-2} + \cdots \qquad (10.57)$$

Using the formulations given in equations (10.56) and (10.57), the forecast error can then be written as

$$e_N(l) = Y_{N+l} - \hat{y}_N(l) = e_{N+l} + \psi_1 e_{N+l-1} + \psi_2 e_{N+l-2} + \cdots + \psi_{l-1} e_{N+1}$$
$$(10.58)$$

and hence

$$\text{Var}[e_N(l)] = (1 + \psi_1^2 + \psi_2^2 + \cdots + \psi_{l-1}^2)\sigma_e^2 \qquad (10.59)$$

since the errors are uncorrelated by assumption. The variance of the forecast error can be used to construct probability limits as follows.

Given the information (that is, the time series) up to and including time N, the forecast origin, Y_{N+l}, is normally distributed (by assumption) about its expected value $\hat{y}_N(l)$ with a variance equal to the variance of the forecast error. Therefore probability limits for a future observation can be constructed

---

[1] If the process is stationary the left-hand side of equation (10.55) is the deviation of Y_t from its mean, $\dot{Y}_t$.

using percentage points of an $N(0, 1)$ distribution. For example, 95-percent probability limits for Y_{N+l} are given by

$$\hat{y}_N(l) \pm 1.96\sqrt{1 + \psi_1^2 + \psi_2^2 + \cdots + \psi_{l-1}^2}\,\sigma_e \tag{10.60}$$

and the general $100(1 - \gamma)$-percent probability limits for Y_{N+l} are

$$\hat{y}_N(l) \pm z_{\gamma/2}\sqrt{1 + \psi_1^2 + \psi_2^2 + \cdots + \psi_{l-1}^2}\,\sigma_e \tag{10.61}$$

where $z_{\gamma/2}$ is the deviate exceeded by a proportion $\gamma/2$ of the standard normal distribution. The probability limits are interpreted in the following manner. Given the information at *time N*, the probability is $1 - \gamma$ that the observed value of Y_{N+l}, when it occurs, will lie between them. Probability limits apply to *individual* forecasts. If probability limits with a fixed probability content are calculated for several lead times, it is *not* true that the series can be expected to lie within *all* the limits simultaneously with probability $1 - \gamma$.

The reader will note that with the convention that $\psi_0 = 1$, equation (10.61) is applicable for any lead time l, and for a given probability level the width of the probability interval increases as the lead time increases since positive terms are added under the radical sign. For stationary series, the contributions of the ψ_j^2 terms are negligible for large j, and the probability limits are eventually horizontal bands. For nonstationary series, however, the magnitudes of the ψ_j^2 terms are not negligible for large j, and the probability limits continue to expand as the lead time increases. This is certainly reasonable since we would expect the uncertainty associated with forecasts to increase as we look farther and farther into the future—particularly if the process being forecasted is nonstationary.

In practice approximate probability limits may be calculated from equation (10.61) by substituting the least squares estimates of the autoregressive and moving average parameters into the expressions for $\psi_1, \psi_2, \ldots, \psi_{l-1}$ to get estimated coefficients $\hat{\psi}_1, \hat{\psi}_2, \ldots, \hat{\psi}_{l-1}$ and using the square root of the residual mean square, $\hat{\sigma}_e$, to estimate σ_e.

Figure 10.13 shows the forecasts of the next six observations and approximate 95-percent probability limits for the refrigerator sales series, the Treasury bill interest rate series, and the unemployment rate series, respectively. In addition the *actual values* for the next six observations for the unemployment rate series are shown.

Updating Forecasts

When new information becomes available in the form of more observations, it can be used to "update" old forecasts. In particular, using the formulation already introduced, it can be shown that

$$\hat{y}_{N+1}(l) = \hat{y}_N(l + 1) + \psi_l e_N(1) \tag{10.62}$$

that is, for any ARIMA model, the N-origin forecast of y_{N+l+1} is updated to

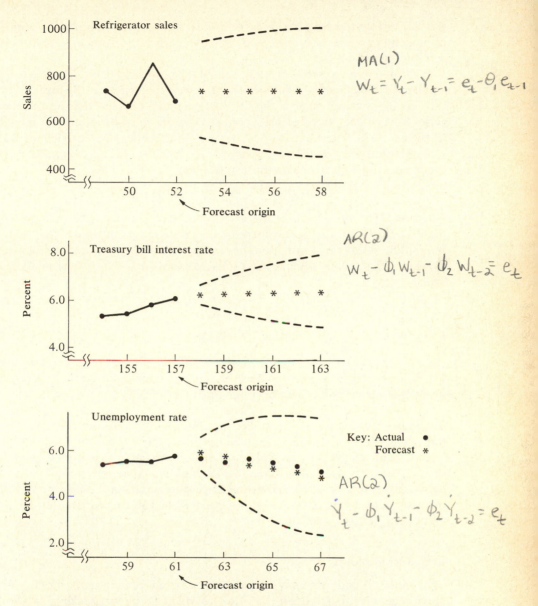

Figure 10.13 Forecasts and 95-percent probability limits

become the $(N + 1)$-origin forecast of the same observation by the addition of a constant multiple of the "one-step-ahead" forecast error $e_N(1) = y_{N+1} - \hat{y}_N(1)$ which, of course, is easily calculated once the *next* observation y_{N+1} is available.

This procedure for updating forecasts has already been demonstrated for the case of a first-order autoregressive process in Chapter 9. Another example of updating forecasts is provided by considering the forecasts of future refrigerator sales.

We showed in Example 10.8 that for IMA(1, 1) processes the forecasts lie along a horizontal straight line whose intercept is the value of the forecast of the *next* observation made at the forecast origin N. For the refrigerator sales series $\hat{y}_N(1) = 742$, and the corresponding straight line of forecasts is shown in Figure 10.13. The forecasts are computed at the forecast origin $N = 52$.

In addition we have shown that $\psi_j = (1 - \theta_1)$, $j = 1, 2, \ldots$, for IMA(1, 1) processes, and therefore for these processes

$$\hat{y}_{N+1}(l) = \hat{y}_N(l + 1) + (1 - \theta_1)e_N(1) \tag{10.63}$$

where $e_N(1) = y_{N+1} - \hat{y}_N(1)$ is the one-step-ahead forecast error at time N. Now we shall suppose that refrigerator sales in the 53 period were 700, that is, $y_{N+1} = 700$. With this information the remaining old forecasts can be updated using equation (10.63) with the least squares estimate $\hat{\theta}_1 = 0.59$ inserted for θ_1. Thus

$$\hat{y}_{N+1}(l) = \hat{y}_N(l + 1) + (1 - 0.59)[y_{N+1} - \hat{y}_N(1)]$$
$$= 742 + 0.41(700 - 742)$$
$$= 724.8, \qquad \text{for all } l$$

since the forecasts, $\hat{y}_N(l)$, made at time N are in this case the same for any lead time; that is, given the new observation, the updated forecasts lie along a horizontal straight line with an intercept of 724.8.

At this point a word of caution is in order. Continual updating from a *previously fitted* time series model may produce misleading results if the underlying stochastic process should change over time. Therefore it is always a good idea, after several new observations are in hand, to apply the model-building strategy once again and either derive a new time series model, or if a new model is not required, to reestimate the parameters in the old model. The time series model should be monitored just as the forecasts are monitored as more data become available.

Comparing Forecasts

In certain situations it may happen that two or more time series models (or, more generally, models of any kind) are not ruled out by the available data and the decision maker may want to discriminate among them on the basis of their forecasting accuracy and structural stability. Various measures for discriminating among time series models have been proposed, and most of them involve some function of the *one-step-ahead* forecast errors. We shall consider one such measure known as the "Janus quotient."

The Janus quotient, proposed originally by Gadd and Wold [29], compares the variance of the one-step-ahead forecast errors in the prediction period to the variance of the errors (residuals) in the observation period. Thus if the entire period of interest is divided up as

$$\left| \begin{array}{c|c} t = 1, 2, \ldots, N & t = N+1, N+2, \ldots, N+m \\ \leftarrow \text{observation period} \rightarrow & \leftarrow \qquad \text{prediction period} \qquad \rightarrow \end{array} \right|$$

the Janus quotient is given by

$$J = \frac{(1/m)\Sigma_{t=N+1}^{N+m}[y_t - \hat{y}_{t-1}(1)]^2}{(1/N)\Sigma_{t=1}^{N}(y_t - \hat{y}_t)^2} \qquad (10.64)$$

The numerator in the Janus quotient is simply the average squared one-step-ahead forecast error[m] (the "variance" of the forecast errors), whereas the denominator is the average squared residual (the "variance" of the fitting errors).

Thus the accuracy of the forecasts can be determined by examining the *numerator* of the Janus quotient, and the structural stability of the time series model can be determined by examining the quotient itself; for if the underlying stochastic process remains stable over time, the magnitude of the Janus quotient should be close to unity. A value for the Janus quotient much larger than unity may indicate an inadequate model[n] since the magnitudes of the forecast errors tend to be larger than the magnitudes of the "fitting errors."

When comparing two models with comparable denominators, the one having the smaller Janus quotient would ordinarily be preferred.

Simple Exponential Smoothing

We pointed out in Section 10.3 that for IMA(1, 1) processes we can express the random variable Y_t entirely in terms of previous Y's and the current error e_t. In particular using equation (10.22) with $t = N + 1$ we have

$$Y_{N+1} = (1 - \theta_1)\{Y_N + \theta_1 Y_{N-1} + \theta_1^2 Y_{N-2} + \cdots\} + e_{N+1} \qquad (10.65)$$

Now at time N, the random variables inside the parentheses in equation (10.65) will have been observed, so that if we evaluate the expected value of Y_{N+1} at the forecast origin N, we obtain

$$E_N(Y_{N+1}) = \hat{y}_N(1) = (1 - \theta_1)y_N + \theta_1(1 - \theta_1)y_{N-1} + \theta_1^2(1 - \theta_1)y_{N-2} + \cdots \qquad (10.66)$$

Because $|\theta_1| < 1$, the coefficients on the right-hand side of this expression decay exponentially, and therefore we see that the optimal forecast for the next observation can be written as an exponentially weighted sum of

---

[m] The one-step-ahead forecasts, $\hat{y}_{t-1}(1)$, appearing in the numerator of the Janus quotient can be computed for any ARIMA model using the updating formula given in equation (10.62).

[n] It may also indicate a changing environment. Therefore some care must be taken in interpreting a Janus quotient.

previous observations. To get the forecast of the next observation we "smooth" (average) the current and previous observations with weights that decay exponentially, and this procedure for computing forecasts is referred to as "simple exponential smoothing."

The exponential smoothing predictor is intuitively appealing because of its weighting pattern, which gives the greatest weight to the most recent observation and yet does not completely ignore the information in the older data. Even though the right-hand side of equation (10.66) is (theoretically) an infinite series, if N is moderately large and $|\theta_1|$ is not close to unity, the weights die out fairly rapidly, and the contribution to the sum of the observations near the beginning of the series can usually be neglected.

The expression for $\hat{y}_N(1)$ in equation (10.66) is not the most convenient form for the exponential smoothing predictor. We have shown that the forecasts made at time N for IMA(1, 1) processes lie along a horizontal line with intercept $\hat{y}_N(1)$; that is, $\hat{y}_N(l) = \hat{y}_N(1)$ for $l \geq 2$. In addition we have noted that the forecasts may be updated using the relationship

$$\hat{y}_{N+1}(l) = \hat{y}_N(l+1) + (1 - \theta_1)e_N(1)$$

where $e_N(1) = y_{N+1} - \hat{y}_N(1)$ is the one-step-ahead forecast error. If we now set $l = 1$ and substitute for $e_N(1)$ in the updating expression, we get

$$\hat{y}_{N+1}(1) = \hat{y}_N(2) + (1 - \theta_1)[y_{N+1} - \hat{y}_N(1)]$$

However $\hat{y}_N(2) = \hat{y}_N(1)$ since the forecasts at time N lie along a horizontal line; therefore after substituting for $\hat{y}_N(2)$ and rearranging terms, we have the basic equation of simple exponential smoothing

$$\hat{y}_{N+1}(1) = (1 - \theta_1)y_{N+1} + \theta_1\hat{y}_N(1) \tag{10.67}$$

Thus the forecast of the next observation at time $N + 1$ is simply a weighted sum of the observation y_{N+1} and the forecast of y_{N+1} made at time N. This method of computing exponentially smoothed forecasts is ideally suited for electronic computers since it is recursive in nature and requires at any time $N + 1$ only the current observation and the previous forecast.

Because of its intuitively appealing nature and its ease of calculation [using equation (10.67)], exponential smoothing predictors are extremely popular and have been employed in a wide variety of situations. We should point out, however, that simple exponential smoothing predictors produce optimal forecasts for a particular stochastic process, the IMA(1, 1) process; and when they are used in situations in which an IMA(1, 1) model is not appropriate, they may no longer be the best forecasts that could be obtained. Optimal forecasts follow directly from the form of the *appropriate* time series model, so the majority of the effort in time series analysis should be directed toward model building; forecasts are immediate by-products.

We note in closing that forecasts of future values may be improved by the use of auxiliary information. For example, movement in the time series of interest (say, refrigerator sales) may be preceded by movement in a related series° (for example, housing starts), and it seems reasonable to assume that the information in the latter series may be combined with the information in the original series to provide better forecasts than may be obtained from the original series itself. In fact there are procedures for expanding the ARIMA models to allow for the dependence of the original process on a related process. The resulting models are called "dynamic-noise" models, and the reader interested in pursuing this topic is referred to references [1] and [17].

10.9   FINAL COMMENTS

We have presented a methodology for time series analysis that has enjoyed considerable success in forecasting and control applications. The availability of computer programs incorporating the "Box-Jenkins" time series methodology considerably eases the computational burdens implied by the identification, fitting, diagnostic checking, and forecasting procedures presented in this chapter.

The time series models and forecasting procedures presented thus far can be easily modified to handle "seasonal" stochastic processes. This is the subject of the next chapter.

° The related series is called a "leading indicator" by economists.

Supplement 10A

STATIONARY STOCHASTIC PROCESSES AND THEIR PROPERTIES

A phenomenon that evolves over time in accordance with certain probability laws is called a "stochastic process." A time series is nothing more than one particular realization of a stochastic process. For example, the sequence of observations in Figure 10.1 (that is, the refrigerator sales data) represents one member of a set of possible sequences that *might* have been observed over the same time period.

Thus in analyzing a time series we regard the observation y_t at a given time t as a realization of a random variable Y_t with probability density function $f(y_t)$. Similarly the observations at any two times t_1 and t_2 may be regarded as realizations of two random variables Y_{t_1} and Y_{t_2} with joint probability density function $f(y_{t_1}, y_{t_2})$. In general the entire sequence of observations $y_{t_1}, y_{t_2}, \ldots, y_{t_N}$ making up the time series may be regarded as a realization of an N-dimensional random variable $(Y_{t_1}, Y_{t_2}, \ldots, Y_{t_N})$ with probability density $f(y_{t_1}, y_{t_2}, \ldots, y_{t_N})$.

Stationary Stochastic Processes

A "stationary time series" is a time series whose statistical properties do not change over time. More precisely we define a stationary time series as a particular realization of a *stochastic process* whose statistical properties do not change over time. A stochastic process in such a state of statistical equilibrium is referred to as a *stationary stochastic process*. We point out that there are various degrees of stationarity. If the statistical properties of the stochastic process are not affected by a change of time origin, the process is said to be strictly stationary. Mathematically we can characterize a strictly stationary stochastic process by the following definition.

Definition 10A1   A *stochastic process* is said to be *strictly stationary* if the joint probability density associated with the N random variables

$Y_{t_1}, Y_{t_2}, \ldots, Y_{t_N}$, for *any* set of times $t_1, \ldots, t_N$, is the same as that associated with the N random variables $Y_{t_1+k}, Y_{t_2+k}, \ldots, Y_{t_N+k}$, where k is any integer.

Thus a discrete stochastic process is strictly stationary if the entire probability structure associated with any set of N observations remains unchanged when the time of observation is shifted either forward or backward by an integer amount k.

Stochastic processes for which only the first- and second-order moments of the random variables $Y_{t_1}, Y_{t_2}, \ldots, Y_{t_N}$ are unaffected by a shift in the time origin are said to be *weakly stationary*, or second-order stationary. In this book we shall assume that the joint density associated with any set of random variables $Y_{t_1}, Y_{t_2}, \ldots, Y_{t_N}$ is the multivariate normal.[a] Since the multivariate normal density is completely specified as soon as its first- and second-order moments are known, second-order stationarity, along with the normality assumption, is enough to ensure strict stationarity, and henceforth we will delete the word "strict" when referring to stationary processes.

A stationary stochastic process, then, can be completely described by its first- and second-order moments, that is, by its mean and covariance function.

The Mean and the Covariance Function

We shall consider the single random variable Y_{t_1} at the particular time point t_1. The conditions for stationarity imply that the probability density function of Y_{t_1}, $f(y_{t_1})$, must be the same as that of the random variable Y_{t_1+k} for any integer k. In particular for $k = 1, f(y_{t_1}) = f(y_{t_1+1})$; that is, the probability density of Y_{t_1} is exactly the same as the probability density of the random variable one time period to the right. Continuing in this way (that is, letting k assume other integer values) it is easy to see that for stationary processes

$$f(y_{t_1}) = f(y_{t_2}) = \cdots = f(y_{t_N})$$

for *any* set of distinct times $t_1, t_2, \ldots, t_N$; that is, each of the Y_t's has exactly the same probability function. Setting $t_2 = t_1 + 1, t_3 = t_1 + 2, \ldots, t_N = t_1 + N$, we have that each member of any set of N *consecutive* random variables has the same probability function. Since, for stationary processes, the probability density $f(y_t)$ is the same for all times t, it follows that the mean and variance of Y_t must be the same for all t. This leads to the following definitions.

Definition 10A2 The *mean of a stationary stochastic process* will be denoted by μ and is given by

$$\mu = E(Y_t) = \int_{-\infty}^{\infty} y_t f(y_t) \, dy_t$$

for any time t.

[a] The multivariate normal distribution is a generalization of the bivariate normal distribution discussed in Section 2.3.

This definition implies that a stationary stochastic process has a constant mean or level about which the process fluctuates.

Definition 10A3   The *variance of a stationary stochastic process* will be denoted by σ_Y^2 and is given by

$$\sigma_Y^2 = E(Y_t - \mu)^2 = \int_{-\infty}^{\infty} (y_t - \mu)^2 f(y_t)\, dy_t$$

for any time t.

The variance is a measure of the fluctuation of the stochastic process about its level. The concepts of mean and variance for a stochastic process are completely analogous to the concepts of mean and variance we introduced earlier for a single random variable. The distinction is that when we talk about a stochastic process we are talking about a *chronological sequence of random variables* and we must define the mean and variance for each random variable in the sequence. If the process is stationary, the means and, similarly, the variances are all the same.

We have pointed out that each observation, y_t, of a time series is just a realization of the corresponding random variable Y_t of the stochastic process, for any time t. For most processes met in practice, given a set of consecutive observations, $y_1, y_2, \ldots, y_N$, the mean and variance of the stochastic process can be *estimated* by

$$\bar{y} = \frac{1}{N} \sum_{t=1}^{N} y_t$$

and

$$\hat{\sigma}_Y^2 = \frac{1}{N} \sum_{t=1}^{N} (y_t - \bar{y})^2$$

The stationarity assumption also implies that the joint density function, $f(y_{t_1}, y_{t_2})$, associated with *any* pair of random variables Y_{t_1}, Y_{t_2} separated by a constant time interval, is the same for all times t_1 and t_2. Setting $t_1 = t$ and $t_2 = t + k$ (k an integer), it follows that a measure of the linear relationship between any two random variables Y_t, Y_{t+k} is provided by the covariance of Y_t, Y_{t+k}. Since the random variables Y_t, Y_{t+k} are part of the same stochastic process—that is, they represent the probabilistic behavior of the same physical mechanism at different time points—the covariance is a measure of the linear relationship existing between pairs of random variables within the stochastic process itself and is called the *autocovariance*.

Definition 10A4   The *autocovariance* of the random variables Y_t, Y_{t+k}, separated by a constant time interval or lag k, is denoted by γ_k and is given by[b]

---

[b] The integral that follows is like a double summation. The double integral can be evaluated by integrating first with respect to y_{t+k} and then with respect to y_t.

$$\gamma_k = E(Y_t - \mu)(Y_{t+k} - \mu) = \int_{-\infty}^{\infty} \int_{-\infty}^{\infty} (y_t - \mu)(y_{t+k} - \mu) f(y_t, y_{t+k}) \, dy_t \, dy_{t+k}$$

for any time t.

The autocovariance is a measure of the linear dependency between two random variables separated by a fixed number of time periods or lags k. For example, γ_1 is a measure of the linear relationship between adjacent random variables.

The concept of autocovariance is completely analogous to the concept of covariance introduced in Chapter 2 for *any* pair of random variables X and Y. In particular it follows that the nature of the autocovariance at lag k can usually be inferred from a scatter diagram constructed using the pairs of observations (y_t, y_{t+k}), $t = 1, \ldots, N - k$, of the time series.

The Autocovariance and Autocorrelation Functions

We have defined the autocovariance γ_k. For $k = 0$, $\gamma_0 = E(Y_t - \mu)^2 = \sigma_Y^2$, the variance of the stochastic process. A plot of γ_k as a function of the lag k is called the *autocovariance function*. It is clear that the autocovariance is measured in the square of the units of the random variables Y_t, Y_{t+k}, so that it is difficult to interpret the magnitude of this number. The reader will recall that a measure of the linear dependence between any two random variables which does not depend on the units of the random variables is provided by the correlation coefficient ρ. In a similar manner we can define a standardized measure of the linear dependence between Y_t and Y_{t+k} known as the *autocorrelation*.

Definition 10A5   The *autocorrelation* of the random variables Y_t, Y_{t+k}, at lag k, is denoted by ρ_k and is given by

$$\rho_k = \frac{\text{Cov}(Y_t, Y_{t+k})}{\sqrt{\text{Var}(Y_t) \, \text{Var}(Y_{t+k})}} = \frac{\gamma_k}{\sqrt{\gamma_0^2}} = \frac{\gamma_k}{\gamma_0}$$

for any time t.

For stationary processes the autocovariance, γ_k, and hence the autocorrelation, ρ_k, is a function only of the time difference or lag k separating the random variables, Y_t and Y_{t+k}, for all t.

When viewed as a function of the lag k, ρ_k is known as the *autocorrelation function*. We note that the autocovariance and hence the autocorrelation function is symmetric about $k = 0$; that is, for $k > 0$, $\gamma_k = \gamma_{-k}$ and thus $\rho_k = \rho_{-k}$. Therefore it is only necessary to plot ρ_k for $k > 0$ to get an indication of the linear dependence between random variables k time units apart.

Under suitable conditions different stationary stochastic processes produce different autocorrelation functions. If a particular autocorrelation structure can be uniquely associated with a particular stationary stochastic process, then knowledge of the autocorrelation function (or an estimate of it) can be used to help identify the process. This result serves as the basis for the identification procedures illustrated in Section 10.5.

Estimation of the Autocovariance and Autocorrelation Functions

Our discussion of autocovariance and autocorrelation up to now has been restricted to the theoretical quantities that characterize a conceptual stationary stochastic process. In practice the theoretical values γ_k and ρ_k are unknown and must be estimated from the available data, that is, from the time series. For example, if we want to estimate γ_1 or ρ_1—a measure of the linear dependence between *adjacent* random variables—it seems reasonable to construct an estimate using the adjacent pairs of observations (y_t, y_{t+1}) of the time series.

Such a measure will now be developed. We denote the N observations of a time series by $y_1, y_2, \ldots, y_N$ (that is, we set the time origin, $t_1 = 1$); then an estimate of $\rho_k, k > 0$, is given by[c]

$$r_k = \frac{c_k}{c_0}, \qquad k = 1, 2, \ldots$$

where

$$c_k = \frac{1}{N} \sum_{t=1}^{N-k} (y_t - \bar{y})(y_{t+k} - \bar{y}), \qquad k = 1, 2, \ldots$$

is an estimate of γ_k, $\bar{y}$ is the arithmetic average, and

$$c_0 = \frac{1}{N} \sum_{t=1}^{N} (y_t - \bar{y})^2$$

The estimated autocorrelations are most easily calculated with the aid of an electronic computer.

For stationary processes the theoretical autocorrelation function tends either to die out with increasing lag or to "cut off" after a particular lag $k = q$ (that is, $\rho_k = 0$ for all $k > q$). To identify a model for the stochastic process being investigated, it will be necessary to have a crude check on whether ρ_k is effectively zero for a given value of k. For this purpose a useful approximation (derived from a result due to M. S. Bartlett [12]) is available.

---

[c] Several estimators of the autocovariance and autocorrelation coefficients have been proposed by statisticians. Studies of the statistical properties of these estimators indicate that the estimates of ρ_k and γ_k given here are the most satisfactory.

Theorem 10A1   If the N observations $y_1, \ldots, y_N$ are realizations from a stationary stochastic process, where any sequence of N random variables making up the process has a multivariate normal probability distribution, then, for N large, the following results hold:

1. If $\rho_k = 0$ *for all* k,

$$\mathrm{Var}(r_k) \doteq \frac{1}{N}, \qquad k = 1, 2, \ldots$$

2. If $\rho_{q+v} = 0$, $v = 1, 2, 3, \ldots$; that is, if the autocorrelation coefficients are zero beyond some lag q $(q = 1, 2, \ldots)$, then

$$\mathrm{Var}(r_k) \doteq \frac{1}{N}\left(1 + 2\sum_{i=1}^{q} \rho_i^2\right), \qquad k > q$$

The results presented in this theorem, used in conjunction with the observed pattern of the estimated autocorrelations r_k, enable us to determine the likely pattern of the theoretical autocorrelations ρ_k and ultimately a suitable time series model.

EXAMPLE 10A1

We shall consider the collection of estimated autocorrelations given below which were calculated from a time series consisting of $N = 100$ observations:

k:	1	2	3	4	5	6	7	8	9	10
r_k:	-0.31	-0.05	0.08	0.01	-0.11	-0.02	0.05	0.00	0.00	-0.07

Suppose we wish to determine whether or not $\rho_k = 0$ for all k, that is, whether this particular time series is completely random.

Assuming $\rho_k = 0$ $(k \geq 1)$, we would expect the r_k's to be distributed about zero for all lags, and use of the theorem gives $\mathrm{Var}(r_k) \doteq 1/N = 1/100 = 0.01$. Hence the standard deviation is $\sqrt{\mathrm{Var}(r_k)} \doteq 0.10$ for all k. A nonzero value for ρ_k is then indicated if $|r_k|$ is "large" when compared with this standard deviation. For the autocorrelations displayed above, we see that $|r_1| = 0.31$ is over three times as large as its standard deviation, whereas the remaining estimated autocorrelations are small. Therefore it appears as if $\rho_1 \neq 0$ and $\rho_k = 0$ $(k \geq 2)$.

We might next ask whether the observed series represents a realization of a stochastic process such that $\rho_1 \neq 0$ and $\rho_k = 0$ $(k \geq 2)$. Under this assumption we have, using part 2 of the theorem with $q = 1$ and substituting r_1 for ρ_1,

$$\mathrm{Var}(r_k) \doteq \frac{1}{N}(1 + 2r_1^2) = \frac{1}{100}\{1 + 2(-0.31)^2\} = 0.012, \qquad k > 1$$

giving an estimated standard deviation of 0.11. Since the estimated autocorrelations for $k > 1$ are small compared with this estimated standard deviation, there is no reason to doubt the hypothesis: $\rho_1 \neq 0$, $\rho_k = 0$ $(k \geq 2)$.

Some care must be taken in the interpretation of individual r_k's because large covariances can exist between neighboring values. This effect can frequently distort the overall appearance of the autocorrelation function. For example, it may not die out as rapidly as expected. However this behavior is not generally harmful, for if ambiguities do arise in the interpretation of the estimated autocorrelations, these ambiguities will be resolved at a later stage of the investigation.

We remark that *estimated* autocorrelations always exist (can always be computed) regardless of whether or not the observed time series represents a realization of a stationary or nonstationary process. However, for nonstationary processes, the estimated autocorrelations fail to die out rapidly, and so calculation of the estimated autocorrelations not only aids in the identification of a particular kind of stationary process, they can also indicate if the underlying stochastic process is, in fact, a stationary process.

Supplement 10B

MORE GENERAL TIME SERIES MODELS

Moving Average Model of Order q

A moving average[a] model of order q, or MA(q) model, is defined by the expression

$$\dot{Y}_t = e_t - \theta_1 e_{t-1} - \theta_2 e_{t-2} - \cdots - \theta_q e_{t-q}$$

This expression is a straightforward generalization of the moving average models introduced in Section 10.3. We note that the current deviation $\dot{Y}_t$ depends on shocks that have entered the system during the preceding q time periods. The coefficients $\theta_1, \theta_2, \ldots, \theta_q$ must satisfy certain restrictions to ensure that the current value of the process variable does not depend overwhelmingly on those in the remote past. The nature of these restrictions is beyond the level of this book, and the interested reader is referred to reference [1].

For a moving average process of order q, the autocorrelations are given by

$$\rho_k = \frac{-\theta_k + \theta_1 \theta_{k+1} + \cdots + \theta_{q-k} \theta_q}{1 + \theta_1^2 + \cdots + \theta_q^2}, \qquad k = 1, 2, \ldots, q$$

$$\rho_k = 0, \qquad k > q$$

that is, the autocorrelations are zero (cut off) beyond the order q of the process.

---

[a] We note at this point that the expression "moving average" generally refers to a linear combination whose coefficients sum to one. This is not the case in "moving average" time series models; however the expression is commonly employed, and we will continue to use it.

Autoregressive Models of Order p

An autoregressive model of order p, or AR(p) model, is defined by the expression

$$\dot{Y}_t = \phi_1 \dot{Y}_{t-1} + \phi_2 \dot{Y}_{t-2} + \cdots + \phi_p \dot{Y}_{t-p} + e_t$$

where the parameters $\phi_1, \phi_2, \ldots, \phi_p$ satisfy certain constraints to ensure stationarity (see reference [1]).

In general a pth-order autoregressive model has the form of a linear regression model, where the dependent variable $\dot{Y}_t$ is regressed on the "independent" variables $\dot{Y}_{t-1}, \dot{Y}_{t-2}, \ldots, \dot{Y}_{t-p}$; that is, the deviations $\dot{Y}_{t-1}, \ldots, \dot{Y}_{t-p}$ for the p previous time periods all have an effect on the current deviation $\dot{Y}_t$.

Generalizing the methods employed for the first- and second-order autoregressive process, it can be shown that the autocorrelations for an AR(p) process satisfy the pth-order difference equation

$$\rho_k - \phi_1 \rho_{k-1} - \phi_2 \rho_{k-2} - \cdots - \phi_p \rho_{k-p} = 0$$

which has a solution of the form

$$\rho_k = A_1 \alpha_1^{\,k} + \cdots + A_p \alpha_p^{\,k}$$

where $\alpha_1, \ldots, \alpha_p$ are roots of $x^p - \phi_1 x^{p-1} - \cdots - \phi_p = 0$ with $|\alpha_i| < 1$, $i = 1, \ldots, p$.

Therefore the autocorrelation function may have a wide variety of appearances. It may exhibit geometric decay, oscillations, sinusoidal behavior, or combinations of these patterns.

Autoregressive-Moving Average Models of Order (p, q)

The autoregressive-moving average model of order (p, q), or ARMA(p, q) model, is defined by the relationship

$$\dot{Y}_t - \phi_1 \dot{Y}_{t-1} - \cdots - \phi_p \dot{Y}_{t-p} = e_t - \theta_1 e_{t-1} - \cdots - \theta_q e_{t-q}$$

where the integers p and q represent the order of the autoregressive and moving average term, respectively. The parameters $\phi_1, \phi_2, \ldots, \phi_p$ are subject to the same restrictions as the parameters of a "pure" autoregressive process of order p (to ensure stationarity), and the parameters $\theta_1, \theta_2, \ldots, \theta_q$ are subject to the same restrictions as the parameters of a "pure" moving average process of order q (to ensure invertibility).

The ARMA (p, q) model and its various special cases [for example, AR(p) and MA(q) models] constitute a large class of stationary time series models. Experience has indicated that these models are capable of representing a wide variety of commonly occurring stationary time series. They are extremely flexible and have the particular advantage of providing parsimonious representations.

General Autoregressive-Integrated-Moving Average Models of Order (p, d, q)

The autoregressive-integrated-moving average model of order p, d, q, or ARIMA(p, d, q) model, is a model for nonstationary series that assumes the dth differences process can be represented by an ARMA(p, q) model. Thus the integers (p, d, q) refer to the order of the autoregressive term, the amount of differencing, and the order of the moving average term, respectively. The term "integrated" has been inserted because although it is actually the differenced process (for example, $W_t = Y_t - Y_{t-1}$, $t = \ldots, -1, 0, 1, \ldots$) which is represented by an ARMA model, an expression for the original process, Y_t, can be obtained by "integrating" (that is, summing).

We shall let the dth difference of Y_t be denoted by $D^d Y_t$ for all t, $d = 0, 1, 2, \ldots$, with the convention that $D^0 Y_t = \dot{Y}_t = Y_t - \mu$. Then, for example,

First difference: $D^1 Y_t = DY_t = Y_t - Y_{t-1}$

Second difference: $D^2 Y_t = D(DY_t) = D(Y_t - Y_{t-1})$
$$= DY_t - DY_{t-1} = Y_t - 2Y_{t-1} + Y_{t-2}$$

Third difference: $D^3 Y_t = D(D^2 Y_t) = Y_t - 3Y_{t-1} + 3Y_{t-2} - Y_{t-3}$

In general, dth difference:

$$D^d Y_t = \binom{d}{0} Y_t - \binom{d}{1} Y_{t-1} + \binom{d}{2} Y_{t-2} - \cdots - (-1)^d \binom{d}{d} Y_{t-d}$$

With this notation the general ARIMA(p, d, q) model becomes

$$D^d Y_t - \phi_1 D^d Y_{t-1} - \cdots - \phi_p D^d Y_{t-p} = e_t - \theta_1 e_{t-1} - \cdots - \theta_q e_{t-q}$$

or with $W_t = D^d Y_t$, $t = \ldots, -1, 0, 1, \ldots$

$$W_t - \phi_1 W_{t-1} - \cdots - \phi_p W_{t-p} = e_t - \theta_1 e_{t-1} - \cdots - \theta_q e_{t-q}$$

The reader will note that with $d = 0$—that is, the original process Y_t is stationary—the expressions above reduce to the ARMA(p, q) model. The restrictions on the parameters $\phi_1, \ldots, \phi_p$, $\theta_1, \ldots, \theta_q$ and the nature of the errors are the same as those for stationary and invertible ARMA processes (see reference [1]).

Supplement 10C

A METHOD FOR DETERMINING THE ψ COEFFICIENTS

It is shown in reference [1] that the coefficients ψ_j, $j = 1, 2, \ldots$ in the expression

$$Y_t = e_t + \psi_1 e_{t-1} + \psi_2 e_{t-2} + \cdots$$

can be determined for any t by equating like powers of X in the polynomial expression

$$(1 - \phi_1 X - \phi_2 X^2 - \cdots - \phi_p X^p)(1 - X)^d (1 + \psi_1 X + \psi_2 X^2 + \cdots)$$
$$= (1 - \theta_1 X - \theta_2 X^2 - \cdots - \theta_q X^q)$$

where p and q are the orders of the autoregressive and moving average terms, respectively, and d is the amount of differencing needed to achieve stationarity. (Note that d may equal zero, in which case the term $(1 - X)^d$ is equal to unity.)

An example will illustrate the method. Further examples are included in the problems at the end of the chapter.

EXAMPLE 10C1

Consider the model

$$(Y_t - Y_{t-1}) - \phi_1(Y_{t-1} - Y_{t-2}) = e_t - \theta_1 e_{t-1}$$

Here $p = d = q = 1$, and hence the ψ_j's can be determined from the expression

$$(1 - \phi_1 X)(1 - X)(1 + \psi_1 X + \psi_2 X^2 + \cdots) = (1 - \theta_1 X)$$

by equating like powers of X. Multiplying the factors on the left-hand side of the equality together gives

$$[1 - (1 + \phi_1)X + \phi_1 X^2] + \psi_1 X[1 - (1 + \phi_1)X + \phi_1 X^2]$$
$$+ \psi_2 X^2[1 - (1 + \phi_1)X + \phi_1 X^2] + \cdots$$

and collecting terms of like powers of X produces

$$1-(1+\phi_1-\psi_1)X-[-\phi_1+(1+\phi_1)\psi_1-\psi_2]X^2-[-\phi_1\psi_1+(1+\phi_1)\psi_2-\psi_3]X^3-\cdots$$

It is clear from the right-hand side of the equality that the coefficient of X must be θ_1 and the coefficients of powers of X higher than the first must be zero. Therefore

$$1+\phi_1-\psi_1=\theta_1$$
$$-\phi_1+(1+\phi_1)\psi_1-\psi_2=0$$
$$-\phi_1\psi_1+(1+\phi_1)\psi_2-\psi_3=0$$

and in general

$$-\phi_1\psi_{j-2}+(1+\phi_1)\psi_{j-1}-\psi_j=0,\qquad j\geq 3$$

Solving the equations above yields

$$\psi_1=(1+\phi_1)-\theta_1$$
$$\psi_2=(1+\phi_1)\psi_1-\phi_1=(1+\phi_1)^2-\theta_1(1+\phi_1)-\phi_1$$

and further ψ_j's can be generated recursively from the expression

$$\psi_j=(1+\phi_1)\psi_{j-1}-\phi_1\psi_{j-2},\qquad j\geq 3$$

PROBLEMS

10.1 (a) Define a time series.
    (b) Give examples of time series likely to be stationary and examples of time series likely to be nonstationary.

10.2 Describe the characteristics of the theoretical autocorrelation function for the following stationary-invertible ARMA process:
    (a) AR(1) process
    (b) AR(2) process
    (c) ARMA(1, 1) process
    (d) MA(1) process
    (e) MA(2) process

10.3 Using the definitions in Supplement 10A and the properties of the error term, evaluate $\gamma_k=E(Y_t-\mu)(Y_{t+k}-\mu)$, $k=0,1,2,\ldots$, for
    (a) MA(1) process
    (b) MA(2) process
    (c) AR(1) process

10.4 Using the results from Problem 10.3, evaluate $\rho_k=\gamma_k/\gamma_0$, $k=0,1,2,\ldots$, for
    (a) MA(1) process
    (b) MA(2) process
    (c) AR(1) process
    Thus verify equations (10.4), (10.7), and (9.5).

10.5 Consider the time series,

$$64\quad 55\quad 41\quad 59\quad 48\quad 71\quad 35\quad 57\quad 40\quad 58$$

Calculate r_1 and r_2. (Note that in practice reliable autocorrelation estimates are only obtained from series consisting of approximately 50 observations or more.)

10.6 Show that the MA(1) model $\dot{Y}_t = e_t - \theta_1 e_{t-1}$ has a representation as an *infinite* autoregressive process with $\phi_1 = -\theta_1$, $\phi_2 = -\theta_1^2$, $\phi_3 = -\theta_1^3, \ldots$
(*Hint*: Note that

$$\dot{Y}_t = e_t - \theta_1 e_{t-1}$$
$$\theta_1 \dot{Y}_{t-1} = \theta_1 e_{t-1} - \theta_1^2 e_{t-2}$$
$$\theta_1^2 \dot{Y}_{t-2} = \theta_1^2 e_{t-2} - \theta_1^3 e_{t-3}$$

.

.

.

Now sum both sides.)

10.7 Show that the AR(1) model $\dot{Y}_t = \phi \dot{Y}_{t-1} + e_t$ has a representation as an infinite moving average process with $\theta_1 = -\phi_1$, $\theta_2 = -\phi_1^2$, $\theta_3 = -\phi_1^3, \ldots$
(*Hint*: Note that

$$\dot{Y}_t - \phi_1 \dot{Y}_{t-1} = e_t$$
$$\phi_1 \dot{Y}_{t-1} - \phi_1^2 \dot{Y}_{t-2} = \phi_1 e_{t-1}$$
$$\phi_1^2 \dot{Y}_{t-2} - \phi_1^3 \dot{Y}_{t-3} = \phi_1^2 e_{t-2}$$

.

.

.

Now sum both sides.)

10.8 Show that ρ_k for an AR(1) process satisfies the difference equation

$$\rho_k - \phi_1 \rho_{k-1} = 0$$

Set $k = 1$ and using the fact that $\rho_0 = 1$, derive ρ_1. With the initial values ρ_0, ρ_1 obtain the general expression for ρ_k.

10.9 Write a computer program to generate a series of $N = 100$ observations using the model

$$Y_t = -0.5 Y_{t-1} + e_t$$

where e_t are $N(0, 1)$ variates. Begin the calculations by setting $y_1 = e_1$, where e_1 is a (pseudo) random $N(0, 1)$ variate. Plot the generated series. Does it appear to be stationary?

10.10 Using the 100 observations generated in Problem 10.9, calculate the sample autocorrelation function r_k for $k = 1, 2, \ldots, 10$. Do the sample autocorrelations indicate an AR(1) process? Explain. (*Note*: It is convenient to use a computer to do these calculations)

10.11   Consider the IMA(1, 1) model $Y_t - Y_{t-1} = e_t - \theta_1 e_{t-1}$.

    (a)   Show how Y_t can be expressed explicitly as an infinite sum of previous e's.

    (b)   Show how Y_t can be expressed as an exponentially weighted sum of previous Y's.

10.12   Consider the time series (read down the columns),

(1)	(2)	(3)	(4)
460	477	490	531
457	479	489	547
452	475	489	551
459	479	485	547
462	476	491	541
459	476	492	545
463	478	494	549
479	479	499	545
493	477	498	549
490	476	500	547
492	475	497	543
498	475	494	540
499	473	495	539
497	474	500	532
496	474	504	517
490	474	513	527
489	465	511	540
478	466	514	542
487	467	510	538
491	471	509	541
487	471	515	541
482	467	519	547
479	473	523	553
478	481	519	559
479	488	523	557

    (a)   Plot the original series. Does it appear as if this series is a realization of a stationary stochastic process? Explain.

    (b)   Plot the series formed by taking first differences. Does it appear as if the first-difference series is stationary? Explain.

10.13   Identify appropriate time series models given the estimated autocorrelations below. Justify your choice using your knowledge of theoretical autocorrelation functions for ARIMA processes.

(a)

k:	1	2	3	4	5	6	7	8
Original series: r_k:	-0.39	0.11	0.05	-0.07	-0.03	0.08	0.02	-0.05
y_t $\sqrt{\text{Vâr}(r_k)}$:	0.13	0.15	0.15	0.15	0.15	0.16	0.16	0.16

$$\bar{y} = 83 \qquad \hat{\sigma}_{\bar{y}} = \hat{\sigma}_Y/\sqrt{N} = 25$$
$$\text{where } \hat{\sigma}_Y^2 = (1/N)\Sigma_{t=1}^N(y_t - \bar{y})^2$$

(b)

k:	1	2	3	4	5	6	7	8
Original series: r_k:	-0.70	0.51	-0.32	0.22	-0.11	0.08	-0.04	0.03
y_t $\sqrt{\text{Vâr}(r_k)}$:	0.16	0.19	0.21	0.23	0.24	0.25	0.25	0.25

$$\bar{y} = 4.37 \qquad \hat{\sigma}_{\bar{y}} = \hat{\sigma}_Y/\sqrt{N} = 1.89$$
$$\text{where } \hat{\sigma}_Y^2 = (1/N)\Sigma_{t=1}^N(y_t - \bar{y})^2$$

(c)

k:	1	2	3	4	5	6	7	8
Original series: r_k:	0.99	0.98	0.98	0.97	0.94	0.91	0.89	0.86
y_t k:	1	2	3	4	5	6	7	8
First-difference series: r_k:	0.45	-0.04	0.12	0.06	-0.18	0.16	0.01	-0.07
$y_t - y_{t-1}$ $\sqrt{\text{Vâr}(r_k)}$:	0.14	0.17	0.17	0.17	0.17	0.18	0.18	0.18

(d)

k:	1	2	3	4	5	6	7	8
Original series: r_k:	0.94	0.93	0.90	0.89	0.87	0.86	0.84	0.81
y_t k:	1	2	3	4	5	6	7	8
First-difference series: r_k:	0.69	0.50	0.33	0.19	0.10	0.08	0.03	0.01
$y_t - y_{t-1}$ $\sqrt{\text{Vâr}(r_k)}$:	0.10	0.15	0.18	0.21	0.23	0.24	0.25	0.25

10.14 Write a computer program (or use an existing program if one is available) to calculate the sample autocorrelations r_k, $k = 1, 2, \ldots, 20$, for the time series given in Problem 10.12. Repeat the calculations for the first-difference series.
  (a) Considering the behavior of the sample autocorrelation functions, does it appear as if the original series or the first-difference series is stationary? Explain. (Note: Your answer should be consistent with the answer to Problem 10.12.)
  (b) Identify an appropriate model for this time series. In particular consider the random walk model, $Y_t - Y_{t-1} = e_t$.

10.15 Suppose the time series model

$$(Y_t - Y_{t-1}) - 0.5(Y_{t-1} - Y_{t-2}) = e_t$$

is an appropriate representation of a given time series. Describe the likely appearance of the sample autocorrelation function(s) leading to the identification of this model. [You may find your answer to Problem 10.13(d) useful.]

10.16 Discuss the likely appearance of the *sample* autocorrelation function for the time series models listed below.

  (a) $\dot{Y}_t - \phi_1\dot{Y}_{t-1} - \phi_2\dot{Y}_{t-2} = e_t$,   where $\dot{Y}_t = Y_t - \mu$
  (b) $\dot{Y}_t = e_t - \theta_1 e_{t-1} - \theta_2 e_{t-2}$,   where $\dot{Y}_t = Y_t - \mu$
  (c) $\dot{Y}_t - \phi_1\dot{Y}_{t-1} = e_t - \theta_1 e_{t-1}$,   where $\dot{Y}_t = Y_t - \mu$
  (d) $W_t = e_t - \theta_1 e_{t-1}$,   where $W_t = Y_t - Y_{t-1}$

(e) $W_t - \phi_1 W_{t-1} - \phi_2 W_{t-2} = e_t,$ where $W_t = Y_t - Y_{t-1}$
(f) $W_t = e_t - \theta_1 e_{t-1} - \cdots - \theta_q e_{t-q},$ where $W_t = Y_t - Y_{t-1}$

10.17   Consider the ARMA(1, 1) model $\dot{Y}_t - \phi_1 \dot{Y}_{t-1} = e_t - \theta_1 e_{t-1}$. Solve for e_t, and indicate how successive e_t's may be calculated recursively. List the initial value(s) required to start the calculation.

10.18   Given the AR(2) model $\dot{Y}_t = \phi_1 \dot{Y}_{t-1} + \phi_2 \dot{Y}_{t-2} + e_t$.
(a)   Solve for e_t, and indicate how successive e's may be calculated.
(b)   Evaluate $E(Y_t|\text{previous } y\text{'s})$, and hence obtain

$$S(\mu, \phi) = \sum_{t=1}^{N} [y_t - E(Y_t|\text{previous } y\text{'s})]^2$$

List the initial values required to calculate $S(\mu, \phi)$ given the time series and values for μ, ϕ_1, and ϕ_2.

10.19   In modeling the weekly sales of a certain commodity over the past six months, the time series model

$$\dot{Y}_t - \phi_1 \dot{Y}_{t-1} = e_t - \theta_1 e_{t-1}$$

was thought to be appropriate.
(a)   Use the five observations

$$y_t: \quad 4 \quad 6 \quad 5 \quad 3 \quad 7$$

to illustrate how to calculate the sum of squares function

$$S(\mu, \phi_1, \theta_1) = \sum [y_t - E(Y_t|\text{previous } y\text{'s})]^2$$

(b)   Suppose the model above was fitted and the autocorrelations of the residuals were

k:	1	2	3	4	5	6	7	8
$r_k(\hat{e})$:	0.50	−0.04	0.03	−0.01	0.01	0.02	0.03	−0.01
$\sqrt{\text{Vâr}[r_k(\hat{e})]}$:	0.08	0.10	0.11	0.11	0.11	0.11	0.11	0.11

Is the assumed model really appropriate? If not, how would you modify the model? Explain.

10.20   Consider the IMA(1, 1) model

$$W_t = Y_t - Y_{t-1} = e_t - \theta_1 e_{t-1}$$

Using the first 10 observations for the *differenced* refrigerator sales series (given in Table 10.6) evaluate the sum of squares function

$$S(\theta_1) = \sum [w_t - E(W_t|\text{previous } w\text{'s})]^2, \quad \text{for } \theta_1 = 0.3, 0.5, 0.7$$

Plot the function, and determine the minimum by inspection for these three parameter values.

10.21   Indicate how the sum of squares function may be obtained for the following time series models. List the starting (initial) values required to calculate the sum of squares for a given set of parameter values.

(a) $\dot{Y}_t = e_t - \theta_1 e_{t-1} - \theta_2 e_{t-2}$,   where $\dot{Y}_t = Y_t - \mu$
(b) $\dot{Y}_t - \phi_1 \dot{Y}_{t-1} - \phi_2 \dot{Y}_{t-2} = e_t - \theta_1 e_{t-1}$,   where $\dot{Y}_t = Y_t - \mu$
(c) $W_t - \phi_1 W_{t-1} = e_t$,   where $W_t = Y_t - Y_{t-1}$
(d) $W_t - \phi_1 W_{t-1} = e_t - \theta_1 e_{t-1}$,   where $W_t = Y_t - Y_{t-1}$

10.22 Discuss in detail the nature of the IMA(1, 1) model $Y_t - Y_{t-1} = e_t - \theta_1 e_{t-1}$. In particular discuss the nature of the forecasts and probability limits produced by this model and its role in "simple exponential smoothing."

10.23 Suppose the model

$$(Y_t - Y_{t-1}) - 0.6(Y_{t-1} - Y_{t-2}) = e_t$$

is appropriate for the time series being analyzed. Given the values $y_{N-2} = 12$, $y_{N-1} = 9$, $y_N = 10$, for the last three observations, forecast the future values y_{N+1} and y_{N+2}. Suppose the future value y_{N+1} is observed and, in fact, $y_{N+1} = 13$; update the forecast of y_{N+2}.

10.24 Consider the ARMA(1, 1) model

$$(Y_t - 60) + 0.3(Y_{t-1} - 60) = e_t - 0.4 e_{t-1}$$

which has been fitted to a time series where the last 10 values are

$$60, 57, 52, 59, 62, 59, 63, 67, 61, 58$$

and the last residual is $\hat{e}_N = -2$.
(a) Calculate the forecasts of the next two observations, and indicate how forecasts can be calculated for lead times greater than two. Show what happens to the forecasts as the lead time becomes arbitrarily large.
(b) Given $\hat{\sigma}_e^2 = 4$, calculate 90-percent probability limits for the next two observations. Interpret these limits.

10.25 Recall that the "random walk" model is defined by the relationship

$$Y_t - Y_{t-1} = e_t$$

(a) Derive a general expression for the N-origin forecast $\hat{y}_N(l)$.
(b) Show that Y_t may be written entirely in terms of the e_t's as

$$Y_t = e_t + e_{t-1} + e_{t-2} + \cdots$$

and hence develop an expression for the variance of the lead l forecast error.
(c) Using the results in parts (a) and (b) indicate how to construct 95-percent probability limits for the future value y_{N+l}.

10.26 The following residual autocorrelations were observed after fitting a time series model to investment in fixed plant and equipment:

k:	1	2	3	4	5	6	7	8
$r_k(\hat{e})$:	-0.19	-0.23	0.08	-0.09	-0.08	0.15	0.07	-0.13
$\sqrt{\hat{\text{Var}}[r_k(\hat{e})]}$:	0.07	0.10	0.12	0.13	0.13	0.13	0.13	0.13

.0156

Given that there were $N = 64$ observations in the series and the behavior of the residual autocorrelations, does it appear as if the original time series model is adequate? Explain. (As part of your explanation include the overall test of model adequacy based on the χ^2 distribution.)

- 10.27 Using the time series generated in Problem 10.9, plot the sum of squares function for values of ϕ_1 in the interval $(-0.9, -0.1)$, and determine the least squares estimate of ϕ_1 by inspection.

10.28 Since the time series referred to in Problem 10.27 is a realization of a first-order autoregressive process with $\mu = 0$, an initial estimate of ϕ_1 is provided by

$$r_1 = \frac{\Sigma y_t y_{t+1}}{\Sigma y_t^2}$$

the lag 1 sample autocorrelation. Evaluate r_1, and compare the value of this initial estimate of ϕ_1 with the least squares value obtained above. Explain any difference in magnitude that may exist. (Note: You may have already calculated r_1 for Problem 10.10.)

10.29 Using the generated first-order autoregressive series from Problem 10.9 and the least squares estimate of ϕ_1 obtained above (Problem 10.27), forecast the next five observations, and calculate 75-percent probability limits for these future values.

10.30 The refrigerator sales series is given in Table 10.6. We have shown that this series is well represented by an IMA(1, 1) model with $\theta_1 \doteq 0.6$.
   (a) Suppose the forecast origin is at the 45th observation, forecast the next observation using the form of the forecast given in equation (10.66). Compare the forecast with the actual value, 751.
   (b) Use the basic equation of simple exponential smoothing [equation (10.67)] to forecast the remaining observations of the series.

10.31 (a) Develop an expression for updating forecasts of future values for the unemployment rate series.
   (b) We have shown that at forecast origin $N = 61$, $\hat{y}_{61}(1) = 5.82$, $\hat{y}_{61}(2) = 5.72$, and $\hat{y}_{61}(3) = 5.54$. Given the actual value $y_{62} = 5.77$, update the forecasts of y_{63} and y_{64}.

10.32 Write out an expression for the forecasts $\hat{y}_N(l)$, $l = 1, 2, 3, 4$, using the fitted Treasury bill interest rates model. Determine the ψ_j weights, and develop an expression for 90-percent probability limits for the future value y_{N+l}.

10.33 A coefficient for judging the accuracy of forecasts has been proposed by H. Theil (see reference [10]). A modification of Theil's coefficient may be expressed as

$$U = \frac{\sqrt{(1/m)\Sigma_{t=1}^m [y_t - \hat{y}_{t-1}(1)]^2}}{\sqrt{(1/m)\Sigma_{t=1}^m y_t^2} + \sqrt{(1/m)\Sigma_{t=1}^m \hat{y}_{t-1}^2(1)}}$$

where m is the number of comparisons between the actual and predicted values.

(a)   Show that $0 \leq U^2 \leq 1$.

(b)   Let D be the denominator in the Theil coefficient and let

$$S_a = \sqrt{(1/m)\Sigma_{t=1}^m (y_t - \bar{y})^2}, \qquad S_p = \sqrt{(1/m)\Sigma_{t=1}^m [\hat{y}_{t-1}(1) - \bar{\hat{y}}]^2}$$

and

$$r_{ap} = \frac{\Sigma_{t=1}^m (y_t - \bar{y})[\hat{y}_{t-1}(1) - \bar{\hat{y}}]}{\sqrt{\Sigma_{t=1}^m (y_t - \bar{y})^2}\sqrt{\Sigma_{t=1}^m [\hat{y}_{t-1}(1) - \bar{\hat{y}}]^2}}$$

where $\bar{y}$ and $\bar{\hat{y}}$ are the averages of the m actual and predicted values, respectively. Show that

$$U^2 = U_m^2 + U_s^2 + U_c^2$$

where

$$U_m^2 = \frac{(\bar{y} - \bar{\hat{y}})^2}{D^2}, \qquad U_s^2 = \frac{(S_a - S_p)^2}{D^2}$$

and

$$U_c^2 = 2S_a S_p \frac{1 - r_{ap}}{D^2}$$

The three components U_m^2, U_s^2, and U_c^2 may be interpreted as follows: U_m^2 is a measure of the error contributed by differences in the mean levels of the actual and predicted values; U_s^2 is a measure of the error due to differences between measures of variation associated with the actual and predicted values; and finally, U_c^2 is a measure of the error due to lack of positive covariation between the actual and predicted values.

11

ANALYSIS OF SEASONAL
TIME SERIES

11.1 INTRODUCTION

A seasonal event is one that occurs with a more or less regular period. Christmas is a seasonal event whose regularity is disturbed only by leap years. Easter, on the other hand, exhibits more widely varying periods from year to year because it is not required to occur on a specific date. A seasonal event like Christmas has important effects on business activities, such as toy, jewelry, and greeting card sales. A seasonal event that reflects in the unemployment rate is the large movement of students into the labor market in the summer months.

Seasonal events may be looked upon as interruptions or disturbances in the "normal flow of events," but because of their regularity, they can be anticipated. If we adopt the perspective of day-to-day operations of a firm, we see immediately that anticipation of and planning for seasonal events is only prudent management. Thus forecasts of the *magnitudes* of the effects of seasonal events should prove valuable to management.

We shall illustrate the modeling of seasonal time series by means of examples. Our approach involves the extension of the analysis presented in Chapters 9 and 10 to seasonal data.

11.2 AN EXAMPLE: INVESTMENTS OF LARGE
NEW YORK CITY BANKS

The data in this example consist of 120 observations on average weekly total investments per month of large New York City banks, from January 1965 through December 1974, as recorded in the *Federal Reserve Bulletin*.[a]

[a] Published monthly by the Division of Administrative Services, Board of Governors of the Federal Reserve System.

Table 11.1   Average Weekly Total Investments per Month of Large New York City Banks, January 1965 through December 1974 (millions of dollars)

Year	Jan.	Feb.	Mar.	Apr.	May	June	July	Aug.	Sept.	Oct.	Nov.	Dec.	Average
						MONTH							
1965	5466.1	4845.9	4671.2	4528.4	4418.1	4461.0	4643.7	4460.0	4260.0	4627.9	4612.4	4759.6	4637.9
1966	4565.7	4193.9	3822.4	4144.9	3823.7	3653.6	3710.4	3705.8	3999.6	3694.6	3855.4	4323.9	3957.8
1967	4475.4	4785.4	4882.0	4648.7	4656.4	4775.3	5017.9	4859.6	5287.4	5794.2	5832.6	5597.6	5051.0
1968	5313.2	5192.7	4807.5	4765.6	4631.0	4727.8	5773.2	5298.9	5909.9	5647.8	5645.1	5965.1	5223.2
1969	5165.4	4580.8	4144.9	4509.0	3984.8	4030.4	4275.2	4291.7	4324.2	4164.4	4615.9	4962.4	4420.8
1970	4550.2	4136.5	4232.1	4936.8	4507.1	4329.7	4500.2	5169.4	4993.0	4879.9	5059.0	5459.8	4729.5
1971	5544.6	5342.9	5106.0	5439.2	4863.7	4657.2	4988.7	4427.2	4562.6	4528.4	5330.7	5263.6	5004.7
1972	5112.2	5031.1	5551.8	5223.9	5004.4	4901.1	4602.1	4746.2	5247.3	4715.1	4862.8	5039.9	5003.2
1973	4931.2	4362.8	4137.6	4316.1	3760.8	4247.1	3831.9	3768.8	4154.2	4257.6	4867.9	5787.9	4485.3
1974	5435.0	5117.0	4984.0	4781.0	3793.0	3679.0	3373.0	4118.0	4679.0	4144.0	4625.0	4932.0	4471.7
Average	5055.90	4758.90	4633.95	4729.36	4344.30	4346.22	4371.63	4484.63	4741.72	4645.39	4930.68	5209.18	

Table 11.1 displays the data along with monthly and yearly averages. A time plot of the data suggests that they fluctuate with relatively constant variability about a mean of 4687.7. An examination of the monthly means suggests that investments tend to be relatively high in the winter and low in the summer.

Figure 11.1 shows the sample autocorrelation function of the data for lags 1 through 12. The sample autocorrelations beyond lag 12 are deemed to be "insignificant" and hence are not shown. Figure 11.1 is suggestive of an AR(1) model with $\phi_1 \doteq 0.75$. Figure 11.2 shows the results of fitting an AR(1) model to the data. The reader will notice the large residual autocorrelations at lags 6 and 12. Such correlations might be interpreted as evidence of "seasonal" variation, that is, semiannual and annual movements in the observations.

The simplest way to model the behavior in Figure 11.2 is to include two moving average terms with lags 6 and 12. In symbols the model has the form

$$(Y_t - \mu) - \phi_1(Y_{t-1} - \mu) = e_t - \theta_6 e_{t-6} - \theta_{12} e_{t-12} \tag{11.1}$$

where the random variables and parameters have the usual interpretations. This is a four-parameter member of the ARMA(p, q) class, and hence the methods developed in Chapters 9 and 10 are applicable to this model also. We now turn to the fitting and diagnostic checking of the model.

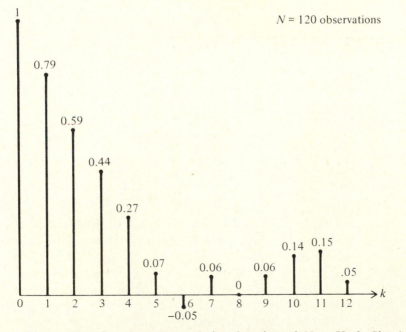

Figure 11.1 Sample autocorrelation function of New York City bank investments

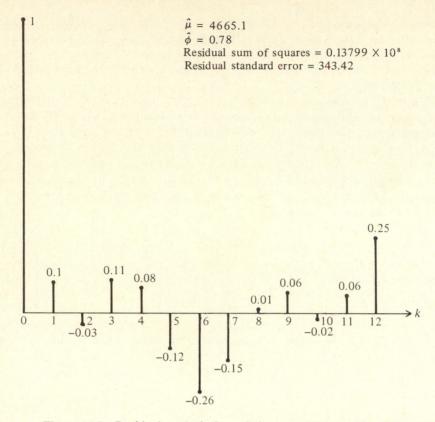

$\hat{\mu} = 4665.1$
$\hat{\phi} = 0.78$
Residual sum of squares $= 0.13799 \times 10^8$
Residual standard error $= 343.42$

Figure 11.2   Residual analysis from fitting an AR(1) model to New York
City bank data

To employ the method of least squares developed in Chapter 10 we solve
equation (11.1) for e_t as a function of past e's, current and past Y's, and the
parameters. We get

$$e_t = (Y_t - \mu) - \phi_1(Y_{t-1} - \mu) + \theta_6 e_{t-6} + \theta_{12} e_{t-12} \qquad (11.2)$$

Now it is clear that for $t = 1, 2, \ldots, 12$, e_t depends on values of e's that
occurred before the observation period began, and furthermore e_1 depends
on Y_0, an unobserved Y. Following the procedure suggested in Chapter 10
we should compute only the values of $e_2, e_3, \ldots, e_{120}$ under the assumption
that all e's occurring before the period of observation are zero. Then
parameter values which minimize $\Sigma_{t=2}^{120} e_t^2$ can be found. Unfortunately this
procedure introduces a fairly substantial number of approximations into
the calculation of the e's and can lead to poor parameter estimates.
     This problem is effectively met by the technique of "back forecasting"

(outlined in reference [1], pp. 315–320). The idea is to extend the series artificially into the past in order to reduce the magnitude of the approximation errors. The procedure involves three steps.

Step 1   Using the crude method of Chapter 10, we estimate the parameters of the model, compute forecasts $\hat{y}_{120}(1), \ldots, \hat{y}_{120}(12)$, and use these forecasts as surrogates for $Y_{121}, \ldots, Y_{132}$.

Step 2   We reverse the order of the extended series to create the series $Y_{132}, Y_{131}, \ldots, Y_1$. We use the crude method to estimate the parameters of equation (11.1) on the basis of the reversed series, and forecast 12 "future" observations that are to be used as surrogates for $Y_{-11}, \ldots, Y_0$.

Step 3   Use the least squares method to estimate the parameters of equation (11.1) on the basis of the series $Y_{-11}, \ldots, Y_0, Y_1, \ldots, Y_{120}$.

Box and Jenkins [1, p. 318] note that the "back forecasting" procedure can be repeated as often as desired, but usually little will be gained from iterations beyond the first.

Table 11.2   Residual Analysis for Fitting Equation (11.1) to the New York City Bank Data

$\hat{\phi}_1 = 0.83$
$\hat{\mu} = 4708.9$
$\hat{\theta}_6 = 0.22$
$\hat{\theta}_{12} = -0.17$
Residual sum of squares $= 0.12666 \times 10^8$
Residual standard error $= 331.87$
Skewness $= 0.03$
Kurtosis $= 3.09$
Chi-square statistic based on 12 lags $= 8.61$ (8 degrees of freedom)

Lag	1	2	3	4	5	6	7	8	9	10	11	12
Residual autocorrelation	0.01	−0.05	0.10	0.09	−0.09	−0.02	−0.16	0	0.10	0	0	0.05

Table 11.2 shows the results of fitting equation (11.1) to the bank investments series. The usual diagnostic checks do not contradict the adequacy of the model.

Forecasts may be generated from equation (11.1) using the algorithm presented in Chapter 10. Letting $t = N + l$ in equation (11.1) and rearranging terms we get

$$Y_{N+l} = \mu + \phi_1(Y_{N+l-1} - \mu) + e_{N+l} - \theta_6 e_{N+l-6} - \theta_{12} e_{N+l-12} \qquad (11.3)$$

Successively setting $l = 1, 2, 3, \ldots$ and computing the conditional expectation $E_N(Y_{N+l})$ the point forecasts are generated. (Of course parameters and

Table 11.3   Forecasts, Forecast Intervals, and Actual Values
for 1975 (New York bank data)

Lead l	Lower 95-Percent Forecast Boundary	Point Forecast	Upper 95-Percent Forecast Boundary	Actual Value
1 (Jan.)	4331.9	4954.1	5577.2	4719.5
2 (Feb.)	3951.5	4759.7	5567.9	4881.0
3 (Mar.)	3736.2	4649.4	5562.6	5624.3
4 (Apr.)	3789.7	4768.2	5746.7	5903.8
5 (May)	3571.5	4592.1	5612.7	6311.0
6 (Jun.)	3523.0	4571.4	5619.8	6930.5
7 (Jul.)	3466.3	4516.3	5566.3	7048.8
8 (Aug.)	3589.4	4640.5	5691.6	6845.5
9 (Sep.)	3677.1	4728.9	5780.1	7452.5
10 (Oct.)	3572.6	4625.0	5677.4	7531.0
11 (Nov.)	3611.1	4663.8	5716.6	8331.8
12 (Dec.)	3635.2	4688.1	5741.1	8369.0

errors must be replaced by estimates in the usual way, so the quantities actually calculated are *estimates* of the forecasts that would be generated if the parameter values and the complete past history of the series were known.) Using the technique in Supplement 10C, the variances of the forecasts can also be computed. Table 11.3 shows point forecasts, the endpoints of the forecast intervals

$$\hat{y}_N(l) \pm 1.96\sqrt{\mathrm{Var}(\hat{y}_N(l))}$$

and the actual values of investments for 1975.

The reader will notice that although the forecasts for January and February are reasonable, the actual values for the rest of the year lie above the upper forecast boundaries. The model based on the past history of the series was not able to forecast the very rapid rise in investments that occurred in 1975. Given this turn of events, how do we make forecasts for 1976? One approach would be to make some subjective judgment as to whether or not investments will maintain the high level achieved in 1975. If our understanding of the economy suggests a future continual rise, fluctuation about a new higher level, or downturn in investments, then our forecast could be made to reflect such understanding.

Another approach, more in keeping with a model building philosophy, could be based on the following line of argument. Although the assumption of stationarity appeared satisfactory during the period 1965–1974, events in 1975 suggest that the series has become more volatile, at least in the short run. Thus a nonstationary model may be more appropriate for representing future values in a period that appears fraught with uncertainty. Equation (11.1) becomes nonstationary if we set $\phi_1 = 1$. In fact, if we fit equation (11.1) to the investments data including the 1975 values, we find the estimate of ϕ_1 is 0.98, and the approximate 95-percent confidence interval for ϕ_1 is (0.90, 1.06), suggesting that a value of $\phi_1 = 1$ is quite plausible. Table 11.4 shows the results of fitting the model

$$Y_t - Y_{t-1} = e_t - \theta_6 e_{t-6} - \theta_{12} e_{t-12} \tag{11.4}$$

Table 11.4   Results of Fitting Equation (11.4) to the New York City Bank Data, 1965–1975

$\hat{\theta}_6 = 0.12$ $\hat{\theta}_{12} = -0.16$
Residual sum of squares $= 0.17015 \times 10^8$
Residual standard error $= 363.18$
Chi-square statistic based on 12 lags $= 6.47$ (10 degrees of freedom)

Lag	1	2	3	4	5	6	7	8	9	10	11	12
Residual autocorrelation	0.03	−0.05	0.09	0.09	−0.10	−0.01	−0.08	0.03	0.10	−0.01	−0.03	0.03

Lead l	Lower 95-Percent Forecast Boundary	Point Forecasts for 1976	Upper 95-Percent Forecast Boundary
1 (Jan.)	7611.1	8322.9	9,034.7
2 (Feb.)	7395.0	8401.7	9,408.4
3 (Mar.)	7225.4	8458.3	9,691.3
4 (Apr.)	7049.9	8473.6	9,897.3
5 (May)	6877.6	8469.3	10,061.1
6 (Jun.)	6818.2	8561.9	10,305.5
7 (Jul.)	6730.7	8584.4	10,438.2
8 (Aug.)	6581.4	8538.9	10,496.6
9 (Sep.)	6580.9	8637.2	10,693.4
10 (Oct.)	6518.5	8668.9	10,819.3
11 (Nov.)	6558.1	8798.7	11,039.3
12 (Dec.)	6485.6	8812.9	11,140.1

to the data including 1975 values. On the basis of the point forecasts, this model predicts a continued gradual upward drift in investments. The continual widening of the forecast intervals is, of course, characteristic of a nonstationary model. The great width of these intervals cautions a person against making very precise statements about the behavior of investments in 1976, a sensible attitude in the face of the demonstrated volatility of the investments series.

11.3 FURTHER EXAMPLES: FOUR WISCONSIN EMPLOYMENT SERIES

Description of Data and Presentation of Models

The data listed in the tables in Supplement 11A are series of monthly numbers of employees in three categories of production and in wholesale and retail trade in the State of Wisconsin from January 1961 through October 1975. The data are published by the State of Wisconsin's Department of Industry, Labor and Human Relations, Bureau of Research and Statistics.

Before we discuss the details of the modeling of these series, let us look at the final models displayed in Table 11.5. The models may be summarized as follows.

Table 11.5 Final Models for Four Wisconsin Series

Category	Model	Residual Root Mean Square
I. Food and kindred products	$W_t = Y_t - Y_{t-12}$ $W_t - 0.75 W_{t-1} = e_t - 0.52 e_{t-12}$	1.17
II. Fabricated metals	$W_t = Y_t - Y_{t-1}$ $W_t - 0.28 W_{t-12} = e_t$	0.68
III. Transportation equipment	$W_t = (Y_t - 40.4) - 0.44(Y_{t-12} - 40.4)$ $W_t - 0.44 W_{t-1} - 0.11 W_{t-2} = e_t + 0.38 e_{t-6}$	2.99
IV. Trade	$W_t = (Y_t - Y_{t-12}) - (Y_{t-1} - Y_{t-13})$ $a_t = e_t - 0.38 e_{t-12}$ $W_t = a_t + 0.18 a_{t-1}$	1.35

Food and Kindred Products The series itself is nonstationary, but the annual seasonal difference, $W_t = Y_t - Y_{t-12}$, is stationary, having an AR(1) component, $W_t - \phi_1 W_{t-1}$, with $\hat{\phi}_1 = 0.75$, and a seasonal moving average component, $e_t - \theta_{12} e_{t-12}$, with $\hat{\theta}_{12} = 0.52$. Substitution of $Y_t - Y_{t-12}$ for W_t yields the following form for the model:

$$Y_t - \phi_1 Y_{t-1} - Y_{t-12} + \phi_1 Y_{t-13} = e_t - \theta_{12} e_{t-12} \tag{11.5}$$

with $\hat{\phi}_1 = 0.75$ and $\hat{\theta}_{12} = 0.52$.

Fabricated Metals The series itself is nonstationary, but the first difference, $W_t = Y_t - Y_{t-1}$, is stationary, having a seasonal autoregressive component, $W_t - \phi_{12} W_{t-12}$, with $\hat{\phi}_{12} = 0.28$. An alternative formula for the model is

$$Y_t - Y_{t-1} - \phi_{12} Y_{t-12} + \phi_{12} Y_{t-13} = e_t \tag{11.6}$$

with $\hat{\phi}_{12} = 0.28$.

Transportation Equipment   The series is stationary with estimated mean $\hat{\mu} = 40.4$. The seasonal pattern is somewhat complicated. First the annual seasonal autoregressive component,

$$W_t = (Y_t - \mu) - \phi_{12}(Y_{t-12} - \mu)$$

with $\hat{\mu} = 40.4$ and $\hat{\phi}_{12} = 0.44$, is formed. The W_t's then follow a model with an AR(2) component,

$$W_t - \phi_1 W_{t-1} - \phi_2 W_{t-2}$$

with $\hat{\phi}_1 = 0.44$ and $\hat{\phi}_2 = 0.11$, and a semiannual moving average component, $e_t - \theta_6 e_{t-6}$, with $\hat{\theta}_6 = -0.38$. An alternative form of the model is

$$\dot{Y}_t - \phi_1 \dot{Y}_{t-1} - \phi_2 \dot{Y}_{t-2} - \phi_{12} \dot{Y}_{t-12} + \phi_1 \phi_{12} \dot{Y}_{t-13} + \phi_2 \phi_{12} \dot{Y}_{t-14} = e_t - \theta_6 e_{t-6}$$

$$(11.7)$$

with $\dot{Y}_t = Y_t - \mu$; $\mu = 40.4$, $\hat{\phi}_1 = 0.44$, $\hat{\phi}_2 = 0.11$, $\hat{\phi}_{12} = 0.44$, and $\hat{\theta}_6 = -0.38$.

Trade   The series itself is nonstationary, but the function

$$W_t = (Y_t - Y_{t-12}) - (Y_{t-1} - Y_{t-12})$$

which is the first difference of the annual seasonal difference, is stationary. The model for w_t is formed as follows. First form the seasonal moving average $a_t = e_t - \theta_{12} e_{t-12}$, where the e's follow the usual assumptions of independence, zero mean, and constant variance, and $\hat{\theta}_{12} = 0.38$. Then the formula for W_t is $W_t = a_t - \theta_1 a_{t-1}$, with $\hat{\theta}_1 = -0.18$. Thus W_t has an MA(1) component involving the seasonal moving average error series a_t. Another way of writing this model is

$$Y_t - Y_{t-1} - Y_{t-12} + Y_{t-13} = e_t - \theta_1 e_{t-1} - \theta_{12} e_{t-12} + \theta_1 \theta_{12} e_{t-13} \qquad (11.8)$$

with $\hat{\theta}_1 = -0.18$ and $\hat{\theta}_{12} = 0.38$. Notice the parallelism between the lags of Y on the left-hand side of equation (11.8) and the lags of e on the right-hand side. The model in equation (11.8) appears rather frequently in modeling seasonal time series.

   Comparison of the four models (11.5) through (11.8) is enlightening. Models (11.5), (11.6), and (11.8) are nonstationary; model (11.7) is stationary. Model (11.5) involves an annual seasonal difference, model (11.6) a regular first difference, and model (11.8) the first difference of an annual seasonal difference. Model (11.5) involves a regular AR(1) component, model (11.6) an annual seasonal autoregressive component, model (11.7) both a regular AR(2) and an annual seasonal autoregressive component, and model (11.8) no autoregressive component at all. Model (11.6) has no moving average component. Models (11.5) and (11.7) contain annual and semiannual seasonal moving average components, respectively. And model (11.8) contains both a regular MA(1) and an annual seasonal moving average component. This summary is suggestive of the great variety of patterns that can occur in modeling seasonal time series.

Some Seasonal Correlation Patterns

The ability to model seasonal time series rests upon the ability to detect certain basic seasonal patterns in sample autocorrelation functions, and a knowledge of the theoretical autocorrelation functions of a few basic seasonal time series models proves to be an invaluable aid in this process. In this section we will present some of the correlation patterns that appear to be most helpful in modeling monthly data. (Similar displays can be made for quarterly and weekly data which are usually simpler and more complicated than monthly data, respectively.) The models are displayed in Table 11.6.

Table 11.6   Useful Seasonal Models and Their Autocorrelation Functions

Model	*Autocorrelation Function (positive lags only)*
I. Semiannual moving average $W_t = e_t - \theta_6 e_{t-6}$	An isolated spike at lag 6 $\rho_6 = -\theta_6/(1 + \theta_6^2)$ All other $\rho_k = 0$
II. Annual moving average $W_t = e_t - \theta_{12} e_{t-12}$	An isolated spike at lag 12 $\rho_{12} = -\theta_{12}/(1 + \theta_{12}^2)$ All other $\rho_k = 0$
III. Combination of MA(1) and annual moving average	
Form A: $W_t = e_t - \theta_1 e_{t-1} - \theta_{12} e_{t-12}$	Spikes at lags 1, 11, 12 $\rho_1 = -\theta_1/(1 + \theta_1^2 + \theta_{12}^2)$, $\rho_{11} = \theta_1\theta_{12}/(1 + \theta_1^2 + \theta_{12}^2)$, $\rho_{12} = -\theta_{12}/(1 + \theta_1^2 + \theta_{12}^2)$ All other $\rho_k = 0$
Form B: $W_t = e_t - \theta_1 e_{t-1} - \theta_{12} e_{t-12} + \theta_1\theta_{12} e_{t-13}$	Spikes at lags 1, 11, 12, and 13 $\rho_1 = -\theta_1/(1 + \theta_1^2)$, $\rho_{12} = -\theta_{12}/(1 + \theta_{12}^2)$ $\rho_{11} = \rho_{13} = \rho_1\rho_{12}$ All other $\rho_k = 0$
IV. Annual autoregressive $W_t - \phi_{12} W_{t-12} = e_t$ $-1 < \phi_{12} < 1$	Exponentially decaying spikes at lags 12, 24, 36, and so on $\rho_{12k} = \phi_{12}^k$, $k = 1, 2, 3, \ldots$ All other $\rho_k = 0$
V. Annual autoregressive with MA(1) $W_t - \phi_{12} W_{t-12} = e_t - \theta_1 e_{t-1}$ $-1 < \phi_{12} < 1$	Spikes at lags 1, 11, 12, 13, 23, 24, 25, 35, 36, 37, and so on $\rho_1 = -\theta_1/(1 + \theta_1^2)$, $\rho_{12k} = \phi_{12}^k$, $k = 1, 2, 3, \ldots$ $\rho_{12k-1} = \rho_{12k+1} = \phi_{12}^k\rho_1$, $k = 1, 2, 3, \ldots$ All other $\rho_k = 0$

Two Examples of Modeling

In this section we shall trace the steps in modeling the Food and Kindred Products and the Trade series. A careful study of these examples should give the reader some hints on how to approach seasonal series. Remember, though, that each series presents its own problems, and successful modeling often requires attention to peculiarities as well as to "standard" patterns in the data. In particular the handling of nonstationarity can lead to difficulties. In this regard, we have noticed in students a tendency to form differences when the fitting of an autoregressive component is more appropriate.

We now turn to the modeling of Food and Kindred Products. Table 11.7 displays the sample autocorrelation function of the series. A plot of the function reveals a sinusoidal pattern with very high peaks at lags 12, 24, and 36. Thus we have elements of AR(2) behavior (sinusoidal) and either autoregressive or nonstationary behavior in the annual seasonal lag 12. One way to check the latter is to fit a seasonal autoregressive model to the data, namely, the model

$$Y_t - \phi_{12} Y_{t-12} = e_t \tag{11.9}$$

When this is done, the estimate of ϕ_{12} is $\hat{\phi}_{12} = 0.982$ with a standard error of 0.02, so a value of 1.0 for ϕ_{12} is plausible. Thus the formation of the annual seasonal differences, $w_t = y_t - y_{t-12}$, seems warranted. Table 11.8 shows the sample autocorrelation function of w_t. The approximate exponential decay in the first two lags suggests fitting an AR(1) model to W_t, but the rather large autocorrelations at later lags may force the fitting of a more complicated model. Starting with the simple AR(1) hypothesis, we come up with the model

$$W_t - \phi_1 W_{t-1} = e_t \tag{11.10}$$

where $\hat{\phi}_1 = 0.66$ with standard error 0.06. The residual autocorrelation function is displayed in Table 11.9. The large, isolated spike at lag 12 shows that we need to add a seasonal moving average component to equation (11.10). Thus we are led to fit the model in equation (11.5), and the residuals from fitting this model suggest no need for further modeling.

We now turn to the Trade series. Table 11.10 shows the sample autocorrelation function of the series. The persistently large spikes suggest strongly that the series is nonstationary, so we form the first differences $x_t = y_t - y_{t-1}$. The sample autocorrelation function of x_t is also shown in Table 11.10. The approximate exponential decay in lags 12, 24, and 36 suggests that X_t may follow a seasonal autoregressive model, but when the model

$$X_t - \phi_{12} X_{t-12} = e_t$$

is fit, we find that $\phi_{12} = 1$ is a plausible hypothesis, and thus that an annual seasonal difference appears warranted. Table 11.10 also displays the sample autocorrelation function of the differences

$$w_t = x_t - x_{t-12} = y_t - y_{t-1} - y_{t-12} + y_{t-13}$$

Here we see the basic pattern of model III, form B, of Table 11.6 with spikes at lags 1, 11, 12, and 13. (There are spikes at other lags, too, but interpreting autocorrelation patterns or seeing the "right" spikes is part of the modeler's

Table 11.7 Sample Autocorrelation Function of the Food and Kindred Products Series

Lags	Autocorrelations											
1–12	0.77	0.32	−0.12	−0.38	−0.52	−0.56	−0.52	−0.38	−0.12	0.29	0.69	0.88
13–24	0.68	0.28	−0.12	−0.38	−0.49	−0.53	−0.49	−0.37	−0.14	0.23	0.61	0.79
25–36	0.62	0.25	−0.12	−0.35	−0.47	−0.50	−0.47	−0.36	−0.15	0.19	0.54	0.72

Table 11.8 Sample Autocorrelation Function of the Annual Seasonal Difference of the Food and Kindred Products Series

Lags	Autocorrelations											
1–12	0.64	0.32	0.20	0.21	0.20	0.18	0.20	0.24	0.29	0.21	0.07	−0.12
13–24	−0.05	0.10	0.20	0.20	0.18	0.19	0.17	0.07	−0.01	0.04	0.10	0.03
25–36	0.00	−0.01	−0.01	0.01	0.00	−0.03	−0.01	0.05	0.13	0.10	−0.06	−0.12

Table 11.9 Residual Autocorrelation Function from Fitting Model (11.10) to the Food and Kindred Products Series

Lags	Autocorrelations											
1–12	0.11	−0.13	−0.13	0.05	0.03	−0.01	0.00	0.03	0.20	0.11	0.07	−0.33

Table 11.10   Sample Autocorrelation Functions of Various Functions of the Trade Series

ORIGINAL SERIES

Autocorrelations

Lags												
1–12	0.97	0.94	0.92	0.90	0.88	0.87	0.85	0.84	0.83	0.82	0.81	0.81
13–24	0.78	0.75	0.72	0.70	0.69	0.67	0.66	0.64	0.63	0.62	0.61	0.60
25–36	0.57	0.54	0.52	0.50	0.49	0.47	0.46	0.44	0.43	0.42	0.41	0.40

FIRST DIFFERENCES

Autocorrelations

Lags												
1–12	0.03	−0.19	−0.32	−0.13	0.01	0.22	0.01	−0.11	−0.30	−0.18	0.04	0.90
13–24	0.02	−0.19	−0.29	−0.12	0.02	0.21	0.01	−0.10	−0.28	−0.16	0.03	0.81
25–36	0.00	−0.17	−0.27	−0.12	0.03	0.19	0.01	−0.09	−0.26	−0.14	0.04	0.75

FIRST DIFFERENCE OF ANNUAL SEASONAL DIFFERENCE

Autocorrelations

Lags												
1–12	0.18	−0.03	−0.01	−0.14	0.00	0.10	0.02	0.09	0.10	0.00	−0.07	−0.28
13–24	−0.10	0.03	0.00	0.14	0.02	−0.10	−0.10	−0.05	−0.02	−0.02	0.05	−0.08
25–36	−0.09	−0.03	−0.08	−0.06	0.04	−0.01	−0.12	−0.05	0.00	0.03	0.07	0.01

art.) We are thus led to fit this model to w_t with the results shown in Table 11.5. In fact the residuals from this fit suggest no reason to be unhappy with it.

11.4   FINAL COMMENTS

In this chapter we have introduced the reader to some of the issues involved in the analysis of seasonal time series. In Section 11.2 we presented an example of modeling a time series that appeared to be generated by a simple seasonal model. We observed, however, that the series was quite volatile and that not much confidence could be placed in forecasts based on the derived model. Such a "negative" example is presented because it illustrates a phenomenon that is not uncommon in practice.

In Section 11.3 we display models of four employment time series. The great variety of correlation patterns exhibited by these models is rather striking. Despite the variety of possible seasonal time series models, certain basic autocorrelation patterns do seem to recur in practice. Some of the more useful patterns are displayed in Table 11.6, along with the models that generate them. Finally, we have considered in some detail the steps in modeling two of the employment series.

Supplement 11A

WISCONSIN EMPLOYMENT DATA

All values are given in units of 1000 employees. Each series contains 178 observations.

Table 11A.1 Food and Kindred Products

Year \ Month	Jan.	Feb.	Mar.	Apr.	May	June	July	Aug.	Sept.	Oct.	Nov.	Dec.
1961	56.3	55.7	55.8	56.3	57.2	59.1	71.5	72.2	72.7	61.5	57.4	56.9
1962	55.3	54.9	54.9	54.9	54.6	57.7	68.2	70.6	71.0	60.0	56.0	54.4
1963	53.3	52.8	53.0	53.4	54.3	58.2	67.4	71.0	69.8	59.4	55.6	54.6
1964	53.4	53.0	53.0	53.2	54.2	58.0	67.5	70.1	68.2	56.6	54.9	54.0
1965	52.9	52.6	52.8	53.0	53.6	56.1	66.1	69.8	69.3	61.2	57.5	54.9
1966	53.4	52.7	53.0	52.9	55.4	58.7	67.9	70.0	68.7	59.3	56.4	54.5
1967	52.8	52.8	53.2	55.3	55.8	58.2	65.3	67.9	68.3	61.7	56.4	53.9
1968	52.6	52.1	52.4	51.6	52.7	57.3	65.1	71.5	69.9	61.9	57.3	55.1
1969	53.6	53.4	53.5	53.3	53.9	52.7	61.0	69.9	70.4	59.4	56.3	54.3
1970	53.5	53.0	53.2	52.5	53.4	56.5	65.3	70.7	66.9	58.2	55.3	53.4
1971	52.1	51.5	51.5	52.4	53.3	55.5	64.2	69.6	69.3	58.5	55.3	53.6
1972	52.3	51.5	51.7	51.5	52.2	57.1	63.6	68.8	68.9	60.1	55.6	53.9
1973	53.3	53.1	53.5	53.5	53.9	57.1	64.7	69.4	70.3	62.6	57.9	55.8
1974	54.8	54.2	54.6	54.3	54.8	58.1	68.1	73.3	75.5	66.4	60.5	57.7
1975	55.8	54.7	55.0	55.6	56.4	60.6	70.8	76.4	74.8	62.2		

Table 11A.2 Fabricated Metals

Month / Year	Jan.	Feb.	Mar.	Apr.	May	June	July	Aug.	Sept.	Oct.	Nov.	Dec.
1961	31.1	29.5	31.0	31.5	32.0	32.8	32.8	33.1	33.6	33.4	33.5	33.5
1962	33.1	33.1	33.4	33.9	34.4	35.2	35.0	35.3	34.6	34.1	33.9	33.8
1963	33.6	33.8	34.0	34.4	34.7	35.4	35.5	35.5	35.2	34.5	34.4	34.3
1964	33.9	34.1	34.5	34.3	35.0	36.0	35.0	37.2	37.4	36.9	37.2	37.3
1965	37.2	37.6	36.0	38.8	38.8	39.5	39.6	40.1	39.4	38.9	39.5	39.2
1966	38.7	39.1	39.5	39.7	40.0	41.4	41.3	39.8	38.5	40.2	41.0	40.9
1967	40.5	40.3	40.4	40.3	40.4	41.8	41.1	42.1	40.8	40.5	40.6	40.7
1968	40.6	40.7	40.7	40.7	40.9	41.6	41.2	42.0	41.8	42.3	43.1	42.9
1969	42.6	43.1	43.3	43.5	43.9	45.7	45.6	46.5	45.3	45.2	45.3	45.0
1970	44.2	44.3	44.4	43.4	42.8	44.3	44.4	44.8	44.4	43.1	42.6	42.4
1971	42.2	41.8	40.1	42.0	42.4	43.1	42.4	43.1	43.2	42.8	43.0	42.8
1972	42.5	42.6	42.3	42.9	43.6	44.7	44.5	45.0	44.8	44.9	45.2	45.2
1973	45.0	45.5	46.2	46.8	47.5	48.3	48.3	49.1	48.9	49.4	50.0	50.0
1974	49.6	49.9	49.6	50.7	50.7	50.9	50.5	51.2	50.7	50.3	49.2	48.1
1975	46.6	45.3	44.6	44.0	43.7	43.8	43.0	43.6	44.0	45.0		

Table 11A.3 Transportation Equipment

Month Year	Jan.	Feb.	Mar.	Apr.	May	June	July	Aug.	Sept.	Oct.	Nov.	Dec.
1961	25.3	23.5	37.9	40.6	43.8	43.8	43.7	35.1	40.5	44.6	44.6	45.2
1962	45.3	45.4	45.7	45.7	45.1	45.5	44.8	35.6	44.1	45.3	46.7	47.8
1963	47.4	47.5	47.7	47.4	47.6	47.7	47.1	41.6	45.5	46.3	47.0	47.4
1964	47.1	44.2	44.3	44.1	43.7	36.3	36.0	40.5	46.2	43.4	48.3	48.7
1965	48.0	45.8	46.1	46.5	46.6	46.6	43.4	43.0	44.9	45.5	45.7	45.3
1966	30.0	43.3	43.2	42.4	42.1	42.1	31.0	34.1	42.7	42.7	43.3	42.6
1967	29.5	29.4	38.0	37.8	37.8	38.1	33.3	30.0	39.2	39.1	39.2	39.4
1968	35.8	38.7	38.8	39.2	38.6	39.2	38.2	31.8	38.2	38.1	38.3	38.4
1969	38.7	38.9	35.6	38.5	32.1	39.3	36.8	35.6	38.6	39.3	30.1	39.6
1970	39.1	34.1	39.1	39.2	38.2	37.8	37.1	32.2	38.7	30.8	30.7	37.9
1971	37.8	29.9	33.1	37.8	37.6	37.6	37.6	36.9	37.7	37.0	37.3	37.1
1972	34.8	37.5	37.7	38.0	38.4	36.4	37.3	37.8	39.8	40.4	40.7	40.7
1973	41.0	42.0	42.6	43.3	44.0	44.7	43.0	43.1	46.3	46.2	46.0	45.3
1974	41.0	40.0	37.9	45.2	46.1	44.6	42.3	37.7	45.2	45.9	46.1	45.6
1975	36.3	38.3	40.3	40.1	40.3	40.7	40.5	43.5	44.2	43.5		

Table 11A.4 Trade

Year / Month	Jan.	Feb.	Mar.	Apr.	May	June	July	Aug.	Sept.	Oct.	Nov.	Dec.
1961	239.6	236.4	236.8	241.5	243.7	246.1	244.1	244.2	244.8	246.6	250.9	261.4
1962	237.6	235.7	236.1	242.6	244.5	246.6	245.7	247.7	248.9	251.4	255.9	263.7
1963	242.2	239.3	239.7	247.2	249.3	252.3	252.8	253.6	254.2	256.1	260.3	268.8
1964	250.1	247.9	249.0	253.8	258.3	261.3	261.3	261.9	263.3	267.3	270.6	281.5
1965	261.9	258.6	259.7	266.0	271.1	274.4	274.0	273.8	274.9	280.0	285.4	295.9
1966	275.4	273.6	275.9	281.1	285.2	289.1	289.2	288.9	291.1	295.3	300.2	310.9
1967	286.9	283.0	286.2	291.5	295.4	299.7	297.9	298.1	300.2	304.8	311.9	320.9
1968	298.3	295.5	297.2	302.7	306.7	309.1	308.7	309.9	310.8	314.7	321.2	329.0
1969	307.6	305.5	308.0	314.4	320.5	323.4	323.0	324.4	326.1	329.3	335.0	341.9
1970	321.8	317.3	318.6	323.4	327.1	327.9	325.3	325.7	330.0	333.5	337.1	341.3
1971	321.6	318.2	319.6	326.2	332.3	334.2	334.5	335.5	335.1	338.2	341.9	347.9
1972	329.5	326.4	329.1	337.2	344.9	349.6	351.0	353.8	354.5	357.4	362.1	367.5
1973	347.9	345.0	348.9	355.3	362.4	366.6	366.0	370.2	370.9	374.5	380.2	384.6
1974	360.6	354.4	357.4	367.0	375.7	381.0	381.2	383.0	384.3	387.0	391.7	396.0
1975	374.0	370.4	373.2	381.1	389.9	394.6	394.0	397.0	397.2	399.4		

PROBLEMS

A number of time series are presented below. Not all the series are seasonal. Derive appropriate time series models for these data.

11.1 Monthly averages of weekly total investments of large New York City banks, January 6, 1965, through December 26, 1973. (Read across.) [Source: *Federal Reserve Bulletin.* 120 observations.]

5466.1	4845.9	4671.2	4528.4	4418.1	4461.0	4643.7	4460.7	4260.0	4627.9
4612.4	4759.6	4565.7	4193.9	3822.4	4144.9	3823.7	3653.6	3710.4	3705.8
3999.6	3694.6	3855.4	4323.9	4475.4	4785.4	4882.0	4648.7	4656.4	4775.3
5017.9	4859.6	5287.4	5794.2	5832.6	5597.6	5313.2	5192.7	4807.5	4765.6
4631.0	4727.8	4773.2	5298.9	5909.9	5647.8	5645.1	5965.1	5165.4	4580.8
4144.9	4509.0	3984.8	4030.4	4275.2	4291.7	4324.2	4164.4	4615.9	4962.4
4550.2	4136.5	4232.1	4936.8	4507.1	4329.7	4500.2	5169.4	4993.0	4879.9
5095.0	5459.8	5544.6	5342.9	5106.0	5439.2	4863.7	4657.2	4988.7	4427.2
4562.6	4528.4	5330.7	5263.6	5112.2	5031.1	5551.8	5223.9	5004.4	4901.1
4602.1	4746.2	5247.3	4715.1	4862.8	5039.9	4931.2	4362.8	4137.6	4316.1
3760.8	4247.1	3831.9	3768.8	4154.2	4257.6	4867.9	5787.9	5435.0	5117.0
4984.0	4781.0	3793.0	3679.0	3373.0	4118.0	4679.0	4144.0	4625.0	4932.0

11.2 Monthly averages of weekly interest rates for three-month U.S. Treasury bills, January 6, 1965, through December 26, 1973. (Read across.) [Source: *Federal Reserve Bulletin.* 120 observations.]

3.9340	4.0000	4.0000	3.9925	3.9560	3.8700	3.8880	3.9400	4.0425	4.1820
4.2375	4.4900	4.7000	4.8250	4.7900	4.7300	4.8125	4.6625	4.8760	5.2225
5.7775	5.6360	5.5925	5.0860	4.6325	4.5950	4.2775	3.9340	3.8025	3.7950
4.6080	4.8000	4.9400	5.0625	5.1875	5.4820	5.2750	5.1300	5.3100	5.4675
5.8075	5.6840	5.4200	5.2400	5.2550	5.3875	5.5780	6.0425	6.2950	6.2875
6.1900	6.1475	6.1380	6.7500	7.1975	7.1940	7.2975	7.2950	7.5460	7.8650
7.8500	7.2225	6.5850	6.4875	7.0400	6.8900	6.5250	6.5400	6.4850	6.2380
5.4600	4.8925	4.5560	3.7700	3.4700	3.9200	4.3200	4.8825	5.5840	5.2725
4.9425	4.6560	4.3775	4.3050	3.7340	3.6200	4.0400	4.2480	4.1200	4.2850
4.5000	4.4525	5.0900	5.1350	5.0725	5.2800	5.5825	5.7700	6.4660	6.5175
6.5600	7.1880	8.0825	8.6425	8.4900	7.3100	7.9100	7.6020	7.7700	7.1800
7.8600	8.2900	8.3300	7.9300	7.5300	8.8000	8.2200	7.3000	7.5800	7.1500

11.3 *Quarterly Financial Times* ordinary share index, second quarter, 1958—fourth quarter, 1966. [Source: Reference [24]. 51 observations. Format (16F5.1).]

148.5165.4178.5187.3195.9205.2191.5183.0184.5181.1173.7185.4201.8198.0168.0161.6
170.2184.5211.0218.3231.7247.4301.9323.8314.1321.0312.9323.7349.3310.4295.8301.2
285.8271.7283.6295.7309.3295.7342.0335.1344.4360.9346.5340.6340.3323.3345.6349.3
359.7320.0299.9

See references [24] and [20].

11.4 *Quarterly Financial Times* commodity index, second quarter, 1958—fourth quarter, 1966. [Source: Reference [24]. 51 observations. Format (16F5.2).]

94.7492.4392.4191.6589.3891.0589.8990.1686.7888.4590.6986.0384.8584.0781.9680.03
79.8080.1980.1380.4282.6782.7882.6182.4781.8679.7077.8977.6178.9079.7278.0877.54
76.9976.2578.1380.3881.7882.8184.9986.3185.9590.7392.4287.1885.2085.4487.8589.95
90.2088.8983.25

See references [24] and [20].

11.5   Quarterly UK car production (seasonally adjusted), second quarter, 1958—
        fourth quarter, 1966. [Source: Reference [24]. 51 observations. Format
        (13F6.0).]

187197195916199253227616215363231728231767211211185200152404156163151567213683
244543523111266580253543261675249407246248293062285809366265374241375764354411
249527206165258410279342264824312983300932323424312780363336378275414457459158
460397462279434255475890439365431666399160449564437555426616399254334587

See references [24] and [20].

11.6   Monthly Bureau of Labor Statistics, all industrial commodities price index
        (1964 = 100), 1957–1966. (Read across.) [Source: Bureau of Labor Statistics. 120
        observations.]

10500	10614	10605	10602	10597	10562	10562	10578	10565	10532	10495	10498
10481	10355	10296	10247	10216	10249	10318	10429	10465	10408	10402	10395
10428	10479	10485	10494	10473	10370	10346	10328	10292	10278	10279	10275
10343	10361	10371	10385	10297	10354	10403	10448	10473	10482	10440	10423
10419	10428	10405	10329	10153	10213	10186	10154	10194	10124	10148	10197
10203	10145	10123	10243	10197	10185	10158	10116	10170	10131	10121	10106
10078	10051	10082	10068	10123	10155	10149	10101	10068	10092	10039	10112
10116	10089	10026	9984	9985	9951	9963	9941	9882	9987	10028	10048
10089	10056	10065	10076	10138	10166	10162	10212	10215	10235	10282	10283
10299	10296	10309	10373	10430	10476	10470	10506	10502	10519	10537	10506

11.7   Monthly sales of company X, January 1965—May 1971. (Read across.) [Source:
        Reference [22]. 77 observations.]

154	96	73	49	36	59	95	169	210	278	298	245	200	118	90	79	78	91	167	169
289	347	375	203	223	104	107	85	75	99	135	211	335	460	488	326	346	261	224	141
148	145	223	272	445	560	612	467	518	404	300	210	196	186	247	343	464	680	711	610
613	392	273	322	189	257	324	404	677	858	895	664	628	308	324	248	272			

See references [22] and [18].

11.8   Weekly average U.S. money supply (demand deposits plus currency), January
        4, 1961—December 30, 1970. (Read across.) [Source: Federal Reserve Bulletin.
        522 observations.]

145.8	145.4	145.6	145.0	144.4	143.7	143.1	141.2	141.0	141.3	142.3
142.2	141.0	141.2	142.9	144.2	143.9	143.2	142.6	141.9	140.5	140.8
141.7	143.4	142.7	140.9	141.3	142.1	142.7	142.6	143.2	143.2	143.2
141.3	141.7	142.8	144.4	145.4	143.1	143.3	144.3	145.8	145.8	146.5
146.8	147.8	146.7	146.8	148.0	149.6	151.4	150.3	151.5	150.5	150.4
149.1	148.0	148.0	147.4	145.3	144.9	145.7	146.4	146.4	144.3	145.1

147.0	148.9	148.4	147.3	146.5	145.9	143.4	143.5	144.9	146.1	146.2
144.0	144.5	145.1	145.9	145.5	146.1	146.3	145.9	143.5	143.6	144.7
146.3	146.8	144.9	145.6	146.2	147.7	147.5	148.5	149.0	149.5	148.7
148.6	149.8	151.4	153.2	152.8	154.1	153.2	153.5	152.3	150.9	151.1
150.4	148.4	147.1	148.3	148.9	149.4	147.1	147.8	149.3	151.9	151.3
150.6	149.7	149.3	146.8	147.1	148.2	149.6	150.1	147.9	148.9	149.3
150.7	150.5	150.7	151.4	150.9	148.7	148.5	150.0	151.4	152.8	150.5
150.7	152.0	153.4	153.3	153.9	155.7	155.9	155.1	154.5	156.2	157.0
158.9	158.2	159.5	159.3	159.0	158.7	156.7	156.7	156.0	153.8	152.1
153.0	153.7	155.0	152.5	153.1	154.3	156.2	157.3	155.4	155.1	154.1
152.1	151.2	153.1	153.7	155.9	154.1	153.6	155.2	156.3	156.3	156.1
157.2	157.0	155.4	154.6	156.2	157.4	159.8	158.4	157.0	158.9	160.1
160.3	160.5	161.9	162.3	162.3	160.8	162.3	163.4	166.2	166.3	165.5
168.1	166.4	165.8	163.7	163.7	162.6	160.9	157.9	160.3	160.4	162.2
159.1	159.1	161.4	163.4	165.0	162.4	161.6	160.1	157.8	156.9	158.8
160.1	162.5	161.4	159.6	161.6	162.5	162.2	161.7	163.2	162.5	161.4
159.9	161.5	162.7	165.3	165.9	163.4	165.5	166.5	167.1	167.1	168.8
168.9	168.9	167.7	168.4	170.5	172.9	174.8	173.9	176.7	175.4	174.4
172.1	172.1	171.2	169.6	166.0	167.5	168.5	170.5	169.0	167.7	170.0
172.6	175.0	172.9	171.4	169.9	167.9	165.8	166.7	169.1	171.1	172.2
168.0	170.0	169.3	169.7	168.1	170.0	168.6	168.8	167.0	168.1	169.9
171.4	173.5	169.3	171.4	171.6	171.9	171.1	172.7	172.6	173.1	172.1
172.3	174.3	175.7	179.9	176.8	179.8	178.2	177.5	174.3	173.6	173.4
172.2	170.6	170.5	171.8	173.5	174.8	172.1	174.3	175.2	176.6	174.0
173.9	172.6	173.1	170.6	172.2	174.6	176.3	177.7	174.2	177.1	177.3
176.6	176.1	178.2	177.6	178.0	176.0	176.7	178.8	180.3	180.9	178.6
181.0	182.0	182.5	181.2	183.0	184.1	184.4	183.4	183.4	186.4	186.9
189.3	189.1	193.2	191.3	191.1	187.2	185.6	186.2	183.9	182.5	180.6
183.6	184.2	184.6	183.4	185.2	186.7	190.8	187.7	186.0	185.9	185.3
183.9	184.2	187.7	188.0	189.2	187.3	191.0	189.9	191.5	188.5	189.8
191.1	190.1	189.0	188.7	191.0	192.7	192.9	190.2	193.4	194.0	195.7
193.2	194.3	198.0	196.6	197.4	197.4	200.4	201.3	203.7	203.6	207.6
208.3	206.1	203.3	199.2	200.2	198.2	197.8	195.5	198.2	198.3	199.0
197.1	199.7	201.6	204.8	202.0	199.4	198.2	198.1	197.3	196.9	198.7
200.8	201.7	198.9	202.0	202.4	202.6	200.1	200.3	201.4	200.4	199.4
197.8	200.2	201.9	203.5	199.3	201.1	203.2	204.1	203.1	202.2	205.6
205.4	205.8	203.8	207.0	207.2	210.1	209.2	213.2	216.5	214.0	211.5
206.2	206.1	203.7	202.7	199.8	204.0	204.3	204.9	203.1	207.2	209.7
211.2	210.0	206.3	207.2	205.7	205.0	203.5	207.1	207.5	210.0	205.3
208.3	210.3	210.5	207.7	207.3	209.4	209.0	209.2	207.2	209.9	211.1
213.9	210.1	210.2	213.7	213.2	213.3	211.3	215.3	214.9	216.3	214.1

See reference [27].

Appendix 1

ELEMENTS OF SET THEORY

Definition A1.1   A *set* is a collection of objects. The objects in a set are called *elements*.

We shall generally denote sets by upper case letters, for example, A, B, S, and Ω: and the elements of the set by lower case letters like a, b, x, and ω. In this book the elements of a set will usually be real numbers; although by definition we can talk about sets of people, or grapes, or functions, and the like.

The principal concept of set theory is that of *belonging*. If x belongs to S (x is an element of the set S), we write

$$x \in S$$

The expression

$$x \notin S$$

means that the element x does *not* belong to the set S. It is important to note that a set itself may be an element of some other set. In addition, a set may contain no elements. The set with no elements is called the *empty* set, or the *void* set, and is usually denoted by ϕ (phi).

There are two ways to describe a set: We can list all of its elements, a particularly trying task if the number of elements is large, or we can specify the attribute(s) possessed by an arbitrary element of the set.

EXAMPLE A1.1

Let A be the set of authors of this book. Then we can easily describe the set A by writing

$$A = \{\text{Miller, Wichern}\}$$

where the braces enclose the elements of the set.

EXAMPLE A1.2

Let S be the set of all real numbers between 0 and 1, inclusive. In this case it is impossible to list all the elements of S, but we can conveniently describe this set by specifying the

447

attribute possessed by an arbitrary element, say, x, of S. Thus we can write

$$S = \{x | 0 \le x \le 1\}$$

which is read: S is the set of all numbers x such that x is between 0 and 1.

It is clear that sets may be *finite* (that is, contain a finite number of elements), as in Example A1.1, or *infinite*, as in Example A1.2.

Definition A1.2   Two sets A and B are equal, denoted by $A = B$, if and only if they have exactly the same elements.

Definition A1.3   The set A is called a *subset* of the set B if every element of A is an element of B. The set A is called a *proper* subset of B if A is a subset of B and there is at least one element of B that is not in A.

A subset A of the set B is denoted symbolically by

$$A \subseteq B$$

If A is a proper subset of B (that is, the case $A = B$ is excluded) we write

$$A \subset B$$

We note that *every* set contains the empty set, ϕ, as a subset.

EXAMPLE A1.3

Consider the set of integers $A = \{1, 2, 3, 4\}$. The sets $B = \{3, 4\}$ and $C = \{1, 2, 4\}$ are proper subsets of A. In fact, *any* set containing fewer than four of the integers in A is a proper subset. Note that the set $X = \{0, 1, 2\}$ is *not* a subset of A since there is an element of X, namely, 0, which is not an element of A.

The set $D = \{3, 1, 2, 4\}$ contains exactly the same elements as A, and therefore $A = D$.

EXAMPLE A1.4

Suppose we toss two coins and observe the outcomes. Before the coins are tossed, the set, S, say, of possible outcomes is a collection of *ordered* pairs of the form (__, __), where the first slot contains the outcome for the first coin (either "heads" or "tails") and the second slot contains the outcome for the second coin. Thus

$$S = \{(H, H), (H, T), (T, H), (T, T)\}$$

It is important to note that the elements (H, T) and (T, H) are distinct since they represent different possible outcomes.

If we let x be the outcome for the first coin and y be the outcome for the second coin, then a possible outcome for the experiment can be represented by the pair (x, y). Now consider the sets:

$$A = \{(x, y) | x \text{ or } y \text{ is a head}\}$$
$$B = \{(x, y) | x \text{ is a head}\}$$
$$C = \{(x, y) | x \text{ and } y \text{ are the same}\}$$

Then A, B, and C can be described alternately by

$$A = \{(H, H), (H, T), (T, H)\}$$
$$B = \{(H, H), (H, T)\}$$
$$C = \{(H, H), (T, T)\}$$

The sets A, B, and C are each proper subsets of S. Moreover $B \subset A$. The set C is not a subset of either A or B.

OPERATIONS WITH SETS

Definition A1.4   The *union* of two sets A and B, denoted by $A \cup B$, is the set C consisting of elements that belong to either A or B; that is,

$$C = A \cup B = \{x \mid x \text{ is either in } A \text{ or } B\}$$

Definition A1.5   The *intersection* of two sets A and B, denoted by $A \cap B$, is the set C consisting of elements that belong to both A and B; that is,

$$C = A \cap B = \{x \mid x \text{ is both in } A \text{ and in } B\}$$

It is fairly easy to see that the union of two sets A and B must be at least as large as the intersection of two sets. This follows since an element in the intersection (which may be the empty set ϕ, see Example A1.5) must necessarily be in either A or B and hence in the union. The converse, however, is not true in general. For example, an element of B, say, x, may not be an element of A; that is, $x \in A \cup B$, but $x \notin A \cap B$ (see Figure A1.2).

Operations with sets, like constructing unions and intersections, are conveniently represented by devices known as "Venn diagrams." Consider Figure A1.1. The outer rectangle, labeled S, represents an arbitrary (and frequently universal) set of elements. The inner circles labeled A, B, and C in the figure represent sets that are subsets of S.

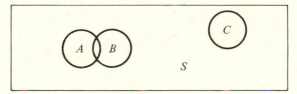

Figure A1.1   A Venn diagram

Note that in Figure A1.1, the sets A and B have some elements in common, indicated by the overlap between the corresponding circles, whereas the sets B and C, for example, do not. Thus we can easily represent the union and intersection of two sets by Venn diagrams, and this is done in Figure A1.2 in which $A \cup B$ and $A \cap B$ are shown.

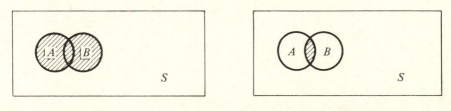

$C = A \cup B$ $C = A \cap B$

Figure A1.2   Venn diagrams of the union and intersection of two sets

If the intersection of two sets is empty, that is, $A \cap B = \phi$, the sets A and B are said to be disjoint (see Figure A1.3).

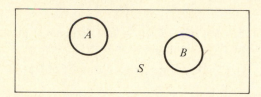

Figure A1.3   Disjoint sets

The notions of union and intersection can be generalized to more than two sets. The extension to three sets and the corresponding notation is depicted in Figure A1.4.

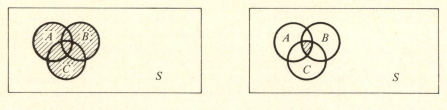

$$C = A \cup B \cup C \qquad\qquad C = A \cap B \cap C$$

Figure A1.4   Venn diagram of the union and intersection of three sets

Result A1.1   Let A, B, and C be three sets, and let ϕ be the empty set. Then the following relationships hold.

Unions

1. $A \cup \phi = A$
2. $A \cup A = A$
3. $A \cup B = B \cup A$
4. $A \cup (B \cup C) = (A \cup B) \cup C$
5. $A \subset B$ if and only if $A \cup B = B$

Intersections

1. $A \cap \phi = \phi$
2. $A \cap A = A$
3. $A \cap B = B \cap A$
4. $A \cap (B \cap C) = (A \cap B) \cap C$
5. $A \subset B$ if and only if $A \cap B = A$

EXAMPLE A1.5

Consider the sets $A = \{1, 2, 3, 4\}$, $B = \{-1, 0, 1\}$, and $C = \{\frac{1}{2}, 3, 5\}$. Then

$$A \cup B = \{-1, 0, 1, 2, 3, 4\}$$

$$B \cup C = \{-1, 0, \tfrac{1}{2}, 1, 3, 5\}$$
$$A \cap B = \{1\}$$
$$B \cap C = \phi \quad \text{(that is, } B \text{ and } C \text{ are disjoint)}$$

Moreover,

$$A \cup B \cup C = \{-1, 0, \tfrac{1}{2}, 1, 2, 3, 4, 5\}$$
$$A \cap B \cap C = \phi$$

Exercise   Using the sets A, B, and C in Example A1.5, verify parts 2–4 in Result A1.1 for both unions and intersections.

EXAMPLE A1.6

Let $A = \{x \,|\, 0 < x < 1\}$ and $B = \{0, 1\}$. Then $A \cap B = \phi$, and $A \cup B = \{x \,|\, 0 \le x \le 1\}$.

Result A1.2   Let A, B, and C be three sets. The operations of union and intersection are connected by the following relations:

1. $(A \cup B) \cap C = (A \cap C) \cup (B \cap C)$
2. $(A \cap B) \cup C = (A \cup C) \cap (B \cup C)$

Exercise   Verify Result A1.2 using the sets in Example A1.5.

Definition A1.6   Let A be a subset of the set S. The *complement* of A, denoted by $\bar{A}$, is the set of elements in S that are *not* in A.
Figure A1.5 illustrates the relationship between a set and its complement.

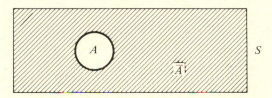

Figure A1.5   A set A and its complement $\bar{A}$

Result A1.3   Let A and B be two subsets of a set S with complements $\bar{A}$ and $\bar{B}$. The following relationships hold:

1. $\overline{(\bar{A})} = A$ (that is, the complement of A complement is A itself)
2. $A \cap \bar{A} = \phi$
3. $A \cup \bar{A} = S$
4. $A \subset B$ if and only if $\bar{B} \subset \bar{A}$
5. $\overline{(A \cup B)} = \bar{A} \cap \bar{B}$
6. $\overline{(A \cap B)} = \bar{A} \cup \bar{B}$
7. $\bar{S} = \phi, \bar{\phi} = S$

Exercise   Verify parts 5 and 6 of Result A1.3 using Venn diagrams. These results are known as the "DeMorgan laws."

EXAMPLE A1.7

Let $S = \{(H, H), (H, T), (T, H), (T, T)\}$ be the set of possible outcomes from a toss of two coins. If $A = \{(H, H), (H, T)\}$ then $\bar{A} = \{(T, H), (T, T)\}$.

EXAMPLE A1.8

Let $S = \{x \mid 0 \leq x \leq 1\}$, $A = \{x \mid 0 \leq x < \frac{1}{2}\}$, and $B = \{x \mid \frac{1}{4} \leq x < \frac{3}{4}\}$. Then

$$\bar{A} = \{x \mid \frac{1}{2} \leq x \leq 1\}$$
$$\bar{B} = \{x \mid 0 \leq x < \frac{1}{4} \quad \text{or} \quad \frac{3}{4} \leq x \leq 1\}$$

Also

$$A \cap B = \{x \mid \frac{1}{4} \leq x < \frac{1}{2}\}$$

and

$$\overline{A \cap B} = \{x \mid 0 \leq x < \frac{1}{4} \quad \text{or} \quad \frac{1}{2} \leq x \leq 1\}$$
$$= \bar{A} \cup \bar{B}$$

Definition A1.7   If A and B are two sets, the *difference* between A and B, denoted by $A - B$, is the set of elements in A that are not in B.

The difference, $A - B$, between two sets is also known as the relative complement of B in A. The difference between two sets is indicated by the shaded region in the Venn diagram of Figure A1.6.

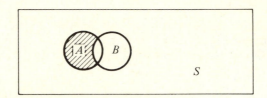

Figure A1.6   The difference set $A - B$

If A and B are subsets of a set S, then it is easy to see (from Figure A1.6, for example) that $A - B = A \cap \bar{B}$.

EXAMPLE A1.9

Let $A = \{-3, 0, \frac{1}{2}, 2\}$ and $B = \{x \mid -\frac{1}{2} \leq x \leq \frac{1}{2}\}$. Then $A - B = \{-3, 2\}$.

EXAMPLE A1.10

Let $A = \{x \mid x$ is a Ph.D. candidate in accounting$\}$ and $B = \{x \mid x$ is a female Ph.D. candidate in any field of business$\}$. We have $A - B = \{x \mid x$ is a male Ph.D. candidate in accounting$\}$. Interchanging the roles of the sets A and B gives $B - A = \{x \mid x$ is a female Ph.D. candidate in any field of business except accounting$\}$.

Appendix 2

CALCULUS REVIEW

We will review briefly the elements of differential and integral calculus used in this book.

LIMITS AND CONTINUITY

Definition A2.1   Let $\{a_n\}$ denote a sequence of numbers. Then L is the limit of the sequence as n tends to infinity $(n \to \infty)$ if $|a_n - L|$ can be made smaller than any preassigned positive number, say, ϵ, for all n sufficiently large. In symbols, $a_n \to L$ (or $\lim_{n \to \infty} a_n = L$) if given any $\epsilon > 0$, there is an integer N such that $|a_n - L| < \epsilon$ whenever $n \geq N$.

EXAMPLE A2.1

The sequence $\{a_n = 1/2^n\}$ has limit $L = 0$. The sequence $\{a_n = 2^n\}$ does not have a finite limit because 2^n grows large without bound as n gets larger and larger (we denote this fact by $\lim_{n \to \infty} 2^n = \infty$). The sequence $\{a_n = (-1)^n\}$ does not have a limit because the values of the sequence oscillate between -1 and $+1$. The sequence $\{a_n = (1 + 1/n)^n\}$ has a limit that is denoted by e and is approximately equal to 2.7. An important generalization of the last result is that $\lim_{n \to \infty} (1 + c/n)^n = e^c$, where c may be any real number.

Definition A2.2   Let $f(x)$ be a real-valued function. Then $f(x)$ is said to have a limit at the point x_0 if for every sequence $\{x_n\}$ such that $x_n \to x_0$ [and $f(x_n)$ is defined for all n] the sequence $\{f(x_n)\}$ has a limit; that is, $f(x_n) \to L$, where L is some number.

EXAMPLE A2.2

Consider the function $f(x) = x^2$, and let $x_0 = 2$. Let $\{x_n\}$ denote any sequence converging to 2. Then we can show that $x_n{}^2 \to x_0{}^2 = 2^2 = 4$; that is,

$$f(x_n) \to f(x_0) = 2^2 = 4$$

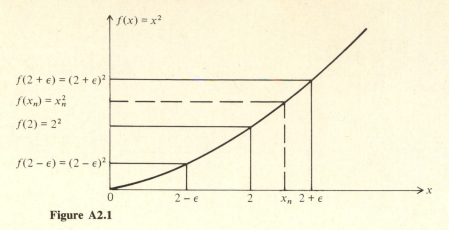

Figure A2.1

This is most easily seen in Figure A2.1. Whenever $|x_n - 2| < \epsilon$, then $f(x_n)$ falls in the interval $[(2 - \epsilon)^2, (2 + \epsilon)^2]$. As ϵ gets closer and closer to zero, $f(x_n)$ must get closer and closer to $f(2) = 2^2 = 4$. Thus $f(x_n) \rightarrow 4 = f(x_0)$ as $n \rightarrow \infty$. A similar argument can be repeated for any value of x_0, showing that $f(x_n) \rightarrow f(x_0)$ whenever $x_n \rightarrow x_0$, no matter what value of x_0 is chosen. This is another way of saying $f(x) = x^2$ is a continuous function of x.

Definition A2.3   A function $f(x)$ is continuous at a point x_0 if for any sequence $\{x_n\}$ such that $x_n \rightarrow x_0$ we have $f(x_n) \rightarrow f(x_0)$. The function is said to be continuous if the above condition holds for any value of x_0.

**DIFFERENTIATION OF FUNCTIONS
OF A SINGLE VARIABLE**

Definition A2.4   Let $f(x)$ be a function. Let x_0 be a fixed point, and let $\{x_n\}$ be any sequence such that $x_n \rightarrow x_0$, but $x_n \neq x_0$ for any value of n. Define the function

$$g(x_n ; x_0) = \frac{f(x_n) - f(x_0)}{x_n - x_0}$$

If the limit of $g(x_n ; x_0)$ as $n \rightarrow \infty$ exists, then that limit is called the *derivative* of $f(x)$ at x_0. The derivative is denoted by $[df(x)]/dx|_{x=x_0}$ or by $f'(x_0)$.
An intuitive notion of derivative can be seen in Figure A2.2. The function

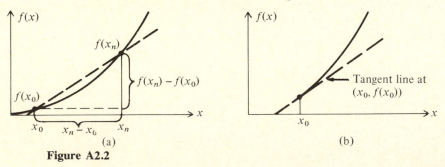

(a)   (b)

Figure A2.2

$$g(x_n ; x_0) = \frac{f(x_n) - f(x_0)}{x_n - x_0}$$

is seen from Figure A2.2(a) to be the *slope* of the line passing through the points $[x_0, f(x_0)]$ and $[x_n, f(x_n)]$. In the limit when x_n "reaches" x_0, $g(x_n ; x_0)$ is the slope of the line *tangent* to the graph of $f(x)$ at the point $[x_0, f(x_0)]$, as shown in Figure A2.2(b).

EXAMPLE A2.3

Let $f(x) = x^2$ and $x_0 = 2$. Then

$$g(x_n ; x_0) = \frac{x_n^2 - x_0^2}{x_n - x_0} = \frac{x_n^2 - 4}{x_n - 2} = \frac{(x_n - 2)(x_n + 2)}{(x_n - 2)} = (x_n + 2)$$

If $x_n \to x_0 = 2$ as $n \to \infty$,

$$g(x_n ; x_0) \to (2 + 2) = 4$$

Thus

$$f'(2) = \frac{dx^2}{dx}\bigg|_{x=2} = 4$$

The same argument applies for any value of x_0; that is, $f'(x_0) = 2x_0$. Dropping the subscript on x_0, we usually write $f'(x) = dx^2/dx = 2x$ for any value of x.

Definition A2.4 can be used to obtain the following formulas for derivatives of some commonly occurring functions. In these results n is an integer, and a and b are real numbers.

Function $f(x)$	Derivative $f'(x)$
x^n	nx^{n-1}
$(x + a)^n$	$n(x + a)^{n-1}$
$(bx + a)^n$	$nb(bx + a)^{n-1}$
e^x	e^x
e^{-x}	$-e^{-x}$
e^{bx+a}	be^{bx+a}
$e^{b(x-a)^2}$	$2b(x - a)e^{b(x-a)^2}$
a	0
$\ln x$	$1/x$

Theorem A2.1   Let $f(x)$ and $g(x)$ be two differentiable functions. Then

1. $\dfrac{d[af(x) + bg(x)]}{dx} = a\dfrac{df(x)}{dx} + b\dfrac{dg(x)}{dx}$

2. $\dfrac{d[f(x)g(x)]}{dx} = f(x)\dfrac{dg(x)}{dx} + \dfrac{df(x)}{dx}g(x)$

Corollary

1. $\dfrac{d[af(x)]}{dx} = a\dfrac{df(x)}{dx}$

2. If

$$f(x) = a_0 + a_1 x + a_2 x^2 + \cdots + a_n x^n$$

then

$$f'(x) = a_1 + 2a_2 x + \cdots + na_n x^{n-1}$$

Definition A2.5 Let $f(x)$ be a function with derivative $f'(x)$. Then the second derivative of $f(x)$, $f''(x)$, is simply the derivative of $f'(x)$; that is,

$$f''(x) = \frac{df'(x)}{dx}$$

Theorem A2.2 Let $f(x)$ be a function that is twice differentiable at all points in the interval (a, b) and attains a maximum or a minimum within the interval. Let x_0 be the point where the maximum or minimum is attained. Then $f'(x_0) = 0$ and $f''(x_0) < 0$ if the maximum is attained, whereas $f'(x_0) = 0$ and $f''(x_0) > 0$ if a minimum is attained (see Figure A2.3).

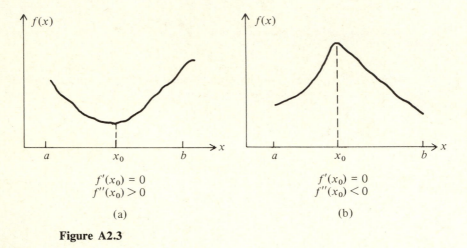

$$f'(x_0) = 0 \qquad\qquad f'(x_0) = 0$$
$$f''(x_0) > 0 \qquad\qquad f''(x_0) < 0$$

(a) \qquad\qquad\qquad (b)

Figure A2.3

EXAMPLE A2.4

Let $f(x) = b(x - a)^2$, $b > 0$. A graph of $f(x)$ will clearly show that the function attains the minimum value of zero at the point $x_0 = a$. Let us confirm this using Theorem A2.2. Now $f'(x) = 2b(x - a)$, so

$$f'(a) = 2b(a - a) = 0$$

Furthermore $f''(x) = 2b$, so

$$f''(a) = 2b > 0$$

Thus we have a minimum point.

EXAMPLE A2.5

Let

$$f(x) = (x - a_1)^2 + (x - a_2)^2$$

Then
$$f'(x) = 2(x - a_1) + 2(x - a_2) = 4x - 2(a_1 + a_2)$$
Setting $f'(x_0) = 0$, we get $4x_0 - 2(a_1 + a_2) = 0$, which implies
$$x_0 = \frac{2(a_1 + a_2)}{4} = \frac{a_1 + a_2}{2}$$
Now $f''(x) = 4$, so $f''(x_0) = 4 > 0$. Hence $f(x)$ attains a minimum at
$$x_0 = \frac{a_1 + a_2}{2}$$

INTEGRATION OF FUNCTIONS OF A SINGLE VARIABLE

The process of integration is analogous to the arithmetic process of addition. The definite integral of a function may be interpreted as an area. These two remarks will be illustrated by simple examples before we formally define integration.

EXAMPLE A2.6

The function $f(x) = 2x$, $0 \le x \le 1$, is graphed in Figure A2.4. The shaded area in the figure is just the area of a right triangle with base 1 and height 2, so the
$$\text{area} = (\tfrac{1}{2})(\text{base})(\text{height}) = (\tfrac{1}{2})(1)(2) = 1$$

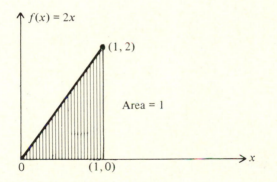

Figure A2.4

Another way to get this area is to divide the interval $[0, 1]$ into two equal parts and erect rectangles over the subintervals $[0, \tfrac{1}{2}]$ and $[\tfrac{1}{2}, 1]$, as shown in Figure A2.5. The left rectangle has a height of $f(\tfrac{1}{4}) = 2(\tfrac{1}{4})$, so its area is
$$(\text{base})(\text{height}) = (\tfrac{1}{2})(2)(\tfrac{1}{4}) = \tfrac{1}{4}$$
The right rectangle has height $f(\tfrac{3}{4}) = 2(\tfrac{3}{4})$, so its area is
$$(\tfrac{1}{2})(2)(\tfrac{3}{4}) = \tfrac{3}{4}$$
The area of the triangle is equal to the *sum of the areas of the rectangles*; that is, $1 = \tfrac{1}{4} + \tfrac{3}{4}$.

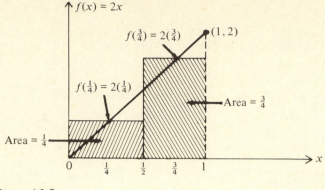

Figure A2.5

EXAMPLE A2.7

Consider the function $f(x) = x^2$, $0 \le x \le 1$. The method of integration can be used to show that the area under this curve is $\frac{1}{3}$. In this example we present a heuristic argument to show how this comes about. Again divide $[0, 1]$ into two subintervals $[0, \frac{1}{2}]$ and $[\frac{1}{2}, 1]$, and construct rectangles as in Example A2.6 (see Figure A2.6). In this case the sum of the areas of

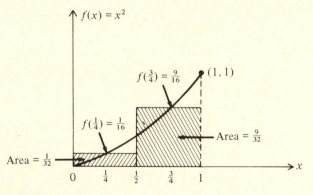

Figure A2.6

the rectangles is $\frac{1}{32} + \frac{9}{32} = \frac{10}{32} = 0.31250$, which is *not* equal to $\frac{1}{3}$, but is accurate to one decimal place. The percentage error in the area calculation is

$$100 \times \frac{0.33333 - 0.31250}{0.33333} = 6.249\%$$

We can decrease the percentage error quite a bit by dividing $[0, 1]$ into more subintervals. For instance, if we use the intervals $[0, \frac{1}{4}]$, $[\frac{1}{4}, \frac{1}{2}]$, $[\frac{1}{2}, \frac{3}{4}]$, and $[\frac{3}{4}, 1]$, the rectangles over these intervals will have heights equal to the values of $f(x)$ at the midpoints of the respective intervals; that is,

$$f\left(\frac{1}{8}\right) = \frac{1}{64}, \quad f\left(\frac{3}{8}\right) = \frac{9}{64}, \quad f\left(\frac{5}{8}\right) = \frac{25}{64}, \quad f\left(\frac{7}{8}\right) = \frac{49}{64}$$

The areas of the rectangles are

$$\left(\frac{1}{4}\right)\left(\frac{1}{64}\right)=\frac{1}{256}, \quad \left(\frac{1}{4}\right)\left(\frac{9}{64}\right)=\frac{9}{256}, \quad \left(\frac{1}{4}\right)\left(\frac{25}{64}\right)=\frac{25}{256}, \quad \left(\frac{1}{4}\right)\left(\frac{49}{64}\right)=\frac{49}{256}$$

The *sum* of these areas is

$$\frac{1+9+25+49}{256}=\frac{84}{256}=0.32812$$

yielding a percentage error of

$$100\times\frac{0.33333-0.32812}{0.33333}=1.563\%$$

(To two decimal places the answer is actually correct.) By using more intervals, we can make the percentage error even smaller. In fact, we can make the percentage error as small as we like by using enough intervals. "In the limit," that is, if we could use an infinite number of "intervals" of infinitely small length, the "sum" of the areas of the "rectangles" over these "intervals" will be equal to the area under the curve, which is $\frac{1}{3}$. This latter quantity is the integral of $f(x)=x^2$ over the interval $[0,1]$. The integral is denoted by $\int_0^1 x^2\,dx$. The symbol $\int$ is called an "integral sign" and is an elongated S to denote that it stands for a special kind of summation.

Definition A2.6   Let $f(x)$ be a function continuous on the interval $[a,b]$. Let

$$a=x_0<x_1<\cdots<x_{n-1}<x_n=b$$

be a collection of points that divide the interval $[a,b]$ into subintervals $[x_0,x_1]$, $[x_1,x_2],\ldots,[x_{n-1},x_n]$. Let $x_1^*,x_2^*,\ldots,x_n^*$ denote the midpoints of these intervals, and consider the sum

$$\sum_{i=1}^{n} f(x_i^*)(x_i-x_{i-1}) \tag{A2.1}$$

which represents the sum of the areas of the rectangles with bases $[x_0,x_1]$, $[x_1,x_2],\ldots,[x_{n-1},x_n]$ and corresponding heights $f(x_1^*),f(x_2^*),\ldots,f(x_n^*)$. The integral of $f(x)$ over $[a,b]$, denoted by $\int_a^b f(x)\,dx$, is defined to be the limit of the sum in equation (A2.1) as $n\to\infty$ in such a way that the lengths of the subintervals tend to zero.

Definition A2.6 can be used to obtain the following formulas for integrals of some commonly occurring functions. In these results, n is an integer, and c and d are real numbers.

Function $f(x)$	*Integral* $\displaystyle\int_a^b f(x)\,dx$
x^n	$(b^{n+1}-a^{n+1})/(n+1)$
$(x+c)^n$	$[(b+c)^{n+1}-(a+c)^{n+1}]/(n+1)$
$(dx+c)^n$	$[(db+c)^{n+1}-(da+c)^{n+1}]/[d(n+1)]$
e^x	e^b-e^a
e^{-x}	$e^{-a}-e^{-b}$
$e^{dx+c},\,d>0$	$(1/d)(e^{db+c}-e^{da+c})$
$e^{-dx+c},\,d>0$	$(1/d)(e^{-da+c}-e^{-db+c})$
$1/x$	$\ln b-\ln a$

Theorem A2.3   Let $f(x)$ and $g(x)$ be continuous functions on $[a,b]$. Then

$$\int_a^b [c_1 f(x)+c_2 g(x)]\,dx = c_1\int_a^b f(x)\,dx + c_2\int_a^b g(x)\,dx \tag{A2.2}$$

Corollary

1. $\displaystyle\int_a^b [c_1f_1(x) + c_2f_2(x) + \cdots + c_nf_n(x)]\, dx$

$$= c_1\int_a^b f_1(x)\, dx + c_2\int_a^b f_2(x)\, dx + \cdots + c_n\int_a^b f_n(x)\, dx$$

2. $\displaystyle\int_a^b (c_0 + c_1x + c_2x^2 + \cdots + c_nx^n)\, dx$

$$= c_0(b - a) + c_1\left(\frac{b^2 - a^2}{2}\right) + \cdots + c_n\left(\frac{b^{n+1} - a^{n+1}}{n + 1}\right)$$

We will now indicate the relationship between differentiation and integration. Let $f(t)$ be a continuous function on $[a, b]$, and let x be any point in the interval (a, b). Then the integral $F(x) = \int_a^x f(t)\, dt$ is the area under the curve over the interval $[a, x]$ and can be thought of as a function of x (see Figure A2.7). The fundamental theorem

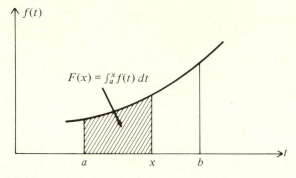

$$F(x) = \int_a^x f(t)\, dt$$

Figure A2.7

of calculus says that

$$F'(x) = \frac{d}{dx}\int_a^x f(t)\, dt = f(x) \tag{A2.3}$$

In the sense stated in equation (A2.3), the derivative of the integral of a function is the function itself. In that sense also, then, differentiation and integration work in exactly opposite directions.

In order to motivate the next definition we present the following example.

EXAMPLE A2.8

We have seen that

$$\int_a^b e^{-x}\, dx = e^{-a} - e^{-b}$$

Letting $a = 0$, we have

$$\int_0^b e^{-x}\, dx = e^0 - e^{-b} = 1 - e^{-b}$$

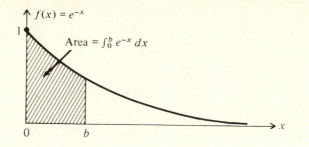

Figure A2.8

In Figure A2.8 this integral is pictured as an area. We ask the questions: What happens to this area as b gets larger and larger? Does it have a limiting value? In symbols, we want the value of

$$\lim_{b \to \infty} \int_0^b e^{-x}\, dx = \lim_{b \to \infty}(1 - e^{-b})$$

It is clear from Figure A2.8 that e^{-b} gets closer and closer to zero as b increases. Hence

$$\lim_{b \to \infty}(1 - e^{-b}) = 1 - 0 = 1$$

and hence

$$\lim_{b \to \infty} \int_0^b e^{-x}\, dx = 1$$

This limit is given the special symbol $\int_0^\infty e^{-x}\, dx$.

Definition A2.7 If $\lim_{b \to \infty} \int_a^b f(x)\,dx$ exists, we denote this limit by the symbol $\int_a^\infty f(x)\, dx$. Similarly,

$$\int_{-\infty}^b f(x)\, dx = \lim_{a \to -\infty} \int_a^b f(x)\, dx \qquad \text{and} \qquad \int_{-\infty}^\infty f(x)\, dx = \lim_{\substack{a \to -\infty \\ b \to +\infty}} \int_a^b f(x)\, dx$$

provided these limits exist.

We conclude this section with some integrals that occur frequently in statistics.

Theorem A2.4

Normal Distribution

1. $\displaystyle \int_{-\infty}^\infty \frac{1}{\sqrt{2\pi}} e^{-(x^2/2)}\, dx = 1$

2. $\displaystyle \int_0^\infty \frac{1}{\sqrt{2\pi}} e^{-(x^2/2)}\, dx = \int_{-\infty}^0 \frac{1}{\sqrt{2\pi}} e^{-(x^2/2)}\, dx = \frac{1}{2}$

(See Figure A2.9)

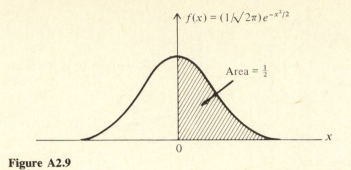

Figure A2.9

PARTIAL DERIVATIVES, MINIMIZATION AND MAXIMIZATION OF FUNCTIONS OF TWO VARIABLES

The purpose of this section is to state conditions under which functions of two variables attain maxima and minima. These conditions are needed for the discussion in Chapter 5, and generalizations of them are needed in Chapter 6. Readers willing to accept the results stated in these chapters on faith may skip this section.

In order to proceed we must first define the concept of partial differentiation. We shall do this informally by means of examples.

EXAMPLE A2.9

Consider the function of two variables

$$f(x, y) = 4x^2 + 3y^2 - 4xy - 8y + 8$$

The partial derivative of $f(x, y)$ with respect to x is found by differentiating $f(x, y)$ as a function of x alone while treating y as a constant. For example, the derivative of the term $4x^2$ is $(4)(2)x = 8x$, whereas the derivative of the term $3y^2$ is zero because this term is treated as a constant. Thus the partial derivative of $f(x, y)$ with respect to x is

$$8x + 0 - 4y - 0 + 0 = 8x - 4y$$

Similarly, the partial derivative of $f(x, y)$ with respect to y (treating x as a constant) is

$$0 + 6y - 4x - 8 + 0 = 6y - 4x - 8$$

The notation for the partial derivative of $f(x, y)$ with respect to x is $[\partial f(x, y)]/\partial x$, and, similarly, the partial derivative of $f(x, y)$ with respect to y is $[\partial f(x, y)]/\partial y$. Using this notation we have

$$\frac{\partial f(x, y)}{\partial x} = 8x - 4y$$

whereas

$$\frac{\partial f(x, y)}{\partial y} = 6y - 4x - 8$$

We can take partial derivatives of partial derivatives to obtain second-order partial derivatives. The second partial derivative of $f(x, y)$ with respect to x is denoted by $[\partial^2 f(x, y)]/\partial x^2$ and is found by

$$\frac{\partial^2 f(x, y)}{\partial x^2} = \frac{\partial}{\partial x}\left[\frac{\partial f(x, y)}{\partial x}\right] = 8 - 0 = 8$$

whereas

$$\frac{\partial^2 f(x, y)}{\partial y^2} = \frac{\partial}{\partial y}\left[\frac{\partial f(x, y)}{\partial y}\right] = 6 - 0 - 0 = 6$$

If we take the partial derivative of $[\partial f(x, y)]/\partial x$ with respect to y, we get the mixed partial derivative

$$\frac{\partial^2 f(x, y)}{\partial y\, \partial x} = \frac{\partial}{\partial y}\left[\frac{\partial f(x, y)}{\partial x}\right] = 0 - 4 = -4$$

and similarly

$$\frac{\partial^2 f(x, y)}{\partial x\, \partial y} = \frac{\partial}{\partial x}\left[\frac{\partial f(x, y)}{\partial y}\right] = 0 - 4 - 0 = -4$$

In this example the two mixed partial derivatives have the same value.

EXAMPLE A2.10

Given the function

$$f(x, y) = x^3 + xy^2 - x^2 y + y^3 - 2$$

we find

$$\frac{\partial f(x, y)}{\partial x} = 3x^2 + y^2 - 2xy$$

$$\frac{\partial f(x, y)}{\partial y} = 2xy - x^2 + 3y^2$$

$$\frac{\partial^2 f(x, y)}{\partial x^2} = 6x - 2y$$

$$\frac{\partial^2 f(x, y)}{\partial y^2} = 2x + 6y$$

$$\frac{\partial^2 f(x, y)}{\partial y\, \partial x} = 2y - 2x$$

and

$$\frac{\partial^2 f(x, y)}{\partial x\, \partial y} = 2y - 2x$$

The procedure for checking for maxima and minima is as follows. First find the partial derivatives of $f(x, y)$ with respect to x and to y. Then set

$$\frac{\partial f(x, y)}{\partial x} = 0 \qquad \text{and} \qquad \frac{\partial f(x, y)}{\partial y} = 0$$

and solve the resulting two equations for x and y. Let x_0 and y_0 denote the solutions. Next find $[\partial^2 f(x, y)]/\partial x^2$, $[\partial^2 f(x, y)]/\partial x\, \partial y$, and $[\partial^2 f(x, y)]/\partial y^2$, and substitute x_0 and y_0 into these functions. Let A, B, and C denote the resulting values. Then if $B^2 - AC < 0$ and $A + C < 0$, (x_0, y_0) is a maximum point; whereas if $B^2 - AC < 0$ and $A + C > 0$, (x_0, y_0) is a minimum point. If $B^2 - AC > 0$, then (x_0, y_0) is neither a maximum nor minimum point; whereas if $B^2 - AC = 0$, the nature of (x_0, y_0) cannot be determined without further analysis. It should be pointed out that this procedure assumes a number of conditions on $f(x, y)$ that we will not delineate. All of the examples in the text satisfy these conditions.

EXAMPLE A2.11

Consider the function

$$f(x, y) = 4x^2 + 3y^2 - 4xy - 8y + 8$$

introduced in Example A2.9. Setting

$$\frac{\partial f(x, y)}{\partial x} = 0 \qquad \text{and} \qquad \frac{\partial f(x, y)}{\partial y} = 0$$

we get

$$8x_0 - 4y_0 = 0$$
$$-4x_0 + 6y_0 = 8$$

which yields $x_0 = 1$ and $y_0 = 2$. Furthermore $A = 8$, $B = -4$, and $C = 6$. This yields

$$B^2 - AC = 16 - (8)(6) = 16 - 48 = -32 < 0$$

and

$$A + C = 8 + 6 = 14 > 0$$

so $(1, 2)$ is a minimum point of the function

$$f(x, y) = 4x^2 + 3y^2 - 4xy - 8y + 8$$

ELEMENTS
OF MATRIX ALGEBRA

Definition A3.1   An $n \times m$ *matrix*, generally denoted by upper-case letters like A, R, Σ, and so forth, is a rectangular array of elements having n rows and m columns.

Examples of matrices follow.

$$A = \begin{bmatrix} -7 & 2 \\ 0 & 1 \\ 3 & 4 \end{bmatrix}, \qquad B = \begin{bmatrix} x & 3 & 0 \\ 4 & -2 & 1/x \\ \sin x & 1 & x^2 \end{bmatrix}, \qquad I = \begin{bmatrix} 1 & 0 & 0 \\ 0 & 1 & 0 \\ 0 & 0 & 1 \end{bmatrix}$$

$$X = \begin{bmatrix} x_1 \\ x_2 \\ x_3 \\ x_4 \end{bmatrix}, \qquad \Sigma = \begin{bmatrix} 1 & 0.5 & -0.3 \\ 0.7 & 0 & 1 \end{bmatrix}, \qquad E = [e_1] \qquad \text{(A3.1)}$$

In our work the matrix elements will be real numbers or functions taking on values in the real numbers.

Definition A3.2   The *dimension* (abbreviated "dim") of an $n \times m$ matrix is the ordered pair $n \times m$; n is the row dimension and m is the column dimension. The dimension of a matrix is frequently indicated in parentheses below the letter representing the matrix. Thus the $n \times m$ matrix A is denoted by

$$\underset{(n \times m)}{A}$$

In the examples above the dimension of the matrix B is 3×3, and this information can be conveyed by writing

$$\underset{(3 \times 3)}{B}$$

An $n \times m$ matrix, say, A, of arbitrary constants can be written

$$\underset{(n \times m)}{A} = \begin{bmatrix} a_{11} & a_{12} & \cdots & a_{1m} \\ a_{21} & a_{22} & \cdots & a_{2m} \\ \cdot & \cdot & & \cdot \\ \cdot & \cdot & & \cdot \\ \cdot & \cdot & & \cdot \\ a_{n1} & a_{n2} & & a_{nm} \end{bmatrix}$$

or more compactly as

$$\underset{(n \times m)}{A} = \{a_{ij}\}$$

where the index i refers to the row and the index j refers to the column.

An $n \times 1$ matrix is referred to as a column *vector*. An $1 \times m$ matrix is referred to as a row *vector*. Unlike more general matrices, vectors are usually denoted by boldface letters.

For example,

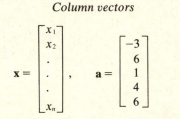

Column vectors	Row vectors

$$\mathbf{x} = \begin{bmatrix} x_1 \\ x_2 \\ \cdot \\ \cdot \\ \cdot \\ x_n \end{bmatrix}, \quad \mathbf{a} = \begin{bmatrix} -3 \\ 6 \\ 1 \\ 4 \\ 6 \end{bmatrix}$$

$$\mathbf{y} = (y_1, y_2, \ldots, y_n)$$
$$\mathbf{b} = (0 \quad 1 \quad 0 \quad 0 \quad -1)$$

A vector (either row or column) consisting of, say, n elements may be regarded as representing a point in n-dimensional space. For example, with $n = 2$, the vector $\mathbf{x} = (x_1, x_2)$ may be regarded as representing the point in the plane with coordinates x_1, x_2 (see Figure A3.1).

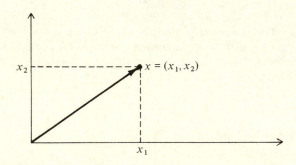

Figure A3.1

Definition A3.3   If an arbitrary matrix A has the *same* number of rows and columns, then A is called a *square* matrix.

In the examples of matrices given by (A3.1) the matrices B, I, and E are square matrices.

Definition A3.4   The $n \times n$ *identity* matrix, denoted by

$$\underset{(n \times n)}{I}$$

is the square matrix with ones on the main (NW–SE) diagonal and zeros elsewhere. A 3×3 identity matrix is included in the collection (A3.1).

Definition A3.5   Consider the $n \times m$ matrix A with arbitrary elements a_{ij}, $i = 1, 2, \ldots, n$, $j = 1, 2, \ldots, m$. The *transpose* of the matrix A, denoted by A', is the $m \times n$ matrix with arbitrary elements a_{ji}, $j = 1, 2, \ldots, m$, $i = 1, 2, \ldots, n$; that is, the transpose of the matrix A is obtained from A by interchanging the rows and columns.

As an example, if

$$\underset{(2 \times 3)}{A} = \begin{bmatrix} 2 & 1 & 3 \\ 7 & -4 & 6 \end{bmatrix}$$

then

$$\underset{(3 \times 2)}{A'} = \begin{bmatrix} 2 & 7 \\ 1 & -4 \\ 3 & 6 \end{bmatrix}$$

Definition A3.6   Two matrices

$$\underset{(n \times m)}{A} = \{a_{ij}\} \qquad \text{and} \qquad \underset{(n \times m)}{B} = \{b_{ij}\}$$

are said to be *identical*, written $A = B$, if $a_{ij} = b_{ij}$, $i = 1, 2, \ldots, n$, $j = 1, 2, \ldots, m$; that is, two matrices are identical if

1. Their dimensionality is the same.
2. Every corresponding element is the same.

Definition A3.7   Let A be an $n \times n$ (square) matrix. The matrix A is said to be *symmetric* if $A = A'$; that is, A is symmetric if $a_{ij} = a_{ji}$, $i = 1, 2, \ldots, n$, $j = 1, 2, \ldots, n$.

Examples of symmetric matrices are

$$\underset{(3 \times 3)}{I} = \begin{bmatrix} 1 & 0 & 0 \\ 0 & 1 & 0 \\ 0 & 0 & 1 \end{bmatrix}, \qquad \underset{(2 \times 2)}{A} = \begin{bmatrix} 2 & 4 \\ 4 & 1 \end{bmatrix}$$

$$\underset{(4 \times 4)}{B} = \begin{bmatrix} a & c & e & f \\ c & b & g & d \\ e & g & c & a \\ f & d & a & d \end{bmatrix}$$

Definition A3.8: Matrix addition   Let the matrices A and B both be of dim $n \times m$ with arbitrary elements a_{ij} and b_{ij}, $i = 1, 2, \ldots, n$, $j = 1, 2, \ldots, m$, respectively. The sum of the matrices A and B is an $n \times m$ matrix C, written $C = A + B$, such that the arbitrary element of C, c_{ij}, is given by

$$c_{ij} = a_{ij} + b_{ij}, \quad \begin{aligned} i &= 1, 2, \ldots, n \\ j &= 1, 2, \ldots, m \end{aligned}$$

Note that the addition of matrices is defined only for matrices of the same dimension. Two examples follow.

$$\begin{bmatrix} 3 & 2 & 3 \\ 4 & 1 & 1 \end{bmatrix} + \begin{bmatrix} 3 & 6 & 7 \\ 2 & -1 & 0 \end{bmatrix} = \begin{bmatrix} 6 & 8 & 10 \\ 6 & 0 & 1 \end{bmatrix}$$

$$\quad A \qquad + \qquad B \qquad = \qquad C$$

$$\begin{bmatrix} 3 \\ -1 \end{bmatrix} + \begin{bmatrix} 1 \\ 0 \end{bmatrix} = \begin{bmatrix} 4 \\ -1 \end{bmatrix}$$

$$\mathbf{x} \quad + \quad \mathbf{y} \quad = \quad \mathbf{z}$$

Definition A3.9: Scalar multiplication   Let α be an arbitrary scalar, that is, a 1×1 matrix, and

$$\underset{(n \times m)}{A} = \{a_{ij}\}$$

Then

$$\underset{(n \times m)}{\alpha A} = \underset{(n \times m)}{A \alpha} = \underset{(n \times m)}{B} = \{b_{ij}\}$$

where $b_{ij} = \alpha a_{ij} = a_{ij} \alpha$, $i = 1, 2, \ldots, n$, $j = 1, 2, \ldots, m$; that is, multiplication of a matrix by a scalar produces a new matrix whose elements are the elements of the original matrix *each* multiplied by the scalar.

For example, $\alpha = 2$

$$2 \begin{bmatrix} 3 & -4 \\ 2 & 6 \\ 0 & 5 \end{bmatrix} = \begin{bmatrix} 3 & -4 \\ 2 & 6 \\ 0 & 5 \end{bmatrix} 2 = \begin{bmatrix} 6 & -8 \\ 4 & 12 \\ 0 & 10 \end{bmatrix}$$

$$\alpha A \qquad = \qquad A \alpha \qquad = \qquad B$$

Result A3.1   For all matrices A, B, and C (*of equal dim*) and scalars α and β,

1. $(A + B) + C = A + (B + C)$
2. $A + B = B + A$
3. $\alpha(A + B) = \alpha A + \alpha B$
4. $(\alpha + \beta)A = \alpha A + \beta A$
5. $(A + B)' = A' + B'$ (that is, the transpose of the sum is equal to the sum of the transposes)
6. $(\alpha \beta)A = \alpha(\beta A)$
7. $(\alpha A)' = \alpha A'$

Definition A3.10: Matrix subtraction   Let

$$\underset{n\times m}{A} = \{a_{ij}\} \quad \text{and} \quad \underset{(n\times m)}{B} = \{b_{ij}\}$$

be two matrices of equal dimension. Then the difference between A and B, written $A - B$, is an $n \times m$ matrix $C = \{c_{ij}\}$ given by

$$C = A - B = A + (-1)B$$

that is, $c_{ij} = a_{ij} + (-1)b_{ij}$, $i = 1, 2, \ldots, n$, $j = 1, 2, \ldots, m$.

Definition A3.11: Matrix multiplication   The product AB of an $n \times m$ matrix $A = \{a_{ij}\}$ and an $m \times p$ matrix $B = \{b_{ij}\}$ is the $n \times p$ matrix C whose element c_{ij} is given by

$$c_{ij} = \sum_{k=1}^{m} a_{ik}b_{kj}, \qquad \begin{array}{l} i = 1, 2, \ldots, n \\ j = 1, 2, \ldots, p \end{array}$$

Note that for the product AB to be defined, the column dimension of A must equal the row dimension of B. Then the row dimension of AB equals the row dimension of A, and the column dimension of AB equals the column dimension of B.

For example, let

$$\underset{(2\times3)}{A} = \begin{bmatrix} 3 & -1 & 2 \\ 4 & 0 & 5 \end{bmatrix} \quad \text{and} \quad \underset{(3\times2)}{B} = \begin{bmatrix} 3 & 4 \\ 6 & -2 \\ 4 & 3 \end{bmatrix}$$

then

$$\underset{(2\times3)}{\begin{bmatrix} 3 & -1 & 2 \\ 4 & 0 & 5 \end{bmatrix}} \underset{(3\times2)}{\begin{bmatrix} 3 & 4 \\ 6 & -2 \\ 4 & 3 \end{bmatrix}} = \begin{bmatrix} 11 & 20 \\ 32 & 31 \end{bmatrix} = \underset{(2\times2)}{\begin{bmatrix} c_{11} & c_{12} \\ c_{21} & c_{22} \end{bmatrix}}$$

where

$c_{11} = (3)(3) + (-1)(6) + (2)(4) = 11$
$c_{12} = (3)(4) + (-1)(-2) + (2)(3) = 20$
$c_{21} = (4)(3) + (0)(6) + (5)(4) = 32$
$c_{22} = (4)(4) + (0)(-2) + (5)(3) = 31$

As an additional example consider the product of two vectors. Let

$$\mathbf{x} = \begin{pmatrix} 1 \\ 0 \\ -2 \\ 3 \end{pmatrix} \quad \text{and} \quad \mathbf{y} = \begin{pmatrix} 2 \\ -3 \\ -1 \\ -8 \end{pmatrix}$$

then

$$\mathbf{x}' = (1 \quad 0 \quad -2 \quad 3) \quad \text{and} \quad \mathbf{x}'\mathbf{y} = (1 \quad 0 \quad -2 \quad 3) \begin{bmatrix} 2 \\ -3 \\ -1 \\ -8 \end{bmatrix} = (-20)$$

Note that the product $\mathbf{xy}$ is undefined since $\mathbf{x}$ is a 4×1 matrix and $\mathbf{y}$ is a 4×1 matrix, and hence the column dim of $\mathbf{x}$, 1, is unequal to the row dim of $\mathbf{y}$, 4.

Result A3.2 For all matrices A, B, and C (of dimensions such that the indicated products are defined) and a scalar α,

1. $\alpha(AB) = (\alpha A)B$
2. $A(BC) = (AB)C$
3. $A(B + C) = AB + AC$
4. $(B + C)A = BA + CA$
5. $(AB)' = B'A'$

There are several important differences between the algebra of matrices and the algebra of real numbers. Two of these differences follow.

Matrix multiplication is, in general, not commutative; that is, in general, $AB \neq BA$. Several examples will illustrate the failure of the commutative law (for matrices).

1.
$$\begin{bmatrix} 3 & -1 \\ 4 & 7 \end{bmatrix}\begin{bmatrix} 0 \\ 2 \end{bmatrix} = \begin{bmatrix} -2 \\ 14 \end{bmatrix}$$

but

$$\begin{bmatrix} 0 \\ 2 \end{bmatrix}\begin{bmatrix} 3 & -1 \\ 4 & 7 \end{bmatrix}$$

is not defined.

2.
$$\begin{bmatrix} 1 & 0 & 1 \\ 2 & -3 & 6 \end{bmatrix}\begin{bmatrix} 7 & 6 \\ -3 & 1 \\ 2 & 4 \end{bmatrix} = \begin{bmatrix} 9 & 10 \\ 35 & 33 \end{bmatrix}$$

but

$$\begin{bmatrix} 7 & 6 \\ -3 & 1 \\ 2 & 4 \end{bmatrix}\begin{bmatrix} 1 & 0 & 1 \\ 2 & -3 & 6 \end{bmatrix} = \begin{bmatrix} 19 & -18 & 43 \\ -1 & -3 & 3 \\ 10 & -12 & 26 \end{bmatrix}$$

3.
$$\begin{bmatrix} 4 & -1 \\ 0 & 1 \end{bmatrix}\begin{bmatrix} 2 & 1 \\ -3 & 4 \end{bmatrix} = \begin{bmatrix} 11 & 0 \\ -3 & 4 \end{bmatrix}$$

but

$$\begin{bmatrix} 2 & 1 \\ -3 & 4 \end{bmatrix}\begin{bmatrix} 4 & -1 \\ 0 & 1 \end{bmatrix} = \begin{bmatrix} 8 & -1 \\ -12 & 7 \end{bmatrix}$$

Let 0 denote the $n \times p$ "zero" matrix, that is, the matrix with zero for every element. In the algebra of real numbers, if the product, ab, say, of two numbers is zero, then $a = 0$ or $b = 0$. In matrix algebra, however, the product of two *nonzero* matrices may be the zero matrix; that is

$$\underset{(n \times m)}{A}\ \underset{(m \times p)}{B} = \underset{(n \times p)}{0}$$

does not imply that $A = 0$ or $B = 0$. For example,

$$\begin{bmatrix} 3 & 1 & 3 \\ 1 & 2 & 2 \end{bmatrix}\begin{bmatrix} 4 \\ 3 \\ -5 \end{bmatrix} = \begin{bmatrix} 0 \\ 0 \end{bmatrix}$$

It is true, however, that if either

$$\underset{(n \times m)}{A} = \underset{(n \times m)}{0} \qquad \text{or} \qquad \underset{(m \times p)}{B} = \underset{(m \times p)}{0}$$

then

$$\underset{(n \times m)}{A} \ \underset{(m \times p)}{B} = \underset{(n \times p)}{0}$$

Definition A3.12   A set of vectors $x_1, x_2, \ldots, x_n$ is said to be *linearly dependent* if there exist n numbers $(a_1, a_2, \ldots, a_n)$ not all zero such that

$$a_1 x_1 + a_2 x_2 + \cdots + a_n x_n = 0$$

Otherwise the set of vectors is said to be *linearly independent*.

For example, let $n = 3$ and

$$x_1 = [1, 1, 1]'$$
$$x_2 = [2, 5, -1]'$$
$$x_3 = [0, 1, -1]'$$

then

$$2x_1 - x_2 + 3x_3 = 0$$

Thus x_1, x_2, x_3 are a linearly dependent set of vectors since any one of them can be written as a linear combination of the others (for example, $x_2 = 2x_1 + 3x_3$). [*Note*: if we change x_3 to $x_3 = [0, 1, 0]'$, for example, then x_1, x_2, x_3 are *linearly independent*.]

Definition A3.13   The *rank* of a set of vectors $x_1, x_2, \ldots, x_n$ is the largest number of linearly independent vectors that can be chosen from the set. In the example above the rank of the set of vectors x_1, x_2, x_3 is two.

Definition A3.14   The *row rank* of a matrix is the rank of its set of rows considered as vectors (that is, row vectors). The *column rank* of a matrix is the rank of its set of columns considered as vectors (that is, column vectors).

For example, let the matrix W be

$$W = \begin{bmatrix} 1 & 1 & 1 \\ 2 & 5 & -1 \\ 0 & 1 & -1 \end{bmatrix}$$

then its row rank was shown above to be two. Note that the column rank of W is also two since

$$-2 \begin{bmatrix} 1 \\ 2 \\ 0 \end{bmatrix} + \begin{bmatrix} 1 \\ 5 \\ 1 \end{bmatrix} + \begin{bmatrix} 1 \\ -1 \\ -1 \end{bmatrix} = \begin{bmatrix} 0 \\ 0 \\ 0 \end{bmatrix}$$

This is no coincidence as the following result indicates.

Result A3.3   The row rank and the column rank of a matrix are equal. Thus the "rank of a matrix" is either the row rank or the column rank.

Definition A3.15   A square matrix

$$\underset{(n \times n)}{A}$$

is *nonsingular* if

$$\underset{(n \times n)}{A} \ \underset{(n \times 1)}{x} = \underset{(n \times 1)}{0} \quad \text{implies} \quad \underset{(n \times 1)}{x} = \underset{(n \times 1)}{0}$$

If a matrix fails to be nonsingular, it is called *singular*. Equivalently a *square* matrix is nonsingular if its rank is equal to the number of rows (or columns).

Definition A3.16   Let A be a square matrix of dim $n \times n$. If B is the unique $n \times n$ matrix such that

$$AB = BA = I$$

where I is the $n \times n$ identity matrix, then B is called the *inverse* of A and is denoted by A^{-1}.

For example,

$$A = \begin{bmatrix} 2 & 3 \\ 1 & 5 \end{bmatrix}, \qquad A^{-1} = \begin{bmatrix} 5/7 & -3/7 \\ -1/7 & 2/7 \end{bmatrix}$$

since

$$\begin{bmatrix} 2 & 3 \\ 1 & 5 \end{bmatrix} \begin{bmatrix} 5/7 & -3/7 \\ -1/7 & 2/7 \end{bmatrix} = \begin{bmatrix} 5/7 & -3/7 \\ -1/7 & 2/7 \end{bmatrix} \begin{bmatrix} 2 & 3 \\ 1 & 5 \end{bmatrix} = \begin{bmatrix} 1 & 0 \\ 0 & 1 \end{bmatrix}$$

Result A3.4   Let A and B be square matrices of the same dimension and let the indicated inverses exist. Then

1. $(A^{-1})' = (A')^{-1}$
2. $(AB)^{-1} = B^{-1}A^{-1}$

Definition A3.17   The *determinant* of the square $n \times n$ matrix $A = \{a_{ij}\}$, denoted by $|A|$, is the scalar

$$|A| = a_{11}, \qquad \text{if } n = 1$$

$$|A| = \sum_{j=1}^{n} a_{1j}|A_{1j}|(-1)^{1+j}, \qquad \text{if } n > 1 \tag{A3.2}$$

where A_{1j} is the $(n-1)\times(n-1)$ matrix obtained by deleting the first row and jth column of A.

Examples of determinants [evaluated using equation (A3.2)]:

1. $\begin{vmatrix} 1 & 3 \\ 6 & 4 \end{vmatrix} = 1|4|(-1)^2 + 3|6|(-1)^3 = 1(4) + 3(6)(-1) = -14$

In general

$$\begin{vmatrix} a_{11} & a_{12} \\ a_{21} & a_{22} \end{vmatrix} = a_{11}a_{22}(-1)^2 + a_{12}a_{21}(-1)^3 = a_{11}a_{22} - a_{12}a_{21}$$

2. $\begin{vmatrix} 3 & 1 & 6 \\ 7 & 4 & 5 \\ 2 & -7 & 1 \end{vmatrix} = 3\begin{vmatrix} 4 & 5 \\ -7 & 1 \end{vmatrix}(-1)^2 + 1\begin{vmatrix} 7 & 5 \\ 2 & 1 \end{vmatrix}(-1)^3 + 6\begin{vmatrix} 7 & 4 \\ 2 & -7 \end{vmatrix}(-1)^4$

$$= 3(39) - 1(-3) + 6(-57) = -222$$

3. $\begin{vmatrix} 1 & 0 & 0 \\ 0 & 1 & 0 \\ 0 & 0 & 1 \end{vmatrix} = 1\begin{vmatrix} 1 & 0 \\ 0 & 1 \end{vmatrix}(-1)^2 + 0\begin{vmatrix} 0 & 0 \\ 0 & 1 \end{vmatrix}(-1)^3 + 0\begin{vmatrix} 0 & 1 \\ 0 & 0 \end{vmatrix}(-1)^4$

$$= 1(1) = 1$$

In general, if I is an $n \times n$ identity matrix, $|I| = 1$.

4.
$$\begin{vmatrix} a_{11} & a_{12} & a_{13} \\ a_{21} & a_{22} & a_{23} \\ a_{31} & a_{32} & a_{33} \end{vmatrix} = a_{11} \begin{vmatrix} a_{22} & a_{23} \\ a_{32} & a_{33} \end{vmatrix} (-1)^2 + a_{12} \begin{vmatrix} a_{21} & a_{23} \\ a_{31} & a_{33} \end{vmatrix} (-1)^3 + a_{13}$$

$$\times \begin{vmatrix} a_{21} & a_{22} \\ a_{31} & a_{32} \end{vmatrix} (-1)^4$$

$$= a_{11}a_{22}a_{33} + a_{12}a_{23}a_{31} + a_{21}a_{32}a_{13}$$
$$- a_{31}a_{22}a_{13} - a_{21}a_{12}a_{33} - a_{32}a_{23}a_{11}$$

that is, the determinant of any 3×3 matrix can be computed by summing the products of elements along the solid lines and subtracting the products along the dashed lines in Figure A3.2. This procedure is *not* valid for matrices of higher dimension, and in general equation (A3.2) can be employed to evaluate the determinant.

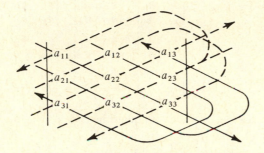

Figure A3.2

Result A3.5   For a square matrix A of dim $n \times n$, the following are equivalent:

1. A is nonsingular.

2. $\underset{(n \times n)}{A} \underset{(n \times 1)}{x} = \underset{(n \times 1)}{0}$ implies $\underset{(n \times 1)}{x} = \underset{(n \times 1)}{0}$.

3. $|A| \neq 0$.

4. There exists a matrix A^{-1} such that $AA^{-1} = A^{-1}A = \underset{(n \times n)}{I}$.

Result A3.6   Let A and B be $n \times n$ square matrices.

1. $|A| = |A'|$.
2. If each element of a row (column) of A is zero, then $|A| = 0$.
3. If any two rows (columns) of A are identical, then $|A| = 0$.
4. If A is nonsingular, then $|A| = 1/|A^{-1}|$; that is, $|A|\,|A^{-1}| = 1$.
5. $|AB| = |A|\,|B|$.

Result A3.7

1. The inverse of the 2×2 matrix

$$A = \begin{bmatrix} a_{11} & a_{12} \\ a_{21} & a_{22} \end{bmatrix}$$

is given by

$$A^{-1} = \frac{1}{|A|} \begin{bmatrix} a_{22} & -a_{12} \\ -a_{21} & a_{11} \end{bmatrix} \tag{A3.3}$$

2. The inverse of a 3×3 matrix

$$A = \begin{bmatrix} a_{11} & a_{12} & a_{13} \\ a_{21} & a_{22} & a_{23} \\ a_{31} & a_{32} & a_{33} \end{bmatrix}$$

is given by

$$A^{-1} = \frac{1}{|A|} \begin{bmatrix} \begin{vmatrix} a_{22} & a_{23} \\ a_{32} & a_{33} \end{vmatrix} & -\begin{vmatrix} a_{12} & a_{13} \\ a_{32} & a_{33} \end{vmatrix} & \begin{vmatrix} a_{12} & a_{13} \\ a_{22} & a_{23} \end{vmatrix} \\ -\begin{vmatrix} a_{21} & a_{23} \\ a_{31} & a_{33} \end{vmatrix} & \begin{vmatrix} a_{11} & a_{13} \\ a_{31} & a_{33} \end{vmatrix} & -\begin{vmatrix} a_{11} & a_{13} \\ a_{21} & a_{23} \end{vmatrix} \\ \begin{vmatrix} a_{21} & a_{22} \\ a_{31} & a_{32} \end{vmatrix} & -\begin{vmatrix} a_{11} & a_{12} \\ a_{31} & a_{32} \end{vmatrix} & \begin{vmatrix} a_{11} & a_{12} \\ a_{21} & a_{22} \end{vmatrix} \end{bmatrix} \tag{A3.4}$$

Definition A3.18   Let A be an $n \times n$ square matrix. A is said to be *orthogonal* if $A^{-1} = A'$.

An example of an orthogonal matrix follows.
Let

$$A = \begin{bmatrix} -\frac{1}{2} & \frac{1}{2} & \frac{1}{2} & \frac{1}{2} \\ \frac{1}{2} & -\frac{1}{2} & \frac{1}{2} & \frac{1}{2} \\ \frac{1}{2} & \frac{1}{2} & -\frac{1}{2} & \frac{1}{2} \\ \frac{1}{2} & \frac{1}{2} & \frac{1}{2} & -\frac{1}{2} \end{bmatrix}$$

then clearly $A = A'$ and $A' = A^{-1}$ since $AA' = A'A = AA$ and

$$\begin{bmatrix} -\frac{1}{2} & \frac{1}{2} & \frac{1}{2} & \frac{1}{2} \\ \frac{1}{2} & -\frac{1}{2} & \frac{1}{2} & \frac{1}{2} \\ \frac{1}{2} & \frac{1}{2} & -\frac{1}{2} & \frac{1}{2} \\ \frac{1}{2} & \frac{1}{2} & \frac{1}{2} & -\frac{1}{2} \end{bmatrix} \begin{bmatrix} -\frac{1}{2} & \frac{1}{2} & \frac{1}{2} & \frac{1}{2} \\ \frac{1}{2} & -\frac{1}{2} & \frac{1}{2} & \frac{1}{2} \\ \frac{1}{2} & \frac{1}{2} & -\frac{1}{2} & \frac{1}{2} \\ \frac{1}{2} & \frac{1}{2} & \frac{1}{2} & -\frac{1}{2} \end{bmatrix} = \begin{bmatrix} 1 & 0 & 0 & 0 \\ 0 & 1 & 0 & 0 \\ 0 & 0 & 1 & 0 \\ 0 & 0 & 0 & 1 \end{bmatrix}$$

$$\qquad\quad A \qquad\qquad\qquad\qquad A \qquad\qquad = \qquad I$$

Appendix 4

TABLES

LIST OF TABLES

Table A   Random Digits

37751	04998	66038	63480	98442	22245	83538	62351	74514	90497
50915	64152	82981	15796	27102	71635	34470	13608	26360	76285
99142	35021	01032	57907	80545	54112	15150	36856	03247	40392
70720	10033	25191	62358	03784	74377	88150	25567	87457	49512
18460	64947	32958	08752	96366	89092	23597	74308	00881	88976
65763	41133	60950	35372	06782	81451	78764	52645	19841	50083
83769	52570	60133	25211	87384	90182	84990	26400	39128	97043
58900	78420	98579	33665	10718	39342	46346	14401	13503	46525
54746	71115	78219	64314	11227	41702	54517	87676	14078	45317
56819	27340	07200	52663	57864	85159	15460	97564	29637	27742

Table A   (*Concluded*)

34990	62122	38223	28526	37006	22774	46026	15981	87291	56946
02269	22795	87593	81830	95383	67823	20196	54850	46779	64519
43042	53600	45738	00261	31100	67239	02004	70698	53597	62617
92565	12211	06868	87786	59576	61382	33972	13161	47208	96604
67424	32620	60841	86848	85000	04835	48576	33884	10101	84129
04015	77148	09535	10743	97871	55919	45274	38304	93125	91847
85226	19763	46105	25289	26714	73253	85922	21785	42624	92741
03360	07457	75131	41209	50451	23472	07438	08375	29312	62264
72460	99682	27970	25632	34096	17656	12736	27476	21938	67305
66960	55780	71778	52629	51692	71442	36130	70425	39874	62035
14824	95631	00697	65462	24815	13930	02938	54619	28909	53950
34001	05618	41900	23303	19928	60755	61404	56947	91441	19299
77718	83830	29781	72917	10840	74182	08293	62588	99625	22088
60930	05091	35726	07414	49211	69586	20226	08274	28167	65279
94180	62151	08112	26646	07617	42954	22521	09395	43561	45692
81073	85543	47650	93830	07377	87995	35084	39386	93141	88309
18467	39689	60801	46828	38670	88243	89042	78452	08032	72566
60643	59399	79740	17295	50094	66436	92677	68345	24025	36489
73372	61697	85728	90779	13235	83114	70728	32093	74306	08325
18395	18482	83245	54942	51905	09534	70839	91073	42193	81199
07261	28720	71244	05064	84873	68020	39037	68981	00670	86291
61679	81529	83725	33269	45958	74265	87460	60525	42539	25605
11815	48679	00556	96871	39835	83055	84949	11681	51687	55896
99007	35050	86440	44280	20320	97527	28138	01088	49037	85430
06446	65608	79291	16624	06135	30622	56133	33998	32308	29434
37913	83900	49166	00249	53178	72307	72190	75931	77613	20172
89444	98195	46733	37201	71901	55023	54570	83126	09462	93979
12582	41940	36060	56756	07999	64138	06492	25815	19518	86938
50494	80008	64774	51382	08059	66448	16437	91579	39197	43798
78301	66128	12840	22254	15193	81210	95747	47344	33660	41707
79457	31686	94486	27386	41641	72199	67265	51794	81521	01556
49337	10475	49588	79338	32156	47732	29464	92835	09498	81902
92540	56528	21200	87462	08924	56993	57330	85069	10903	80904
17729	61914	74616	20433	59474	21270	96406	13090	94308	02072
24003	80475	19793	71578	52010	72216	15692	96689	80452	46312
16129	49245	21693	20946	60873	82451	32516	23823	30046	06870
05453	03060	83621	43443	17082	04401	15299	64642	73497	88426
67711	70526	46700	00171	55077	11440	95932	91116	17259	19645
76306	39287	31026	49379	30267	68885	98147	70311	43856	37376
81300	17782	76403	00972	12558	46140	19818	20440	83967	61036

SOURCE: *A Million Random Digits with* 100,000 *Normal Deviates*, The RAND Corporation, 1955.

Table B Binomial Probabilities

This table gives binomial probabilities,

$$P(R = r \mid n, p) = \binom{n}{r} p^r (1 - p)^{n-r},$$

for $n = 1(1)20$, $r = 0(1)n$, and $p = .05(.05).50$. For $p > .50$, take
$P(r \mid n, p) = P(n - r \mid n, 1 - p)$.

Examples: $P(R = 3 \mid n = 8, p = .25) = .2076$,
and $P(R = 2 \mid n = 5, p = .60) = P(R = 3 \mid n = 5, p = .40) = .2304$.

n	r	.05	.10	.15	.20	.25	.30	.35	.40	.45	.50
1	0	.9500	.9000	.8500	.8000	.7500	.7000	.6500	.6000	.5500	.5000
	1	.0500	.1000	.1500	.2000	.2500	.3000	.3500	.4000	.4500	.5000
2	0	.9025	.8100	.7225	.6400	.5625	.4900	.4225	.3600	.3025	.2500
	1	.0950	.1800	.2550	.3200	.3750	.4200	.4550	.4800	.4950	.5000
	2	.0025	.0100	.0225	.0400	.0625	.0900	.1225	.1600	.2025	.2500
3	0	.8574	.7290	.6141	.5120	.4219	.3430	.2746	.2160	.1664	.1250
	1	.1354	.2430	.3251	.3840	.4219	.4410	.4436	.4320	.4084	.3750
	2	.0071	.0270	.0574	.0960	.1406	.1890	.2389	.2880	.3341	.3750
	3	.0001	.0010	.0034	.0080	.0156	.0270	.0429	.0640	.0911	.1250
4	0	.8145	.6561	.5220	.4096	.3164	.2401	.1785	.1296	.0915	.0625
	1	.1715	.2916	.3685	.4096	.4219	.4116	.3845	.3456	.2995	.2500
	2	.0135	.0486	.0975	.1536	.2109	.2646	.3105	.3456	.3675	.3750
	3	.0005	.0036	.0115	.0256	.0469	.0756	.1115	.1536	.2005	.2500
	4	.0000	.0001	.0005	.0016	.0039	.0081	.0150	.0256	.0410	.0625
5	0	.7738	.5905	.4437	.3277	.2373	.1681	.1160	.0778	.0503	.0312
	1	.2036	.3280	.3915	.4096	.3955	.3602	.3124	.2592	.2059	.1562
	2	.0214	.0729	.1382	.2048	.2637	.3087	.3364	.3456	.3369	.3125
	3	.0011	.0081	.0244	.0512	.0879	.1323	.1811	.2304	.2757	.3125
	4	.0000	.0004	.0022	.0064	.0146	.0284	.0488	.0768	.1128	.1562
	5	.0000	.0000	.0001	.0003	.0010	.0024	.0053	.0102	.0185	.0312
6	0	.7351	.5314	.3771	.2621	.1780	.1176	.0754	.0467	.0277	.0156
	1	.2321	.3543	.3993	.3932	.3560	.3025	.2437	.1866	.1359	.0938
	2	.0305	.0984	.1762	.2458	.2966	.3241	.3280	.3110	.2780	.2344
	3	.0021	.0146	.0415	.0819	.1318	.1852	.2355	.2765	.3032	.3125
	4	.0001	.0012	.0055	.0154	.0330	.0595	.0951	.1382	.1861	.2344
	5	.0000	.0001	.0004	.0015	.0044	.0102	.0205	.0369	.0609	.0938
	6	.0000	.0000	.0000	.0001	.0002	.0007	.0018	.0041	.0083	.0156
7	0	.6983	.4783	.3206	.2097	.1335	.0824	.0490	.0280	.0152	.0078
	1	.2573	.3720	.3960	.3670	.3115	.2471	.1848	.1306	.0872	.0547
	2	.0406	.1240	.2097	.2753	.3115	.3177	.2985	.2613	.2140	.1641
	3	.0036	.0230	.0617	.1147	.1730	.2269	.2679	.2903	.2918	.2734
	4	.0002	.0026	.0109	.0287	.0577	.0972	.1442	.1935	.2388	.2734
	5	.0000	.0002	.0012	.0043	.0115	.0250	.0466	.0774	.1172	.1641
	6	.0000	.0000	.0001	.0004	.0013	.0036	.0084	.0172	.0320	.0547
	7	.0000	.0000	.0000	.0000	.0001	.0002	.0006	.0016	.0037	.0078

Table B   (*Continued*)

n	r	.05	.10	.15	.20	.25	.30	.35	.40	.45	.50
8	0	.6634	.4305	.2725	.1678	.1001	.0576	.0319	.0168	.0084	.0039
	1	.2793	.3826	.3847	.3355	.2760	.1977	.1373	.0896	.0548	.0312
	2	.0515	.1488	.2376	.2936	.3115	.2965	.2587	.2090	.1569	.1094
	3	.0054	.0331	.0839	.1468	.2076	.2541	.2786	.2787	.2568	.2188
	4	.0004	.0046	.0185	.0459	.0865	.1361	.1875	.2322	.2627	.2734
	5	.0000	.0004	.0026	.0092	.0231	.0467	.0808	.1239	.1719	.2188
	6	.0000	.0000	.0002	.0011	.0038	.0100	.0217	.0413	.0703	.1094
	7	.0000	.0000	.0000	.0001	.0004	.0012	.0033	.0079	.0164	.0312
	8	.0000	.0000	.0000	.0000	.0000	.0001	.0002	.0007	.0017	.0039
9	0	.6302	.3874	.2316	.1342	.0751	.0404	.0277	.0101	.0046	.0020
	1	.2985	.3874	.3679	.3020	.2253	.1556	.1004	.0605	.0339	.0176
	2	.0629	.1722	.2597	.3020	.3003	.2668	.2162	.1612	.1110	.0703
	3	.0077	.0446	.1069	.1762	.2336	.2668	.2716	.2508	.2119	.1641
	4	.0006	.0074	.0283	.0661	.1168	.1715	.2194	.2508	.2600	.2461
	5	.0000	.0008	.0050	.0165	.0389	.0735	.1181	.1672	.2128	.2461
	6	.0000	.0001	.0006	.0028	.0087	.0210	.0424	.0743	.1160	.1641
	7	.0000	.0000	.0000	.0003	.0012	.0039	.0098	.0212	.0407	.0703
	8	.0000	.0000	.0000	.0000	.0001	.0004	.0013	.0035	.0083	.0176
	9	.0000	.0000	.0000	.0000	.0000	.0000	.0001	.0003	.0008	.0020
10	0	.5987	.3487	.1969	.1074	.0563	.0282	.0135	.0060	.0025	.0010
	1	.3151	.3874	.3474	.2684	.1877	.1211	.0725	.0403	.0207	.0098
	2	.0746	.1937	.2759	.3020	.2816	.2335	.1757	.1209	.0763	.0439
	3	.0105	.0574	.1298	.2013	.2503	.2668	.2522	.2150	.1665	.1172
	4	.0010	.0112	.0401	.0881	.1460	.2001	.2377	.2508	.2384	.2051
	5	.0001	.0015	.0085	.0264	.0584	.1029	.1536	.2007	.2340	.2461
	6	.0000	.0001	.0012	.0055	.0162	.0368	.0689	.1115	.1596	.2051
	7	.0000	.0000	.0001	.0008	.0031	.0090	.0212	.0425	.0746	.1172
	8	.0000	.0000	.0000	.0001	.0004	.0014	.0043	.0106	.0229	.0439
	9	.0000	.0000	.0000	.0000	.0000	.0001	.0005	.0016	.0042	.0098
	10	.0000	.0000	.0000	.0000	.0000	.0000	.0000	.0001	.0003	.0010
11	0	.5688	.3138	.1673	.0859	.0422	.0198	.0088	.0036	.0014	.0005
	1	.3293	.3835	.3248	.2362	.1549	.0932	.0518	.0266	.0125	.0054
	2	.0867	.2131	.2866	.2953	.2581	.1998	.1395	.0887	.0513	.0269
	3	.0137	.0710	.1517	.2215	.2581	.2568	.2254	.1774	.1259	.0806
	4	.0014	.0158	.0536	.1107	.1721	.2201	.2428	.2365	.2060	.1611
	5	.0001	.0025	.0132	.0388	.0803	.1231	.1830	.2207	.2360	.2256
	6	.0000	.0003	.0023	.0097	.0268	.0566	.0985	.1471	.1931	.2256
	7	.0000	.0000	.0003	.0017	.0064	.0173	.0379	.0701	.1128	.1611
	8	.0000	.0000	.0000	.0002	.0011	.0037	.0102	.0234	.0462	.0806
	9	.0000	.0000	.0000	.0000	.0001	.0005	.0018	.0052	.0126	.0269
	10	.0000	.0000	.0000	.0000	.0000	.0000	.0002	.0007	.0021	.0054
	11	.0000	.0000	.0000	.0000	.0000	.0000	.0000	.0000	.0002	.0005
12	0	.5404	.2824	.1422	.0687	.0317	.0138	.0057	.0022	.0008	.0002
	1	.3413	.3766	.3012	.2062	.1267	.0712	.0368	.0174	.0075	.0029
	2	.0988	.2301	.2924	.2835	.2323	.1678	.1088	.0639	.0339	.0161
	3	.0173	.0852	.1720	.2362	.2581	.2397	.1954	.1419	.0923	.0537
	4	.0021	.0213	.0683	.1329	.1936	.2311	.2367	.2128	.1700	.1208
	5	.0002	.0038	.0193	.0532	.1032	.1585	.2039	.2270	.2225	.1934
	6	.0000	.0005	.0040	.0155	.0401	.0792	.1281	.1766	.2124	.2256

Table B (*Continued*)

n	r	.05	.10	.15	.20	.25	.30	.35	.40	.45	.50
12	7	.0000	.0000	.0006	.0033	.0115	.0291	.0591	.1009	.1489	.1934
	8	.0000	.0000	.0001	.0005	.0024	.0078	.0199	.0420	.0762	.1208
	9	.0000	.0000	.0000	.0001	.0004	.0015	.0048	.0125	.0277	.0537
	10	.0000	.0000	.0000	.0000	.0000	.0002	.0008	.0025	.0068	.0161
	11	.0000	.0000	.0000	.0000	.0000	.0000	.0001	.0003	.0010	.0029
	12	.0000	.0000	.0000	.0000	.0000	.0000	.0000	.0000	.0001	.0002
13	0	.5133	.2542	.1209	.0550	.0238	.0097	.0037	.0013	.0004	.0001
	1	.3512	.3672	.2774	.1787	.1029	.0540	.0259	.0113	.0045	.0016
	2	.1109	.2448	.2937	.2680	.2059	.1388	.0836	.0453	.0220	.0095
	3	.0214	.0997	.1900	.2457	.2517	.2181	.1651	.1107	.0660	.0349
	4	.0028	.0277	.0838	.1535	.2097	.2337	.2222	.1845	.1350	.0873
	5	.0003	.0055	.0266	.0691	.1258	.1803	.2154	.2214	.1989	.1571
	6	.0000	.0008	.0063	.0230	.0559	.1030	.1546	.1968	.2169	.2095
	7	.0000	.0001	.0011	.0058	.0186	.0442	.0833	.1312	.1775	.2095
	8	.0000	.0000	.0001	.0011	.0047	.0142	.0336	.0656	.1089	.1571
	9	.0000	.0000	.0000	.0001	.0009	.0034	.0101	.0243	.0495	.0873
	10	.0000	.0000	.0000	.0000	.0001	.0006	.0022	.0065	.0162	.0349
	11	.0000	.0000	.9000	.0000	.0000	.0001	.0003	.0012	.0036	.0095
	12	.0000	.0000	.0000	.0000	.0000	.0000	.0000	.0001	.0005	.0016
	13	.0000	.0000	.0000	.0000	.0000	.0000	.0000	.0000	.0000	.0001
14	0	.4877	.2288	.1028	.0440	.0178	.0068	.0024	.0008	.0002	.0001
	1	.3593	.3559	.2539	.1539	.0832	.0407	.0181	.0073	.0027	.0009
	2	.1229	.2570	.2912	.2501	.1802	.1134	.0634	.0317	.0141	.0056
	3	.0259	.1142	.2056	.2501	.2402	.1943	.1366	.0845	.0462	.0222
	4	.0037	.0349	.0998	.1720	.2202	.2290	.2022	.1549	.1040	.0611
	5	.0004	.0078	.0352	.0860	.1468	.1963	.2178	.2066	.1701	.1222
	6	.0000	.0013	.0093	.0322	.0734	.1262	.1759	.2066	.2088	.1833
	7	.0000	.0002	.0019	.0092	.0280	.0618	.1082	.1574	.1952	.2095
	8	.0000	.0000	.0003	.0020	.0082	.0232	.0510	.0918	.1398	.1833
	9	.0000	.0000	.0000	.0003	.0018	.0066	.0183	.0408	.0762	.1222
	10	.0000	.0000	.0000	.0000	.0003	.0014	.0049	.0136	.0312	.0611
	11	.0000	.0000	.0000	.0000	.0000	.0002	.0010	.0033	.0093	.0222
	12	.0000	.0000	.0000	.0000	.0000	.0000	.0001	.0005	.0019	.0056
	13	.0000	.0000	.0000	.0000	.0000	.0000	.0000	.0001	.0002	.0009
	14	.0000	.0000	.0000	.0000	.0000	.0000	.0000	.0000	.0000	.0001
15	0	.4633	.2059	.0874	.0352	.0134	.0047	.0016	.0005	.0001	.0000
	1	.3658	.3432	.2312	.1319	.0668	.0305	.0126	.0047	.0016	.0005
	2	.1348	.2669	.2856	.2309	.1559	.0916	.0476	.0219	.0090	.0032
	3	.0307	.1285	.2184	.2501	.2252	.1700	.1110	.0634	.0318	.0139
	4	.0049	.0428	.1156	.1876	.2252	.2186	.1792	.1268	.0780	.0417
	5	.0006	.0105	.0449	.1032	.1651	.2061	.2123	.1859	.1404	.0916
	6	.0000	.0019	.0132	.0430	.0917	.1472	.1906	.2066	.1914	.1527
	7	.0000	.0003	.0030	.0138	.0393	.0811	.1319	.1771	.2013	.1964
	8	.0000	.0000	.0005	.0035	.0131	.0348	.0710	.1181	.1647	.1964
	9	.0000	.0000	.0001	.0007	.0034	.0116	.0298	.0612	.1048	.1527
	10	.0000	.0000	.0000	.0001	.0007	.0030	.0096	.0245	.0515	.0916
	11	.0000	.0000	.0000	.0000	.0001	.0006	.0024	.0074	.0191	.0417
	12	.0000	.0000	.0000	.0000	.0000	.0001	.0004	.0016	.0052	.0139
	13	.0000	.0000	.0000	.0000	.0000	.0000	.0001	.0003	.0010	.0032
	14	.0000	.0000	.0000	.0000	.0000	.0000	.0000	.0000	.0001	.0005
	15	.0000	.0000	.0000	.0000	.0000	.0000	.0000	.0000	.0000	.0000

Table B   (Continued)

n	r	.05	.10	.15	.20	.25	.30	.35	.40	.45	.50
16	0	.4401	.1853	.0743	.0281	.0100	.0033	.0010	.0003	.0001	.0000
	1	.3706	.3294	.2097	.1126	.0535	.0228	.0087	.0030	.0009	.0002
	2	.1463	.2745	.2775	.2111	.1336	.0732	.0353	.0150	.0056	.0018
	3	.0359	.1423	.2285	.2463	.2079	.1465	.0888	.0468	.0215	.0085
	4	.0061	.0514	.1311	.2001	.2252	.2040	.1553	.1014	.0572	.0278
	5	.0008	.0137	.0555	.1201	.1802	.2099	.2008	.1623	.1123	.0667
	6	.0001	.0028	.0180	.0550	.1101	.1649	.1982	.1983	.1684	.1222
	7	.0000	.0004	.0045	.0197	.0524	.1010	.1524	.1889	.1969	.1746
	8	.0000	.0001	.0009	.0055	.0197	.0487	.0923	.1417	.1812	.1964
	9	.0000	.0000	.0001	.0012	.0058	.0185	.0442	.0840	.1318	.1746
	10	.0000	.0000	.0000	.0002	.0014	.0056	.0167	.0392	.0755	.1222
	11	.0000	.0000	.0000	.0000	.0002	.0013	.0049	.0142	.0337	.0667
	12	.0000	.0000	.0000	.0000	.0000	.0002	.0011	.0040	.0115	.0278
	13	.0000	.0000	.0000	.0000	.0000	.0000	.0002	.0008	.0029	.0085
	14	.0000	.0000	.0000	.0000	.0000	.0000	.0000	.0001	.0005	.0018
	15	.0000	.0000	.0000	.0000	.0000	.0000	.0000	.0000	.0001	.0002
	16	.0000	.0000	.0000	.0000	.0000	.0000	.0000	.0000	.0000	.0000
17	0	.4181	.1668	.0631	.0225	.0075	.0023	.0007	.0002	.0000	.0000
	1	.3741	.3150	.1893	.0957	.0426	.0169	.0060	.0019	.0005	.0001
	2	.1575	.2800	.2673	.1914	.1136	.0581	.0260	.0102	.0035	.0010
	3	.0415	.1556	.2359	.2393	.1893	.1245	.0701	.0341	.0144	.0052
	4	.0076	.0605	.1457	.2093	.2209	.1868	.1320	.0796	.0411	.0182
	5	.0010	.0175	.0668	.1361	.1914	.2081	.1849	.1379	.0875	.0472
	6	.0001	.0039	.0236	.0680	.1276	.1784	.1991	.1839	.1432	.0944
	7	.0000	.0007	.0065	.0267	.0668	.1201	.1685	.1927	.1841	.1484
	8	.0000	.0001	.0014	.0084	.0279	.0644	.1143	.1606	.1883	.1855
	9	.0000	.0000	.0003	.0021	.0093	.0276	.0611	.1070	.1540	.1855
	10	.0000	.0000	.0000	.0004	.0025	.0095	.0263	.0571	.1008	.1484
	11	.0000	.0000	.0000	.0001	.0005	.0026	.0090	.0242	.0525	.0944
	12	.0000	.0000	.0000	.0000	.0001	.0006	.0024	.0081	.0215	.0472
	13	.0000	.0000	.0000	.0000	.0000	.0001	.0005	.0021	.0068	.0182
	14	.0000	.0000	.0000	.0000	.0000	.0000	.0001	.0004	.0016	.0052
	15	.0000	.0000	.0000	.0000	.0000	.0000	.0000	.0001	.0003	.0010
	16	.0000	.0000	.0000	.0000	.0000	.0000	.0000	.0000	.0000	.0001
	17	.0000	.0000	.0000	.0000	.0000	.0000	.0000	.0000	.0000	.0000
18	0	.3972	.1501	.0536	.0180	.0056	.0016	.0004	.0001	.0000	.0000
	1	.3763	.3002	.1704	.0811	.0338	.0126	.0042	.0012	.0003	.0001
	2	.1683	.2835	.2556	.1723	.0958	.0458	.0190	.0069	.0022	.0006
	3	.0473	.1680	.2406	.2297	.1704	.1046	.0547	.0246	.0095	.0031
	4	.0093	.0700	.1592	.2153	.2130	.1681	.1104	.0614	.0291	.0117
	5	.0014	.0218	.0787	.1507	.1988	.2017	.1664	.1146	.0666	.0327
	6	.0002	.0052	.0310	.0816	.1436	.1873	.1941	.1655	.1181	.0708
	7	.0000	.0010	.0091	.0350	.0820	.1376	.1792	.1892	.1657	.1214
	8	.0000	.0002	.0022	.0120	.0376	.0811	.1327	.1734	.1864	.1669
	9	.0000	.0000	.0004	.0033	.0139	.0386	.0794	.1284	.1694	.1855
	10	.0000	.0000	.0001	.0008	.0042	.0149	.0385	.0771	.1248	.1669
	11	.0000	.0000	.0000	.0001	.0010	.0046	.0151	.0374	.0742	.1214
	12	.0000	.0000	.0000	.0000	.0002	.0012	.0047	.0145	.0354	.0708

Table B   (*Concluded*)

n	r	.05	.10	.15	.20	.25	.30	.35	.40	.45	.50
						p					
18	13	.0000	.0000	.0000	.0000	.0000	.0002	.0012	.0045	.0134	.0327
	14	.0000	.0000	.0000	.0000	.0000	.0000	.0002	.0011	.0039	.0117
	15	.0000	.0000	.0000	.0000	.0000	.0000	.0000	.0002	.0009	.0031
	16	.0000	.0000	.0000	.0000	.0000	.0000	.0000	.0000	.0001	.0006
	17	.0000	.0000	.0000	.0000	.0000	.0000	.0000	.0000	.0000	.0001
	18	.0000	.0000	.0000	.0000	.0000	.0000	.0000	.0000	.0000	.0000
19	0	.3774	.1351	.0456	.0144	.0042	.0011	.0003	.0001	.0000	.0000
	1	.3774	.2852	.1529	.0685	.0268	.0093	.0029	.0008	.0002	.0000
	2	.1787	.2852	.2428	.1540	.0803	.0358	.0138	.0046	.0013	.0003
	3	.0533	.1796	.2428	.2182	.1517	.0869	.0422	.0175	.0062	.0018
	4	.0112	.0798	.1714	.2182	.2023	.1491	.0909	.0467	.0203	.0074
	5	.0018	.0266	.0907	.1636	.2023	.1916	.1468	.0933	.0497	.0222
	6	.0002	.0069	.0374	.0955	.1574	.1916	.1844	.1451	.0949	.0518
	7	.0000	.0014	.0122	.0443	.0974	.1525	.1844	.1797	.1443	.0961
	8	.0000	.0002	.0032	.0166	.0487	.0981	.1489	.1797	.1771	.1442
	9	.0000	.0000	.0007	.0051	.0198	.0514	.0980	.1464	.1771	.1762
	10	.0000	.0000	.0001	.0013	.0066	.0220	.0528	.0976	.1449	.1762
	11	.0000	.0000	.0000	.0003	.0018	.0077	.0233	.0532	.0970	.1442
	12	.0000	.0000	.0000	.0000	.0004	.0022	.0083	.0237	.0529	.0961
	13	.0000	.0000	.0000	.0000	.0001	.0005	.0024	.0085	.0233	.0518
	14	.0000	.0000	.0000	.0000	.0000	.0001	.0006	.0024	.0082	.0222
	15	.0000	.0000	.0000	.0000	.0000	.0000	.0001	.0005	.0022	.0074
	16	.0000	.0000	.0000	.0000	.0000	.0000	.0000	.0001	.0005	.0018
	17	.0000	.0000	.0000	.0000	.0000	.0000	.0000	.0000	.0001	.0003
	18	.0000	.0000	.0000	.0000	.0000	.0000	.0000	.0000	.0000	.0000
	19	.0000	.0000	.0000	.0000	.0000	.0000	.0000	.0000	.0000	.0000
20	0	.3585	.1216	.0388	.0115	.0032	.0008	.0002	.0000	.0000	.0000
	1	.3774	.2702	.1368	.0576	.0211	.0068	.0020	.0005	.0001	.0000
	2	.1887	.2852	.2293	.1369	.0669	.0278	.0100	.0031	.0008	.0002
	3	.0596	.1901	.2428	.2054	.1339	.0716	.0323	.0123	.0040	.0011
	4	.0133	.0898	.1821	.2182	.1897	.1304	.0738	.0350	.0139	.0046
	5	.0022	.0319	.1028	.1746	.2023	.1789	.1272	.0746	.0365	.0148
	6	.0003	.0089	.0454	.1091	.1686	.1916	.1712	.1244	.0746	.0370
	7	.0000	.0020	.0160	.0545	.1124	.1643	.1844	.1659	.1221	.0739
	8	.0000	.0004	.0046	.0222	.0609	.1144	.1614	.1797	.1623	.1201
	9	.0000	.0001	.0011	.0074	.0271	.0654	.1158	.1597	.1771	.1602
	10	.0000	.0000	.0002	.0020	.0099	.0308	.0686	.1171	.1593	.1762
	11	.0000	.0000	.0000	.0005	.0030	.0120	.0336	.0710	.1185	.1602
	12	.0000	.0000	.0000	.0001	.0008	.0039	.0136	.0355	.0727	.1201
	13	.0000	.0000	.0000	.0000	.0002	.0010	.0045	.0146	.0366	.0739
	14	.0000	.0000	.0000	.0000	.0000	.0002	.0012	.0049	.0150	.0370
	15	.0000	.0000	.0000	.0000	.0000	.0000	.0003	.0013	.0049	.0148
	16	.0000	.0000	.0000	.0000	.0000	.0000	.0000	.0003	.0013	.0046
	17	.0000	.0000	.0000	.0000	.0000	.0000	.0000	.0000	.0002	.0011
	18	.0000	.0000	.0000	.0000	.0000	.0000	.0000	.0000	.0000	.0002
	19	.0000	.0000	.0000	.0000	.0000	.0000	.0000	.0000	.0000	.0000
	20	.0000	.0000	.0000	.0000	.0000	.0000	.0000	.0000	.0000	.0000

SOURCE: From *Handbook of Probability and Statistics with Tables* (2d ed.). Copyright © 1970 by McGraw-Hill, Inc. Used by permission of McGraw-Hill Book Company.

Table C   Poisson Probabilities

This table gives Poisson probabilities,

$$P(R = r|\lambda, t) = \frac{e^{-\lambda t}(\lambda t)^r}{r!},$$

for $\lambda t = 1(.1)10(1)20$ and suitable values of r.

Example: $P(R = 2|\lambda = 1.5, t = 5) = .0156.$

λt

r	0.1	0.2	0.3	0.4	0.5	0.6	0.7	0.8	0.9	1.0
0	.9048	.8187	.7408	.6703	.6065	.5488	.4966	.4493	.4066	.3679
1	.0905	.1637	.2222	.2681	.3033	.3293	.3476	.3595	.3659	.3679
2	.0045	.0164	.0333	.0536	.0758	.0988	.1217	.1438	.1647	.1839
3	.0002	.0011	.0033	.0072	.0126	.0198	.0284	.0383	.0494	.0613
4	.0000	.0001	.0002	.0007	.0016	.0030	.0050	.0077	.0111	.0153
5	.0000	.0000	.0000	.0001	.0002	.0004	.0007	.0012	.0020	.0031
6	.0000	.0000	.0000	.0000	.0000	.0000	.0001	.0002	.0003	.0005
7	.0000	.0000	.0000	.0000	.0000	.0000	.0000	.0000	.0000	.0001

λt

r	1.1	1.2	1.3	1.4	1.5	1.6	1.7	1.8	1.9	2.0
0	.3329	.3012	.2725	.2466	.2231	.2019	.1827	.1653	.1496	.1353
1	.3662	.3614	.3543	.3452	.3347	.3230	.3106	.2975	.2842	.2707
2	.2014	.2169	.2303	.2417	.2510	.2584	.2640	.2678	.2700	.2707
3	.0738	.0867	.0998	.1128	.1255	.1378	.1496	.1607	.1710	.1804
4	.0203	.0260	.0324	.0395	.0471	.0551	.0636	.0723	.0812	.0902
5	.0045	.0062	.0084	.0111	.0141	.0176	.0216	.0260	.0309	.0361
6	.0008	.0012	.0018	.0026	.0035	.0047	.0061	.0078	.0098	.0120
7	.0001	.0002	.0003	.0005	.0008	.0011	.0015	.0020	.0027	.0034
8	.0000	.0000	.0001	.0001	.0001	.0002	.0003	.0005	.0006	.0009
9	.0000	.0000	.0000	.0000	.0000	.0000	.0001	.0001	.0001	.0002

λt

r	2.1	2.2	2.3	2.4	2.5	2.6	2.7	2.8	2.9	3.0
0	.1225	.1108	.1003	.0907	.0821	.0743	.0672	.0608	.0550	.0498
1	.2572	.2438	.2306	.2177	.2052	.1931	.1815	.1703	.1596	.1494
2	.2700	.2681	.2652	.2613	.2565	.2510	.2450	.2384	.2314	.2240
3	.1890	.1966	.2033	.2090	.2138	.2176	.2205	.2225	.2237	.2240
4	.0992	.1082	.1169	.1254	.1336	.1414	.1488	.1557	.1622	.1680
5	.0417	.0476	.0538	.0602	.0668	.0735	.0804	.0872	.0940	.1008
6	.0146	.0174	.0206	.0241	.0278	.0319	.0362	.0407	.0455	.0540
7	.0044	.0055	.0068	.0083	.0099	.0118	.0139	.0163	.0188	.0216
8	.0011	.0015	.0019	.0025	.0031	.0038	.0047	.0057	.0068	.0081
9	.0003	.0004	.0005	.0007	.0009	.0011	.0014	.0018	.0022	.0027
10	.0001	.0001	.0001	.0002	.0002	.0003	.0004	.0005	.0006	.0008
11	.0000	.0000	.0000	.0000	.0000	.0001	.0001	.0001	.0002	.0002
12	.0000	.0000	.0000	.0000	.0000	.0000	.0000	.0000	.0000	.0001

Table C (*Continued*)

λt

r	3.1	3.2	3.3	3.4	3.5	3.6	3.7	3.8	3.9	4.0
0	.0450	.0408	.0369	.0344	.0302	.0273	.0247	.0224	.0202	.0183
1	.1397	.1304	.1217	.1135	.1057	.0984	.0915	.0850	.0789	.0733
2	.2165	.2087	.2008	.1929	.1850	.1771	.1692	.1615	.1539	.1465
3	.2237	.2226	.2209	.2186	.2158	.2125	.2087	.2046	.2001	.1954
4	.1734	.1781	.1823	.1858	.1888	.1912	.1931	.1944	.1951	.1954
5	.1075	.1140	.1203	.1264	.1322	.1377	.1429	.1477	.1522	.1563
6	.0555	.0608	.0662	.0716	.0771	.0826	.0881	.0936	.0989	.1042
7	.0246	.0278	.0312	.0348	.0385	.0425	.0466	.0508	.0551	.0595
8	.0095	.0111	.0129	.0148	.0169	.0191	.0215	.0241	.0269	.0298
9	.0033	.0040	.0047	.0056	.0066	.0076	.0089	.0102	.0116	.0132
10	.0010	.0013	.0016	.0019	.0023	.0028	.0033	.0039	.0045	.0053
11	.0003	.0004	.0005	.0006	.0007	.0009	.0011	.0013	.0016	.0019
12	.0001	.0001	.0001	.0002	.0002	.0003	.0003	.0004	.0005	.0006
13	.0000	.0000	.0000	.0000	.0001	.0001	.0001	.0001	.0002	.0002
14	.0000	.0000	.0000	.0000	.0000	.0000	.0000	.0000	.0000	.0001

λt

r	4.1	4.2	4.3	4.4	4.5	4.6	4.7	4.8	4.9	5.0
0	.0166	.0150	.0136	.0123	.0111	.0101	.0091	.0082	.0074	.0067
1	.0679	.0630	.0583	.0540	.0500	.0462	.0427	.0395	.0365	.0337
2	.1393	.1323	.1254	.1188	.1125	.1063	.1005	.0948	.0894	.0842
3	.1904	.1852	.1798	.1743	.1687	.1631	.1574	.1517	.1460	.1404
4	.1951	.1944	.1933	.1917	.1898	.1875	.1849	.1820	.1789	.1755
5	.1600	.1633	.1662	.1687	.1708	.1725	.1738	.1747	.1753	.1755
6	.1093	.1143	.1191	.1237	.1281	.1323	.1362	.1398	.1432	.1462
7	.0640	.0686	.0732	.0778	.0824	.0869	.0914	.0959	.1002	.1044
8	.0328	.0360	.0393	.0428	.0463	.0500	.0537	.0575	.0614	.0653
9	.0150	.0168	.0188	.0209	.0232	.0255	.0280	.0307	.0334	.0363
10	.0061	.0071	.0081	.0092	.0104	.0118	.0132	.0147	.0164	.0181
11	.0023	.0027	.0032	.0037	.0043	.0049	.0056	.0064	.0073	.0082
12	.0008	.0009	.0011	.0014	.0016	.0019	.0022	.0026	.0030	.0034
13	.0002	.0003	.0004	.0005	.0006	.0007	.0008	.0009	.0011	.0013
14	.0001	.0001	.0001	.0001	.0002	.0002	.0003	.0003	.0004	.0005
15	.0000	.0000	.0000	.0000	.0001	.0001	.0001	.0001	.0001	.0002

λt

r	5.1	5.2	5.3	5.4	5.5	5.6	5.7	5.8	5.9	6.0
0	.0061	.0055	.0050	.0045	.0041	.0037	.0033	.0030	.0027	.0025
1	.0311	.0287	.0265	.0244	.0225	.0207	.0191	.0176	.0162	.0149
2	.0793	.0746	.0701	.0659	.0618	.0580	.0544	.0509	.0477	.0446
3	.1348	.1293	.1239	.1185	.1133	.1082	.1033	.0985	.0938	.0892
4	.1719	.1681	.1641	.1600	.1558	.1515	.1472	.1428	.1383	.1339
5	.1753	.1748	.1740	.1728	.1714	.1697	.1678	.1656	.1632	.1606
6	.1490	.1515	.1537	.1555	.1571	.1584	.1594	.1601	.1605	.1606
7	.1086	.1125	.1163	.1200	.1234	.1267	.1298	.1326	.1353	.1377
8	.0692	.0731	.0771	.0810	.0849	.0887	.0925	.0962	.0998	.1033
9	.0392	.0423	.0454	.0486	.0519	.0552	.0586	.0620	.0654	.0688

Table C   (Continued)

λt

r	5.1	5.2	5.3	5.4	5.5	5.6	5.7	5.8	5.9	6.0
10	.0200	.0220	.0241	.0262	.0285	.0309	.0334	.0359	.0386	.0413
11	.0093	.0104	.0116	.0129	.0143	.0157	.0173	.0190	.0207	.0225
12	.0039	.0045	.0051	.0058	.0065	.0073	.0082	.0092	.0102	.0113
13	.0015	.0018	.0021	.0024	.0028	.0032	.0036	.0041	.0046	.0052
14	.0006	.0007	.0008	.0009	.0011	.0013	.0015	.0017	.0019	.0022
15	.0002	.0002	.0003	.0003	.0004	.0005	.0006	.0007	.0008	.0009
16	.0001	.0001	.0001	.0001	.0001	.0002	.0002	.0002	.0003	.0003
17	.0000	.0000	.0000	.0000	.0000	.0001	.0001	.0001	.0001	.0001

λt

r	6.1	6.2	6.3	6.4	6.5	6.6	6.7	6.8	6.9	7.0
0	.0022	.0020	.0018	.0017	.0015	.0014	.0012	.0011	.0010	.0009
1	.0137	.0126	.0116	.0106	.0098	.0090	.0082	.0076	.0070	.0064
2	.0417	.0390	.0364	.0340	.0318	.0296	.0276	.0258	.0240	.0223
3	.0848	.0806	.0765	.0726	.0688	.0652	.0617	.0584	.0552	.0521
4	.1294	.1249	.1205	.1162	.1118	.1076	.1034	.0992	.0952	.0912
5	.1579	.1549	.1519	.1487	.1454	.1420	.1385	.1349	.1314	.1277
6	.1605	.1601	.1595	.1586	.1575	.1562	.1546	.1529	.1511	.1490
7	.1399	.1418	.1435	.1450	.1462	.1472	.1480	.1486	.1489	.1490
8	.1066	.1099	.1130	.1160	.1188	.1215	.1240	.1263	.1284	.1304
9	.0723	.0757	.0791	.0825	.0858	.0891	.0923	.0954	.0985	.1014
10	.0441	.0469	.0498	.0528	.0558	.0588	.0618	.0649	.0679	.0710
11	.0245	.0265	.0285	.0307	.0330	.0353	.0377	.0401	.0426	.0452
12	.0124	.0137	.0150	.0164	.0179	.0194	.0210	.0227	.0245	.0264
13	.0058	.0065	.0073	.0081	.0089	.0098	.0108	.0119	.0130	.0142
14	.0025	.0029	.0033	.0037	.0041	.0046	.0052	.0058	.0064	.0071
15	.0010	.0012	.0014	.0016	.0018	.0020	.0023	.0026	.0029	.0033
16	.0004	.0005	.0005	.0006	.0007	.0008	.0010	.0011	.0013	.0014
17	.0001	.0002	.0002	.0002	.0003	.0003	.0004	.0004	.0005	.0006
18	.0000	.0001	.0001	.0001	.0001	.0001	.0001	.0002	.0002	.0002
19	.0000	.0000	.0000	.0000	.0000	.0000	.0000	.0001	.0001	.0001

λt

r	7.1	7.2	7.3	7.4	7.5	7.6	7.7	7.8	7.9	8.0
0	.0008	.0007	.0007	.0006	.0006	.0005	.0005	.0004	.0004	.0003
1	.0059	.0054	.0049	.0045	.0041	.0038	.0035	.0032	.0029	.0027
2	.0208	.0194	.0180	.0167	.0156	.0145	.0134	.0125	.0116	.0107
3	.0492	.0464	.0438	.0413	.0389	.0366	.0345	.0324	.0305	.0286
4	.0874	.0836	.0799	.0764	.0729	.0696	.0663	.0632	.0602	.0573
5	.1241	.1204	.1167	.1130	.1094	.1057	.1021	.0986	.0951	.0916
6	.1468	.1445	.1420	.1394	.1367	.1339	.1311	.1282	.1252	.1221
7	.1489	.1486	.1481	.1474	.1465	.1454	.1442	.1428	.1413	.1396
8	.1321	.1337	.1351	.1363	.1373	.1382	.1388	.1392	.1395	.1396
9	.1042	.1070	.1096	.1121	.1144	.1167	.1187	.1207	.1224	.1241
10	.0740	.0770	.0800	.0829	.0858	.0887	.0914	.0941	.0967	.0993
11	.0478	.0504	.0531	.0558	.0585	.0613	.0640	.0667	.0695	.0722

Table C (*Continued*)

λt

r	7.1	7.2	7.3	7.4	7.5	7.6	7.7	7.8	7.9	8.0
12	.0283	.0303	.0323	.0344	.0366	.0388	.0411	.0434	.0457	.0481
13	.0154	.0168	.0181	.0196	.0211	.0227	.0243	.0260	.0278	.0296
14	.0078	.0086	.0095	.0104	.0113	.0123	.0134	.0145	.0157	.0169
15	.0037	.0041	.0046	.0051	.0057	.0062	.0069	.0075	.0083	.0090
16	.0016	.0019	.0021	.0024	.0026	.0030	.0033	.0037	.0041	.0045
17	.0007	.0008	.0009	.0010	.0012	.0013	.0015	.0017	.0019	.0021
18	.0003	.0003	.0004	.0004	.0005	.0006	.0006	.0007	.0008	.0009
19	.0001	.0001	.0001	.0002	.0002	.0002	.0003	.0003	.0003	.0004
20	.0000	.0000	.0001	.0001	.0001	.0000	.0001	.0001	.0001	.0002
21	.0000	.0000	.0000	.0000	.0000	.0000	.0000	.0000	.0001	.0001

λt

r	8.1	8.2	8.3	8.4	8.5	8.6	8.7	8.8	8.9	9.0
0	.0003	.0003	.0002	.0002	.0002	.0002	.0002	.0002	.0001	.0001
1	.0025	.0023	.0021	.0019	.0017	.0016	.0014	.0013	.0012	.0011
2	.0100	.0092	.0086	.0079	.0074	.0068	.0063	.0058	.0054	.0050
3	.0269	.0252	.0237	.0222	.0208	.0195	.0183	.0171	.0160	.0150
4	.0544	.0517	.0491	.0466	.0443	.0420	.0398	.0377	.0357	.0337
5	.0882	.0849	.0816	.0784	.0752	.0722	.0692	.0663	.0635	.0607
6	.1191	.1160	.1128	.1097	.1066	.1034	.1003	.0972	.0941	.0911
7	.1378	.1358	.1338	.1317	.1294	.1271	.1247	.1222	.1197	.1171
8	.1395	.1392	.1388	.1382	.1375	.1366	.1356	.1344	.1332	.1318
9	.1256	.1269	.1280	.1290	.1299	.1306	.1311	.1315	.1317	.1318
10	.1017	.1040	.1063	.1084	.1104	.1123	.1140	.1157	.1172	.1186
11	.0749	.0776	.0802	.0828	.0853	.0878	.0902	.0925	.0948	.0970
12	.0505	.0530	.0555	.0579	.0604	.0629	.0654	.0679	.0703	.0728
13	.0315	.0334	.0354	.0374	.0395	.0416	.0438	.0459	.0481	.0504
14	.0182	.0196	.0210	.0225	.0240	.0256	.0272	.0289	.0306	.0324
15	.0098	.0107	.0116	.0126	.0136	.0147	.0158	.0169	.0182	.0194
16	.0050	.0055	.0060	.0066	.0072	.0079	.0086	.0093	.0101	.0109
17	.0024	.0026	.0029	.0033	.0036	.0040	.0044	.0048	.0053	.0058
18	.0011	.0012	.0014	.0015	.0017	.0019	.0021	.0024	.0026	.0029
19	.0005	.0005	.0006	.0007	.0008	.0009	.0010	.0011	.0012	.0014
20	.0002	.0002	.0002	.0003	.0003	.0004	.0004	.0005	.0005	.0006
21	.0001	.0001	.0001	.0001	.0001	.0002	.0002	.0002	.0002	.0003
22	.0000	.0000	.0000	.0000	.0001	.0001	.0001	.0001	.0001	.0001

λt

r	9.1	9.2	9.3	9.4	9.5	9.6	9.7	9.8	9.9	10
0	.0001	.0001	.0001	.0001	.0001	.0001	.0001	.0001	.0001	.0000
1	.0010	.0009	.0009	.0008	.0007	.0007	.0006	.0005	.0005	.0005
2	.0046	.0043	.0040	.0037	.0034	.0031	.0029	.0027	.0025	.0023
3	.0140	.0131	.0123	.0115	.0107	.0100	.0093	.0087	.0081	.0076
4	.0319	.0302	.0285	.0269	.0254	.0240	.0226	.0213	.0201	.0189

Table C   (*Continued*)

					λt					
r	9.1	9.2	9.3	9.4	9.5	9.6	9.7	9.8	9.9	10
5	.0581	.0555	.0530	.0506	.0483	.0460	.0439	.0418	.0398	.0378
6	.0881	.0851	.0822	.0793	.0764	.0736	.0709	.0682	.0656	.0631
7	.1145	.1118	.1091	.1064	.1037	.1010	.0982	.0955	.0928	.0901
8	.1302	.1286	.1269	.1251	.1232	.1212	.1191	.1170	.1148	.1126
9	.1317	.1315	.1311	.1306	.1300	.1293	.1284	.1274	.1263	.1251
10	.1198	.1210	.1219	.1228	.1235	.1241	.1245	.1249	.1250	.1251
11	.0991	.1012	.1031	.1049	.1067	.1083	.1098	.1112	.1125	.1137
12	.0752	.0776	.0799	.0822	.0844	.0866	.0888	.0908	.0928	.0948
13	.0526	.0549	.0572	.0594	.0617	.0640	.0662	.0685	.0707	.0729
14	.0342	.0361	.0380	.0399	.0419	.0439	.0459	.0479	.0500	.0521
15	.0208	.0221	.0235	.0250	.0265	.0281	.0297	.0313	.0330	.0347
16	.0118	.0127	.0137	.0147	.0157	.0168	.0180	.0192	.0204	.0217
17	.0063	.0069	.0075	.0081	.0088	.0095	.0103	.0111	.0119	.0128
18	.0032	.0035	.0039	.0042	.0046	.0051	.0055	.0060	.0065	.0071
19	.0015	.0017	.0019	.0021	.0023	.0026	.0028	.0031	.0034	.0037
20	.0007	.0008	.0009	.0010	.0011	.0012	.0014	.0015	.0017	.0019
21	.0003	.0003	.0004	.0004	.0005	.0006	.0006	.0007	.0008	.0009
22	.0001	.0001	.0002	.0002	.0002	.0002	.0003	.0003	.0004	.0004
23	.0000	.0001	.0001	.0001	.0001	.0001	.0001	.0001	.0002	.0002
24	.0000	.0000	.0000	.0000	.0000	.0000	.0000	.0001	.0001	.0001

					λt					
r	11	12	13	14	15	16	17	18	19	20
0	.0000	.0000	.0000	.0000	.0000	.0000	.0000	.0000	.0000	.0000
1	.0002	.0001	.0000	.0000	.0000	.0000	.0000	.0000	.0000	.0000
2	.0010	.0004	.0002	.0001	.0000	.0000	.0000	.0000	.0000	.0000
3	.0037	.0018	.0008	.0004	.0002	.0001	.0000	.0000	.0000	.0000
4	.0102	.0053	.0027	.0013	.0006	.0003	.0001	.0001	.0000	.0000
5	.0224	.0127	.0070	.0037	.0019	.0010	.0005	.0002	.0001	.0001
6	.0411	.0255	.0152	.0087	.0048	.0026	.0014	.0007	.0004	.0002
7	.0646	.0437	.0281	.0174	.0104	.0060	.0034	.0018	.0010	.0005
8	.0888	.0655	.0457	.0304	.0194	.0120	.0072	.0042	.0024	.0013
9	.1085	.0874	.0661	.0473	.0324	.0213	.0135	.0083	.0050	.0029
10	.1194	.1048	.0859	.0663	.0486	.0341	.0230	.0150	.0095	.0058
11	.1194	.1144	.1015	.0844	.0663	.0496	.0355	.0245	.0164	.0106
12	.1094	.1144	.1099	.0984	.0829	.0661	.0504	.0368	.0259	.0176
13	.0926	.1056	.1099	.1060	.0956	.0814	.0658	.0509	.0378	.0271
14	.0728	.0905	.1021	.1060	.1024	.0930	.0800	.0655	.0541	.0387
15	.0534	.0724	.0885	.0989	.1024	.0992	.0906	.0786	.0650	.0516
16	.0367	.0543	.0719	.0866	.0960	.0992	.0963	.0884	.0772	.0646
17	.0237	.0383	.0550	.0713	.0847	.0934	.0963	.0936	.0863	.0760
18	.0145	.0256	.0397	.0554	.0706	.0830	.0909	.0936	.0911	.0844
19	.0084	.0161	.0272	.0409	.0557	.0699	.0814	.0887	.0911	.0888
20	.0046	.0097	.0177	.0286	.0418	.0559	.0692	.0798	.0866	.0888
21	.0024	.0055	.0109	.0191	.0299	.0426	.0560	.0684	.0783	.0846
22	.0012	.0030	.0065	.0121	.0204	.0310	.0433	.0560	.0676	.0769
23	.0006	.0016	.0037	.0074	.0133	.0216	.0320	.0438	.0559	.0669
24	.0003	.0008	.0020	.0043	.0083	.0144	.0226	.0328	.0442	.0557

Table C (*Concluded*)

	λt									
r	11	12	13	14	15	16	17	18	19	20
25	.0001	.0004	.0010	.0024	.0050	.0092	.0154	.0237	.0336	.0446
26	.0000	.0002	.0005	.0013	.0029	.0057	.0101	.0164	.0246	.0343
27	.0000	.0001	.0002	.0007	.0016	.0034	.0063	.0109	.0173	.0254
28	.0000	.0000	.0001	.0003	.0009	.0019	.0038	.0070	.0117	.0181
29	.0000	.0000	.0001	.0002	.0004	.0011	.0023	.0044	.0077	.0125
30	.0000	.0000	.0000	.0001	.0002	.0006	.0013	.0026	.0049	.0083
31	.0000	.0000	.0000	.0000	.0001	.0003	.0007	.0015	.0030	.0054
32	.0000	.0000	.0000	.0000	.0001	.0001	.0004	.0009	.0018	.0034
33	.0000	.0000	.0000	.0000	.0000	.0001	.0002	.0005	.0010	.0020
34	.0000	.0000	.0000	.0000	.0000	.0000	.0001	.0002	.0006	.0012
35	.0000	.0000	.0000	.0000	.0000	.0000	.0000	.0001	.0003	.0007
36	.0000	.0000	.0000	.0000	.0000	.0000	.0000	.0001	.0002	.0004
37	.0000	.0000	.0000	.0000	.0000	.0000	.0000	.0000	.0001	.0002
38	.0000	.0000	.0000	.0000	.0000	.0000	.0000	.0000	.0000	.0001
39	.0000	.0000	.0000	.0000	.0000	.0000	.0000	.0000	.0000	.0001

SOURCE: From *Handbook of Probability and Statistics with Tables* (2d ed.). Copyright © 1970 by McGraw-Hill, Inc. Used by permission of McGraw-Hill Book Company.

Table D   Standard Normal Distribution

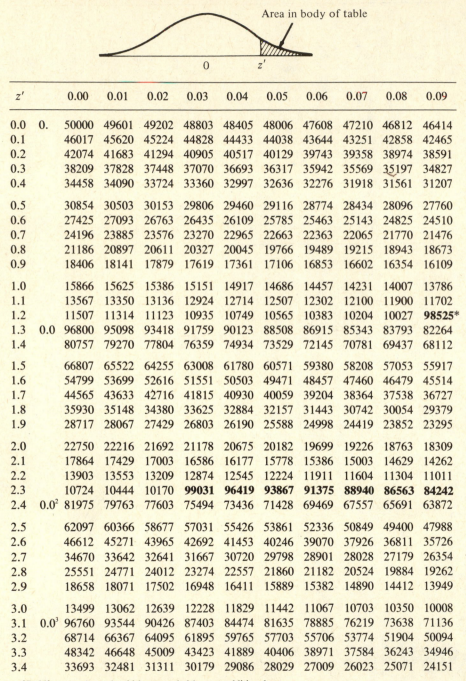

Area in body of table

z'		0.00	0.01	0.02	0.03	0.04	0.05	0.06	0.07	0.08	0.09
0.0	0.	50000	49601	49202	48803	48405	48006	47608	47210	46812	46414
0.1		46017	45620	45224	44828	44433	44038	43644	43251	42858	42465
0.2		42074	41683	41294	40905	40517	40129	39743	39358	38974	38591
0.3		38209	37828	37448	37070	36693	36317	35942	35569	35197	34827
0.4		34458	34090	33724	33360	32997	32636	32276	31918	31561	31207
0.5		30854	30503	30153	29806	29460	29116	28774	28434	28096	27760
0.6		27425	27093	26763	26435	26109	25785	25463	25143	24825	24510
0.7		24196	23885	23576	23270	22965	22663	22363	22065	21770	21476
0.8		21186	20897	20611	20327	20045	19766	19489	19215	18943	18673
0.9		18406	18141	17879	17619	17361	17106	16853	16602	16354	16109
1.0		15866	15625	15386	15151	14917	14686	14457	14231	14007	13786
1.1		13567	13350	13136	12924	12714	12507	12302	12100	11900	11702
1.2		11507	11314	11123	10935	10749	10565	10383	10204	10027	**98525***
1.3	0.0	96800	95098	93418	91759	90123	88508	86915	85343	83793	82264
1.4		80757	79270	77804	76359	74934	73529	72145	70781	69437	68112
1.5		66807	65522	64255	63008	61780	60571	59380	58208	57053	55917
1.6		54799	53699	52616	51551	50503	49471	48457	47460	46479	45514
1.7		44565	43633	42716	41815	40930	40059	39204	38364	37538	36727
1.8		35930	35148	34380	33625	32884	32157	31443	30742	30054	29379
1.9		28717	28067	27429	26803	26190	25588	24998	24419	23852	23295
2.0		22750	22216	21692	21178	20675	20182	19699	19226	18763	18309
2.1		17864	17429	17003	16586	16177	15778	15386	15003	14629	14262
2.2		13903	13553	13209	12874	12545	12224	11911	11604	11304	11011
2.3		10724	10444	10170	**99031**	**96419**	**93867**	**91375**	**88940**	**86563**	**84242**
2.4	0.0^2	81975	79763	77603	75494	73436	71428	69469	67557	65691	63872
2.5		62097	60366	58677	57031	55426	53861	52336	50849	49400	47988
2.6		46612	45271	43965	42692	41453	40246	39070	37926	36811	35726
2.7		34670	33642	32641	31667	30720	29798	28901	28028	27179	26354
2.8		25551	24771	24012	23274	22557	21860	21182	20524	19884	19262
2.9		18658	18071	17502	16948	16411	15889	15382	14890	14412	13949
3.0		13499	13062	12639	12228	11829	11442	11067	10703	10350	10008
3.1	0.0^3	96760	93544	90426	87403	84474	81635	78885	76219	73638	71136
3.2		68714	66367	64095	61895	59765	57703	55706	53774	51904	50094
3.3		48342	46648	45009	43423	41889	40406	38971	37584	36243	34946
3.4		33693	32481	31311	30179	29086	28029	27009	26023	25071	24151

*Boldface numbers should be preceded by one additional zero.

Table D (*Concluded*)

z'		0.00	0.01	0.02	0.03	0.04	0.05	0.06	0.07	0.08	0.09
3.5		23263	22405	21577	20778	20006	19262	18543	17849	17180	16534
3.6		15911	15310	14730	14171	13632	13112	12611	12128	11662	11213
3.7		10780	10363	**99611**	**95740**	**92010**	**88417**	**84957**	**81624**	**78414**	**75324**
3.8	0.0^4	72348	69483	66726	64072	61517	59059	56694	54418	52228	50122
3.9		48096	46148	44274	42473	40741	39076	37475	35936	34458	33037
4.0		31671	30359	29099	27888	26726	25609	24536	23507	22518	21569
4.1		20658	19783	18944	18138	17365	16624	15912	15230	14575	13948
4.2		13346	12769	12215	11685	11176	10689	10221	**97736**	**93447**	**89337**
4.3	0.0^5	85399	81627	78015	74555	71241	68069	65031	62123	59340	56675
4.4		54125	51685	49350	47117	44979	42935	40980	39110	37322	35612
4.5		33977	32414	30920	29492	28127	26823	25577	24386	23249	22162
4.6		21125	20133	19187	18283	17420	16597	15810	15060	14344	13660
4.7		13008	12386	11792	11226	10686	10171	**96796**	**92113**	**87648**	**83391**
4.8	0.0^6	79333	75465	71779	68267	64920	61731	58693	55799	53043	50418
4.9		47918	45538	43272	41115	39061	37107	35247	33476	31792	30190

SOURCE: Table D is taken from Fisher and Yates, *Statistical Tables for Biological, Agricultural, and Medical Research*, (6th edition, 1974) published by Longman Group Ltd., London (previously published by Oliver & Boyd, Edinburgh), and by permission of the authors and publishers.

*Boldface numbers should be preceded by one additional zero.

Table E Percentage Points of the *t* Distribution

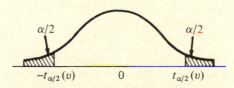

Area in both tails (α)

v	0.1	0.05	0.02	0.01
1	6.314	12.706	31.821	63.657
2	2.920	4.303	6.965	9.925
3	2.353	3.182	4.541	5.841
4	2.132	2.776	3.747	4.604
5	2.015	2.571	3.365	4.032
6	1.943	2.447	3.143	3.707
7	1.895	2.365	2.998	3.499
8	1.860	2.306	2.896	3.355
9	1.833	2.262	2.821	3.250
10	1.812	2.228	2.764	3.169

Table E (*Concluded*)

v	0.1	0.05	0.02	0.01
11	1.796	2.201	2.718	3.106
12	1.782	2.179	2.681	3.055
13	1.771	2.160	2.650	3.012
14	1.761	2.145	2.624	2.977
15	1.753	2.131	2.602	2.947
16	1.746	2.120	2.583	2.921
17	1.740	2.110	2.567	2.898
18	1.734	2.101	2.552	2.878
19	1.729	2.093	2.539	2.861
20	1.725	2.086	2.528	2.845
21	1.721	2.080	2.518	2.831
22	1.717	2.074	2.508	2.819
23	1.714	2.069	2.500	2.807
24	1.711	2.064	2.492	2.797
25	1.708	2.060	2.485	2.787
26	1.706	2.056	2.479	2.779
27	1.703	2.052	2.473	2.771
28	1.701	2.048	2.467	2.763
29	1.699	2.045	2.462	2.756
30	1.697	2.042	2.457	2.750
40	1.684	2.021	2.423	2.704
60	1.671	2.000	2.390	2.660
120	1.658	1.980	2.358	2.617
∞	1.645	1.960	2.326	2.576

SOURCE: Table E is abridged from Table III of Fisher and Yates, *Statistical Tables for Biological, Agricultural, and Medical Research*, (6th edition, 1974) published by Longman Group Ltd., London (previously published by Oliver & Boyd, Edinburgh), and by permission of the authors and publishers.

Table F Percentage Points of the Chi-square Distribution

v	0.99	0.98	0.95	0.90	0.80	0.70	0.50	0.30	0.20	0.10	0.05	0.02	0.01	0.001
1	0.0^3157	0.0^3628	0.00393	0.0158	0.0642	0.148	0.455	1.074	1.642	2.706	3.841	5.412	6.635	10.827
2	0.0201	0.0404	0.103	0.211	0.446	0.713	1.386	2.408	3.219	4.605	5.991	7.824	9.210	13.815
3	0.115	0.185	0.352	0.584	1.005	1.424	2.366	3.665	4.642	6.251	7.815	9.837	11.345	16.266
4	0.297	0.429	0.711	1.064	1.649	2.195	3.357	4.878	5.989	7.779	9.488	11.668	13.277	18.467
5	0.554	0.752	1.145	1.610	2.343	3.000	4.351	6.064	7.289	9.236	11.070	13.388	15.086	20.515
6	0.872	1.134	1.635	2.204	3.070	3.828	5.348	7.231	8.558	10.645	12.592	15.033	16.812	22.457
7	1.239	1.564	2.167	2.833	3.822	4.671	6.346	8.383	9.803	12.017	14.067	16.622	18.475	24.322
8	1.646	2.032	2.733	3.490	4.594	5.527	7.344	9.524	11.030	13.362	15.507	18.168	20.090	26.125
9	2.088	2.532	3.325	4.168	5.380	6.393	8.343	10.656	12.242	14.684	16.919	19.679	21.666	27.877
10	2.558	3.059	3.940	4.865	6.179	7.267	9.342	11.781	13.442	15.987	18.307	21.161	23.209	29.588
11	3.053	3.609	4.575	5.578	6.989	8.148	10.341	12.899	14.631	17.275	19.675	22.618	24.725	31.264
12	3.571	4.178	5.226	6.304	7.807	9.034	11.340	14.011	15.812	18.549	21.026	24.054	26.217	32.909
13	4.107	4.765	5.892	7.042	8.634	9.926	12.340	15.119	16.985	19.812	22.362	25.472	27.688	34.528
14	4.660	5.368	6.571	7.790	9.467	10.821	13.339	16.222	18.151	21.064	23.685	26.873	29.141	36.123
15	5.229	5.985	7.261	8.547	10.307	11.721	14.339	17.322	19.311	22.307	24.996	28.259	30.578	37.697

Table F (*Concluded*)

v	0.99	0.98	0.95	0.90	0.80	0.70	0.50	0.30	0.20	0.10	0.05	0.02	0.01	0.001
16	5.812	6.614	7.962	9.312	11.152	12.624	15.338	18.418	20.465	23.542	26.296	29.633	32.000	39.252
17	6.408	7.255	8.672	10.085	12.002	13.531	16.338	19.511	21.615	24.769	27.587	30.995	33.409	40.790
18	7.015	7.906	9.390	10.865	12.857	14.440	17.338	20.601	22.760	25.989	28.869	32.346	34.805	42.312
19	7.633	8.567	10.117	11.651	13.716	15.352	18.338	21.689	23.900	27.204	30.144	33.687	36.191	43.820
20	8.260	9.237	10.851	12.443	14.578	16.266	19.337	22.775	25.038	28.412	31.410	35.020	37.566	45.315
21	8.897	9.915	11.591	13.240	15.445	17.182	20.337	23.858	26.171	29.615	32.671	36.343	38.932	46.797
22	9.542	10.600	12.338	14.041	16.314	18.101	21.337	24.939	27.301	30.813	33.924	37.659	40.289	48.268
23	10.196	11.293	13.091	14.848	17.187	19.021	22.337	26.018	28.429	32.007	35.172	38.968	41.638	49.728
24	10.856	11.992	13.848	15.659	18.062	19.943	23.337	27.096	29.553	33.196	36.415	40.270	42.980	51.179
25	11.524	12.697	14.611	16.473	18.940	20.867	24.337	28.172	30.675	34.382	37.652	41.566	44.314	52.620
26	12.198	13.409	15.379	17.292	19.820	21.792	25.336	29.246	31.795	35.563	38.885	42.856	45.642	54.052
27	12.879	14.125	16.151	18.114	20.703	22.719	26.336	30.319	32.912	36.741	40.113	44.140	46.963	55.476
28	13.565	14.847	16.928	18.939	21.588	23.647	27.336	31.391	34.027	37.916	41.337	45.419	48.278	56.893
29	14.256	15.574	17.708	19.768	22.475	24.577	28.336	32.461	35.139	39.087	42.557	46.693	49.588	58.302
30	14.953	16.306	18.493	20.599	23.364	25.508	29.336	33.530	36.250	40.256	43.773	47.962	50.892	59.703
32	16.362	17.783	20.072	22.271	25.148	27.373	31.336	35.665	38.466	42.585	46.194	50.487	53.486	62.487
34	17.789	19.275	21.664	23.952	26.938	29.242	33.336	37.795	40.676	44.903	48.602	52.995	56.061	65.247
36	19.233	20.783	23.269	25.643	28.735	31.115	35.336	39.922	42.879	47.212	50.999	55.489	58.619	67.985
38	20.691	22.304	24.884	27.343	30.537	32.992	37.335	42.045	45.076	49.513	53.384	57.969	61.162	70.703
40	22.164	23.838	26.509	29.051	32.345	34.872	39.335	44.165	47.269	51.805	55.759	60.436	63.691	73.402

42	23.650	25.383	28.144	30.765	34.157	36.755	41.335	46.282	49.456	54.090	58.124	62.892	66.206	76.084
44	25.148	26.939	29.787	32.487	35.974	38.641	43.335	48.396	51.639	56.369	60.481	65.337	68.710	78.750
46	26.657	28.504	31.439	34.215	37.795	40.529	45.335	50.507	53.818	58.641	62.830	67.771	71.201	81.400
48	28.177	30.080	33.098	35.949	39.621	42.420	47.335	52.616	55.993	60.907	65.171	70.197	73.683	84.037
50	29.707	31.664	34.764	37.689	41.449	44.313	49.335	54.723	58.164	63.167	67.505	72.613	76.154	86.661
52	31.246	33.256	36.437	39.433	43.281	46.209	51.335	56.827	60.332	65.422	69.832	75.021	78.616	89.272
54	32.793	34.856	38.116	41.183	45.117	48.106	53.335	58.930	62.496	67.673	72.153	77.422	81.069	91.872
56	34.350	36.464	39.801	42.937	46.955	50.005	55.335	61.031	64.658	69.919	74.468	79.815	83.513	94.461
58	35.913	38.078	41.492	44.696	48.797	51.906	57.335	63.129	66.816	72.160	76.778	82.201	85.950	97.039
60	37.485	39.699	43.188	46.459	50.641	53.809	59.335	65.227	68.972	74.397	79.082	84.580	88.379	99.607
62	39.063	41.327	44.889	48.226	52.487	55.714	61.335	67.322	71.125	76.630	81.381	86.953	90.802	102.166
64	40.649	42.960	46.595	49.996	54.336	57.620	63.335	69.416	73.276	78.860	83.675	89.320	93.217	104.716
66	42.240	44.599	48.305	51.770	56.188	59.527	65.335	71.508	75.424	81.085	85.965	91.681	95.626	107.258
68	43.838	46.244	50.020	53.548	58.042	61.436	67.335	73.600	77.571	83.308	88.250	94.037	98.028	109.791
70	45.442	47.893	51.739	55.329	59.898	63.346	69.334	75.689	79.715	85.527	90.531	96.388	100.425	112.317

SOURCE: Table F is taken from Fisher and Yates, *Statistical Tables for Biological, Agricultural, and Medical Research*, (6th edition, 1974) published by Longman Group Ltd., London (previously published by Oliver & Boyd, Edinburgh), and by permission of the authors and publishers.

Table G Percentage Points of the F Distribution

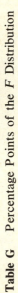

$F_{0.05}(v_1, v_2)$ $F_{0.01}(v_1, v_2)$

Roman type — Bold face type

v_1

v_2	1	2	3	4	5	6	7	8	9	10	11	12	14	16	20	24	30	40	50	75	100	200	500	∞
1	161	200	216	225	230	234	237	239	241	242	243	244	245	246	248	249	250	251	252	253	253	254	254	254
	4,052	**4,999**	**5,403**	**5,625**	**5,764**	**5,859**	**5,928**	**5,981**	**6,022**	**6,056**	**6,082**	**6,106**	**6,142**	**6,169**	**6,208**	**6,234**	**6,261**	**6,286**	**6,302**	**6,323**	**6,334**	**6,352**	**6,361**	**6,366**
2	18.51	19.00	19.16	19.25	19.30	19.33	19.36	19.37	19.38	19.39	19.40	19.41	19.42	19.43	19.44	19.45	19.46	19.47	19.47	19.48	19.49	19.49	19.50	19.50
	98.49	**99.00**	**99.17**	**99.25**	**99.30**	**99.33**	**99.36**	**99.37**	**99.39**	**99.40**	**99.41**	**99.42**	**99.43**	**99.44**	**99.45**	**99.46**	**99.47**	**99.48**	**99.48**	**99.49**	**99.49**	**99.49**	**99.50**	**99.50**
3	10.13	9.55	9.28	9.12	9.01	8.94	8.88	8.84	8.81	8.78	8.76	8.74	8.71	8.69	8.66	8.64	8.62	8.60	8.58	8.57	8.56	8.54	8.54	8.53
	34.12	**30.82**	**29.46**	**28.71**	**28.24**	**27.91**	**27.67**	**27.49**	**27.34**	**27.23**	**27.13**	**27.05**	**26.92**	**26.83**	**26.69**	**26.60**	**26.50**	**26.41**	**26.35**	**26.27**	**26.23**	**26.18**	**26.14**	**26.12**
4	7.71	6.94	6.59	6.39	6.26	6.16	6.09	6.04	6.00	5.96	5.93	5.91	5.87	5.84	5.80	5.77	5.74	5.71	5.70	5.68	5.66	5.65	5.64	5.63
	21.20	**18.00**	**16.69**	**15.98**	**15.52**	**15.21**	**14.98**	**14.80**	**14.66**	**14.54**	**14.45**	**14.37**	**14.24**	**14.15**	**14.02**	**13.93**	**13.83**	**13.74**	**13.69**	**13.61**	**13.57**	**13.52**	**13.48**	**13.46**
5	6.61	5.79	5.41	5.19	5.05	4.95	4.88	4.82	4.78	4.74	4.70	4.68	4.64	4.60	4.56	4.53	4.50	4.46	4.44	4.42	4.40	4.38	4.37	4.36
	16.26	**13.27**	**12.06**	**11.39**	**10.97**	**10.67**	**10.45**	**10.29**	**10.15**	**10.05**	**9.96**	**9.89**	**9.77**	**9.68**	**9.55**	**9.47**	**9.38**	**9.29**	**9.24**	**9.17**	**9.13**	**9.07**	**9.04**	**9.02**
6	5.99	5.14	4.76	4.53	4.39	4.28	4.21	4.15	4.10	4.06	4.03	4.00	3.96	3.92	3.87	3.84	3.81	3.77	3.75	3.72	3.71	3.69	3.68	3.67
	13.74	**10.92**	**9.78**	**9.15**	**8.75**	**8.47**	**8.26**	**8.10**	**7.98**	**7.87**	**7.79**	**7.72**	**7.60**	**7.52**	**7.39**	**7.31**	**7.23**	**7.14**	**7.09**	**7.02**	**6.99**	**6.94**	**6.90**	**6.88**

Column labels (top, left → right): 7 8 9 10 11 12 13 14 15 16 17 18

Each cell is given as *upper value / lower value*.

row																								
7	3.23 / 5.65	3.24 / 5.67	3.25 / 5.70	3.28 / 5.75	3.29 / 5.78	3.32 / 5.85	3.34 / 5.90	3.38 / 5.98	3.41 / 6.07	3.44 / 6.15	3.49 / 6.27	3.52 / 6.35	3.57 / 6.47	3.60 / 6.54	3.63 / 6.62	3.68 / 6.71	3.73 / 6.84	3.79 / 7.00	3.87 / 7.19	3.97 / 7.46	4.12 / 7.85	4.35 / 8.45	4.74 / 9.55	5.59 / 12.25
8	2.93 / 4.86	2.94 / 4.88	2.96 / 4.91	2.98 / 4.96	3.00 / 5.00	3.03 / 5.06	3.05 / 5.11	3.08 / 5.20	3.12 / 5.28	3.15 / 5.36	3.20 / 5.48	3.23 / 5.56	3.28 / 5.67	3.31 / 5.74	3.34 / 5.82	3.39 / 5.91	3.44 / 6.03	3.50 / 6.19	3.58 / 6.37	3.69 / 6.63	3.84 / 7.01	4.07 / 7.59	4.46 / 8.65	5.32 / 11.26
9	2.71 / 4.31	2.72 / 4.33	2.73 / 4.36	2.76 / 4.41	2.77 / 4.45	2.80 / 4.51	2.82 / 4.56	2.86 / 4.64	2.90 / 4.73	2.93 / 4.80	2.98 / 4.92	3.02 / 5.00	3.07 / 5.11	3.10 / 5.18	3.13 / 5.26	3.18 / 5.35	3.23 / 5.47	3.29 / 5.62	3.37 / 5.80	3.48 / 6.06	3.63 / 6.42	3.86 / 6.99	4.26 / 8.02	5.12 / 10.56
10	2.54 / 3.91	2.55 / 3.93	2.56 / 3.96	2.59 / 4.01	2.61 / 4.05	2.64 / 4.12	2.67 / 4.17	2.70 / 4.25	2.74 / 4.33	2.77 / 4.41	2.82 / 4.52	2.86 / 4.60	2.91 / 4.71	2.94 / 4.78	2.97 / 4.85	3.02 / 4.95	3.07 / 5.06	3.14 / 5.21	3.22 / 5.39	3.33 / 5.64	3.48 / 5.99	3.71 / 6.55	4.10 / 7.56	4.96 / 10.04
11	2.40 / 3.60	2.41 / 3.62	2.42 / 3.66	2.45 / 3.70	2.47 / 3.74	2.50 / 3.80	2.53 / 3.86	2.57 / 3.94	2.61 / 4.02	2.65 / 4.10	2.70 / 4.21	2.74 / 4.29	2.79 / 4.40	2.82 / 4.46	2.86 / 4.54	2.90 / 4.63	2.95 / 4.74	3.01 / 4.88	3.09 / 5.07	3.20 / 5.32	3.36 / 5.67	3.59 / 6.22	3.98 / 7.20	4.84 / 9.65
12	2.30 / 3.36	2.31 / 3.38	2.32 / 3.41	2.35 / 3.46	2.36 / 3.49	2.40 / 3.56	2.42 / 3.61	2.46 / 3.70	2.50 / 3.78	2.54 / 3.86	2.60 / 3.98	2.64 / 4.05	2.69 / 4.16	2.72 / 4.22	2.76 / 4.30	2.80 / 4.39	2.85 / 4.50	2.92 / 4.65	3.00 / 4.82	3.11 / 5.06	3.26 / 5.41	3.49 / 5.95	3.88 / 6.93	4.75 / 9.33
13	2.21 / 3.16	2.22 / 3.18	2.24 / 3.21	2.26 / 3.27	2.28 / 3.30	2.32 / 3.37	2.34 / 3.42	2.38 / 3.51	2.42 / 3.59	2.46 / 3.67	2.51 / 3.78	2.55 / 3.85	2.60 / 3.96	2.63 / 4.02	2.67 / 4.10	2.72 / 4.19	2.77 / 4.30	2.84 / 4.44	2.92 / 4.62	3.02 / 4.86	3.18 / 5.20	3.41 / 5.74	3.80 / 6.70	4.67 / 9.07
14	2.13 / 3.00	2.14 / 3.02	2.16 / 3.06	2.19 / 3.11	2.21 / 3.14	2.24 / 3.21	2.27 / 3.26	2.31 / 3.34	2.35 / 3.43	2.39 / 3.51	2.44 / 3.62	2.48 / 3.70	2.53 / 3.80	2.56 / 3.86	2.60 / 3.94	2.65 / 4.03	2.70 / 4.14	2.77 / 4.28	2.85 / 4.46	2.96 / 4.69	3.11 / 5.03	3.34 / 5.56	3.74 / 6.51	4.60 / 8.86
15	2.07 / 2.87	2.08 / 2.89	2.10 / 2.92	2.12 / 2.97	2.15 / 3.00	2.18 / 3.07	2.21 / 3.12	2.25 / 3.20	2.29 / 3.29	2.33 / 3.36	2.39 / 3.48	2.43 / 3.56	2.48 / 3.67	2.51 / 3.73	2.55 / 3.80	2.59 / 3.89	2.64 / 4.00	2.70 / 4.14	2.79 / 4.32	2.90 / 4.56	3.06 / 4.89	3.29 / 5.42	3.68 / 6.36	4.54 / 8.68
16	2.01 / 2.75	2.02 / 2.77	2.04 / 2.80	2.07 / 2.86	2.09 / 2.98	2.13 / 2.96	2.16 / 3.01	2.20 / 3.10	2.24 / 3.18	2.28 / 3.25	2.33 / 3.37	2.37 / 3.45	2.42 / 3.55	2.45 / 3.61	2.49 / 3.69	2.54 / 3.78	2.59 / 3.89	2.66 / 4.03	2.74 / 4.20	2.85 / 4.44	3.01 / 4.77	3.24 / 5.29	3.63 / 6.23	4.49 / 8.53
17	1.96 / 2.65	1.97 / 2.67	1.99 / 2.70	2.02 / 2.76	2.04 / 2.79	2.08 / 2.86	2.11 / 2.92	2.15 / 3.00	2.19 / 3.08	2.23 / 3.16	2.29 / 3.27	2.33 / 3.35	2.38 / 3.45	2.41 / 3.52	2.45 / 3.59	2.50 / 3.68	2.55 / 3.79	2.62 / 3.93	2.70 / 4.10	2.81 / 4.34	2.96 / 4.67	3.20 / 5.18	3.59 / 6.11	4.45 / 8.40
18	1.92 / 2.57	1.93 / 2.59	1.95 / 2.62	1.98 / 2.68	2.00 / 2.71	2.04 / 2.78	2.07 / 2.83	2.11 / 2.91	2.15 / 3.00	2.19 / 3.07	2.25 / 3.19	2.29 / 3.27	2.34 / 3.37	2.37 / 3.44	2.41 / 3.51	2.46 / 3.60	2.51 / 3.71	2.58 / 3.85	2.77 / 4.01	2.77 / 4.25	2.93 / 4.58	3.16 / 5.09	3.55 / 6.01	4.41 / 8.28

Row labels (bottom): 7 8 9 10 11 12 13 14 15 16 17 18

Table G (Concluded)

v_2		1	2	3	4	5	6	7	8	9	10	11	12	14	16	20	24	30	40	50	75	100	200	500	∞
19		4.38	3.52	3.13	2.90	2.74	2.63	2.55	2.48	2.43	2.38	2.34	2.31	2.26	2.21	2.15	2.11	2.07	2.02	2.00	1.96	1.94	1.91	1.90	1.88
		8.18	**5.93**	**5.01**	**4.50**	**4.17**	**3.94**	**3.77**	**3.63**	**3.52**	**3.43**	**3.36**	**3.30**	**3.19**	**3.12**	**3.00**	**2.92**	**2.84**	**2.76**	**2.70**	**2.63**	**2.60**	**2.54**	**2.51**	**2.49**
20		4.35	3.49	3.10	2.87	2.71	2.60	2.52	2.45	2.40	2.35	2.31	2.28	2.23	2.18	2.12	2.08	2.04	1.99	1.96	1.92	1.90	1.87	1.85	1.84
		8.10	**5.85**	**4.94**	**4.43**	**4.10**	**3.87**	**3.71**	**3.56**	**3.45**	**3.37**	**3.30**	**3.23**	**3.13**	**3.05**	**2.94**	**2.86**	**2.77**	**2.69**	**2.63**	**2.56**	**2.53**	**2.47**	**2.44**	**2.42**
21		4.32	3.47	3.07	2.84	2.68	2.57	2.49	2.42	2.37	2.32	2.28	2.25	2.20	2.15	2.09	2.05	2.00	1.96	1.93	1.89	1.87	1.84	1.82	1.81
		8.02	**5.78**	**4.87**	**4.37**	**4.04**	**3.81**	**3.65**	**3.51**	**3.40**	**3.31**	**3.24**	**3.17**	**3.07**	**2.99**	**2.88**	**2.80**	**2.72**	**2.63**	**2.58**	**2.51**	**2.47**	**2.42**	**2.38**	**2.36**
22		4.30	3.44	3.05	2.82	2.66	2.55	2.47	2.40	2.35	2.30	2.26	2.23	2.18	2.13	2.07	2.03	1.98	1.93	1.91	1.87	1.84	1.81	1.80	1.78
		7.94	**5.72**	**4.82**	**4.31**	**3.99**	**3.76**	**3.59**	**3.45**	**3.35**	**3.26**	**3.18**	**3.12**	**3.02**	**2.94**	**2.83**	**2.75**	**2.67**	**2.58**	**2.53**	**2.46**	**2.42**	**2.37**	**2.33**	**2.31**
23		4.28	3.42	3.03	2.80	2.64	2.53	2.45	2.38	2.32	2.28	2.24	2.20	2.14	2.10	2.04	2.00	1.96	1.91	1.88	1.84	1.82	1.79	1.77	1.76
		7.88	**5.66**	**4.76**	**4.26**	**3.94**	**3.71**	**3.54**	**3.41**	**3.30**	**3.21**	**3.14**	**3.07**	**2.97**	**2.89**	**2.78**	**2.70**	**2.62**	**2.53**	**2.48**	**2.41**	**2.37**	**2.32**	**2.28**	**2.26**
24		4.26	3.40	3.01	2.78	2.62	2.51	2.43	2.36	2.30	2.26	2.22	2.18	2.13	2.09	2.02	1.98	1.94	1.89	1.86	1.82	1.80	1.76	1.74	1.73
		7.82	**5.61**	**4.72**	**4.22**	**3.90**	**3.67**	**3.50**	**3.36**	**3.25**	**3.17**	**3.09**	**3.03**	**2.93**	**2.85**	**2.74**	**2.66**	**2.58**	**2.49**	**2.44**	**2.36**	**2.33**	**2.27**	**2.23**	**2.21**
25		4.24	3.38	2.99	2.76	2.60	2.49	2.41	2.34	2.28	2.24	2.20	2.16	2.11	2.06	2.00	1.96	1.92	1.87	1.84	1.80	1.77	1.74	1.72	1.71
		7.77	**5.57**	**4.68**	**4.18**	**3.86**	**3.63**	**3.46**	**3.32**	**3.21**	**3.13**	**3.05**	**2.99**	**2.89**	**2.81**	**2.70**	**2.62**	**2.54**	**2.45**	**2.40**	**2.32**	**2.29**	**2.23**	**2.19**	**2.17**
26		4.22	3.37	2.98	2.74	2.59	2.47	2.39	2.32	2.27	2.22	2.18	2.15	2.10	2.05	1.99	1.95	1.90	1.85	1.82	1.78	1.76	1.72	1.70	1.69
		7.72	**5.53**	**4.64**	**4.14**	**3.82**	**3.59**	**3.42**	**3.29**	**3.17**	**3.09**	**3.02**	**2.96**	**2.86**	**2.77**	**2.66**	**2.58**	**2.50**	**2.41**	**2.36**	**2.28**	**2.25**	**2.19**	**2.15**	**2.13**

v_1

SOURCE: Reprinted by permission from *Statistical Methods* by George W. Snedecor and William G. Cochran, sixth edition © 1967 by Iowa State University Press, Ames, Iowa.

Table H Transformation of r to $z = \frac{1}{2}\ln[(1+r)/(1-r)]$

z	0.00	0.01	0.02	0.03	0.04	0.05	0.06	0.07	0.08	0.09
0.0	0.0000	0.0100	0.0200	0.0300	0.0400	0.0500	0.0599	0.0699	0.0798	0.0898
0.1	0.0997	0.1096	0.1194	0.1293	0.1391	0.1489	0.1586	0.1684	0.1781	0.1877
0.2	0.1974	0.2070	0.2165	0.2260	0.2355	0.2449	0.2543	0.2636	0.2729	0.2821
0.3	0.2913	0.3004	0.3095	0.3185	0.3275	0.3364	0.3452	0.3540	0.3627	0.3714
0.4	0.3800	0.3885	0.3969	0.4053	0.4136	0.4219	0.4301	0.4382	0.4462	0.4542
0.5	0.4621	0.4699	0.4777	0.4854	0.4930	0.5005	0.5080	0.5154	0.5227	0.5299
0.6	0.5370	0.5441	0.5511	0.5580	0.5649	0.5717	0.5784	0.5850	0.5915	0.5980
0.7	0.6044	0.6107	0.6169	0.6231	0.6291	0.6351	0.6411	0.6469	0.6527	0.6584
0.8	0.6640	0.6696	0.6751	0.6805	0.6858	0.6911	0.6963	0.7014	0.7064	0.7114
0.9	0.7163	0.7211	0.7259	0.7306	0.7352	0.7398	0.7443	0.7487	0.7531	0.7574
1.0	0.7616	0.7658	0.7699	0.7739	0.7779	0.7818	0.7857	0.7895	0.7932	0.7969
1.1	0.8005	0.8041	0.8076	0.8110	0.8144	0.8178	0.8210	0.8243	0.8275	0.8306
1.2	0.8337	0.8367	0.8397	0.8426	0.8455	0.8483	0.8511	0.8538	0.8565	0.8591
1.3	0.8617	0.8643	0.8668	0.8692	0.8717	0.8741	0.8764	0.8787	0.8810	0.8832
1.4	0.8854	0.8875	0.8896	0.8917	0.8937	0.8957	0.8977	0.8996	0.9015	0.9033
1.5	0.9051	0.9069	0.9087	0.9104	0.9121	0.9138	0.9154	0.9170	0.9186	0.9201
1.6	0.9217	0.9232	0.9246	0.9261	0.9275	0.9289	0.9302	0.9316	0.9329	0.9341
1.7	0.9354	0.9366	0.9379	0.9391	0.9402	0.9414	0.9425	0.9436	0.9447	0.9458
1.8	0.94681	0.94783	0.94884	0.94983	0.95080	0.95175	0.95268	0.95359	0.95449	0.95537
1.9	0.95624	0.95709	0.95792	0.95873	0.95953	0.96032	0.96109	0.96185	0.96259	0.96331
2.0	0.96403	0.96473	0.96541	0.96609	0.96675	0.96739	0.96803	0.96865	0.96926	0.96986
2.1	0.97045	0.97103	0.97159	0.97215	0.97269	0.97323	0.97375	0.97426	0.97477	0.97526
2.2	0.97574	0.97622	0.97668	0.97714	0.97759	0.97803	0.97846	0.97888	0.97929	0.97970
2.3	0.98010	0.98049	0.98087	0.98124	0.98161	0.98197	0.98233	0.98267	0.98301	0.98335
2.4	0.98367	0.98399	0.98431	0.98462	0.98492	0.98522	0.98551	0.98579	0.98607	0.98635
2.5	0.98661	0.98688	0.98714	0.98739	0.98764	0.98788	0.98812	0.98835	0.98858	0.98881
2.6	0.98903	0.98924	0.98945	0.98966	0.98987	0.99007	0.99026	0.99045	0.99064	0.99083
2.7	0.99101	0.99118	0.99136	0.99153	0.99170	0.99186	0.99202	0.99218	0.99233	0.99248
2.8	0.99263	0.99278	0.99292	0.99306	0.99320	0.99333	0.99346	0.99359	0.99372	0.99384
2.9	0.99396	0.99408	0.99420	0.99431	0.99443	0.99454	0.99464	0.99475	0.99485	0.99495

	0.0	0.1	0.2	0.3	0.4	0.5	0.6	0.7	0.8	0.9
3	0.99505	0.99595	0.99668	0.99728	0.99777	0.99818	0.99851	0.99878	0.99900	0.99918
4	0.99933	0.99945	0.99955	0.99963	0.99970	0.99975	0.99980	0.99983	0.99986	0.99989

SOURCE: Table H is taken from Fisher and Yates, *Statistical Tables for Biological, Agricultural, and Medical Research*, (6th edition, 1974) published by Longman Group Ltd., London (previously published by Oliver & Boyd, Edinburgh), and by permission of the authors and publishers.

Table I   Durbin-Watson Test Critical Values

| | LEVEL OF SIGNIFICANCE $\alpha = 0.05$ | | | | | | | | | |
| | $p = 1$ | | $p = 2$ | | $p = 3$ | | $p = 4$ | | $p = 5$ | |
n	d_L	d_U	d_L	d_U	d_L	d_U	d_L	d_U	d_L	d_U
15	1.08	1.36	0.95	1.54	0.82	1.75	0.69	1.97	0.56	2.21
16	1.10	1.37	0.98	1.54	0.86	1.73	0.74	1.93	0.62	2.15
17	1.13	1.38	1.02	1.54	0.90	1.71	0.78	1.90	0.67	2.10
18	1.16	1.39	1.05	1.53	0.93	1.69	0.82	1.87	0.71	2.06
19	1.18	1.40	1.08	1.53	0.97	1.68	0.86	1.85	0.75	2.02
20	1.20	1.41	1.10	1.54	1.00	1.68	0.90	1.83	0.79	1.99
21	1.22	1.42	1.13	1.54	1.03	1.67	0.93	1.81	0.83	1.96
22	1.24	1.43	1.15	1.54	1.05	1.66	0.96	1.80	0.86	1.94
23	1.26	1.44	1.17	1.54	1.08	1.66	0.99	1.79	0.90	1.92
24	1.27	1.45	1.19	1.55	1.10	1.66	1.01	1.78	0.93	1.90
25	1.29	1.45	1.21	1.55	1.12	1.66	1.04	1.77	0.95	1.89
26	1.30	1.46	1.22	1.55	1.14	1.65	1.06	1.76	0.98	1.88
27	1.32	1.47	1.24	1.56	1.16	1.65	1.08	1.76	1.01	1.86
28	1.33	1.48	1.26	1.56	1.18	1.65	1.10	1.75	1.03	1.85
29	1.34	1.48	1.27	1.56	1.20	1.65	1.12	1.74	1.05	1.84
30	1.35	1.49	1.28	1.57	1.21	1.65	1.14	1.74	1.07	1.83
31	1.36	1.50	1.30	1.57	1.23	1.65	1.16	1.74	1.09	1.83
32	1.37	1.50	1.31	1.57	1.24	1.65	1.18	1.73	1.11	1.82
33	1.38	1.51	1.32	1.58	1.26	1.65	1.19	1.73	1.13	1.81
34	1.39	1.51	1.33	1.58	1.27	1.65	1.21	1.73	1.15	1.81
35	1.40	1.52	1.34	1.58	1.28	1.65	1.22	1.73	1.16	1.80
36	1.41	1.52	1.35	1.59	1.29	1.65	1.24	1.73	1.18	1.80
37	1.42	1.53	1.36	1.59	1.31	1.66	1.25	1.72	1.19	1.80
38	1.43	1.54	1.37	1.59	1.32	1.66	1.26	1.72	1.21	1.79
39	1.43	1.54	1.38	1.60	1.33	1.66	1.27	1.72	1.22	1.79
40	1.44	1.54	1.39	1.60	1.34	1.66	1.29	1.72	1.23	1.79
45	1.48	1.57	1.43	1.62	1.38	1.67	1.34	1.72	1.29	1.78
50	1.50	1.59	1.46	1.63	1.42	1.67	1.38	1.72	1.34	1.77
55	1.53	1.60	1.49	1.64	1.45	1.68	1.41	1.72	1.38	1.77
60	1.55	1.62	1.51	1.65	1.48	1.69	1.44	1.73	1.41	1.77
65	1.57	1.63	1.54	1.66	1.50	1.70	1.47	1.73	1.44	1.77
70	1.58	1.64	1.55	1.67	1.52	1.70	1.49	1.74	1.46	1.77
75	1.60	1.65	1.57	1.68	1.54	1.71	1.51	1.74	1.49	1.77
80	1.61	1.66	1.59	1.69	1.56	1.72	1.53	1.74	1.51	1.77
85	1.62	1.67	1.60	1.70	1.57	1.72	1.55	1.75	1.52	1.77
90	1.63	1.68	1.61	1.70	1.59	1.73	1.57	1.75	1.54	1.78
95	1.64	1.69	1.62	1.71	1.60	1.73	1.58	1.75	1.56	1.78
100	1.65	1.69	1.63	1.72	1.61	1.74	1.59	1.76	1.57	1.78

Table I   (*Concluded*)

	LEVEL OF SIGNIFICANCE $\alpha = 0.01$								
	$p = 1$		$p = 2$		$p = 3$		$p = 4$		$p = 5$	
n	d_L	d_U	d_L	d_U	d_L	d_U	d_L	d_U	d_L	d_U
15	0.81	1.07	0.70	1.25	0.59	1.46	0.49	1.70	0.39	1.96
16	0.84	1.09	0.74	1.25	0.63	1.44	0.53	1.66	0.44	1.90
17	0.87	1.10	0.77	1.25	0.67	1.43	0.57	1.63	0.48	1.85
18	0.90	1.12	0.80	1.26	0.71	1.42	0.61	1.60	0.52	1.80
19	0.93	1.13	0.83	1.26	0.74	1.41	0.65	1.58	0.56	1.77
20	0.95	1.15	0.86	1.27	0.77	1.41	0.68	1.57	0.60	1.74
21	0.97	1.16	0.89	1.27	0.80	1.41	0.72	1.55	0.63	1.71
22	1.00	1.17	0.91	1.28	0.83	1.40	0.75	1.54	0.66	1.69
23	1.02	1.19	0.94	1.29	0.86	1.40	0.77	1.53	0.70	1.67
24	1.04	1.20	0.96	1.30	0.88	1.41	0.80	1.53	0.72	1.66
25	1.05	1.21	0.98	1.30	0.90	1.41	0.83	1.52	0.75	1.65
26	1.07	1.22	1.00	1.31	0.93	1.41	0.85	1.52	0.78	1.64
27	1.09	1.23	1.02	1.32	0.95	1.41	0.88	1.51	0.81	1.63
28	1.10	1.24	1.04	1.32	0.97	1.41	0.90	1.51	0.83	1.62
29	1.12	1.25	1.05	1.33	0.99	1.42	0.92	1.51	0.85	1.61
30	1.13	1.26	1.07	1.34	1.01	1.42	0.94	1.51	0.88	1.61
31	1.15	1.27	1.08	1.34	1.02	1.42	0.96	1.51	0.90	1.60
32	1.16	1.28	1.10	1.35	1.04	1.43	0.98	1.51	0.92	1.60
33	1.17	1.29	1.11	1.36	1.05	1.43	1.00	1.51	0.94	1.59
34	1.18	1.30	1.13	1.36	1.07	1.43	1.01	1.51	0.95	1.59
35	1.19	1.31	1.14	1.37	1.08	1.44	1.03	1.51	0.97	1.59
36	1.21	1.32	1.15	1.38	1.10	1.44	1.04	1.51	0.99	1.59
37	1.22	1.32	1.16	1.38	1.11	1.45	1.06	1.51	1.00	1.59
38	1.23	1.33	1.18	1.39	1.12	1.45	1.07	1.52	1.02	1.58
39	1.24	1.34	1.19	1.39	1.14	1.45	1.09	1.52	1.03	1.58
40	1.25	1.34	1.20	1.40	1.15	1.46	1.10	1.52	1.05	1.58
45	1.29	1.38	1.24	1.42	1.20	1.48	1.16	1.53	1.11	1.58
50	1.32	1.40	1.28	1.45	1.24	1.49	1.20	1.54	1.16	1.59
55	1.36	1.43	1.32	1.47	1.28	1.51	1.25	1.55	1.21	1.59
60	1.38	1.45	1.35	1.48	1.32	1.52	1.28	1.56	1.25	1.60
65	1.41	1.47	1.38	1.50	1.35	1.53	1.31	1.57	1.28	1.61
70	1.43	1.49	1.40	1.52	1.37	1.55	1.34	1.58	1.31	1.61
75	1.45	1.50	1.42	1.53	1.39	1.56	1.37	1.59	1.34	1.62
80	1.47	1.52	1.44	1.54	1.42	1.57	1.39	1.60	1.36	1.62
85	1.48	1.53	1.46	1.55	1.43	1.58	1.41	1.60	1.39	1.63
90	1.50	1.54	1.47	1.56	1.45	1.59	1.43	1.61	1.41	1.64
95	1.51	1.55	1.49	1.57	1.47	1.60	1.45	1.62	1.42	1.64
100	1.52	1.56	1.50	1.58	1.48	1.60	1.46	1.63	1.44	1.65

SOURCE: Reprinted, with permission, from J. Durbin and G. S. Watson, "Testing for Serial Correlation in Least Squares Regression. II," *Biometrika*, Vol. 38 (1951), pp. 159–78.

Table J   Chart to Find Estimates of the Parameters of AR(1) and AR(2) Models

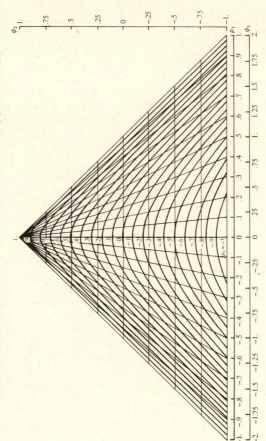

SOURCE: Reproduced by permission from C. M. Stralkowski's, "Lower Order Autoregressive-Moving Average Stochastic Models and Their Use for the Characterization of Abrasive Cutting Tools." Ph.D. Thesis, University of Wisconsin, 1968.

Table K   Chart to Find Estimates of the Parameters of MA(1) and MA(2) Models

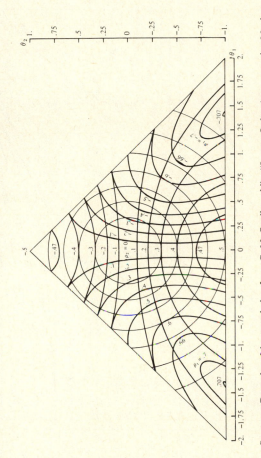
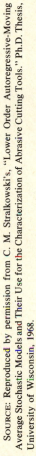

SOURCE: Reproduced by permission from C. M. Stralkowski's, "Lower Order Autoregressive-Moving Average Stochastic Models and Their Use for the Characterization of Abrasive Cutting Tools." Ph.D. Thesis, University of Wisconsin, 1968.

Table L   Chart to Find Estimates of the Parameters of
ARMA(1, 1) Models

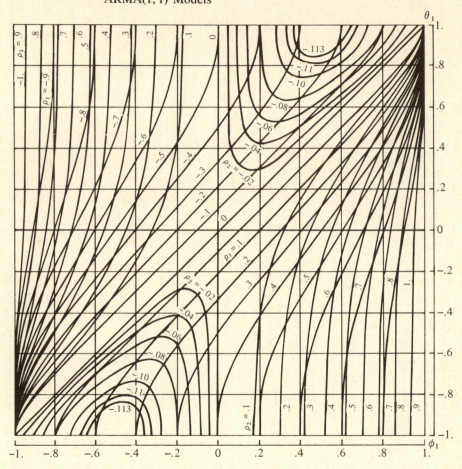

SOURCE: Reproduced by permission from C. M. Stralkowski's, "Lower Order Autoregressive-Moving Average Stochastic Models and Their Use for the Characterization of Abrasive Cutting Tools." Ph.D. Thesis, University of Wisconsin, 1968.

Table M Squares and Square Roots

n	n^2	$\sqrt{n}$	$\sqrt{10n}$	n	n^2	$\sqrt{n}$	$\sqrt{10n}$
1	1	1.00 000	3.16 228	**50**	2 500	7.07 107	22.36 07
2	4	1.41 421	4.47 214	51	2 601	7.14 143	22.58 32
3	9	1.73 205	5.47 723	52	2 704	7.21 110	22.80 35
4	16	2.00 000	6.32 456	53	2 809	7.28 011	23.02 17
5	25	2.23 607	7.07 107	54	2 916	7.34 847	23.23 79
				55	3 025	7.41 620	23.45 21
6	36	2.44 949	7.74 597				
7	49	2.64 575	8.36 660	**56**	3 136	7.48 331	23.66 43
8	64	2.82 843	8.94 427	57	3 249	7.54 983	23.87 47
9	81	3.00 000	9.48 683	58	3 364	7.61 577	24.08 32
10	100	3.16 228	10.00 00	59	3 481	7.68 115	24.28 99
				60	3 600	7.74 597	24.49 49
11	121	3.31 662	10.48 81				
12	144	3.46 410	10.95 45	**61**	3 721	7.81 025	24.69 82
13	169	3.60 555	11.40 18	62	3 844	7.87 401	24.89 98
14	196	3.74 166	11.83 22	63	3 969	7.93 725	25.09 98
15	225	3.87 298	12.24 74	64	4 096	8.00 000	25.29 82
				65	4 225	8.06 226	25.49 51
16	256	4.00 000	12.64 91				
17	289	4.12 311	13.03 84	**66**	4 356	8.12 404	25.69 05
18	324	4.24 264	13.41 64	67	4 489	8.18 535	25.88 44
19	361	4.35 890	13.78 40	68	4 624	8.24 621	26.07 68
20	400	4.47 214	14.14 21	69	4 761	8.30 662	26.26 79
				70	4 900	8.36 660	26.45 75
21	441	4.58 258	14.49 14				
22	484	4.69 042	14.83 24	**71**	5 041	8.42 615	26.64 58
23	529	4.79 583	15.16 58	72	5 184	8.48 528	26.83 28
24	576	4.89 898	15.49 19	73	5 329	8.54 400	27.01 85
25	625	5.00 000	15.81 14	74	5 476	8.60 233	27.20 29
				75	5 625	8.66 025	27.38 61
26	676	5.09 902	16.12 45				
27	729	5.19 615	16.43 17	**76**	5 776	8.71 780	27.56 81
28	784	5.29 150	16.73 32	77	5 929	8.77 496	27.74 89
29	841	5.38 516	17.02 94	78	6 084	8.83 176	27.92 85
30	900	5.47 723	17.32 05	79	6 241	8.88 819	28.10 69
				80	6 400	8.94 427	28.28 43
31	961	5.56 776	17.60 68				
32	1 024	5.65 685	17.88 85	**81**	6 561	9.00 000	28.46 05
33	1 089	5.74 456	18.16 59	82	6 724	9.05 539	28.63 56
34	1 156	5.83 095	18.43 91	83	6 889	9.11 043	28.80 97
35	1 225	5.91 608	18.70 83	84	7 056	9.16 515	28.98 28
				85	7 225	9.21 954	29.15 48
36	1 296	6.00 000	18.97 37				
37	1 369	6.08 276	19.23 54	**86**	7 396	9.27 362	29.32 58
38	1 444	6.16 441	19.49 36	87	7 569	9.32 738	29.49 58
39	1 521	6.24 500	19.74 84	88	7 744	9.38 083	29.66 48
40	1 600	6.32 456	20.00 00	89	7 921	9.43 398	29.83 29
				90	8 100	9.48 683	30.00 00
41	1 681	6.40 312	20.24 85				
42	1 764	6.48 074	20.49 39	**91**	8 281	9.53 939	30.16 62
43	1 849	6.55 744	20.73 64	92	8 464	9.59 166	30.33 15
44	1 936	6.63 325	20.97 62	93	8 649	9.64 365	30.49 59
45	2 025	6.70 820	21.21 32	94	8 836	9.69 536	30.65 94
				95	9 025	9.74 679	30.82 21
46	2 116	6.78 233	21.44 76				
47	2 209	6.85 565	21.67 95	**96**	9 216	9.79 796	30.98 39
48	2 304	6.92 820	21.90 89	97	9 409	9.84 886	31.14 48
49	2 401	7.00 000	22.13 59	98	9 604	9.89 949	31.30 50
50	2 500	7.07 107	22.36 07	99	9 801	9.94 987	31.46 43
				100	10 000	10.00 000	31.62 28
n	n^2	$\sqrt{n}$	$\sqrt{10n}$	n	n^2	$\sqrt{n}$	$\sqrt{10n}$

Table M   (*Continued*)

n	n^2	$\sqrt{n}$	$\sqrt{10n}$		n	n^2	$\sqrt{n}$	$\sqrt{10n}$
100	10 000	10.00 00	31.62 28		**150**	22 500	12.24 74	38.72 98
101	10 201	10.04 99	31.78 05		151	22 801	12.28 82	38.85 87
102	10 404	10.09 95	31.93 74		152	23 104	12.32 88	38.98 72
103	10 609	10.14 89	32.09 36		153	23 409	12.36 93	39.11 52
104	10 816	10.19 80	32.24 90		154	23 716	12.40 97	39.24 28
105	11 025	10.24 70	32.40 37		155	24 025	12.44 99	39.37 00
106	11 236	10.29 56	32.55 76		**156**	24 336	12.49 00	39.49 68
107	11 449	10.34 41	32.71 09		157	24 649	12.53 00	39.62 32
108	11 664	10.39 23	32.86 34		158	24 964	12.56 98	39.74 92
109	11 881	10.44 03	33.01 51		159	25 281	12.60 95	39.87 48
110	12 100	10.48 81	33.16 62		160	25 600	12.64 91	40.00 00
111	12 321	10.53 57	33.31 67		**161**	25 921	12.68 86	40.12 48
112	12 544	10.58 30	33.46 64		162	26 244	12.72 79	40.24 92
113	12 769	10.63 01	33.61 55		163	26 569	12.76 71	40.37 33
114	12 996	10.67 71	33.76 39		164	26 896	12.80 62	40.49 69
115	13 225	10.72 38	33.91 16		165	27 225	12.84 52	40.62 02
116	13 456	10.77 03	34.05 88		**166**	27 556	12.88 41	40.74 31
117	13 689	10.81 67	34.20 53		167	27 889	12.92 28	40.86 56
118	13 924	10.86 28	34.35 11		168	28 224	12.96 15	40.98 78
119	14 161	10.90 87	34.49 64		169	28 561	13.00 00	41.10 96
120	14 400	10.95 45	34.64 10		170	28 900	13.03 84	41.23 11
121	14 641	11.00 00	34.78 51		**171**	29 241	13.07 67	41.35 21
122	14 884	11.04 54	34.92 85		172	29 584	13.11 49	41.47 29
123	15 129	11.09 05	35.07 14		173	29 929	13.15 29	41.59 33
124	15 376	11.13 55	35.21 36		174	30 276	13.19 09	41.71 33
125	15 625	11.18 03	35.35 53		175	30 625	13.22 88	41.83 30
126	15 876	11.22 50	35.49 65		**176**	30 976	13.26 65	41.95 24
127	16 129	11.26 94	35.63 71		177	31 329	13.30 41	42.07 14
128	16 384	11.31 37	35.77 71		178	31 684	13.34 17	42.19 00
129	16 641	11.35 78	35.91 66		179	32 041	13.37 91	42.30 84
130	16 900	11.40 18	36.05 55		180	32 400	13.41 64	42.42 64
131	17 161	11.44 55	36.19 39		**181**	32 761	13.45 36	42.54 41
132	17 424	11.48 91	36.33 18		182	33 124	13.49 07	42.66 15
133	17 689	11.53 26	36.46 92		183	33 489	13.52 77	42.77 85
134	17 956	11.57 58	36.60 60		184	33 856	13.56 47	42.89 52
135	18 225	11.61 90	36.74 23		185	34 225	13.60 15	43.01 16
136	18 496	11.66 19	36.87 82		**186**	34 596	13.63 82	43.12 77
137	18 769	11.70 47	37.01 35		187	34 969	13.67 48	43.24 35
138	19 044	11.74 73	37.14 84		188	35 344	13.71 13	43.35 90
139	19 321	11.78 98	37.28 27		189	35 721	13.74 77	43.47 41
140	19 600	11.83 22	37.41 66		190	36 100	13.78 40	43.58 90
141	19 881	11.87 43	37.55 00		**191**	36 481	13.82 03	43.70 35
142	20 164	11.91 64	37.68 29		192	36 864	13.85 64	43.81 78
143	20 449	11.95 83	37.81 53		193	37 249	13.89 24	43.93 18
144	20 736	12.00 00	37.94 73		194	37 636	13.92 84	44.04 54
145	21 025	12.04 16	38.07 89		195	38 025	13.96 42	44.15 88
146	21 316	12.08 30	38.20 99		**196**	38 416	14.00 00	44.27 19
147	21 609	12.12 44	38.34 06		197	38 809	14.03 57	44.38 47
148	21 904	12.16 55	38.47 08		198	39 204	14.07 12	44.49 72
149	22 201	12.20 66	38.60 05		199	39 601	14.10 67	44.60 94
150	22 500	12.24 74	38.72 98		200	40 000	14.14 21	44.72 14
n	n^2	$\sqrt{n}$	$\sqrt{10n}$		n	n^2	$\sqrt{n}$	$\sqrt{10n}$

Table M (*Continued*)

n	n²	√n	√10n
200	40 000	14.14 21	44.72 14
201	40 401	14.17 74	44.83 30
202	40 804	14.21 27	44.94 44
203	41 209	14.24 78	45.05 55
204	41 616	14.28 29	45.16 64
205	42 025	14.31 78	45.27 69
206	42 436	14.35 27	45.38 72
207	42 849	14.38 75	45.49 73
208	43 264	14.42 22	45.60 70
209	43 681	14.45 68	45.71 65
210	44 100	14.49 14	45.82 58
211	44 521	14.52 58	45.93 47
212	44 944	14.56 02	46.04 35
213	45 369	14.59 45	46.15 19
214	45 796	14.62 87	46.26 01
215	46 225	14.66 29	46.36 81
216	46 656	14.69 69	46.47 58
217	47 089	14.73 09	46.58 33
218	47 524	14.76 48	46.69 05
219	47 961	14.79 86	46.79 74
220	48 400	14.83 24	46.90 42
221	48 841	14.86 61	47.01 06
222	49 284	14.89 97	47.11 69
223	49 729	14.93 32	47.22 29
224	50 176	14.96 66	47.32 86
225	50 625	15.00 00	47.43 42
226	51 076	15.03 33	47.53 95
227	51 529	15.06 65	47.64 45
228	51 984	15.09 97	47.74 93
229	52 441	15.13 27	47.85 39
230	52 900	15.16 58	47.95 83
231	53 361	15.19 87	48.06 25
232	53 824	15.23 15	48.16 64
233	54 289	15.26 43	48.27 01
234	54 756	15.29 71	48.37 35
235	55 225	15.32 97	48.47 68
236	55 696	15.36 23	48.57 98
237	56 169	15.39 48	48.68 26
238	56 644	15.42 72	48.78 52
239	57 121	15.45 96	48.88 76
240	57 600	15.49 19	48.98 98
241	58 081	15.52 42	49.09 18
242	58 564	15 55 63	49.19 35
243	59 049	15.58 85	49.29 50
244	59 536	15.62 05	49.39 64
245	60 025	15.65 25	49.49 75
246	60 516	15.68 44	49.59 84
247	61 009	15.71 62	49.69 91
248	61 504	15.74 80	49.79 96
249	62 001	15.77 97	49.89 99
250	62 500	15.81 14	50.00 00
n	n²	√n	√10n

n	n²	√n	√10n
250	62 500	15.81 14	50.00 00
251	63 001	15.84 30	50.09 99
252	63 504	15.87 45	50.19 96
253	64 009	15.90 60	50.29 91
254	64 516	15.93 74	50.39 84
255	65 025	15.96 87	50.49 75
256	65 536	16.00 00	50.59 64
257	66 049	16.03 12	50.69 52
258	66 564	16.06 24	50.79 37
259	67 081	16.09 35	50.89 20
260	67 600	16.12 45	50.99 02
261	68 121	16.15 55	51.08 82
262	68 644	16.18 64	51.18 59
263	69 169	16.21 73	51.28 35
264	69 696	16.24 81	51.38 09
265	70 225	16.27 88	51.47 82
266	70 756	16.30 95	51.57 52
267	71 289	16.34 01	51.67 20
268	71 824	16.37 07	51.76 87
269	72 361	16.40 12	51.86 52
270	72 900	16.43 17	51.96 15
271	73 441	16.46 21	52.05 77
272	73 984	16.49 24	52.15 36
273	74 529	16.52 27	52.24 94
274	75 076	16.55 29	52.34 50
275	75 625	16.58 31	52.44 04
276	76 176	16.61 32	52.53 57
277	76 729	16.64 33	52.63 08
278	77 284	16.67 33	52.72 57
279	77 841	16.70 33	52.82 05
280	78 400	16.73 32	52.91 50
281	78 961	16.76 31	53.00 94
282	79 524	16.79 29	53.10 37
283	80 089	16.82 26	53.19 77
284	80 656	16.85 23	53.29 17
285	81 225	16.88 19	53.38 54
286	81 796	16.91 15	53.47 90
287	82 369	16.94 11	53.57 24
288	82 944	16.97 06	53.66 56
289	83 521	17.00 00	53.75 87
290	84 100	17.02 94	53.85 16
291	84 681	17.05 87	53.94 44
292	85 264	17.08 80	54.03 70
293	85 849	17.11 72	54.12 95
294	86 436	17.14 64	54.22 18
295	87 025	17.17 56	54.31 39
296	87 616	17.20 47	54.40 59
297	88 209	17.23 37	54.49 77
298	88 804	17.26 27	54.58 94
299	89 401	17.29 16	54.68 09
300	90 000	17.32 05	54.77 23
n	n²	√n	√10n

Table M (*Continued*)

n	n²	√n	√10n	n	n²	√n	√10n
300	90 000	17.32 05	54.77 23	**350**	122 500	18.70 83	59.16 08
301	90 601	17.34 94	54.86 35	351	123 201	18.73 50	59.24 53
302	91 204	17.37 81	54.95 45	352	123 904	18.76 17	59.32 96
303	91 809	17.40 69	55.04 54	353	124 609	18.78 83	59.41 38
304	92 416	17.43 56	55.13 62	354	125 316	18.81 49	59.49 79
305	93 025	17.46 42	55.22 68	355	126 025	18.84 14	59.58 19
306	93 636	17.49 29	55.31 73	**356**	126 736	18.86 80	59.66 57
307	94 249	17.52 14	55.40 76	357	127 449	18.89 44	59.74 95
308	94 864	17.54 99	55.49 77	358	128 164	18.92 09	59.83 31
309	95 481	17.57 84	55.58 78	359	128 881	18.94 73	59.91 66
310	96 100	17.60 68	55.67 76	360	129 600	18.97 37	60.00 00
311	96 721	17.63 52	55.76 74	**361**	130 321	19.00 00	60.08 33
312	97 344	17.66 35	55.85 70	362	131 044	19.02 63	60.16 64
313	97 969	17.69 18	55.94 64	363	131 769	19.05 26	60.24 95
314	98 596	17.72 00	56.03 57	364	132 496	19.07 88	60.33 24
315	99 225	17.74 82	56.12 49	365	133 225	19.10 50	60.41 52
316	99 856	17.77 64	56.21 39	**366**	133 956	19.13 11	60.49 79
317	100 489	17.80 45	56.30 28	367	134 689	19.15 72	60.58 05
318	101 124	17.83 26	56.39 15	368	135 424	19.18 33	60.66 30
319	101 761	17.86 06	56.48 01	369	136 161	19.20 94	60.74 54
320	102 400	17.88 85	56.56 85	370	136 900	19.23 54	60.82 76
321	103 041	17.91 65	56.65 69	**371**	137 641	19.26 14	60.90 98
322	103 684	17.94 44	56.74 50	372	138 384	19.28 73	60.99 18
323	104 329	17.97 22	56.83 31	373	139 129	19.31 32	61.07 37
324	104 976	18.00 00	56.92 10	374	139 876	19.33 91	61.15 55
325	105 625	18.02 78	57.00 88	375	140 625	19.36 49	61.23 72
326	106 276	18.05 55	57.09 64	**376**	141 376	19.39 07	61.31 88
327	106 929	18.08 31	57.18 39	377	142 129	19.41 65	61.40 03
328	107 584	18.11 08	57.27 13	378	142 884	19.44 22	61.48 17
329	108 241	18.13 84	57.35 85	379	143 641	19.46 79	61.56 30
330	108 900	18.16 59	57.44 56	380	144 400	19.49 36	61.64 41
331	109 561	18.19 34	57.53 26	**381**	145 161	19.51 92	61.72 52
332	110 224	18.22 09	57.61 94	382	145 924	19.54 48	61.80 61
333	110 889	18.24 83	57.70 62	383	146 689	19.57 04	61.88 70
334	111 556	18.27 57	57.79 27	384	147 456	19.59 59	61.96 77
335	112 225	18.30 30	57.87 92	385	148 225	19.62 14	62.04 84
336	112 896	18.33 03	57.96 55	**386**	148 996	19.64 69	62.12 89
337	113 569	18.35 76	58.05 17	387	149 769	19.67 23	62.20 93
338	114 244	18.38 48	58.13 78	388	150 544	19.69 77	62.28 96
339	114 921	18.41 20	58.22 37	389	151 321	19.72 31	62.36 99
340	115 600	18 43 91	58.30 95	390	152 100	19.74 84	62.45 00
341	116 281	18.46 62	58.39 52	**391**	152 881	19.77 37	62.53 00
342	116 964	18.49 32	58.48 08	392	153 664	19.79 90	62.60 99
343	117 649	18.52 03	58.56 62	393	154 449	19.82 42	62.68 97
344	118 336	18.54 72	58.65 15	394	155 236	19.84 94	62.76 94
345	119 025	18.57 42	58.73 67	395	156 025	19.87 46	62.84 90
346	119 716	18.60 11	58.82 18	**396**	156 816	19.89 97	62.92 85
347	120 409	18.62 79	58.90 67	397	157 609	19.92 49	63.00 79
348	121 104	18.65 48	58.99 15	398	158 404	19.94 99	63.08 72
349	121 801	18.68 15	59.07 62	399	159 201	19.97 50	63.16 64
350	122 500	18.70 83	59.16 08	400	160 000	20.00 00	63.24 56
n	n²	√n	√10n	n	n²	√n	√10n

Table M (*Continued*)

n	n²	√n	√10n
400	160 000	20.00 00	63.24 56
401	160 801	20.02 50	63.32 46
402	161 604	20.04 99	63.40 35
403	162 409	20.07 49	63.48 23
404	163 216	20.09 98	63.56 10
405	164 025	20.12 46	63.63 96
406	164 836	20.14 94	63.71 81
407	165 649	20.17 42	63.79 66
408	166 464	20.19 90	63.87 49
409	167 281	20.22 37	63.95 31
410	168 100	20.24 85	64.03 12
411	168 921	20.27 31	64.10 93
412	169 744	20.29 78	64.18 72
413	170 569	20.32 24	64.26 51
414	171 396	20.34 70	64.34 28
415	172 225	20.37 15	64.42 05
416	173 056	20.39 61	64.49 81
417	173 889	20.42 06	64.57 55
418	174 724	20.44 50	64.65 29
419	175 561	20.46 95	64.73 02
420	176 400	20.49 39	64.80 74
421	177 241	20.51 83	64.88 45
422	178 084	20.54 26	64.96 15
423	178 929	20.56 70	65.03 85
424	179 776	20.59 13	65.11 53
425	180 625	20.61 55	65.19 20
426	181 476	20.63 98	65.26 87
427	182 329	20.66 40	65.34 52
428	183 184	20.68 82	65.42 17
429	184 041	20.71 23	65.49 81
430	184 900	20.73 64	65.57 44
431	185 761	20.76 05	65.65 06
432	186 624	20.78 46	65.72 67
433	187 489	20.80 87	65.80 27
434	188 356	20.83 27	65.87 87
435	189 225	20.85 67	65.95 45
436	190 096	20.88 06	66.03 03
437	190 969	20.90 45	66.10 60
438	191 844	20.92 84	66.18 16
439	192 721	20.95 23	66.25 71
440	193 600	20.97 62	66.33 25
441	194 481	21.00 00	66.40 78
442	195 364	21.02 38	66.48 31
443	196 249	21.04 76	66.55 82
444	197 136	21.07 13	66.63 33
445	198 025	21.09 50	66.70 83
446	198 916	21.11 87	66.78 32
447	199 809	21.14 24	66.85 81
448	200 704	21.16 60	66.93 28
449	201 601	21.18 96	67.00 75
450	202 500	21.21 32	67.08 20
n	n²	√n	√10n

n	n²	√n	√10n
450	202 500	21.21 32	67.08 20
451	203 401	21.23 68	67.15 65
452	204 304	21.26 03	67.23 09
453	205 209	21.28 38	67.30 53
454	206 116	21.30 73	67.37 95
455	207 025	21.33 07	67.45 37
456	207 936	21.35 42	67.52 78
457	208 849	21.37 76	67.60 18
458	209 764	21.40 09	67.67 57
459	210 681	21.42 43	67.74 95
460	211 600	21.44 76	67.82 33
461	212 521	21.47 09	67.89 70
462	213 444	21.49 42	67.97 06
463	214 369	21.51 74	68.04 41
464	215 296	21.54 07	68.11 75
465	216 225	21.56 39	68.19 09
466	217 156	21.58 70	68.26 42
467	218 089	21.61 02	68.33 74
468	219 024	21.63 33	68.41 05
469	219 961	21.65 64	68.48 36
470	220 900	21.67 95	68.55 65
471	221 841	21.70 25	68.62 94
472	222 784	21.72 56	68.70 23
473	223 729	21.74 86	68.77 50
474	224 676	21.77 15	68.84 77
475	225 625	21.79 45	68.92 02
476	226 576	21.81 74	68.99 28
477	227 529	21.84 03	69.06 52
478	228 484	21.86 32	69.13 75
479	229 441	21.88 61	69.20 98
480	230 400	21.90 89	69.28 20
481	231 361	21.93 17	69.35 42
482	232 324	21.95 45	69.42 62
483	233 289	21.97 73	69.49 82
484	234 256	22.00 00	69.57 01
485	235 225	22.02 27	69.64 19
486	236 196	22.04 54	69.71 37
487	237 169	22.06 81	69.78 54
488	238 144	22.09 07	69.85 70
489	239 121	22.11 33	69.92 85
490	240 100	22.13 59	70.00 00
491	241 081	22.15 85	70.07 14
492	242 064	22.18 11	70.14 27
493	243 049	22.20 36	70.21 40
494	244 036	22.22 61	70.28 51
495	245 025	22.24 86	70.35 62
496	246 016	22.27 11	70.42 73
497	247 009	22.29 35	70.49 82
498	248 004	22.31 59	70.56 91
499	249 001	22.33 83	70.63 99
500	250 000	22.36 07	70.71 07
n	n²	√n	√10n

Table M (*Continued*)

n	n²	√n	√10n		n	n²	√n	√10n
500	250 000	22.36 07	70.71 07		**550**	302 500	23.45 21	74.16 20
501	251 001	22.38 30	70.78 14		551	303 601	23.47 34	74.22 94
502	252 004	22.40 54	70.85 20		552	304 704	23.49 47	74.29 67
503	253 009	22.42 77	70.92 25		553	305 809	23.51 60	74.36 40
504	254 016	22.44 99	70.99 30		554	306 916	23.53 72	74.43 12
505	255 025	22.47 22	71.06 34		555	308 025	23.55 84	74.49 83
506	256 036	22.49 44	71.13 37		**556**	309 136	23.57 97	74.56 54
507	257 049	22.51 67	71.20 39		557	310 249	23.60 08	74.63 24
508	258 064	22.53 89	71.27 41		558	311 364	23.62 20	74.69 94
509	259 081	22.56 10	71.34 42		559	312 481	23.64 32	74.76 63
510	260 100	22.58 32	71.41 43		560	313 600	23.66 43	74.83 31
511	261 121	22.60 53	71.48 43		**561**	314 721	23.68 54	74.89 99
512	262 144	22.62 74	71.55 42		562	315 844	23.70 65	74.96 67
513	263 169	22.64 95	71.62 40		563	316 969	23.72 76	75.03 33
514	264 196	22.67 16	71.69 38		564	318 096	23.74 87	75.09 99
515	265 225	22.69 36	71.76 35		565	319 225	23.76 97	75.16 65
516	266 256	22.71 56	71.83 31		**566**	320 356	23.79 08	75.23 30
517	267 289	22.73 76	71.90 27		567	321 489	23.81 18	75.29 94
518	268 324	22.75 96	71.97 22		568	322 624	23.83 28	75.36 58
519	269 361	22.78 16	72.04 17		569	323 761	23.85 37	75.43 21
520	270 400	22.80 35	72.11 10		570	324 900	23.87 47	75.49 83
521	271 441	22.82 54	72.18 03		**571**	326 041	23.89 56	75.56 45
522	272 484	22.84 73	72.24 96		572	327 184	23.91 65	75.63 07
523	273 529	22.86 92	72.31 87		573	328 329	23.93 74	75.69 68
524	274 576	22.89 10	72.38 78		574	329 476	23.95 83	75.76 28
525	275 625	22.91 29	72.45 69		575	330 625	23.97 92	75.82 88
526	276 676	22.93 47	72.52 59		**576**	331 776	24.00 00	75.89 47
527	277 729	22.95 65	72.59 48		577	332 929	24.02 08	75.96 05
528	278 784	22.97 83	72.66 36		578	334 084	24.04 16	76.02 63
529	279 841	23.00 00	72.73 24		579	335 241	24.06 24	76.09 20
530	280 900	23.02 17	72.80 11		580	336 400	24.08 32	76.15 77
531	281 961	23.04 34	72.86 97		**581**	337 561	24.10 39	76.22 34
532	283 024	23.06 51	72.93 83		582	338 724	24.12 47	76.28 89
533	284 089	23.08 68	73.00 68		583	339 889	24.14 54	76.35 44
534	285 156	23.10 84	73.07 53		584	341 056	24.16 61	76.41 99
535	286 225	23.13 01	73.14 37		585	342 225	24.18 68	76.48 53
536	287 296	23.15 17	73.21 20		**586**	343 396	24.20 74	76.55 06
537	288 369	23.17 33	73.28 03		587	344 569	24.22 81	76.61 59
538	289 444	23.19 48	73.34 85		588	345 744	24.24 87	76.68 12
539	290 521	23.21 64	73.41 66		589	346 921	24.26 93	76.74 63
540	291 600	23.23 79	73.48 47		590	348 100	24.28 99	76.81 15
541	292 681	23.25 94	73.55 27		**591**	349 281	24.31 05	76.87 65
542	293 764	23.28 09	73.62 06		592	350 464	24.33 11	76.94 15
543	294 849	23.30 24	73.68 85		593	351 649	24.35 16	77.00 65
544	295 936	23.32 38	73.75 64		594	352 836	24.37 21	77.07 14
545	297 025	23.34 52	73.82 41		595	354 025	24.39 26	77.13 62
546	298 116	23.36 66	73.89 18		**596**	355 216	24.41 31	77.20 10
547	299 209	23.38 80	73.95 94		597	356 409	24.43 36	77.26 58
548	300 304	23.40 94	74.02 70		598	357 604	24.45 40	77.33 05
549	301 401	23.43 07	74.09 45		599	358 801	24.47 45	77.39 51
550	302 500	23.45 21	74.16 20		600	360 000	24.49 49	77.45 97
n	n²	√n	√10n		n	n²	√n	√10n

Table M   (*Continued*)

n	n²	√n	√10n
600	360 000	24.49 49	77.45 97
601	361 201	24.51 53	77.52 42
602	362 404	24.53 57	77.58 87
603	363 609	24.55 61	77.65 31
604	364 816	24.57 64	77.71 74
605	366 025	24.59 67	77.78 17
606	367 236	24.61 71	77.84 60
607	368 449	24.63 74	77.91 02
608	369 664	24.65 77	77.97 44
609	370 881	24.67 79	78.03 85
610	372 100	24.69 82	78.10 25
611	373 321	24.71 84	78.16 65
612	374 544	24.73 86	78.23 04
613	375 769	24.75 88	78.29 43
614	376 996	24.77 90	78.35 82
615	378 225	24.79 92	78.42 19
616	379 456	24.81 93	78.48 57
617	380 689	24.83 95	78.54 93
618	381 924	24.85 96	78.61 30
619	383 161	24.87 97	78.67 66
620	384 400	24.89 98	78.74 01
621	385 641	24.91 99	78.80 36
622	386 884	24.93 99	78.86 70
623	388 129	24.96 00	78.93 03
624	389 376	24.98 00	78.99 37
625	390 625	25.00 00	79.05 69
626	391 876	25.02 00	79.12 02
627	393 129	25.04 00	79.18 33
628	394 384	25.05 99	79.24 65
629	395 641	25.07 99	79.30 95
630	396 900	25.09 98	79.37 25
631	398 161	25.11 97	79.43 55
632	399 424	25.13 96	79.49 84
633	400 689	25.15 95	79.56 13
634	401 956	25.17 94	79.62 41
635	403 225	25.19 92	79.68 69
636	404 496	25.21 90	79.74 96
637	405 769	25.23 89	79.81 23
638	407 044	25.25 87	79.87 49
639	408 321	25.27 84	79.93 75
640	409 600	25.29 82	80.00 00
641	410 881	25.31 80	80.06 25
642	412 164	25.33 77	80.12 49
643	413 449	25.35 74	80.18 73
644	414 736	25.37 72	80.24 96
645	416 025	25.39 69	80.31 19
646	417 316	25.41 65	80.37 41
647	418 609	25.43 62	80.43 63
648	419 904	25.45 58	80.49 84
649	421 201	25.47 55	80.56 05
650	422 500	25.49 51	80.62 26
n	n²	√n	√10n

n	n²	√n	√10n
650	422 500	25.49 51	80.62 26
651	423 801	25.51 47	80.68 46
652	425 104	25.53 43	80.74 65
653	426 409	25.55 39	80.80 84
654	427 716	25.57 34	80.87 03
655	429 025	25.59 30	80.93 21
656	430 336	25.61 25	80.99 38
657	431 649	25.63 20	81.05 55
658	432 964	25.65 15	81.11 72
659	434 281	25.67 10	81.17 88
660	435 600	25.69 05	81.24 04
661	436 921	25.70 99	81.30 19
662	438 244	25.72 94	81.36 34
663	439 569	25.74 88	81.42 48
664	440 896	25.76 82	81.48 62
665	442 225	25.78 76	81.54 75
666	443 556	25.80 70	81.60 88
667	444 889	25.82 63	81.67 01
668	446 224	25.84 57	81.73 13
669	447 561	25.86 50	81.79 24
670	448 900	25.88 44	81.85 35
671	450 241	25.90 37	81.91 46
672	451 584	25.92 30	81.97 56
673	452 929	25.94 22	82.03 66
674	454 276	25.96 15	82.09 75
675	455 625	25.98 08	82.15 84
676	456 976	26.00 00	82.21 92
677	458 329	26.01 92	82.28 00
678	459 684	26.03 84	82.34 08
679	461 041	26.05 76	82.40 15
680	462 400	26.07 68	82.46 21
681	463 761	26.09 60	82.52 27
682	465 124	26.11 51	82.58 33
683	466 489	26.13 43	82.64 38
684	467 856	26.15 34	82.70 43
685	469 225	26.17 25	82.76 47
686	470 596	26.19 16	82.82 51
687	471 969	26.21 07	82.88 55
688	473 344	26.22 98	82.94 58
689	474 721	26.24 88	83.00 60
690	476 100	26.26 79	83.06 62
691	477 481	26.28 69	83.12 64
692	478 864	26.30 59	83.18 65
693	480 249	26.32 49	83.24 66
694	481 636	26.34 39	83.30 67
695	483 025	26.36 29	83.36 67
696	484 416	26.38 18	83.42 66
697	485 809	26.40 08	83.48 65
698	487 204	26.41 97	83.54 64
699	488 601	26.43 86	83.60 62
700	490 000	26.45 75	83.66 60
n	n²	√n	√10n

Table M (*Continued*)

n	n^2	$\sqrt{n}$	$\sqrt{10n}$	n	n^2	$\sqrt{n}$	$\sqrt{10n}$
700	490 000	26.45 75	83.66 60	**750**	562 500	27.38 61	86.60 25
701	491 401	26.47 64	83.72 57	751	564 001	27.40 44	86.66 03
702	492 804	26.49 53	83.78 54	752	565 504	27.42 26	86.71 79
703	494 209	26.51 41	83.84 51	753	567 009	27.44 08	86.77 56
704	495 616	26.53 30	83.90 47	754	568 516	27.45 91	86.83 32
705	497 025	26.55 18	83.96 43	755	570 025	27.47 73	86.89 07
706	498 436	26.57 07	84.02 38	**756**	571 536	27.49 55	86.94 83
707	499 849	26.58 95	84.08 33	757	573 049	27.51 36	87.00 57
708	501 264	26.60 83	84.14 27	758	574 564	27.53 18	87.06 32
709	502 681	26.62 71	84.20 21	759	576 081	27.55 00	87.12 06
710	504 100	26.64 58	84.26 15	760	577 600	27.56 81	87.17 80
711	505 521	26.66 46	84.32 08	**761**	579 121	27.58 62	87.23 53
712	506 944	26.68 33	84.38 01	762	580 644	27.60 43	87.29 26
713	508 369	26.70 21	84.43 93	763	582 169	27.62 25	87.34 99
714	509 796	26.72 08	84.49 85	764	583 696	27.64 05	87.40 71
715	511 225	26.73 95	84.55 77	765	585 225	27.65 86	87.46 43
716	512 656	26.75 82	84.61 68	**766**	586 756	27.67 67	87.52 14
717	514 089	26.77 69	84.67 59	767	588 289	27.69 48	87.57 85
718	515 524	26.79 55	84.73 49	768	589 824	27.71 28	87.63 56
719	516 961	26.81 42	84.79 39	769	591 361	27.73 08	87.69 26
720	518 400	26.83 28	84.85 28	770	592 900	27.74 89	87.74 96
721	519 841	26.85 14	84.91 17	**771**	594 441	27.76 69	87.80 66
722	521 284	26.87 01	84.97 06	772	595 984	27.78 49	87.86 35
723	522 729	26.88 87	85.02 94	773	597 529	27.80 29	87.92 04
724	524 176	26.90 72	85.08 82	774	599 076	27.82 09	87.97 73
725	525 625	26.92 58	85.14 69	775	600 625	27.83 88	88.03 41
726	527 076	26.94 44	85.20 56	**776**	602 176	27.85 68	88.09 09
727	528 529	26.96 29	85.26 43	777	603 729	27.87 47	88.14 76
728	529 984	26.98 15	85.32 29	778	605 284	27.89 27	88.20 43
729	531 441	27.00 00	85.38 15	779	606 841	27.91 06	88.26 10
730	532 900	27.01 85	85.44 00	780	608 400	27.92 85	88.31 76
731	534 361	27.03 70	85.49 85	**781**	609 961	27.94 64	88.37 42
732	535 824	27.05 55	85.55 70	782	611 524	27.96 43	88.43 08
733	537 289	27.07 40	85.61 54	783	613 089	27.98 21	88.48 73
734	538 756	27.09 24	85.67 38	784	614 656	28.00 00	88.54 38
735	540 225	27.11 09	85.73 21	785	616 225	28.01 79	88.60 02
736	541 696	27.12 93	85.79 04	**786**	617 796	28.03 57	88.65 66
737	543 169	27.14 77	85.84 87	787	619 369	28.05 35	88.71 30
738	544 644	27.16 62	85.90 69	788	620 944	28.07 13	88.76 94
739	546 121	27.18 46	85.96 51	789	622 521	28.08 91	88.82 57
740	547 600	27.20 29	86.02 33	790	624 100	28.10 69	88.88 19
741	549 081	27.22 13	86.08 14	**791**	625 681	28.12 47	88.93 82
742	550 564	27.23 97	86.13 94	792	627 264	28.14 25	88.99 44
743	552 049	27.25 80	86.19 74	793	628 849	28.16 03	89.05 05
744	553 536	27.27 64	86.25 54	794	630 436	28.17 80	89.10 67
745	555 025	27.29 47	86.31 34	795	632 025	28.19 57	89.16 28
746	556 516	27.31 30	86.37 13	**796**	633 616	28.21 35	89.21 88
747	558 009	27.33 13	86.42 92	797	635 209	28.23 12	89.27 49
748	559 504	27.34 96	86.48 70	798	636 804	28.24 89	89.33 08
749	561 001	27.36 79	86.54 48	799	638 401	28.26 66	89.38 68
750	562 500	27.38 61	86.60 25	800	640 000	28.28 43	89.44 27
n	n^2	$\sqrt{n}$	$\sqrt{10n}$	n	n^2	$\sqrt{n}$	$\sqrt{10n}$

Table M (*Continued*)

n	n²	√n	√10n
800	640 000	28.28 43	89.44 27
801	641 601	28.30 19	89.49 86
802	643 204	28.31 96	89.55 45
803	644 809	28.33 73	89.61 03
804	646 416	28.35 49	89.66 60
805	648 025	28.37 25	89.72 18
806	649 636	28.39 01	89.77 75
807	651 249	28.40 77	89.83 32
808	652 864	28.42 53	89.88 88
809	654 481	28.44 29	89.94 44
810	656 100	28.46 05	90.00 00
811	657 721	28.47 81	90.05 55
812	659 344	28.49 56	90.11 10
813	660 969	28.51 32	90.16 65
814	662 596	28.53 07	90.22 19
815	664 225	28.54 82	90.27 74
816	665 856	28.56 57	90.33 27
817	667 489	28.58 32	90.38 81
818	669 124	28.60 07	90.44 34
819	670 761	28.61 82	90.49 86
820	672 400	28.63 56	90.55 39
821	674 041	28.65 31	90.60 91
822	675 684	28.67 05	90.66 42
823	677 329	28.68 80	90.71 93
824	678 976	28.70 54	90.77 44
825	680 625	28.72 28	90.82 95
826	682 276	28.74 02	90.88 45
827	683 929	28.75 76	90.93 95
828	685 584	28.77 50	90.99 45
829	687 241	28.79 24	91.04 94
830	688 900	28.80 97	91.10 43
831	690 561	28.82 71	91.15 92
832	692 224	28.84 44	91.21 40
833	693 889	28.86 17	91.26 88
834	695 556	28.87 91	91.32 36
835	697 225	28.89 64	91.37 83
836	698 896	28.91 37	91.43 30
837	700 569	28.93 10	91.48 77
838	702 244	28.94 82	91.54 23
839	703 921	28.96 55	91.59 69
840	705 600	28.98 28	91.65 15
841	707 281	29.00 00	91.70 61
842	708 964	29.01 72	91.76 06
843	710 649	29.03 45	91.81 50
844	712 336	29.05 17	91.86 95
845	714 025	29.06 89	91.92 39
846	715 716	29.08 61	91.97 83
847	717 409	29.10 33	92.03 26
848	719 104	29.12 04	92.08 69
849	720 801	29.13 76	92.14 12
850	722 500	29.15 48	92.19 54

n	n²	√n	√10n
850	722 500	29.15 48	92.19 54
851	724 201	29.17 19	92.24 97
852	725 904	29.18 90	92.30 38
853	727 609	29.20 62	92.35 80
854	729 316	29.22 33	92.41 21
855	731 025	29.24 04	92.46 62
856	732 736	29.25 75	92.52 03
857	734 449	29.27 46	92.57 43
858	736 164	29.29 16	92.62 83
859	737 881	29.30 87	92.68 23
860	739 600	29.32 58	92.73 62
861	741 321	29.34 28	92.79 01
862	743 044	29.35 98	92.84 40
863	744 769	29.37 69	92.89 78
864	746 496	29.39 39	92.95 16
865	748 225	29.41 09	93.00 54
866	749 956	29.42 79	93.05 91
867	751 689	29.44 49	93.11 28
868	753 424	29.46 18	93.16 65
869	755 161	29.47 88	93.22 02
870	756 900	29.49 58	93.27 38
871	758 641	29.51 27	93.32 74
872	760 384	29.52 96	93.38 09
873	762 129	29.54 66	93.43 45
874	763 876	29.56 35	93.48 80
875	765 625	29.58 04	93.54 14
876	767 376	29.59 73	93.59 49
877	769 129	29.61 42	93.64 83
878	770 884	29.63 11	93.70 17
879	772 641	29.64 79	93.75 50
880	774 400	29.66 48	93.80 83
881	776 161	29.68 16	93.86 16
882	777 924	29.69 85	93.91 49
883	779 689	29.71 53	93.96 81
884	781 456	29.73 21	94.02 13
885	783 225	29.74 89	94.07 44
886	784 996	29.76 58	94.12 76
887	786 769	29.78 25	94.18 07
888	788 544	29.79 93	94.23 38
889	790 321	29.81 61	94.28 68
890	792 100	29.83 29	94.33 98
891	793 881	29.84 96	94.39 28
892	795 664	29.86 64	94.44 58
893	797 449	29.88 31	94.49 87
894	799 236	29.89 98	94.55 16
895	801 025	29.91 66	94.60 44
896	802 816	29.93 33	94.65 73
897	804 609	29.95 00	94.71 01
898	806 404	29.96 66	94.76 29
899	808 201	29.98 33	94.81 56
900	810 000	30.00 00	94.86 83

Table M   (Concluded)

n	n^2	$\sqrt{n}$	$\sqrt{10n}$	n	n^2	$\sqrt{n}$	$\sqrt{10n}$
900	810 000	30.00 00	94.86 83	**950**	902 500	30.82 21	97.46 79
901	811 801	30.01 67	94.92 10	951	904 401	30.83 83	97.51 92
902	813 604	30.03 33	94.97 37	952	906 304	30.85 45	97.57 05
903	815 409	30.05 00	95.02 63	953	908 209	30.87 07	97.62 17
904	817 216	30.06 66	95.07 89	954	910 116	30.88 69	97.67 29
905	819 025	30.08 32	95.13 15	955	912 025	30.90 31	97.72 41
906	820 836	30.09 98	95.18 40	**956**	913 936	30.91 92	97.77 53
907	822 649	30.11 64	95.23 65	957	915 849	30.93 54	97.82 64
908	824 464	30.13 30	95.28 90	958	917 764	30.95 16	97.87 75
909	826 281	30.14 96	95.34 15	959	919 681	30.96 77	97.92 85
910	828 100	30.16 62	95.39 39	960	921 600	30.98 39	97.97 96
911	829 921	30.18 28	95.44 63	**961**	923 521	31.00 00	98.03 06
912	831 744	30.19 93	95.49 87	962	925 444	31.01 61	98.08 16
913	833 569	30.21 59	95.55 10	963	927 369	31.03 22	98.13 26
914	835 396	30.23 24	95.60 33	964	929 296	31.04 83	98.18 35
915	837 225	30.24 90	95.65 56	965	931 225	31.06 44	98.23 44
916	839 056	30.26 55	95.70 79	**966**	933 156	31.08 05	98.28 53
917	840 889	30.28 20	95.76 01	967	935 089	31.09 66	98.33 62
918	842 724	30.29 85	95.81 23	968	937 024	31.11 27	98.38 70
919	844 561	30.31 50	95.86 45	969	938 961	31.12 88	98.43 78
920	846 400	30.33 15	95.91 66	970	940 900	31.14 48	98.48 86
921	848 241	30.34 80	95.96 87	**971**	942 841	31.16 09	98.53 93
922	850 084	30.36 45	96.02 08	972	944 784	31.17 69	98.59 01
923	851 929	30.38 09	96.07 29	973	946 729	31.19 29	98.64 08
924	853 776	30.39 74	96.12 49	974	948 676	31.20 90	98.69 14
925	855 625	30.41 38	96.17 69	975	950 625	31.22 50	98.74 21
926	857 476	30.43 02	96.22 89	**976**	952 576	31.24 10	98.79 27
927	859 329	30.44 67	96.28 08	977	954 529	31.25 70	98.84 33
928	861 184	30.46 31	96.33 28	978	956 484	31.27 30	98.89 39
929	863 041	30.47 95	96.38 46	979	958 441	31.28 90	98.94 44
930	864 900	30.49 59	96.43 65	980	960 400	31.30 50	98.99 49
931	866 761	30.51 23	96.48 83	**981**	962 361	31.32 09	99.04 54
932	868 624	30.52 87	96.54 01	982	964 324	31.33 69	99.09 59
933	870 489	30.54 50	96.59 19	983	966 289	31.35 28	99.14 64
934	872 356	30.56 14	96.64 37	984	968 256	31.36 88	99.19 68
935	874 225	30.57 78	96.69 54	985	970 225	31.38 47	99.24 72
936	876 096	30.59 41	96.74 71	**986**	972 196	31.40 06	99.29 75
937	877 969	30.61 05	96.79 88	987	974 169	31.41 66	99.34 79
938	879 844	30.62 68	96.85 04	988	976 144	31.43 25	99.39 82
939	881 721	30.64 31	96.90 20	989	978 121	31.44 84	99.44 85
940	883 600	30.65 94	96.95 36	990	980 100	31.46 43	99.49 87
941	885 481	30.67 57	97.00 52	**991**	982 081	31.48 02	99.54 90
942	887 364	30.69 20	97.05 67	992	984 064	31.49 60	99.59 92
943	889 249	30.70 83	97.10 82	993	986 049	31.51 19	99.64 94
944	891 136	30.72 46	97.15 97	994	988 036	31.52 78	99.69 95
945	893 025	30.74 09	97.21 11	995	990 025	31.54 36	99.74 97
946	894 916	30.75 71	97.26 25	**996**	992 016	31.55 95	99.79 98
947	896 809	30.77 34	97.31 39	997	994 009	31.57 53	99.84 99
948	898 704	30.78 96	97.36 53	998	996 004	31.59 11	99.89 99
949	900 601	30.80 58	97.41 66	999	998 001	31.60 70	99.95 00
950	902 500	30.82 21	97.46 79	1000	1000 000	31.62 28	100.00 00
n	n^2	$\sqrt{n}$	$\sqrt{10n}$	n	n^2	$\sqrt{n}$	$\sqrt{10n}$

BIBLIOGRAPHY

Books

1. Box, G. E. P., and G. M. Jenkins, *Time Series Analysis: Forecasting and Control.* San Francisco: Holden-Day, 1970.
2. Draper, N. R., and H. Smith, *Applied Regression Analysis.* New York: Wiley, 1966.
3. Goldberger, A. S., *Topics in Regression Analysis.* New York: Macmillan, 1968.
4. Graybill, F. A., *An Introduction to Linear Statistical Models.* New York: McGraw-Hill, 1961.
5. Jenkins, G. M., and D. G. Watts, *Spectral Analysis and Its Applications.* San Francisco: Holden-Day, 1968.
6. Nelson, C. R., *Applied Time Series Analysis for Managerial Forecasting.* San Francisco: Holden-Day, 1973.
7. Neter, J., and W. Wasserman, *Applied Linear Statistical Models.* Homewood, Ill.: Irwin, 1974.
8. Scheffé, H., *The Analysis of Variance.* New York: Wiley, 1959.
9. Searle, S. R., *Linear Models.* New York: Wiley, 1971.
10. Theil, H., *Applied Economic Forecasting.* Amsterdam: North-Holland, 1965.
11. Theil, H., *Principles of Econometrics.* New York: Wiley, 1971.

Articles

12. Bartlett, M. S., On the theoretical specification of sampling properties of autocorrelated time series. *Journal of the Royal Statistical Society (B)*, **8**, 27, 1946.
13. Box, G. E. P., Normality and tests on variances. *Biometrika*, **40**, 318, 1953.
14. Box, G. E. P., Some theorems on quadratic forms applied in the study of analysis of variance problems: I. Effect of inequality of variance in the one-way classification. *Annals of Mathematical Statistics*, **25**, 290, 1954.
15. Box, G. E. P., Some theorems on quadratic forms applied in analysis of variance

problems: II. Effects of inequality of variance and of correlation of errors in the two-way classification. *Annals of Mathematical Statistics*, **25**, 484, 1954.

16. Box, G. E. P., and D. R. Cox, An analysis of transformations. *Journal of the Royal Statistical Society (B)*, **26**, 211, 1964.

17. Box, G. E. P., and G. M. Jenkins, Some recent advances in forecasting and control. *Applied Statistics*, **17**, 91, 1968.

18. Box, G. E. P., and G. M. Jenkins, Some comments on a paper by Chatfield and Prothero and a review by Kendall. *Journal of the Royal Statistical Society (A)*, **136**, 337, 1973.

19. Box, G. E. P., and M. E. Muller, A note on the generation of normal deviates. *Annals of Mathematical Statistics*, **28**, 610, 1958.

20. Box, G. E. P., and P. Newbold, Some comments on a paper of Coen, Gomme, and Kendall. *Journal of the Royal Statistical Society (A)*, **134**, 229, 1971.

21. Box, G. E. P., and D. A. Pierce, Distribution of residual autocorrelations in autoregressive-integrated-moving average time series models. *Journal of the American Statistical Association*, **65**, 1509, 1970.

22. Chatfield, C., and D. L. Prothero, Box-Jenkins seasonal forecasting: Problems in a case study. *Journal of the Royal Statistical Society (A)*, **136**, 295, 1973.

23. Cleveland, W. S., The inverse autocorrelations of a time series and their applications. *Technometrics*, **14**, 277, 1972.

24. Coen, P. J., E. D. Gomme, and M. G. Kendall, Lagged relationships in economic forecasting. *Journal of the Royal Statistical Society (A)*, **132**, 133, 1969.

25. Cox, D. R., and E. J. Snell, The choice of variables in observational studies. *Applied Statistics*, **23**, 51, 1974.

26. Cramer, E., Significance tests and tests of models in multiple regression. *The American Statistician*, **26**(4), 26, 1972.

27. Cramer, R. H., and R. B. Miller, Development of a deposit forecasting procedure for use in bank financial management. *Journal of Bank Research*, **4**, 74, 1973.

28. Durbin, J., and G. S. Watson, Testing for serial correlation in least squares regression. *Biometrika*, **38**, 173, 1951.

29. Gadd, A., and H. Wold, The Janus quotient: A measure for the accuracy of prediction. In H. Wold (ed.), *Econometric Model Building*. Amsterdam: North-Holland, 1964.

30. Hartley, H. O., The modified Gauss-Newton method for the fitting of nonlinear regression functions by least squares. *Technometrics*, **3**, 269, 1969.

31. Hoerl, A. E., and R. W. Kennard, Ridge regression: Biased estimation for nonorthogonal problems. *Technometrics*, **12**, 55, 1970.

32. Hoerl, A. E., and R. W. Kennard, Ridge regression: Applications to nonorthogonal problems. *Technometrics*, **12**, 69, 1970.

33. Hsu, D. A., R. B. Miller, and D. W. Wichern, On the stable paretian behavior of stock market prices. *Journal of the American Statistical Association*, **69**, 108, 1974.

34. Kowalski, C. J., On the effects of nonnormality on the distribution of the sample product moment correlation coefficient. *Applied Statistics*, **21**, 1, 1972.

35. Kruskal, J. B., Special problems of statistical analysis: II. Transformations of data. *International Encyclopedia of Social Sciences*, **15**, 182–193. New York: Macmillan and Free Press, 1968.

36. Layard, M. W. J., Robust large sample tests for homogeneity of variance. *Journal of the American Statistical Association*, **68**, 195, 1973.
37. Marquardt, D. W., and R. D. Snee, Ridge regression in practice. *The American Statistician*, **29**(1), 3, 1975.
38. Schlesselman, J. J., Data transformations in two-way analysis of variance. *Journal of the American Statistical Association*, **68**, 369, 1973.
39. Shapiro, S. S., and M. B. Wilk, An analysis of variance test for normality (complete samples). *Biometrika*, **52**, 591, 1965.
40. Shapiro, S. S., M. B. Wilk, and H. J. Chen, A comparative study of various tests for normality. *Journal of the American Statistical Association*, **63**, 1343, 1968.
41. Stralkowski, C. M., *Lower-order autoregressive-moving average stochastic models and their characterization of abrasive cutting tools*. Ph.D. thesis (Mechanical Engineering), University of Wisconsin, 1968.
42. Swindel, B. F., Instability of regression coefficients illustrated. *The American Statistician*, **28**(2), 63, 1974.
43. Thompson, H. E., and G. C. Tiao, Analysis of telephone data: A case study of forecasting seasonal time series. *The Bell Journal of Economics and Management Science*, **2**, 515, 1971.
44. Tukey, J. W., One degree of freedom for nonadditivity. *Biometrika*, **5**, 232, 1949.
45. Welch, B. L., The significance of the difference between two means when the population variances are unequal. *Biometrika*, **29**, 350, 1938.
46. Yule, G. U., On a method of investigating periodicities in disturbed series, with special reference to Wölfer's sunspot numbers. *Philosophical Transactions A*, **226**, 269, 1927.
47. Zar, J. H., Significance testing of the Spearman rank correlation coefficient. *Journal of the American Statistical Association*, **67**, 578, 1972.

Index

Index